CLAYS IN THE MINERALS PROCESSING VALUE CHAIN

Clays are increasingly becoming a major problem in the mining, extraction and value-adding processes for a wide range of commodity raw materials. Clays can impact negatively on virtually every unit process within the mining and minerals processing sector, having long-term environmental implications that go well beyond the lifetime of the mining operation. This book is the first to compile, explain and evaluate the effects of clays in the minerals processing value chain, from mining to minerals processing, and finally, tailings disposal. Focusing on topics from the chemistry and rheology of clays to their detection and dissolution behaviour, this book provides comprehensive coverage of the effects on processes such as settling, preg-robbing, flotation and comminution. It is an excellent reference for professional mineralogists and geologists, industrial engineers and researchers interested in clays and clay minerals.

MARKUS GRÄFE is the National Coordinator of the Department of Soil and Water Management at Ecuador's National Agricultural Research Institute (INIAP), and an Associate Editor for the *Journal of Environmental Quality*. His research focuses on Ecuador's soil mineralogy and chemistry, stretching from the Amazon Basin across the Andes to the Pacific Ocean and Galapagos.

CRAIG KLAUBER came to mineral processing research from a background in fundamental surface chemistry, and spent several decades as a research scientist at CSIRO. His work principally focuses on utilizing X-ray photoelectron and Auger spectroscopies in order to solve problems in mineral processing systems. He is currently an Adjunct Professor at Curtin University.

ANGUS J. MCFARLANE is a Senior Geoscientist for CSIRO Mineral Resources, with expertise in applied mineralogy and geometallurgy. He is currently working on projects that enable scientists and mining professionals to link ore properties and their behaviour during mining processes via improved characterization and data analytics techniques.

DAVID J. ROBINSON is the Group Leader for Base and Precious Metals Processing within CSIRO, where he has been since 2007 following a 20-year career at Anglo Platinum and Anglo American. He leads a number of research- and pilot-scale projects in base metals hydrometallurgy in areas such as mineral characterization, geometallurgy, leaching, refining and solvent extraction.

CLAYS IN THE MINERALS PROCESSING VALUE CHAIN

Edited by

MARKUS GRÄFE, CRAIG KLAUBER,
ANGUS J. MCFARLANE AND DAVID J. ROBINSON
Commonwealth Scientific Industrial Research Organisation (CSIRO)

CSIRO

CAMBRIDGE
UNIVERSITY PRESS

CAMBRIDGE
UNIVERSITY PRESS

University Printing House, Cambridge CB2 8BS, United Kingdom

One Liberty Plaza, 20th Floor, New York, NY 10006, USA

477 Williamstown Road, Port Melbourne, VIC 3207, Australia

4843/24, 2nd Floor, Ansari Road, Daryaganj, Delhi – 110002, India

79 Anson Road, #06-04/06, Singapore 079906

Cambridge University Press is part of the University of Cambridge.

It furthers the University's mission by disseminating knowledge in the pursuit of education, learning, and research at the highest international levels of excellence.

www.cambridge.org
Information on this title: www.cambridge.org/9781107157323
DOI: 10.1017/9781316661888

First published 2017

Printed in the United Kingdom by TJ International Ltd. Padstow Cornwall

A catalogue record for this publication is available from the British Library.

Library of Congress Cataloging-in-Publication Data
Names: Gräfe, Markus, editor.
Title: Clays in the minerals processing value chain / edited by Markus Gräfe, Craig Klauber, Angus J. McFarlane and David J. Robinson, Commonwealth Scientific Industrial Research Organisation.
Description: Cambridge: Cambridge University Press, 2017. | Includes bibliographical references and index.
Identifiers: LCCN 2017011180 | ISBN 9781107157323 (hardback: paper)
Subjects: LCSH: Clay. | Clay minerals.
Classification: LCC QE471.3.C536 2017 | DDC 553.6/1–dc23
LC record available at https://lccn.loc.gov/2017011180

ISBN 978-1-107-15732-3 Hardback

Contents

Contributors

Warren Bruckard, Mineral Resources, CSIRO

Andrew Chryss, Mineral Resources, CSIRO

Phillip Fawell, Mineral Resources, CSIRO

Elizaveta Forbes, Mineral Resources, CSIRO

Markus Gräfe, Mineral Resources, CSIRO;
Instituto Nacional de Investigaciones Agropecuarias (INIAP) del Ecuador

Craig Klauber, Mineral Resources, CSIRO

Carsten Laukamp, Mineral Resources, CSIRO

Jian Li, Mineral Resources, CSIRO

Mark Ma, Mineral Resources, CSIRO

Angus J. McFarlane, Mineral Resources, CSIRO

Mark Pownceby, Mineral Resources, CSIRO

Erick Ramanaidou, Mineral Resources, CSIRO

Mark Raven, Land and Water, CSIRO

David J. Robinson, Mineral Resources, CSIRO

Peter Self, Land and Water, CSIRO

Peter Smith, Mineral Resources, CSIRO

Martin Wells, Mineral Resources, CSIRO

Kai Yang, Mineral Resources, CSIRO

Reviewers

The editors of this work gratefully acknowledge the free and voluntary expertise lent to the chapters of this book by the following reviewers:

Stephen J. Adkins
Global Mining Technical Manager, SNF-Mining

Professor Dee Bradshaw
Department of Chemical Engineering, University of Cape Town

Dr Laurence Dyer
Department of Mining Engineering and Metallurgical Engineering, Western Australian School of Mines, Curtin University of Technology

Professor George Franks
Department of Chemical and Biomolecular Engineering, University of Melbourne

Dr Will Gates
Institute of Frontier Materials, Deakin University

Dr Kathryn Hadler
Faculty of Engineering, Department of Earth Science & Engineering, Imperial College London

Dr Robert Hewson
School of Mathematical & Geospatial Sciences, RMIT University

Dr Steve Hillier
Environmental and Biochemical Sciences, The James Hutton Institute

Professor Christian Ihle
Ingenería de Minas, Facultad de Ciencias Físicas y Matemáticas, Universidad de Chile

Dr Aleksandar Nikoloski
School of Engineering and Information Technology, Murdoch University

Dr Louisa O'Connor
Department of Mining Engineering & Metallurgical Engineering, Western Australian School of Mines, Curtin University of Technology

Professor Gamini Senanayake
School of Engineering and Information Technology, Murdoch University

Paul Slatter
Principal Engineer, Rheology and Slurry Engineering, ATC Williams

Foreword

This text successfully addresses a major gap and a major need within the minerals industry. To my knowledge, there is no other text that tackles clays from a minerals-centric focus while looking at impacts across the whole minerals processing value chain and across a wide range of mineral and metal commodities.

Clays have a widespread and a very significant economic impact on the minerals industry. I do not have to work hard to find many examples from my industry background to support that contention – they come to mind readily. I can testify that it is very disappointing to see a new mine and process plant be permanently closed down not long after completion of construction and at a time when commissioning was planned to have delivered full production – mainly because clay–sand mixtures at various points in the operation did not behave as intended in settling, in consolidation or in geotechnical strength. The opposite feeling comes from knowing that in an existing operation, treating a sulphide ore containing large amounts of problematic serpentine minerals, a well-structured technical development programme was able to increase flotation recovery by a massive 10 per cent. In yet another example, the challenge was to extract value from approximately 10 per cent of mined ore, which was so high in talc content that it had to be routinely stockpiled. The solution came from an innovative cross-disciplinary approach. Good intention and good fortune came together in a copper heap leaching expansion project, when the feasibility study included trials at a scale sufficient to identify that clays/fines would cause blinding (loss of heap permeability) and hence serious loss of production. The engineering solution involved size separation of the feedstock followed by a separate tank leach of the 'fines' and heap leach of the coarse materials. A very different, but no less impactful, situation occurred at a porphyry copper site located in a region in which water supply was strictly regulated by permit. Water efficiency (usage per tonne of product) was found to have dropped unexpectedly and persistently. When water availability controls the production of concentrate and hence revenue, such a trend is expected to attract serious attention. Variations in the volume of mined clays leading to variations in water retention in the tailings dam became the prime suspect. At a different site, a tank leach process was the victim of repeated, serious disruptions through apparently capricious step changes in slurry rheology. The tank contents would unexpectedly turn to 'porridge'. Detailed studies revealed that this was associated with the variable abundances of specific clays in the ore delivered from the mine to the process plant.

Clay-related issues have also arisen when: considering 'reactive silica' as an important ore characteristic used in the evaluation of bauxite deposits and in subsequent process design; taking account of the effect of moisture on materials handling and wet weather operations in nickel laterite locations; developing a new tar sands extraction process and seeking not to add to the enormous volume of unconsolidated caustic tailings that have been an unfortunate hallmark of that industry; and, finally, evaluating options for managing aluminium content in the production and marketing of iron ore as part of short-term and long-term planning.

Undoubtedly, other colleagues have many additional examples. Even so, to industry practitioners, clay issues can appear to have a sporadic occurrence. The differences between individual deposits and the differences in individual plant designs contribute to a sense of uniqueness. Typically, clays do not present themselves in a 'pure and unadulterated' form, amenable to simple 'plug and play' treatment or handling solutions. They require specific study combining knowledge of the materials and the operational context.

I really like the approach taken in this text. It is written with the academic rigour expected of an authoritative work, but in an accessible style. More especially, in every chapter the importance of the end game is kept in view – the need to contribute to positive industry outcomes. This work is not presented as a set of iconoclastic technical silos where the reader is left to do their best to join the dots. The authors have recognized the need to emphasize overlaps across technical boundaries, moving us (at least part way) towards a sense of cross-disciplinary coherence, which is so important in practice in the minerals context.

When a problem arises – and I have a great deal of sympathy for those who find themselves having to deal with such problems in the field – we tend to call in the expert in that part of the process giving us a headache – perhaps pumping, flotation, leaching, tailings management or whatever. This pattern of dealing with silos of expertise, whose practitioners are partially rather than deeply knowledgeable of clays, is often reflected in the world of technology development as well. This is not how it should be.

When we look to ore-body information, either retrospectively or proactively, to understand the potential for future or ongoing clay-related events, and wish to build this into our short- and long-term planning, we will often find that the focus has been on mapping value elements. Mapping of the sometimes equally important value-destroying constituents, such as clays, is often underdone.

Whether we are addressing issues proactively or reactively in evaluation, design, planning or operating domains, we would like to know what technical 'levers' are available to us so we can move forward reliably in the presence of clays. We would also like to believe that the best possible 'levers' have been developed through an innovation process in which cross-disciplinary R&D has been well targeted and appropriately supported. We still have a distance to travel in that respect.

I congratulate the authors of this text and those companies that had the foresight to support them in this endeavour. It is both original and adventurous. If we are to move to a future in which we can manage clays more effectively and with greater certainty, where technical development in this field is better targeted and supported, where industry experience is better documented and disseminated, where best practice becomes a continuously improving frontier, where we have a chance of developing a holistic view across the value chain and across commodities, then we will need to change our approach to innovation – away from the opportunistic, occasional and siloed to one that reflects a more deliberate strategic intent.

This text is a first step. It provides a view of the current state of knowledge in the minerals industry context. It is impressive in the capability, experience, and insight it represents. On its own, though, it is not enough. I would like to see a further step in which stakeholders come together to consider developing a centre with the mix of industrial experience, specialist knowledge and research focus capable of making a difference and delivering value.

Brian Smith
Visiting Scientist, CSIRO Mineral Resources,
Melbourne, Victoria, Australia

Preface

The review 'Clays in the Minerals Resources Value Chain' was initially prepared in 2013 in response to a clear industry need for a comprehensive treatise covering aspects of clays in mining, processing of ores and tailings disposal. While there exists a wealth of fundamental and applied research that directly or indirectly considers aspects within the topic of clays and their impact on the minerals resources value chain, this information has not been condensed into a useful reference form. Until now the challenge for many would have been to find the publication(s) that most closely answer the many and varied questions that are associated with this broad topic.

The various chapters have therefore attempted to cover more fundamental topics such as definitions, identification, characterization, quantification and some of their properties, along with more applied topics on occurrence, geology, deposit types and the impact clays have on minerals processing and standard unit operations such as comminution, flotation, leaching, solution chemistry, solid–liquid separations and tailings handling.

Each chapter has been written in an accessible manner for non-experts, while also providing relevant source references and background information for a more detailed discourse with each topic. Both the editors and authors alike anticipate that this book will become a useful reference document for practitioners in the various components of the minerals value chain, research and technology developers and in academia.

The original review was made possible thanks to the financial support of the Australian Commonwealth Scientific and Industrial Research Organisation (CSIRO), the Minerals and Energy Research Institute of Western Australia (MERIWA) and nine industry leaders (in alphabetical order): Antofagasta Minerals SA, BASF SE, BHP Billiton Minerals Pty Ltd, Newcrest Mining Limited, Newmont USA Limited, Resource Mining Corporation Limited, Technological Resources Pty Ltd, Teck Resources Ltd and Total E&P Canada Ltd.

The original review was produced by 18 authors with expertise and technical interests in their subject area, and was subsequently updated in 2015/2016 by the same group of authors to include the most recent public domain findings. Gaps in our combined knowledge became evident while reviewing this significant body of literature of some 1100 primary references plus additional references for preg-robbing (75) and characterization (400) summary tables (Chapter 1). Significant effort was put into ensuring that the review was of value to its original industry sponsors; their feedback and indeed contributions in some specific areas is greatly appreciated and has made this book more relevant. With an aim to stimulate research and improve the understanding of the impact clays have on the minerals resources value chain, the final chapter addresses some 20 specific opportunities to better focus research efforts. In particular, towards a better understanding in two general areas:

(1) the chemistry of clays (see Chapters 1 and 3), including metal–clay interactions and selective dissolution; and
(2) particle–particle interactions (see Chapters 2, 7 and 8), including the fundamental science, flotation and the impact of clays on froth rheology, and the impact of clays on the condensing of aggregates.

The editors gratefully acknowledge the support and encouragement of Dr Brian Smith (Visiting Fellow, CSIRO), Dr Jonathan Law (CSIRO, Mineral Resources Director) and Dr Chris Vernon (CSIRO, Mineral Resources, Programme Director).

1

Clays and the Minerals Processing Value Chain (MPVC)

Markus Gräfe, Angus McFarlane and Craig Klauber

1.1 Introduction

Clays represent the raw material for mankind's first transformed product, namely pottery (e.g. Schroeder and Erickson, 2014). While remaining important as a raw material on their own, they are increasingly seen as a major problem in the mining, extraction and value-adding processes for a wide range of commodity raw materials. Clays can impact negatively on virtually every unit process within the mining and minerals processing sector. The extent of this impact is shown schematically in Fig. 1.1.

These impacts can vary from trivial problems such as cleaning mining machinery through to serious mining and materials handling safety issues. The common issues can be summarized as either relating to materials handling (rheological consequences and 'stickiness'), rheological or chemical interference in a beneficiation (or upgrading) process and impacting upon general solid–liquid separation operations. The latter in particular is not only germane to many parts of the mineral processing operation (e.g. S/L in Fig. 1.1), but almost invariably is the last operation prior to tailings disposal. Moreover, these issues continue to operate, often with negative consequences, for tailings areas pre- and post-closure. Presence of clay(s) and clay behaviour can thus have long-term environmental implications that go well beyond the lifetime of the mining and refining operations.

Mining and mineral processing are very energy intensive – in comparison to the agricultural sector, the energy intensity (per dollar added to the economy) is typically 2–3 times greater (Rábago *et al.*, 2001). Whether the commodity is a high-value item such as gold (usually found in parts per million (ppm) concentrations) or an almost pure but low-value bulk commodity such as iron ore, typical operations handle many millions of tonnes of ore material each year. Every time material is handled and for every change in physical state (size range, temperature, pressure, etc.) or chemical environment (pH, Eh, ionic strength, solution composition, etc.) there is a cost penalty in terms of energy, time, raw material inputs and often additional labour and capital expenditure, and the recurring cost thereby entailed. When present in the ore-body, clays cannot be easily removed in the mining phase because of their intimate association with other gangue minerals. Minimizing their impact along the value chain is critical to reducing costs and improving overall efficiencies, hence

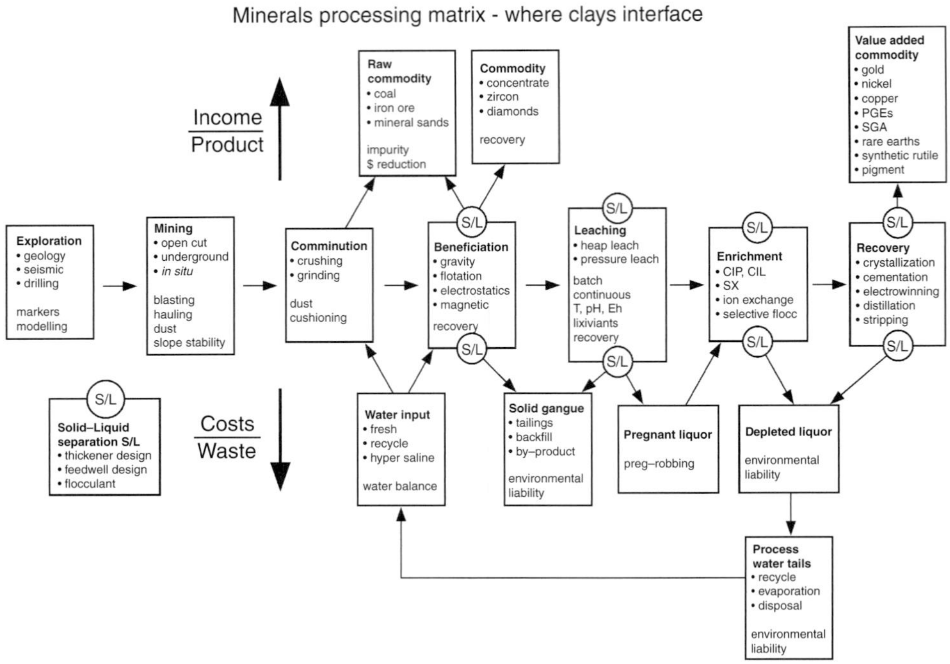

Figure 1.1. Schematic representation of the extent of impact of clays within the mining and minerals processing industry. The problems caused by clays can add substantially to the costs of commodity recovery and waste management.

an understanding of clay behaviour is fundamental to achieving greater overall cost and handling efficiencies.

In the MPVC context there are seven key areas of clay behaviour:

(1) water absorption/desorption (expansion behaviour) and capacity;
(2) suspension rheology;
(3) suspension settling;
(4) species adsorption/sorption/desorption (including cation and anion exchange);
(5) dissolution/precipitation
(6) clay–mineral adhesion/particle–particle interactions; and
(7) clay–polymer adhesion.

These areas in turn are primarily governed by five factors:

(1) clay type (including interlayer structure) – group, subgroup and species;
(2) particle morphology (size, aspect ratio, specific surface area (SSA));
(3) surface charge and double-layer interactions;
(4) acid–base response; and
(5) solution composition of the mineral slurry.

While clay type is a principal driver of clay behaviour, the particular aluminosilicate structures are dynamic and quasi-reversible systems that reflect both the clay's history and its current environment. In the specific context of mining and mineral processing this is especially important:

> we need to know the properties of the clay mineral in question, and those of the interacting compound (water, polymer, iron, organic/inorganic molecule), [so] research into the clay–environment interaction requires collaboration between several disciplines.
>
> (Bergaya and Lagaly, 2006)

A topical search on the Web of Knowledge^SM (Thomson Reuters), one of the search engines commonly used in science, using the Boolean search criteria 'clay*' returned more than 86 000 entries. A subsequent search within those citations for 'hydromet*' returned 76 articles, 'metallurg*' 86 articles, while the Web of Knowledge catalogued only 718 citations against 'Mining Mineral Processing'. Thus, not even 1 per cent of research published on clays since 1900 (the earliest citation on record in the Web of Knowledge) is *directly* associated with mining, hydrometallurgy, and/or minerals processing, i.e. the minerals process value chain (MPVC). In comparison, agronomy (4006) and soil science (11 350) make up nearly 20 per cent of the published work, disregarding some additional 3000–4000 entries in related disciplines. Even though the number of published works on clay indirectly relating to hydrometallurgy, minerals processing and mining exploration is likely to be higher, the numbers reflect the need for a review drawing on the expertise of hydrometallurgists, mining experts and clay scientists into a single body, particularly as clays are recognized by operators to cause significant problems along the MPVC (mine to product/ mine to tails). The intention of this review was thus to place properties and behaviours of clays into the context of minerals processing. As noted above, a review for the generic MPVC must necessarily be cross-disciplinary, hence research activity in the modified clays area (clays modified for another technical application) can assist with providing insight and alternative technical approaches for clay problems in the hydrometallurgical context. Indeed, it is also hoped that by stepping outside the limitations imposed by technical disciplines, unit process specialization or commodity interest, common core principles become evident while critical research gaps could be defined.

1.2 Geology

Economic mineral deposits bearing clay minerals exist in a wide variety of geological contexts in time and space, and thus generalizations are fraught with exceptions and objections, especially where ore-genesis is concerned. Clay minerals are ubiquitous in regions of the crust where alteration and weathering have taken place at low temperatures and pressures, and they are therefore present in most ore deposits that either formed under such conditions or that have become exposed to them through removal of overlying rock over time. Although all mineable ores reside at relatively shallow depths (due to both exploration and mining economics considerations), there are many examples where clay minerals are not likely to be present – for example, orogenic gold deposits formed deep in the crust or lead–zinc deposits formed at shallow depth that subsequently metamorphosed at higher temperatures and pressures after deposition.

Classification of ore deposits into igneous-, sedimentary- and weathering-hosted domains illustrates the breadth of associations between different commodities and clay minerals (Fig. 1.2). The relevance of clay minerals to the geology of deposits and the associated process of exploration are examined in detail in Chapters 5 and 6. For the benefit of those readers more concerned with the post-mining impact of clays, it is nonetheless useful to touch upon some of the geochemistry of various ore bodies containing clays.

For example, much of the world's copper is extracted from 'porphyry' systems, whereby volcanic rocks (of intermediate composition) have been altered by magmatic and meteoric fluids. Along with deposition of copper sulphides, fluids successively alter the original rock mass to leave a characteristic signature corresponding to the system's temperature and composition. Argillic alteration is characterized by the presence of kaolinite, while smectite and interlayered illite–smectite may be present in propylitic (alkaline), argillic and phyllic alteration zones (Inoue, 1995). Depending on the timing of the copper mineralization relative to the alteration event(s), clay minerals will be present in the economic zones of the deposits. Porphyry copper ore bodies are often capped by weathered and oxidized rock where clay minerals are present in addition to those formed through alteration (Berger *et al.*, 2008).

Porphyry copper deposits may be mined through underground or open pit methods, and the presence of clay minerals is of particular concern to mining engineers in designing safe stopes and slopes. Where the main copper mineralization is sulphidic (e.g. chalcopyrite or bornite), the preferred processing route is by flotation to form a saleable concentrate, whereas ore-grade oxidized copper is extracted via leaching (see also Chapter 3). The presence of clays needs to be managed such that: (1) the flotation process is unimpeded by entrainment or rheology-related mixing problems; (2) the leaching process is not constrained by poor permeability; and (3) the tailings are dewatered and stored efficiently and effectively. In each unit process (flotation, leaching, tailings disposal), knowledge of the potential impacts of clays is essential to obtain optimum economic value from the available commodity, in this case copper. Other ore bodies formed through shallow intrusion related to hydrothermal activity and which are associated with clay minerals include porphyry-hosted gold (Au) and molybdenum (Mo), and volcanogenic massive sulphide hydrothermal (VMSH) lead (Pb), zinc (Zn) and copper (Cu).

Extensive weathering of stable continental crust has produced thick (>100 m) layers of weathered and transported regolith which, while an obstacle to exploration, also hosts economic concentrations of residual U, Au, Ni and Al in lateritic ores (Freyssinet *et al.*, 2005) (see also Chapter 5). Formation of laterite profiles takes place when (1) the rate of weathering is greater than the rate of erosion and (2) soluble cations (such as sodium (Na), magnesium (Mg), calcium (Ca) and potassium (K)) are removed preferentially to less soluble ions (such as iron (Fe), aluminium (Al) and silicon (Si)), usually due to excess precipitation (over-evaporation) some time over the lifetime of the surface soil–saprolite–parent rock continuum. As the parent rock is dissolved, clays can form through either crystallization

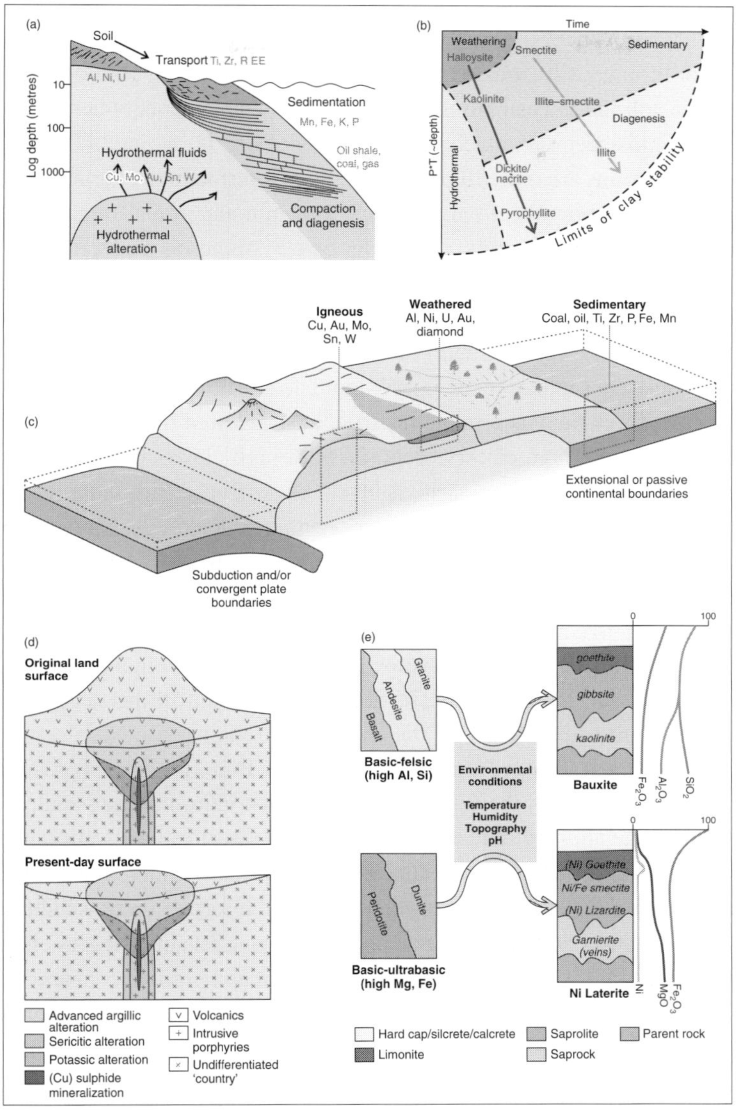

Figure 1.2. Clay minerals are hosted in a variety of geological settings and are likewise associated with numerous metallic ore deposits. In igneous-hosted deposits (such as Cu, Mo, Au, Sn and W), clays can form via both hydrothermal alteration and weathering of the volcanoclastic sediment pile. Lateritic ores can form in stable intracontinental settings, through the interaction of suitable host rocks and favourable environmental circumstances, resulting in clays associated with Au, Ni, Al and U concentrations. Clays are also common in sedimentary hosted ore deposits formed at extensional or passive continental boundaries through erosion/deposition (e.g. Ti, Zr, P) and diagenetic (e.g. coal, oil and gas) processes. Regardless of the formation conditions, ongoing tectonic evolution can modify the clay species through both prograde and retrograde modification. An overview of the geological context of clay mineral formation and ore types they may be associated with: (a) clays and ore types as a function of depth (based on Velde 1995); (b) clay stability as a function of pressure/temperature (depth and geothermal gradient) and time (based on Velde, 1995); (c) ore deposits in their tectonic context; (d) evolution of clay minerals in a porphyry copper deposit; (e) evolution of clay minerals in lateritic Al and Ni deposits. A black and white version of this figure will appear in some formats. For the colour version, please refer to the plate section.

from solution or through diffusion and replacement of pre-existing phyllosilicate structures (Galán, 2006). Kaolinite is typically formed in acid environments, whereas smectites are more commonly present in alkaline conditions where Mg is retained. The precursor primary minerals of kaolins are more likely to form from intermediate to felsic magmas, while the precursor primary minerals of smectites form from mafic and ultramafic magmas. The dissolution process that initiates the formation of clay minerals in weathering terrains is also responsible for the concentration of valuable elements: soluble components of the parent rock are leached away to leave higher grades of the insoluble elements. Clay minerals are therefore intimately associated with ores from these environments.

While gravity-based separation of coarse gold from clay-rich ores is relatively easy, leaching may be less efficient as a result of partial adsorption of the $Au(CN)_2^-$ complex onto clay particles. Uranium (U) in calcrete deposits with high clay (smectite) content presents particular problems to processing, as high levels of geogenic carbonate require leaching in concentrated sodium- and bicarbonate solutions that promote Ca–Na exchange, thereby leaving Na-saturated smectite with very different rheology from the Ca-smectites. Lateritic ores represent the majority of the world's known nickel (Ni) reserves but have been notoriously difficult to mine economically. Clay minerals are minor components of limonitic and saprolitic accumulations of Ni, but may impact on processing through reagent consumption or rheological problems. In some deposits Ni may be present in the structure of smectite minerals, which requires dissolution of the clay structure to release Ni into solution, while minimizing rheological issues and reagent consumption. Bauxites have been the world's primary source of Al for virtually the entire history of industrial use of the metal, and the close association of gibbsite and kaolinite in bauxite horizons requires desilication prior to Bayer refining in order to minimize Si contamination of the pregnant liquor, and minimize reagent (NaOH) and product ($Al(OH)_4^-$) consumption through the formation of sodalite and/or cancrinite. Thus, while extraction aims to be as selective as possible, physical or chemical methods are used to mitigate kaolinite dissolution and reagent and product consumption, because sodalite/cancrinite have long-lasting impacts on tailings rehabilitation (vegetation) and dust management in dry and hot climates such as those encountered in Western Australia.

While clay mineral composition can aid interpretation of coal diagenesis in sedimentary basins (Pevear, 1999), clays are generally problematic for coal production. As clay seams are exposed during mining, their exposure to the atmosphere and removal of structural support produces a zone of weakness that can be a locus of mass movement (Hatherly *et al.*, 2013; Ward *et al.*, 2005). The close association of clays with coal seams also leads to dilution, which impacts upon the final energy rating and ash content, while the associated moisture content requires control to ensure efficient and safe handling during transportation and shipping (Holuszko and Laskowski, 2004). Beneficiation is often employed to remove clays; however, this can lead to tailings management problems and may add moisture to the product unless dry processing methods are employed (de Kretser *et al.*, 1997; Quigley *et al.*, 1982). Further examples of the interplay between the geological setting and metallurgical implications of clays are expanded upon for other sedimentary-hosted commodities in Chapter 5.

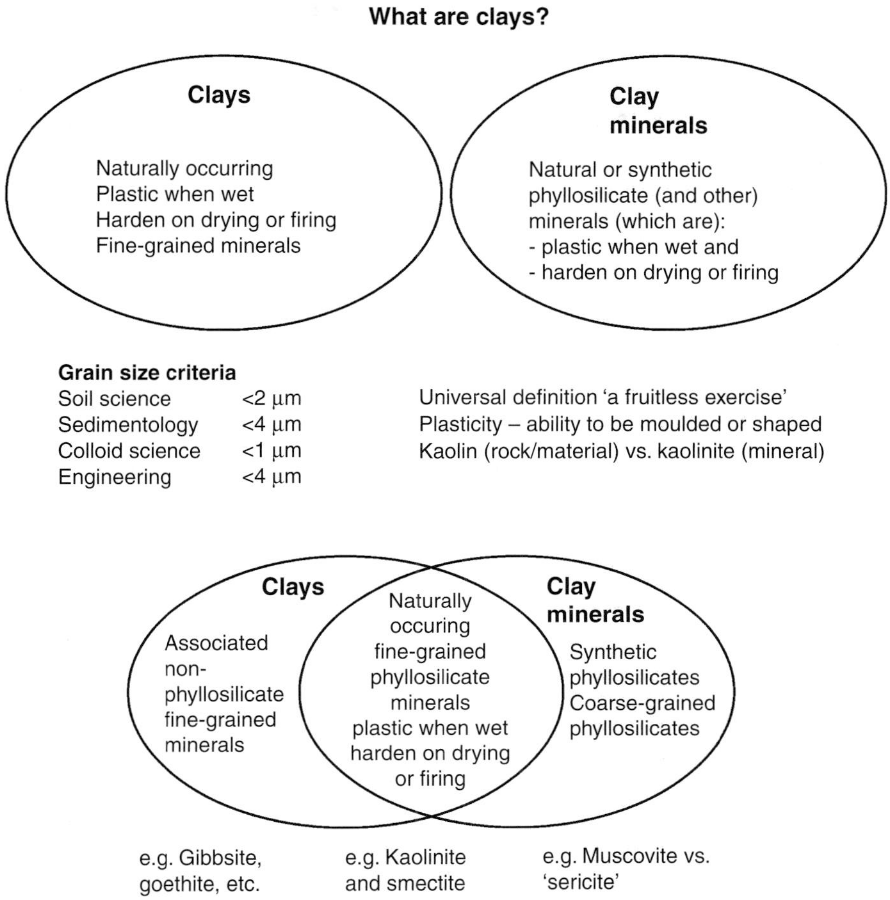

Figure 1.3. The clays or clay minerals that we refer to in this book have the properties in the overlap region. The classification context is shown in Fig. 1.4.

1.3 Defining Clay

Clays have been variously described as particles of any composition below a certain size and/or as the phyllosilicate mineral group or part thereof and/or as materials that display certain physical properties such as plasticity when wet or hardening upon drying. As schematically illustrated in Fig. 1.3, for the purposes of this book the focus is:

- The clay minerals of kaolinite and smectite, due to their frequent occurrence in the fine fraction of mineral processing slurries. Most references discussing these minerals with respect to the MPVC are likely to be relevant.
- Other phyllosilicate minerals where they occur in the fine fraction of ores to the extent that they are associated with kaolinite or smectite. While micas, chlorites, talc and serpentine minerals are present in ores and potentially problematic, they are not necessarily 'clay-like' in behaviour for

various reasons, and hence do not justify the broad attention generally demanded by kaolinite and smectite.

- Other fine-grained non-phyllosilicates and their frequent association with 'fines' will require discussion (but they could be treated as 'non-clay' fines).

For further discussion of these distinctions the reader is referred to chapter 1 of the *Handbook of Clay Science* (Bergaya and Lagaly, 2006; and references therein). Definitions of the various mineral species are taken from the current criteria given by the Clay Mineral Society and International Mineral Association (Birch *et al.*, 2012), although some flexibility is useful to allow a less cumbersome discussion of the more loosely defined terminology.

Many of the characteristics that define clays are problematic rather than beneficial to mining and mineral processing. Their inherently fine size (attributed to extensive defect concentrations in the crystal structure which then poisons crystal growth and favours the nucleation of new crystals (Meunier, 2006)), leads to permeability challenges which slow unit processes, increase reagent consumption and/or adsorb soluble metal ions of value. The plasticity of clays, which is related to their colloidal size and influenced by their shape and surface charge characteristics, creates handling problems that can impede throughput and mixing of slurries, and trap water in tailings disposal streams. The clay minerals that are responsible for these observed behaviours, even when present in relatively minor amounts in ores, are difficult to characterize and quantify, requiring multiple methods to identify and quantify all relevant clay parameters (type, concentration, size distribution, etc.), making anticipation and pre-emptive control of problematic ore parcels cumbersome and expensive.

1.4 Clay Mineral Structure and Classification

Clay mineral structures and their classification are extensively covered in a number of texts (Bailey, 1980; Bergaya and Lagaly, 2006; Brigatti *et al.*, 2006; Murray, 2006), as well as in Chapter 4 of this work. Our brief discourse on this subject matter draws largely on those reviews.

Clay minerals are taxonomically organized over four levels, namely: (1) type, (2) group, (3) subgroup and (4) species (Table 1.1).

All four levels of organization have structurally and/or physicochemically distinguishing characteristics. In terms of simple structural layer representation they are most typically or traditionally displayed as shown in Fig. 1.4. These four levels and their resultant characteristic endmember states are graphically represented as a 'Clay Tunnel' (Fig. 1.5), so as to illustrate the transition between clay minerals.

Type specifies whether a clay mineral is composed of a 1:1 or 2:1 arrangement of tetrahedral (T) to octahedral (O) sheets; these arrangements describe the basic layering of clay minerals (Fig. 1.6a). One-to-one (1:1, TO) and 2:1 (TOT) clay mineral types are *grouped* according to the extent of their permanent negative charge and *subgrouped* further on

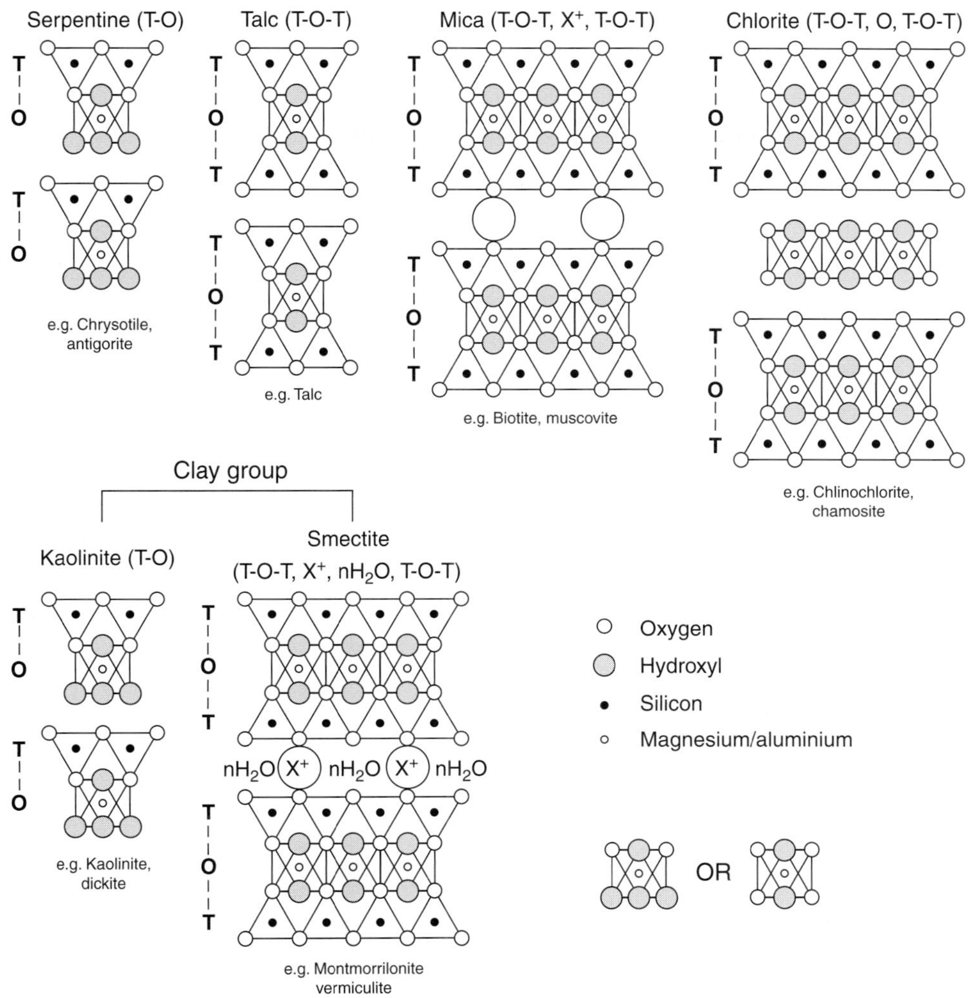

Figure 1.4. Schematic representation of the conventional classification of phyllosilicate minerals.

whether the octahedral sheet is di-octahedral (filled with trivalent metal centres: M^{3+} – e.g. Al^{3+}, Fe^{3+}) or tri-octahedral (filled with divalent metal centres: M^{2+} – e.g. Mg^{2+}, Zn^{2+}). At the *species* level additional, unique structural changes, which cannot be generalized into a further level, are captured, e.g. the difference between kaolinite and halloysite, 1:1 type clay minerals of nearly no permanent negative charge with di-octahedral sheets which are bridged to one another by H-bonding or a water layer in kaolinite or halloysite, respectively. Thus, the classification of clay minerals (type, group, subgroup, specie) may be recaptured with more practical key words, such as: layer-organization, charge, M^{3+} or M^{2+}, and the species' particular(s).

Table 1.1. *Classification of clay minerals, after (Bailey, 1980) based on AIEPA recommendations*

Type	Group	Layer charge	Subgroup	Species (examples)
1:1	Kaolinite–serpentine	~0	kaolinite	
2:1	Talc–pyrophyllite	~0	(Di-octahedral) pyrophyllite	Pyrophyllite
			(Tri-octahedral) talc	Talc
	Smectite	0.2–0.6	Di-octahedral	Montmorillonite
			Tri-octahedral	Saponite
	Vermiculite	0.6–0.9		
	'Illite'	0.9–1.0		
	Mica	1.0–2.0+	Di-octahedral	Muscovite
			Tri-octahedral	Biotite

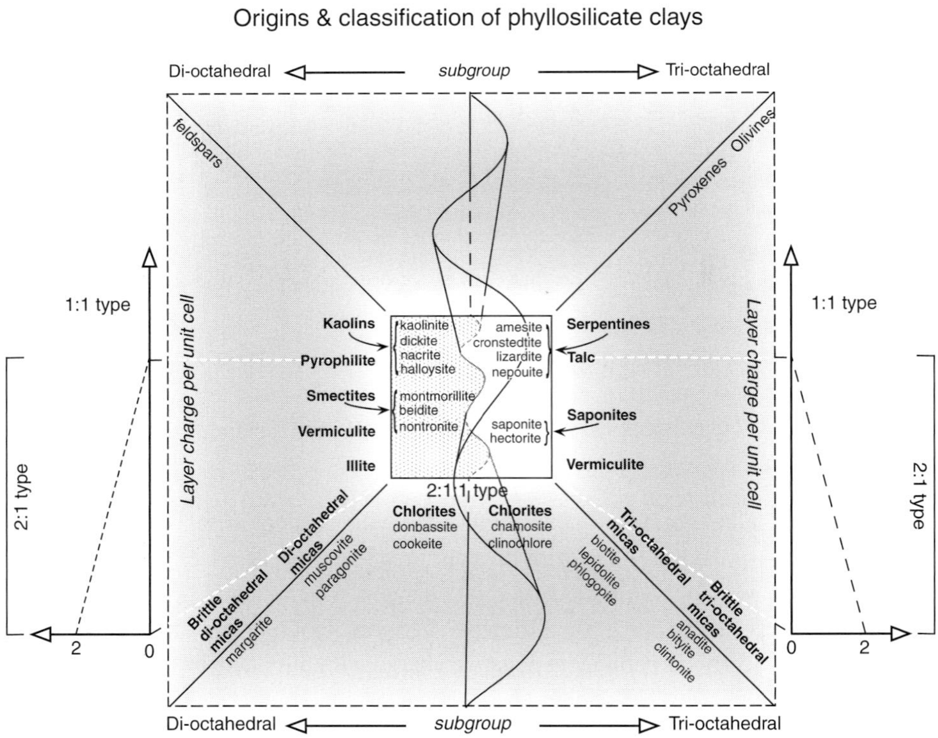

Figure 1.5. Origins and classifications of phyllosilicate clays, represented as a pictorial tunnel. Not all genesis pathways for clays are shown in this image. For more detail the reader is referred to (Galán, 2006; Galán and Ferrell, 2013).

1.4.1 Type: Tetrahedral and Octahedral Sheets

The tetrahedral sheet is composed of an approximately hexagonal (di-trigonal) array of tetrahedra, which propagate infinitely in two directions thus creating a sheet or plane (Fig. 1.6b) in which tetrahedra are uniformly oriented in the same direction: apical O (O_a) atoms pointing towards the octahedral sheet, and the basal (O_b) oxygen atoms pointing away from the octahedral sheet (Fig. 1.6a). A single tetrahedron may thus be described as having three O_b atoms and one O_a atom, $T(O_b)_3(O_a)$, in which every O_b atom is shared between two tetrahedral units, and the O_a atom is shared with a metal centre in the octahedral sheet, and T is the central cation. Silica (Si^{4+}) most commonly occupies the tetrahedral cavity ($T = Si^{4+}$); however, Al^{3+} and Fe^{3+} may substitute for Si^{4+}, which is one mechanism that gives rise to permanent negative charge properties in clay minerals.

In the octahedral sheet, six ligands (usually O, OH and/or F) are coordinated by a central metal cation (*M*), whose charge is either 2^+ or 3^+. Each octahedron is coordinated to two other octahedra via edge-sharing involving one O_a from the tetrahedral sheet and one OH group, which is only shared between the two octahedra (Fig. 1.6c). The OH groups point approximately into the vacant di-trigonal cavity of the tetrahedral sheet (Figs. 1.6a and 1.6d). The symmetry of the conjoined octahedral units is, similarly to the tetrahedral arrangement, approximately hexagonal (di-trigonal); however, because the two 'hexagonal' (T and O) sheets do not align perfectly, a distortion on both sheets is required, which permits sharing of the common ligand moieties (tet O_a) and thus physical connectivity between the two sheets. For more details on the structural details of tetrahedral and octahedral sheets, the reader should consult Brigatti *et al.* (2006).

1.4.2 Groups and Subgroups: Permanent Negative Charge in Di- and Tri-O_h Sheets

Although it is at the subgroup level at which the dominant nature of the octahedral sheet becomes relevant, it is useful to examine the di- and tri-octahedral concept prior to elaborating on the origins and extents of permanent charge (which is covered at the group level). In Fig. 1.5 (the clay tunnel) endmember states for 1:1 and 2:1 type clay minerals are based on the dominant metal occupancy in the octahedral sheet. A clay mineral is deemed di-octahedral if the dominant metal character (in the octahedral sheet) is trivalent, e.g. Al^{3+}, and tri-octahedral if the dominant metal character is divalent, e.g. Mg^{2+}. In the half unit cell of a *di*-octahedral clay, *two* out of three octahedral sites will necessarily be occupied by trivalent metals (Fig. 1.6c, total charge = 6^+), whereas in the half unit cell of a *tri*-octahedral clay mineral, all *three* sites will be occupied by divalent metals (Fig. 1.6e, total charge = 6^+). Perfectly di-octahedral and tri-octahedral clay minerals possess no permanent charge, and this is depicted by the layer-charge delineation in Fig. 1.6 (group level) for the 1:1 type clay endmembers kaolinite/saponite (di-/tri-) and for the 2:1 type clay endmembers pyrophyllite/talc.

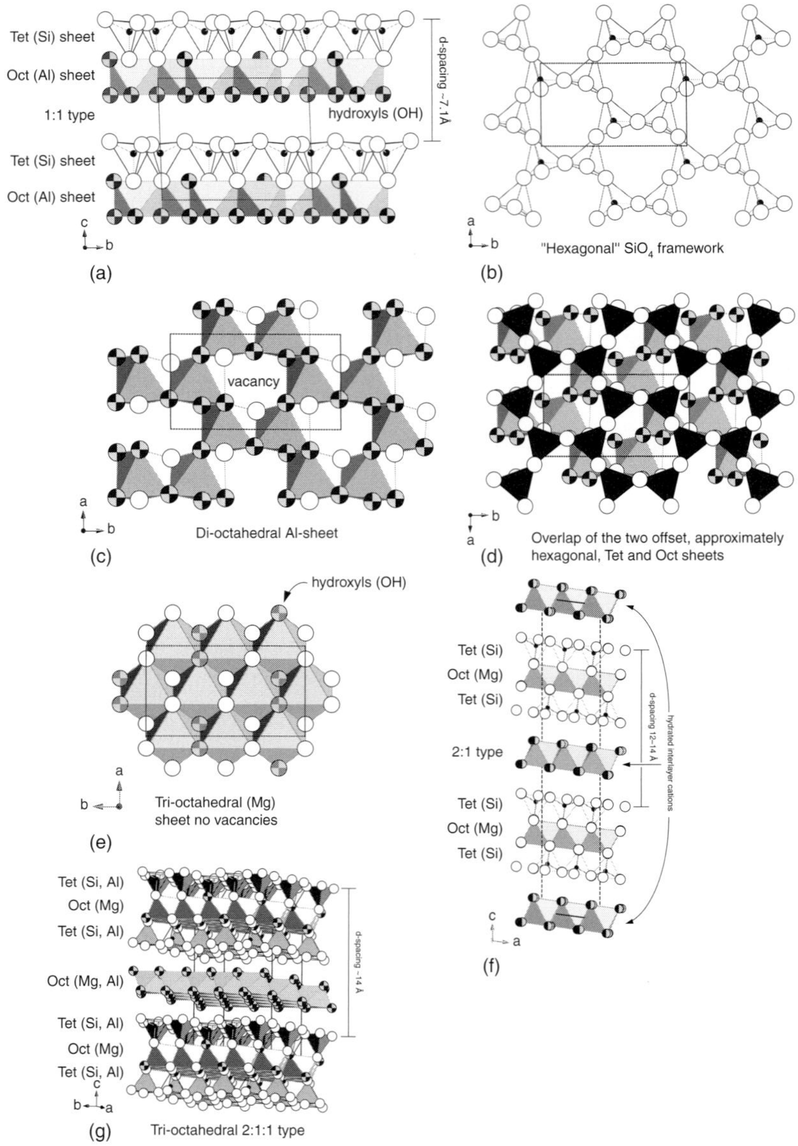

Figure 1.6. Structural aspects of clays that lead to their classification. (a) 1:1 type (kaolinite) clay with H^+ on oxide groups (i.e. hydroxyls, see beach balls) in the interlayer connecting layers together. (b–d) Individual tetrahedral and di-octahedral sheets (note the vacancy in the unit cell) and their superimposition as seen along the c-axis. Note that the hydroxyl groups are approximately aligned at the centres of the di-trigonal (distorted hexagonal) cavities of the tetrahedral sheet. (e) The tri-octahedral sheet of a vermiculite. Note the absence of a vacancy in the unit cell when the octahedral sheet is composed of M^{2+} cations like Mg^{2+}. (f) 2:1 type, tri-octahedral vermiculite with hydrated Mg^{2+} ions in the interlayer. Upon dehydration or increase in pH, these Mg^{2+} ions may form a continuous brucite ($Mg(OH)_2$) interlayer. (g) Clinochlore, which is a tri-octahedral 2:1:1 type chlorite mineral. Note the mixed composition of the tetrahedral sheet and the interlayer octahedral sheet. In the former net negative charge is produced in the 2:1 layer, while in the interlayer octahedral sheet, the substitution of Al^{3+} for Mg^{2+} results in (balancing) positive charge.

Permanent negative charge is created when a metal of lower charge substitutes for a metal of higher charge in parts of the same crystallographic position and there is no change in the crystal form. The latter is referred to as isomorphous substitution. Permanent negative charge can also be created when an octahedral or tetrahedral site remains unfilled, which has ramifications for pH-dependent charge, because such vacancies or defects create Lewis adducts. Permanent positive charge is created when a metal of higher valence (e.g. Al^{3+}) substitutes for a metal of lower valence (e.g. Mg^{2+}) and that may occur in the octahedral sheets ($Mg_6Al_2(OH)_{16}{}^{2+}$) in the interlayers of chlorites (see section below).

Isomorphous substitution may occur in the tetrahedral and octahedral sheets and is not mutually exclusive to any substitution involving discrete metal pairs (e.g. Al and Si). In the tetrahedral sheet, Si^{4+} is the dominant 'metal' centre and is often replaced by Al^{3+} and/or Fe^{3+}. In di-octahedral sheets (dominantly of M^{3+} character) substitution of Al^{3+}, Fe^{3+} and/or Cr^{3+} by Mg^{2+}, Mn^{2+}, Fe^{2+}, Co^{2+} or Ni^{2+} will lead to permanent negative charge. In tri-octahedral sheets (dominantly M^{2+} character), substitution of (usually) Mg^{2+} for a monovalent metal cation (e.g. Li^{+}) will create permanent negative charge (e.g. hectorite). The extent of permanent negative charge is directly proportional to the extent of isomorphous substitution, the number of structural vacancies, and the mitigation by substitution of higher-for-lower valence metals (permanent positive charge). In addition, the negative charge is exacerbated when the valence discrepancy between two substituting metals is greater than 1, e.g. Fe^{2+} substitution for Si^{4+} in the tetrahedral sheet (Guven, 2009).

1.4.3 Chlorites: 2:1:1 Type Clay Minerals

The chlorite minerals are structurally similar to 2:1 type clay minerals but additionally have a positively charged octahedral sheet in the interlayer (Brigatti *et al.*, 2006). The positive charge in the interlayer octahedral sheet is caused by substitution of an M^{3+} cation (e.g. Al^{3+}) for a M^{2+} cation (e.g. Mg^{2+}). The interlayer octahedral sheet bonds to the basal planes of neighbouring 2:1 layers via H-bonding (Brigatti *et al.*, 2006) through the hydroxyl groups of the interlayer octahedral sheet (Fig. 1.6g) and makes the interlayer sheet non-exchangeable – an important distinction to hydroxy-interlayered vermiculites (HIVs), which are classified under 2:1 type clay minerals.

Chlorites are classified at the subgroup level according to the nature of the octahedral sheets in the 2:1 and interlayer structure (Figs 1.5 and 1.6). The most commonly occurring chlorites are those in which both octahedral sheets are tri-octahedral (e.g. clinochlore). The nomenclature used in the literature identifies chlorites as tri- or di-octahedral (both octahedral sheets are either tri- or di-octahedral, respectively) or as di-tri-octahedral or tri-di-octahedral, where the nature of the 2:1 octahedral sheet is mentioned before that of the interlayer. For more details on chlorite minerals, see Chapters 2 and 4 and/or consult Brigatti *et al.* (2006).

1.4.4 *Edge Charge*

In addition to the permanent charge due to isomorphous substitution, a second form of charge develops at 'broken' edge sites, steps and defects along the mineral surface where terminal S–OH and S–OH_2 groups (Lewis adducts) create the mineral–water interface (Schoonheydt and Johnston, 2006). Terminal OH/OH_2 groups are under-coordinated and may carry negative, positive or zero charge. Above the point of zero charge (pzc), the edge charge is negative while below the pzc it is positive. The extent of positive or negative charge is dependent on the coordinating metal cation (in the octahedral sheet: e.g. Al^{3+}, Mg^{2+}, Fe^{2+}/Fe^{3+}) and the degree of protonation/deprotonation that is regulated by the pH of the ambient solution. Hence, edge charge is often referred to in the literature as pH-dependent or variable charge. It is not uncommon that a clay particle may exhibit net negative charge along its basal planes due to isomorphous substitution, while the charge along its edge sites may be positive because the solution pH is below the pzc. The net charging behaviour of phyllosilicate groups such as kaolin/serpentine and pyrophyllite/talc may be largely regulated by solution pH, because they have little or no isomorphous substitution and therefore little or no permanent negative charge. Point of zero charge determinations are therefore an important defining clay metric in order to understand the attraction (preg-robbing/sorption) or repulsion (negative sorption) of processing reagents, as well as particle–particle interactions (see Chapters 2 and 3).

1.5 Methods to Characterize Clays

A large number of physical, chemical, microscopic, structural and spectroscopic methods have been applied to characterize clay systems and their behaviour. Many of these methods characterize the *modified* clay, and often target the non-clay component, as opposed to the original material. This is particularly the case with more sophisticated instrumental approaches. Modification of clays covers everything from the dispersed colloid through to catalyst supports, nano-particles and polymer–clay composites. More traditional clay research aligned with geology, soil science or fired product technology tends to use a narrower range of characterization tools. From the viewpoint of the MPVC and the impact of clays, intrinsic clay behaviour is most important and characterizing the first four of the five clay factors indicated in Section 1.1 – i.e. (1) clay type, (2) morphology, (3) surface charge and (4) acid–base response – is the first step to better understanding the seven key areas of behaviour, as also outlined in Section 1.1. Invariably clays are just one of many components within a more complex mineral processing system.

Between 1980 and 2012, more than 60 different methods were applied to gain insights into intrinsic and modified clay systems (see Table 1.2). With the exception of the specific analytical methods such as powder X-ray diffraction (XRD) that can differentiate both between clay types and other minerals, most characterization methods will only have value on relatively pure systems. In that respect they are largely the domain of fundamental clay studies and outside the scope of this book. Even for methods such as XRD it is

Table 1.2. *Summary of main techniques for determining the structure and properties of clays. Data compiled from approximately 400 publications (unlisted) over the period 1980–2012. ● indicates good-quality information (direct determination or deduction of the property of interest), ○ indicates partial or supplementary information (that might aid in understanding the property of interest).*

							information									
method	% usgae	clay type	particle size / distribution	particle morphology	particle charge	surface area	surface area comment	pore size / distribution	interlayer dimensions	water response	chemical composition	chemical states	site coordination	functional groups	pore water	mass fraction
adsorption																
BET – surface area	5					●	within 20%									
EGME – ethylene glycol monoethyl ether						○	mixed clays									
methylene blue	1					○	relative SSA									
other dyes						○	relative SSA									
indicator ions																
reaction																
CEC – cation exchange capacity	3	○			○									○		○
acid response	2													●		
base response														●		
alkylammonium	2															
physical response																
plasticity			●		●					●						
rheometry	2		●	●	●											
He pycnometry								●								
Hg intrusion	1							●								
hydraulic conductivity	1							●							○	
hydroacoustic																
thermal response																
TGA – thermogravimetric analysis	13									●	○	○		○		○
DTA – differential thermal analysis										●	○					○
electrical response																
AC impedance																
Zeta potential	1				●											
ESA – electrokinetic sonic amplitude																
microscopy																
optical (various)	1			●												
SEM – scanning electron microscopy	8	○	●	●		●	direct	○								
TEM – transmission electron microscopy	7		●	●					●							
AFM – atomic force microscopy	2			●		●	direct									
diffraction																
XRD – X-ray powder diffraction	23	●							●							
Rietveld refinement	1	●							●							●
SAXS – small angle scattering	1		●	○												
WAXD – wide angle X-ray diffraction	1															
laser diffraction	1		●													
spectroscopy																
MS – Mössbauer spectroscopy	1											○				
XAS – X-ray absorption spectroscopy	1										●	●	●	●		
XPS – X-ray photoelectron spectroscopy											●	●		●		
FTIR / IR – Fourier transform infrared	9													●	○	○
NMR – nuclear magnetic resonance	3												●	●		
EPR – electron paramagnetic resonance ESR – electron spin resonance	2										○	○				
EDAX – energy-dispersive X-ray spectroscopy		○									●					
FT Raman														●		
PALS – positron annihilation lifetime spectroscopy																
XRF – X-ray fluorescence	4	○									●					
UV/Vis	1										○			○		
reflectance spectroscopy																
other																
neutron activation																
REM – radiometric emanation method																
ultrasonics																

very useful to physically isolate the clay component prior to analysis (see Chapter 4), in order to (1) concentrate the clays, and (2) manipulate them so as to differentiate among clay minerals (e.g. hydrazine intercalation, see Chapter 4). In terms of the MPVC, the keys to minimizing impact and costs are to first understand which clays are present, under what (environmental) conditions, what their behaviour and impact will be on the unit process (e.g. flotation), and then to consider what can be done to alter/negate that behaviour or remove the clay component altogether. With that in mind, outside of Chapter 4, only a small subset of the most useful (pure system) techniques are summarized below by way of introduction (Table 1.2). For more extensive information the reader is directed to recent publications in the area such as Bergaya and Lagaly (2006) and Murray (2006).

1.5.1 Crystal Structure

1.5.1.1 X-ray Diffraction

Powder X-ray diffraction, particularly sample preparation and approaches to interpretation, is dealt with in detail in Chapter 4. X-ray diffraction is the single most valuable technique in determining clay type and relative abundance; moreover, it can yield information on the extent of crystallinity (or alternatively crystal disorder), strains, polytypes, isomorphous substitution, and others. The key to obtaining this level of information is (clay) sample preparation. The classification of the sheet silicates is primarily dependent on the basal *d*-spacing determination, for which XRD is particularly successful. For non-sheet clays, e.g. allophane and imogolite, classification is more usefully based on Al/Si ratios (Parfitt, 2009), and while XRD plays a role in characterization, the XRD patterns are often diffuse and thus of limited diagnostic value (Abidin *et al.*, 2008; Farmer *et al.*, 1983; van der Gaast *et al.*, 1985). Indeed, the imogolite structure, consisting of 10 Å internal (aluminol groups) and 20 Å external (silanol groups) diameter nanotubes, required both electron diffraction and chemical methods for differentiating silicate anions for its complete structure determination (Cradwick *et al.*, 1972). These types of clays have good short-range order and are not strictly amorphous as is typically the case for glasses. Outside of XRD all the characterization methods are based on specific diagnostic methods that recognize a structural or functional group or short-range bonding environment. Imogolite is a useful illustration because of the unusual tubular structure. Gualtieri *et al.* (2008) demonstrate the extent to which structural characterization of clay can be taken by combining experimental diffraction data with various structure simulations by using the illite–1M polytype as an example, including proposing a position for the interlayer water molecule.

1.5.1.2 Nuclear Magnetic Resonance

In general, nuclides that *do not* have an even number of both protons and neutrons will exhibit nuclear magnetic resonance (NMR) absorption, i.e. having an odd number of protons or an odd number of neutrons. The extent of an NMR shift for the absorbing nuclei can enable identification of specific site environments and in particular complements XRD in increasingly disordered systems. From the viewpoint of clay characterization, the most useful isotopes are those that also have a high isotopic abundance. Hence ^{1}H (99.985 per cent), ^{23}Na (100 per cent), ^{25}Mg (10.0 per cent), ^{27}Al (100 per cent), ^{29}Si (4.7 per cent) and ^{43}Ca (0.14 per cent) can all be utilized as probes for intrinsic clay structure or interlayer cations, whereas the main ^{16}O isotope cannot. With 100 per cent natural isotopic abundance ^{27}Al in particular has been commonly employed in clay studies. For example, with magic angle spinning (MAS), the octahedral (O_h) Al in kaolinite displays an NMR chemical shift of 2 to 4 ppm compared to the aqueous solution aluminate $Al(H_2O)_6{}^{3+}$ reference species (i.e. 2–4 ppm of the shift in resonant frequency of the sample compared to the reference, divided by the reference frequency). This shift can be easily distinguished from a tetrahedral (T_d) Al site at 53–57 ppm, which is induced by disordering kaolinite through mechanical grinding (MacKenzie *et al.*, 2004). With Fe^{3+} substitution in O_h Al sites of montmorillonites (up to 5.6 per cent overall iron content), ^{27}Al information enables the deduction of non-uniform clustering of the iron content and the presence of three different types of O_h sites, including Brønsted/Lewis acid edge sites (Morris *et al.*, 1990).

1.5.1.3 Electron Paramagnetic Resonance

Electron paramagnetic resonance (EPR) is a technique analogous to NMR that investigates unpaired electron spins instead of atomic nuclei. The requirement for the unpaired spins limits the scope of EPR, and although it can be very specific, the interpretation is not always straightforward. Gaite *et al.* (1997) propose that Fe^{3+} EPR can be used as a probe for the degree of structural disorder in kaolinites, and Balan *et al.* (2000) have developed a method for measuring absolute concentrations of paramagnetic Fe^{3+}. The concentration of isolated structural Fe^{3+} (due to substitution of Al^{3+}) represented less than half of the total iron within the kaolinites examined. A related method enables the quantitative determination of VO^{2+} species in clays, relevant as the conversion of V to VO^{2+} is an indicator of the clay's geochemical history.

1.5.1.4 X-ray Absorption Spectroscopy

X-ray absorption spectroscopy (XAS) measures quantum (energy) and physical (geometric) states of the element of interest based on the transfer (absorption) of energy from incident X-rays (photons) to core electrons of the element (Kelly *et al.*, 2008). The key advantages of this technique are element specificity, independence from crystallinity and physical form of the sample (solid, liquid, gas), and minimal sample preparation (minimal possibility of altering the original state of the sample). These measurements provide information about the oxidation state, symmetry and identity of the coordinating ligand environment, and information about more distant neighbouring atoms. For example, for a sample in which Zn is adsorbed on montmorillonite, a Zn K-edge XAFS spectrum could provide information of whether Zn is four- or six-fold coordinated, whether Zn is adsorbed as an outer-sphere complex (e.g. through ion exchange in the interlayer) or is further coordinated by the clay's edge surface. Churakov and Dähn (2012) demonstrate that for Zn adsorption on montmorillonite, three different states can be identified. Structurally incorporated Zn^{2+}

substitutes for Al^{3+} on the trans-symmetric sites of the octahedral layer, whereas at low loadings Zn is incorporated into (010) and (110) edge octahedra and at intermediate loading it attaches as mono- and bidentate surface complexes.

X-ray absorption spectra collected on samples suspected of having a range of chemical states or species for the target element can be interpreted as multi-component systems using abstract factor analysis (principal component analysis, target factor analysis, linear least-square combination fitting). Contributions from spectra of standard materials are indicative of the relative amount of the reference material. Thus, for a sample from a Zn mine site where Zn may be suspected of occurring as the native ore mineral ZnS (sphalerite) and some weathering product other than sphalerite (e.g. Zn adsorbed on montmorillonite or kaolinite), linear combination fitting with XAFS spectra from a reference sphalerite sample and Zn adsorbed on montmorillonite will yield a binary solution of spectral contributions indicative of the approximate composition of the Zn domains in the sample. If the fit residual shows significant structure and the concomitant least squares goodness of fit parameter is poor, then the operator knows that Zn exists in additional states. In that case, a third and possibly fourth reference spectrum will be required. This necessarily requires that the spectra of standard reference materials are well characterized. Series of related XAS spectra representing a multi-component system can be interrogated using factor analysis (principal component and target factor analyses). Both the principal component and target analysis provide the operator with important information about the likely number of total components (Zn states) and which specific Zn states (e.g. sphalerite, Zn adsorbed on montmorillonite, Zn–kerolite, etc.) appear to be present. This in turn tailors the linear least squares fit approach to the most likely reference materials. In the soil and environmental sciences, XAS is considered a key tool to directly determine the chemical states of elements of interest. The information gleaned from these studies is commonly used as input data for geo- and solution-chemical modelling, environmental risk assessments and to study the effects of changing one or more physical or chemical parameters on the speciation of the element of interest. In the MPVC, XAS would therefore lend itself to preg-robbing studies of commodity metals, mineral transformation studies due to processing parameters, and/or environmental risk assessment of penalty elements such as arsenic.

1.5.1.5 Mössbauer Absorption Spectrometry

Mössbauer absorption spectrometry (MAS) is commonly used as an ancillary technique in combination with other characterization methods (Wolters *et al.*, 2009) rather than as the principal technique. A brief review of the technique in the context of clay minerals is given by Murad (2006). Mössbauer absorption spectroscopy measures the extent of gamma-ray resonance absorption by a sample as a specific gamma-ray source is accelerated through a range of velocities (in the order of mm s^{-1}). If the nuclide absorbing the gamma-ray photon is in a different chemical/lattice environment to the gamma-ray source, resonance absorption will not occur unless the gamma-ray energy is varied slightly via the Doppler Effect. The absorption spectrum structure, as a function of the source velocity, is very sensitive to that chemical environment, especially oxidation state and ligand field symmetry, e.g.

for a series of allophanes, Kaufhold *et al.* (2010) find that the relative abundance of the iron content in terms of the Fe^{2+}, Fe^{3+}, hematite and goethite phases can be determined. The major disadvantage of MAS is the limited number of nuclides that can be examined, the most common being ^{57}Fe spectra using a ^{57}Co source. Although Mössbauer spectra can be obtained for elements other than Fe, clay studies appear restricted to that nuclide (e.g. Abdrakhimova and Abdrakhimov, 2006); and for the most part it is for clays and their behaviour as a transformed product.

1.5.1.6 Infrared

The application of infrared (IR) and Raman spectroscopy to characterizing clay minerals has been comprehensively reviewed (e.g. Kloprogge, 2005). The IR methods are particularly useful for following clay transformations *in situ*. Infrared is now almost exclusively undertaken via Fourier Transform Infrared (FTIR) due to the speed of data acquisition. This can either be via absorbance or emission, with the emission approach having advantages for dehydration and dehydroxylation transformations (Kloprogge and Frost, 2005). At temperatures approaching the dehydroxylation point of kaolinite (Fripiat and Toussaint, 1960), the hydroxyl IR vibration bands coalesce indicating complete hydroxyl proton delocalization and thus homogeneous dehydroxylation. By comparison, IR shows the related clay minerals halloysite and dickite lose the outer hydroxyl groups before the inner (Frost and Vassallo, 1996). FTIR can run into ambiguities if the clay system is complex. For example, Cuadros and Dudek (2006) investigate a mixed-layer kaolinite–smectite clay and its progressive kaolinization with heating. Although it is possible to decompose the vibrational spectrum into individual metal–OH–metal hydroxyl bending bands (Al–OH–Al, Al–OH–Fe, Al–OH–Mg and Fe–OH–Fe) and observe the progressive dehydroxylation of the kaolinite domains, evidence for the sequential OH loss for differing cation environments is yet to be established. The size and shape of clay particles modifies the spectral response induced by the chemical bonds alone (Balan *et al.*, 2014). Additional care is therefore needed in calibrating spectral data to identify or quantify various clay species or subspecies. Where size and shape characteristics of minerals affect their behaviour during mining and mineral processing, which is often the case, the additional information present in spectral data might be useful in generating predictive models of ore behaviour.

Apart from laboratory-based applications of vibrational spectroscopy for detailed functional group identification, IR data play a pivotal role in rapid geological mapping via reflectance spectroscopy of drill cores. This technique enables detailed characterization of entire ore deposits (see also Chapter 6), e.g. the Al-smectite and kaolin group distribution in iron ore (Haest *et al.*, 2012).

1.5.1.7 Thermogravimetric Analysis

The common thermal methods are thermogravimetric analysis (TGA), differential thermal analysis (DTA) and differential scanning calorimetry (DSC). Thermogravimetric analysis measures absolute sample mass loss as a function of increasing temperature, and it may be coupled with mass spectrometry to identify the volatilized components. It is commonly employed to characterize water desorption and dehydroxylation ($2OH^- \rightarrow H_2O + O^{2-}$) reactions. Differential thermal analysis applies a thermal cycle to both the sample and a reference material and records temperature differences between the two, thereby detecting exothermic and endothermic transitions (reactions or phase changes) that might not involve any mass changes. Thermogravimetric analysis and DTA can usually be done on the same instrument. Differential scanning calorimetry differs from DTA in that sample and reference material are maintained at the same temperature and the heat requirements for that maintenance are measured over time. For clay samples, DSC measures primarily dehydration and dehydroxylation processes, but with modified clays other processes can also be followed.

Combined TGA/DSC is recognized as a quantitative tool for clay fractions in soils (Karathanasis and Hajek, 1982) and is also useful as an adjunct to EPR, XRD and other techniques. For example, Gehring *et al.* (1993) correlate V^{4+} EPR signal of VO^{2+} species in octahedral mica sites to thermal transformations of transient tetragonal-pyramidal sites. The thermal methods are particularly powerful when used in conjunction with XRD as changes in basal plane spacing are easily followed, as is well illustrated with variably cation-exchanged montmorillonite. The decrease in basal spacing (Δd_{001}) upon dehydration is dependent on the exchanged interlayer cation with $Mg^{2+} > Al^{3+} > Li^+ > Na^+ > NH_4^+ > Cs^+$ and Δd_{001} ranging from 0.448 down to 0.094 nm (Balek *et al.*, 2008). The interlayer cation's polarizing ability influences both how much interlayer water is retained and the dehydration temperature. While a fully dehydrated Na- montmorillonite has a basal spacing of 0.96 nm, water may expand the spacing by up to 57 per cent (Abramova *et al.*, 2007).

Variants of the thermal methods provide interesting additional information, e.g. surface acidity of an acid-activated montmorillonite characterized by microcalorimetry for molar enthalpy of ammonia adsorption can yield acid strength profiles as a function of ammonia surface coverage (Felix *et al.*, 2005). Such fundamental information could find direct application in the MPVC issues to obtain fundamental information about molar enthalpy of reagent adsorption on the whole ore sample or isolated components, e.g. the clay fraction.

1.5.2 Chemical Behaviour

1.5.2.1 Charge: CEC/AEC

The two most important attributes of clays are the physical swelling in the presence of water (in particular smectites) and their ability to electrostatically exchange ions. In terms of their role in the functionality of soils, the ion exchange behaviour is particularly important for plant nutrition (Verburg and Baveye, 1994) and thus much of what is known about this aspect of clays arises from agronomy and more specifically soil science. The impact on the MPVC is not only direct in terms of the exchange affecting liquor chemistry, but the exchange can also lead to changes in macroscopic clay particle behaviour, e.g. aggregation or disaggregation, with obvious consequences for rheological behaviour and solid–liquid separation. Exchange capacity is normally expressed as centimoles (10^{-2} moles) of charge

(+, –) per kilogram (IUPAC), which is numerically equivalent to milliequivalents (meq) of hydrogen (H^+) per 100 g. Unlike some ion exchange systems familiar within mineral processing, e.g. resins, the clay behaviour tends to be less selective. Helfferich (1962) developed and explained a selectivity sequence for soils in general on considerations of Coulomb's Law and the general ability of cations (or metals) to polarize anionic charge. The smaller the hydrated ion radius of the cation/metal, the greater its polarization, and the higher the cation's/metal's valence, the stronger the polarization power: $Ba^{2+} > Pb^{2+} > Sr^{2+} > Ca^{2+} > Ni^{2+} > Cd^{2+} > Cu^{2+} > Co^{2+} > Zn^{2+} > Mg^{2+} > Ag^+ > Cs^+ > Rb^+ > K^+ > NH_4^+ > Na^+ > Li^+$ (Helfferich, 1962).

The ion exchange capacity for the 2:1 clays is significantly higher than the 1:1 clays due to the volumetric interlayer capacity (San Miguel *et al.*, 2006). It also plays a very important role in interlayer behaviour with respect to occupancy of multiple cations/metals and other species capable of co-existence in the interlayer region. As pointed out in Section 1.4.2, it is the isomorphous substitution or defect structure that leads to the permanent negative charge for the smectite group. Directly related to the cation exchange capacity (CEC) is thus the actual layer charge, which is expressed as charge per half unit cell $[O_{10}(OH)_2]$ after Lagaly (1994). The CEC depends both on this as well as pH-dependent edge charges (Mermut and Lagaly, 2001), whereby pH-dependent edge charge for smectites may contribute ~20 per cent. Hydrated metal cations (monovalent, divalent and trivalent) can easily exchange but can be fully dehydrated by 200 °C (Balek *et al.*, 2008). The interlayer cations can also coexist with other polar or charged molecules besides water. As the interlayer spacing only has a restriction to a critical expansion point (that would represent dispersion) there is no predefined constraint to the make-up of the occupants of the interlayer region. In the modified clays used for various industrial applications this behaviour is widely exploited for beneficial purposes. While unlikely to be beneficial, from the perspective of the MPVC, it is essential to recognize that the variety of possible exchangeable species are essentially without limit, from simple co-molecules such as urea $(NH_2)_2CO$ (Abramova *et al.*, 2007) through to the larger dye molecules in conjunction with cations, or large polymeric ions. Model examples of the latter are the primary *n*-alkylammonium salts $C_nH_{2n+1}NH_3^+Cl^-$ (Mermut and Lagaly, 2001) through to quaternary ammonium cations $[(CH_3)_3NR]^+$ or $[(CH_3)_2NR_2]^+$ where R could be an alkyl or aromatic group (Lee and Kim, 2002). The induced basal spacing here can be up to 2.1 nm or even ~3 nm when distorted, with the longer chain length cations taking on altered packing arrangements (e.g. monolayer to bilayer to tilted paraffinic structure) to satisfy charge requirements (Mermut and Lagaly, 2001). This type of clay behaviour vividly demonstrates that not only can clays directly interlayer absorb/exchange a variety of metal species, but also take up charged organic molecules. The reversibility may depend upon conditions the clay is subjected to; for example, in the case of urea (Abramova *et al.*, 2007) heating can lead to a number of polymeric transformations. Organics that fail to evaporate may fully carbonize *in situ*.

In terms of fundamental clay research, much of the effort on exchange capacity is about clay classification as well as behaviour. For example, within the montmorillonite–beidellite series, 96 structural variations theoretically exist (Emmerich *et al.*, 2009). These compositional variations control charge distributions and capacities. For montmorillonites a CEC of 70–120 cmol(+) kg^{-1} (Mermut and Lagaly, 2001) equates, for an organic singly charged cation of molecular weight 270, to a nominal exchange/sorption mass capacity of 190–325 g kg^{-1}. Clearly the capacity of clays to influence MPVC chemistries should not be underestimated. For a commodity metal, e.g. divalent Co or Cu, this could represent a nominal metal exchange/sorption capacity of 20–38 g kg^{-1} and would thus have a substantial economic impact.

1.5.2.2 Dyes

Water-soluble dye molecules are generically classified as acidic/anionic or basic/cationic and can be either organic or inorganic. A special subset of the cation exchange capability of clays is the use of cationic dyes in lieu of simple metallic cations or other small cationic species. By nature of their absorption spectrum, the adsorption response of dyes can be easily followed *in situ*, so the popularity of dye characterization is understandable. Schoonheydt (1994) reviews the cationic clays in the context of clay minerals and importantly points out that on the two-dimensional surface of clays the dye molecules have a strong tendency to agglomerate. In the case of the commonly used dye methylene blue $C_{16}H_{18}N_3S^+$, Yu *et al.* (2000) use molecular dynamics simulations across a variety of clay surfaces as a function of water content. The simulations indicate a variety of aggregate structures; the ions can lie flat, be tilted or arrange in irregular clusters. That study was unable to confirm earlier work by Hähner *et al.* (1996); who used near-edge X-ray absorption fine structure (NEXAFS) spectroscopy at the carbon and nitrogen 1s edges to conclude that the dye molecules adsorb tilted 65–70° with respect to the surface, although in that case the samples were examined *ex situ* in vacuum, so the conditions were not comparable. Yu *et al.* (2000) suggest that methylene blue adsorption is a poor method for SSA (see below) due to the variety of possible orientations, but that it remains useful for measuring relative areas and that the variation of surface orientations could be an advantage for CEC evaluation. Even with the ordered orientation model, SSAs could thus be overestimated using methylene blue (Hähner *et al.*, 1996). The apparent molecular area of 66 $Å^2$ probably has little value. Applying the same *ex situ* methods of NEXAFS and X-ray photoelectron spectroscopy (XPS), Fischer *et al.* (1998) have investigated the adsorption of crystal violet $[C(C_6H_4N(CH_3)_2)_3]^+$ and the related malachite green $[C_6H_5C(C_6H_4N(CH_3)_2)_2]^+$ on model mica surfaces (both cation exchange processes). Despite the large dimensions of the dye molecules they attach to the mica surfaces through a single group inclined at 55–60°.

Kahr and Madsen (1995) point out that the CEC of bentonites can only be determined using methylene blue if the bentonite is in the sodium form. To obtain a reasonable surface area, one must assume a surface area for the dye of 130 $Å^2$. Thus, use of dye methods for estimating SSAs of clays should be undertaken with caution, especially if the mechanism is not understood. For example, Congo red is an anionic dye molecule that would appear to occupy the interlayer space of montmorillonite by a mechanism other than ion exchange (Yermiyahu *et al.*, 2002) and hence would represent a poor choice of dye.

1.5.2.3 Functional Groups, Acid–Base Functionality

The acid–base functionality of clay systems is important in their role in soils, as environmental buffers and industrial catalysts, and would thus be expected to play a role in clay behaviour within the MPVC. Accurate proton affinities, most usefully expressed as a frequency distribution of acidity constants, pK_a, can be easily obtained by titration with a normalized base (Jagiello *et al.*, 1995). As mineral processing unit operations are most often hydrometallurgical in nature (as opposed to non-aqueous solvent processing), the acidity from a water-based viewpoint is the most appropriate. By contrast, in the industrial application of modified clays, e.g. to catalysis, it may be more appropriate to consider acidity in terms of Lewis and Brønsted type sites. This can be done by direct spectroscopy, e.g. pyridine adsorption, then selective temperature-dependent desorption, followed by FTIR (Carvalho *et al.*, 2003). Acid activation (Komadel and Madejová, 2006) is not a specific issue in the MPVC except for the changes in clay behaviour induced by the acid exposure or the unintended dissolution of the clay (see Chapter 3).

1.5.3 Physical Morphology

1.5.3.1 Specific Surface Area and Porosity

The meaning of SSA is dependent on the method by which the area is determined, especially for high SSA materials. Such materials are usually high surface area due to small particle size and/or an extensive pore structure. Typically high surface area materials are measured via the BET (Brunauer–Emmett–Teller) method, relying on the physisorption of gas molecules (e.g. N_2) with a known adsorption cross-section. Molecular nitrogen (cross-section 0.162 nm^2 assuming close-packed monolayer adsorption) is the most suitable (Sing *et al.*, 1985), but other gases can be employed as well. In the case of smectites, Carter *et al.* (1986) note that N_2 adsorbs weakly and may not penetrate the interlayer region, thereby approximating external rather than total surface area. Equally, the relevance of the BET equation for H_2O vapour adsorption is questionable since the molecules cluster around cation sites (Quirk, 1955). The accuracy of BET is within 20 per cent, dependent on the surface packing, but of more concern is the *accessible* pore volume being lower than the actual internal surface area. For micropores (<2.0 nm) the adsorbent–adsorbate interaction is much stronger than in the mesopore region (2–50 nm), where it is possible to derive a description of the pore size distribution. Clays represent an additional complexity over most high surface area systems (e.g. activated carbons, molecular sieves, etc.) as the surface area is not static, but can depend on the extent of dispersion and the extent of ion exchange. This is especially true for clays within the MPVC as most mineral processing occurs in the presence of water. In the case of pure clay systems the SSA may mean the total specific surface area, which is method dependent, but it may also be described in terms of the edge surface area (ESA) and basal surface area (BSA). The porosity of pillared clay catalysts has been reviewed in detail by Gil *et al.* (2008) and represents the key issues for determining the porosity of clay minerals. Liu *et al.* (2011) give a direct approach through SEM image processing.

Clearly a BET value for a dry clay and a clay in a saturated environment can be very different, e.g. an easily dispersed smectite. Blum and Eberl (2004) have demonstrated that polyvinylpyrrolidone (PVP-55K) has a consistent adsorption density across a wide range of smectites, illites and kaolinites of ~0.99 mg m^{-2} (55K is the average molecular weight, so 55K would be ~500 monomer units long), enabling a viable method for smectite differentiation by gravimetric (or FTIR; Blum *et al.*, 2011)) adsorption. Despite the polymer functionality, it is insensitive to surface chemistry, charge and pH, and hence enables a dispersed surface area to be determined with PVP. This can also be assessed by dye cation adsorption, but the result depends critically upon the cation exchange state. Methylene blue chloride $C_{16}H_{18}N_3SCl$ has terminal cationic dimethylamino ligands and so will adsorb on clays via cation exchange, but the values of CEC and surface area are only reliable for bentonites when the clay is fully sodium exchanged and the pH is neutral. The requirement is that the clay's surface area per unit charge approximately matches the physical area of the molecule (Kahr and Madsen, 1995).

Ethylene glycol monoethyl ether (EGME), or 2-ethoxyethanol, $C_2H_5OC_2H_4OH$, is a water miscible, polar, low vapour pressure ether. Its gravimetric adsorption to saturation of the solid, followed by vacuum evaporation to constant mass in order to measure surface area, is a simple and robust method (Cihacek and Bremner, 1979; Heilman *et al.*, 1965). Originally developed for SSA of soils, Quirk and Murray (1999) established that the original assumption of the glycol forming a monolayer on each opposing basal plane within a montmorillonite is approximately correct. The EGME method primarily correlates clay content and clay CEC, so is insensitive to the extent of any ion exchange, and hence represents a useful surface area method for mixed systems.

Attempts have been made to quantify the surface area components for clays using the direct measurement capability afforded by atomic force microscopy (AFM) (Bickmore *et al.*, 2002) and then comparing it to other methods such as N_2 BET and EGME (Macht *et al.*, 2011). The statistical limit of the AFM approach is less of an issue than the method of dispersion (to make the method viable) that subsequently resulted in delamination (Macht *et al.*, 2011), leading to large SSA values from the AFM for high surface area montmorillonite.

1.5.3.2 Scanning and Transmission Electron Microscopy

Scanning and transmission electron microscopy (SEM and TEM, respectively) are the most widely utilized tools for clay characterization after XRD. Scanning electron microscopy in particular is easy to apply and gives immediate feedback on the morphology of the clay (see cover image of this work). For example, even without clay identification, knowing whether the clay has a plate-like structure or a fibrous structure can have major implications for clay behaviour in the MPVC, especially in terms of beneficiation strategies.

SEM, coupled with electron dispersive X-ray spectroscopy (EDS), becomes a particularly useful adjunct to quantitative powder XRD (QXRD) as it can assist the identification of minor components and confirm individual mineral compositions. For example, out of 47 participants in the 3rd Reynolds Cup (2006), nine used SEM/EDS as a supplementary identification method or for quantification (Omotoso *et al.*, 2006). Interestingly, those that used it for quantification had a poorer overall outcome. Applying image processing to SEM micrographs (Liu *et al.*, 2011) enables a full characterization of clay microporosity by describing the pore structure in terms of the three major components of pore directionality, pore area distribution and pore form factor variation. The use of other SEM-based analysis methods such as electron back-scatter diffraction (EBSD) can be particularly useful in discerning the crystallography of individual clay platelets (Kohyama, 1992), although the method may be restricted to relatively large particles (Balan *et al.*, 2014).

The application of TEM to clay minerals, especially in the context of soil systems, has been reviewed by Elsass *et al.* (2008) and more generally elsewhere (Elsass, 2006; Guthrie George and Veblen David, 1990). Transmission electron microscopy, as well as providing morphology information for clays, can directly image the clay lattice fringe systems as exemplified by De Souza Santos (1993), and with selected area electron diffraction (SAED) and XRD can provide a very complete structural characterization. Having additional functions to the TEM, such as EDS in general, can be particularly useful (Kohyama, 1992) although, for example, TEM has been used to elucidate the nature of 'copper clays', demonstrating that copper resides as nanocrystalline oxide phases, an observation otherwise unresolvable with XRD and SEM/EDS (Kelm *et al.*, 2013).

1.6 Clay-Interface Chemistry

1.6.1 Clay Minerals and Their Surface Chemistry

The surface chemistry of clay minerals is distinctly influenced by (1) their chemical make-up; (2) the nature of the surface atoms that create an interface between the clay mineral and another phase (solution, gas or both); (3) special sites along the clay mineral that are marked by defects (missing metal centres); (4) the nature and origin (octahedral vs tetrahedral sheet) of layer charge (isomorphous substitution); and (5) the cations involved in charge compensation (Schoonheydt and Johnston, 2006).

The extraction and refining of ores usually involves interaction between clay mineral grains and water. Since clay mineral grains are usually small (<1 μm), their behaviour is usually governed by colloidal forces that are in turn dependent on the mineral's surface characteristics. The nature of clay surfaces is therefore pivotal in linking macroscopic behaviours of mineral processing slurries with the underlying crystal structure of the clay mineral. Geologically derived changes (e.g. edge-basal face ratios of clays influenced by how external (mineral, organic matter, other) environments/conditions act on crystal growth) to clay surfaces can produce unforeseen and potentially problematic modifications to their expected processing behaviour. Understanding the surface chemistry of clay systems is also an opportunity to judiciously avoid or enhance the behaviour in their respective process streams.

Interactions between simple colloidal particles (i.e. homogeneously charged, monodispersed spheres) comprise the sum of van der Waal's forces (attractive) and electrostatic forces (repulsive). Van der Waal's forces are essentially constant regardless of the solution conditions; thus the net force between particles depends on changes to the electrostatic forces. In aqueous media, charged particles modify the ion distribution close to their surfaces relative to the bulk solution. When interaction of these ion distributions takes place at distances beyond the influence of van der Waal's forces, particles are prevented from contacting one another (i.e. coagulating) and remain separated (dispersed). When the concentration of ions in bulk solution is high, the ion distribution adjacent to surfaces is compressed, allowing particles to approach close enough to one another for van der Waal's forces to dominate, inducing coagulation. The origin of surface charge in many colloidal materials is the result of pH-mediated equilibria. Below a critical pH, chemical equilibrium favours adsorption of excess H^+ ions, inducing a positive surface charge. Above this pH the reverse occurs, removing H^+ ions from the surface into solution and inducing a negative surface charge. In the pH range close to the critical pH, the surface charge is close to zero, thus interacting particles do not undergo strong repulsion and are therefore more likely to coagulate. This critical pH is commonly known as the pzc and relates only to the H^+/OH^- behaviour; the isoelectric point (iep) is a zero net charge situation that may be complicated by other sources of charge so may not be strictly equivalent or occur at the same pH.

1.6.2 Consequences of Permanent Charge

There are distinct physical consequences of the layers (1:1 or 2:1 layer units) not being charge neutral (a universal requirement of closed systems is to have net zero charge). In the first instance, the sheets develop the so-called electric double layers that act repulsively on each other. The only way in which any two negatively charged layers would approach each other is if considerable work (e.g. pressure → energy) is expended to bring the layers closer together or if the charge is compensated by a charge of opposite sign, i.e. countercharge. One form in which repelling layers can thus be brought closer together is through the charge compensation by interlayer cations, metals and other positively charged species. In low charge, 1:1 type clays, e.g. kaolinite, this is accomplished through H^+ ions that bridge the basal tetrahedral sheet to the octahedral sheet; in halloysite the same sheets are bridged by a layer of H_2O molecules. In smectites, notably montmorillonite, monovalent alkali (Li–Cs) and divalent alkaline (Mg–Ba) earth metals fulfil this role; however, third row transition metals, notably Mn through to Zn, heavier metals such as Ag, Cd, Au, Pb and U (as uranyl, UO_2^{2+}) will either replace ions from the exchange complex or adsorb specifically at terminal OH groups as well. The mineral processing consequence of liquor 'preg-robbing' is considered in Section 1.8.1. The ability to attract dissolved ions of positive charge to maintain overall neutrality may also contribute to reagent consumption.

In aqueous environments, the presence of alkali and alkaline earth metals acting as countercharge, for otherwise repelling clay mineral layers, has profound implications on the rheology of such systems because of the differences in solvation energies associated with the different cations. Moreover, the ability of clay mineral layers to approach one another is a function of the metal cations acting as the countercharge. Ultimately, the extent and origin (e.g. smectites vs vermiculites; both tetrahedral vs octahedral sheet) of the permanent negative charge and the physicochemical properties of hydrated (metal) cations acting as countercharges will determine to what extent clay mineral layers will overcome repulsion. It is this inherent nature of clay minerals to repel each other that contributes to many of the rheological problems in hydrometallurgical circuits.

1.7 Variability Within Clay Mineral Structures

There are a number of ways in which kaolinite and smectite clays vary from their ideal structure and/or particle morphology. Many of these differences can affect the surface chemistry of the particles with possible implications for their macroscopic behaviour.

1.7.1 Kaolinite

The kaolinite crystal structure is not considered capable of significant isomorphous substitution, hence known substantive differences that occur between kaolinites from different sources must necessarily relate to the original transformation of primary minerals. These include:

(1) 'Group level' differences in crystallography between halloysite ('hydrated' kaolinite), nacrite and dickite (highly ordered 'kaolinite') and kaolinite *sensu stricto*. These differences are relatively easy to identify using XRD. Halloysite, nacrite and dickite are rarely mentioned in mining/mineral processing literature.
(2) A more subtle crystallographic variation has been observed between highly ordered and poorly ordered kaolinites. Various indices based on XRD patterns (e.g. Hinckley Index) have been proposed to classify the extent of disorder. Differences in macroscopic behaviour such as minimum viscosity, brightness and equilibrium (settled) solids concentrations have been attributed to differences in structural order. Minor differences in Fe substitution are postulated as a root cause of the differences in structural order (Mestdagh *et al.*, 1980; Murray and Lyons, 1955; Murray and Lyons, 1960).
(3) Interlayered kaolinite–smectite are reported to produce higher cation exchange capacity in kaolinites, potentially inducing smectite-like behaviours (e.g. greater delamination and higher equilibrium solids content (Ma and Eggleton, 1999). The nature of the smectite interlayering may prevent its detection with routine XRD analysis. Interlayered kaolinite, along with other clays, are the primary problematic minerals in mineral sands tailings (Kasperski and Mikula, 2011).
(4) Possibly related to the crystallographic variation, physical differences between kaolinites (e.g. particle size, surface area and aspect ratio) can drastically alter handling properties. Kaolinite sourced from a single deposit can vary significantly in surface area (e.g. respective surface areas of 19.7 and 27.1 m^2/g for Birdwood kaolinite in Żbik (2006) and McFarlane *et al.* (2005). Further changes in physical characteristics can occur as a result of mechanical agitation under dry (Aglietti *et al.*, 1986) or wet (Kim

et al., 2004) conditions, with further implications for processing behaviour such as slurry rheology.

(5) Studies of commercial kaolinites ('kaolins') typically identify a range of accessory minerals present at up to 20 wt% (e.g. Teh *et al.*, 2009). Such contaminants need to be accounted for in efforts to understand kaolin behaviour in mineral processing through either isolated experiments on the individual mineral species present or, more commonly, through careful analysis of behavioural trends with respect to changes in bulk composition. Individual muscovite, hematite, gibbsite or silica suspensions, for example, settle to significantly higher equilibrium solids content than typical kaolinite suspensions.

1.7.2 Smectite

Many of the variations to the mineralogical ideal listed for kaolinite are also possible for smectites, including interlayering and mineral associations. However, since smectite clays can differ in extent and type of isomorphous substitution, additional considerations are needed for assessing them:

(1) The smectite mineral 'group' can be divided according to two broad criteria (Guven, 2009). The first, based on the nature of the octahedral sheet as tri- ($3 \times M^{2+}$) or di- ($2 \times M^{3+}$) octahedral is not considered to have a strong influence on macroscopic behaviour. Unlike the analogous difference between kaolinite and serpentines, both tri-octahedral and di-octahedral smectites are likely to exhibit clay-like characteristics, whereas serpentines tend not to be clay-like. The second criterion, based on the location of the charge deficiency in either the tetrahedral or octahedral sheets, is likely to produce specific or diffuse interlayer absorption, respectively, and alter macroscopic behaviours as a result.

(2) The extent of charge deficiency in smectites can range from 0.2 to 0.6 charge equivalents per half unit cell. Somewhat counter-intuitively, some studies have demonstrated that higher charge deficiency is related to lower magnitude rheological parameters, due to the presence of interstratified illite layers and/or stronger interlayer cation adsorption (Christidis *et al.*, 2006; Su and Shen, 2005).

(3) The oxidation state of Fe in either the tetrahedral or octahedral sheets is shown to cause variation in the properties of smectite-bearing materials due to differences in the magnitude of the charge deficiency. As an example, biological reduction of octahedral Fe^{3+} in smectites has been demonstrated to reduce the availability of nutrients such as NH_4^+ in soils while reducing the effective swelling pressure (Kostka *et al.*, 1999; Scherer and Zhang, 2002). Thus, the increased charge deficiency of the Fe^{2+} smectite has a greater capacity for holding interlayer cations, such as value-metals dissolved from ores in a mineral processing context.

(4) The nature of the interlayer cation plays a key role in modifying the behaviour of smectites. This modification can take place naturally, with the extreme case of geological

conversion from smectite to interlayered illite–smectite resulting in decreased value for industrial applications (e.g. reduced swelling) but presumably more favourable behaviour during mineral processing (Christidis and Huff, 2009). In most cases, judicious application of saline solutions, preferably rich in Ca^{2+}, can suppress smectite swelling and induce more manageable handling properties (Lim *et al.*, 2009).

1.8 Implications for Clay Mineral Behaviours in Mineral Processing

1.8.1 Preg-Robbing

The industry term 'preg-robbing' loosely refers to losses of commodity metals from pregnant process solutions prior to the commodity recovery step, resulting in <100 per cent recovery due to sorption processes encompassing ion exchange, surface adsorption, precipitation and other reactions with components in the process liquor. Thus, where mining operations utilize alkaline or acid leaching prior to a further concentration step, any metal sorption represents a process, and therefore also economic, loss. Originally used in the context of carbon-in-pulp (CIP) processing for gold ores containing carbonaceous gangue thought to rob or remove aurocyanide from the leach solution, while not a particularly scientific term (Habashi and van Deventer, 2001) 'preg-robbing' is well understood in the minerals processing industry. In terms of impact on the MPVC, as well as indirect processing impact in areas such as materials handling, rheology, particulate beneficiation, solid–liquid separation and tailings management, the clays can themselves directly impact on the commodity solution chemistry. In terms of economics it is essential that the nature and extent of such preg-robbing are understood. The table in the Appendix elaborates in detail the extent of potential 'preg-robbing' for various clays and ligand/ion interaction across the commodities of Au, Cu, Ni, Co, Zn, Pb and U as a function of pH, ionic strength, electrolyte composition, temperature, metal concentration, and competing ligands and quantified/qualified through metal adsorption, pH of maximum sorption, pH_{50} (pH at which 50 per cent sorption occurs) and other relevant factors (e.g. specific surface area, acid activation, among others). This level of understanding largely derives from the environmental chemistry literature and, while the table is extensive, the values cited are likely limited in validity to the conditions and parameters of the sorption study, including those not reported (NR) by the study authors. It is evident (see Appendix) and well known that sorption reactions of metals/cations on clays is strongly dependent on any number of these factors in combination and isolation (McBride, 1994; Sparks, 2003; Stumm, 1992). Hence, any change may not only alter the extent of sorption but also the outcome in the nature and stability (reversibility) of the surface complex(es). Furthermore, molecular-level details (e.g. specific surface adsorption geometry between sorbate and sorbent) are difficult to impossible to obtain from macroscopic sorption experiments only, thus bulk solution measurements reflect all underlying sorption mechanisms while specific clay features also remain unspecified (Cygan and Tazaki, 2014). Overall there is little published work for preg-robbing studies of commodity metals under industrially relevant conditions: high temperature, extreme pH, high ionic strength, etc. Preg-robbing in the industrial leach circuit is discussed in Chapter 3 for Au and U.

1.8.2 Settling Behaviour of Kaolinite Suspensions

The study by Zbik and Frost (2009) provides an example of how some of the mineralogical differences between kaolinites listed in Section 1.7 impact upon particle interactions and subsequently the rate and extent of settling. In contrast to the Georgia kaolinite sample (KGa-1), a sample of Birdwood kaolinite displays a greater magnitude of (negative) zeta potential across the pH range, a lower crystallinity index and lower equilibrium solids content at equivalent pH. Microscopic examination of cryogenically immobilized suspensions suggest the particles in the Georgia kaolinite suspension dispersion coagulate in a compact 'house of cards' configuration via electrostatic attraction between edge and basal faces. Birdwood kaolinite particles, on the other hand, appear to have an extensive edge–edge configuration; the apparent absence of repulsive force between the negatively charged basal faces appears to be the result of hydrocarbon-induced hydrophobicity.

Two particular strengths of the study by Zbik and Frost (2009) are, first, the multi-level characterization of the clay samples examined, enabling the crystallographic and surface chemistry of the clays to be mechanistically linked with macroscopic behaviours. Second, the use of a clay standard from the Clay Minerals Society Source Clays for which background physical characterization and comparison studies are available (see Costanzo (2001) and accompanying papers). This comprehensive approach is unfortunately rarely taken in clay studies relevant to minerals processing, and therefore limits the extent to which research findings can be compared to one another, including within industrial contexts.

1.8.3 Rheological Behaviour of Different Smectite Species

Expanding on the effect of layer charge differences noted in Section 1.7.2, different studies have reported apparent contradictions with respect to rheological behaviour. Su and Shen (2005) thermally modified a high-charge smectite to produce lower-charge varieties through insertion of Li into vacant octahedral spaces. With decreased charge, the resultant slurries show decreased swelling pressure and decreased tactoid size, apparently due to reduced electrostatic repulsion. Conversely Christidis *et al.* (2006) examined the swelling pressure of naturally formed smectites of different layer charge extents and found the opposite trend. They surmise that the increased swelling pressure is the result of reduced charge localization and non-specific cation adsorption to siloxane cavities in the low-charge smectites, while higher-charge smectites are also more likely to contain interstratified, non-swelling, illitic layers. Understanding whether one, or both, of these studies reflects the likely variability of rheological response (assuming swelling pressure to be an analogous property) is yet to be applied to smectite variability in metalliferous ore bodies.

These examples illustrate that there are many aspects of fundamental and applied clay science aspects yet to be investigated with direct potential benefit to handling clays in the MPVC. Hence, this presents a challenge to mineral processing operations, since even rudimentary clay characterization is often considered impractical or irrelevant, despite new techniques becoming available that make this task easier. It is also a challenge to mineral processing researchers to recognize the subtlety of clay mineral systems and ensure that fundamental and applied studies are sufficiently sophisticated in their method and conclusions.

1.9 Summary

The extensive physical and chemical complexity of clays translates to a host of MPVC problems. Seven key areas of clay behaviour are relevant to the MPVC: (1) water absorption/desorption (expansion behaviour) and net water 'holding' capacity; (2) suspension rheology; (3) suspension settling; (4) species adsorption/sorption/desorption (cation and anion exchange: preg-robbing); (5) dissolution/formation/precipitation; (6) clay–mineral adhesion/particle–particle interaction; and (7) clay–polymer adhesion. Five factors govern most of these behaviours: (1) clay type (including interlayer structure), group, subgroup and species; (2) particle morphology (size, aspect ratio, SSA); (3) surface charge, its distribution and double-layer interactions; (4) acid–base response; and (5) solution composition of the mineral slurry. In the context of the MPVC, the effects of clays on any unit process thus require a dual analytical approach: (1) testing one or more relevant behaviours (e.g. settling) of clay and (2) a factorial approach (variation and combination of one, two or more factors) to understand/quantify the variance in clay behaviour for that unit process.

Not nearly enough hydrometallurgical–clay research has yet been conducted. Of more than 86 000 entries covering research in clays, less than 1 per cent (since 1900) is *directly* associated with mining, hydrometallurgy and/or minerals processing. Other disciplines, e.g. agronomy and related sciences (e.g. soil) make up more than 20 per cent of accumulated knowledge. Metal–clay interactions via ion exchange, surface adsorption or any other mechanism satisfying *preg-robbing* has been extensively researched in environmental and agronomic contexts at low temperature, low ionic strengths (0.001–1.0) and pH (3–10), but not at elevated temperatures, in highly acidic or caustic pH regimes (<3, >10) and/or at high salinity ($I \gg 1.0$). A consequential shortfall is that variations in preg-robbing behaviour by clays as a function of any one or combination of these factors (temperature, pH, ionic strength, etc.) is not understood and cannot be readily used as a potential mitigation tool.

A large and extensive set of clay characterization techniques is available, but which combination produces the best result to understand/predict/know how clays impact a certain unit process? Careful consideration of the relevant clay behaviour and apparent factors will provide initial guidance to the appropriate tool(s); however, understanding what each technique indeed *measures* and the robustness of information that is *interpreted* from the data needs to be equally well understood. For example, SEM/EDAX does not measure the presence of a particular mineral, but rather infers its presence based on the elements co-occurring within an excitation volume (effectively an image pixel) of the sample. Consequently, QEMScan™ analysis cannot differentiate between rutile (TiO_2) and anatase (also TiO_2), and therefore specific ratios measured of Al, Si and K can no more than infer either whether the mineral conforms more to a 2:1 or 1:1 type clay mineral – direct evidence of clay species can only be gathered from probes that measure structural aspects, e.g. XRD, IR or XAS, and often a secondary tool (e.g. a TGA or SSA-EGME measurement) is required to confirm or strengthen the assignment of the first (including and especially inference-based assignments).

In the broader context of resource evaluation, including early exploration, where vast volumes have to be assessed, careful consideration needs to be given to the speed vs depth/extent of information. With ongoing improvements in the quality of semi-automated spectral and diffraction data, the next hurdle to improved ore-body understanding becomes integration and appropriate interpretation of large datasets. The abundance of data should allow improved selection of representative samples for both high-quality diagnostic analysis of clay content and metallurgical tests for proxy or pilot-scale testing. Some sense of the level of detail required in these studies can be gleaned from the accompanying chapters in which known clay behaviours in different processing contexts are described. However, the many subtleties of clay mineralogy may necessitate increasing the depth of analytical detail and mineralogical discrimination in order to enable adequate forecasting and control of processing behaviours. While clays will remain problematic for many aspects of mineral processing, the new technologies available to identify and quantify clay minerals, and the emerging understanding of their idiosyncrasies, promises an exciting and empowering experience for both operators and researchers seeking to understand them.

References

Abdrakhimova, E. S. & Abdrakhimov, V. Z. 2006. A Mossbauer spectroscopy study of the transformation of iron compounds in clay materials. *Russian Journal of Physical Chemistry*, 80 (7), 1077–1082.

Abidin, Z., Matsue, N. & Henmi, T. 2008. A new method for nano tube imogolite synthesis. *Japanese Journal of Applied Physics*, 47, 5079–5082.

Abramova, E., Lapides, I. & Yariv, S. 2007. Thermo-XRD investigation of monoionic montmorillonites mechanochemically treated with urea. *Journal of Thermal Analysis and Calorimetry*, 90 (1), 99–106.

Aglietti, E. F., Porto Lopez, J. M. & Pereira, E. 1986. Kinetic aspects of kaolinite acid dissolution: II. Mineral after mechanochemical treatment. *Reactivity of Solids*, 2 (1–2), 35–44.

Bailey, S. W. 1980. Summary of recommendations of AIPEA Nomenclature Committee. *Clay Minerals*, 15 (1), 85–93.

Balan, E., Allard, T., Boizot, B., Morin, G. & Muller, J. P. 2000. Quantitative measurement of paramagnetic Fe^{3+} in kaolinite. *Clays and Clay Minerals*, 48 (4), 439–445.

Balan, E., Calas, G. & Bish, D. L. 2014. Kaolin-group minerals: From hydrogen-bonded layers to environmental recorders. *Elements*, 10 (3), 183–188.

Balek, V., Benes, M., Subrt, J., *et al.* 2008. Thermal characterization of montmorillonite clays saturated with various cations. *Journal of Thermal Analysis and Calorimetry*, 92 (1), 191–197.

Bergaya, F. & Lagaly, G. 2006. General introduction: Clays, clay minerals, and clay science. In: Bergaya, F., Theng, B. K. G. & Lagaly, G. (eds) *Developments in Clay Science: Volume 1. Handbook of Clay Science.* New York: Elsevier.

Berger, B. R., Ayuso, R. A., Wynn, J. C. & Seal, R. R. 2008. Preliminary model of porphyry copper deposits. Open-File Report 2008-1321. U.S. Geological Survey.

Bickmore, B. R., Nagy, K. L., Sandlin, P. E. & Crater, T. S. 2002. Quantifying surface areas of clays by atomic force microscopy. *American Mineralogist*, 87 (5–6), 780–783.

Birch, W. D., Ciriotti, M. E., Downs, R. T., *et al.* 2012. The new IMA list of minerals: A work in progress. Update: November 2012. International Mineral Association.

Blum, A. E. & Eberl, D. D. 2004. Measurement of clay surface areas by polyvinylpyrrolidone (PVP) sorption and its use for quantifying illite and smectite abundance. *Clays Clay Minerals*, 52 (5), 589–602.

Blum, A., Lee, L. & Eberl, D. 2011. Measurement of clay surface areas by polyvinylpyrrolidone (PVP) sorption: A new method for quantifying illite and smectite abundance. *Clays and Clay Minerals*, 59 (2), 212–213.

Brigatti, M. F., Galan, E. & Theng, B. K. G. 2006. Structures and mineralogy of clay minerals. In: Bergaya, F., Theng, B. K. G. & Lagaly, G. (eds) *Developments in Clay Science: Volume 1. Handbook of Clay Science.* New York: Elsevier.

Carter, D. L., Mortland, M. M. & Kemper, W. D. 1986. Specific surface. In: Klute, A. (ed.) *Methods of Soil Analysis*, 2nd edition, Madison, WI: American Society of Agronomy and Soil Science Society of America.

Carvalho, A. P., Martins, A., Silva, J. M., *et al.* 2003. Characterization of the acidity of Al- and Zr-pillared clays. *Clays and Clay Minerals*, 51 (3), 340–349.

Christidis, G. E. & Huff, W. D. 2009. Geological aspects and genesis of bentonites. *Elements*, 5 (2), 93–98.

Christidis, G. E., Blum, A. E. & Eberl, D. D. 2006. Influence of layer charge and charge distribution of smectites on the flow behaviour and swelling of bentonites. *Applied Clay Science*, 34 (1–4), 125–138.

Churakov, S. V. & Dähn, R. 2012. Zinc adsorption on clays inferred from atomistic simulations and EXAFS spectroscopy. *Environmental Science & Technology*, 46 (11), 5713–5719.

Cihacek, L. J. & Bremner, J. M. 1979. A simplified ethylene glycol monoethyl ether procedure for assessment of soil surface area. *Soil Science Society of America Journal*, 43 (4), 821–822.

Costanzo, P. M. 2001. Baseline studies of the clay minerals society source clays: Introduction. *Clays and Clay Minerals*, 49 (5), 372–373.

Cradwick, P. D. G., Farmer, V. C., Russell, J. D., *et al.* 1972. Imogolite, a hydrated aluminum silicate of tubular structure. *Nature Physical Science*, 240 (104), 187–189.

Cuadros, J. & Dudek, T. 2006. FTIR investigation of the evolution of the octahedral sheet of kaolinite–smectite with progressive kaolinization. *Clays and Clay Minerals*, 54 (1), 1–11.

Cygan, R. T. & Tazaki, K. 2014. Interactions of kaolin minerals in the environment. *Elements*, 10 (3), 195–200.

de Kretser, R., Scales, P. J. & Boger, D. V. 1997. Improving clay-based tailings disposal: Case study on coal tailings. *American Institute of Chemical Engineers Journal*, 43 (7), 1894–1903.

De Souza Santos, P. 1993. The use of clay particle morphology studies to characterize industrial clay deposits: Examples from Brazil. *Clay Minerals*, 28 (4), 539–53.

Elsass, F. 2006. Transmission electron microscopy. In: Bergaya, F., Theng, B. K. G. & Lagaly, G. (eds) *Developments in Clay Science: Volume 1. Handbook of Clay Science.* New York: Elsevier.

Elsass, F., Chenu, C. & Tessier, D. 2008. Transmission electron microscopy for soil samples: Preparation methods and use. In: Ulery, A. L. & Richard Drees, L. (eds) *Methods of Soil Analysis. Part 5. Mineralogical Methods.* Madison, WI: Soil Science Society of America.

Emmerich, K., Wolters, F., Kahr, G. & Lagaly, G. 2009. Clay profiling: The classification of montmorillonites. *Clays and Clay Minerals*, 57 (1), 104–114.

Farmer, V. C., Adams, M. J., Fraser, A. R. & Palmieri, F. 1983. Synthetic imogolite: Properties, synthesis, and possible applications. *Clay Minerals*, 18 (4), 459–472.

Felix, S. P., Savill-Jowitt, C. & Brown, D. R. 2005. Base adsorption calorimetry for characterising surface acidity: A comparison between pulse flow and conventional 'static' techniques. *Thermochimica Acta*, 433 (1–2), 59–65.

Fischer, D., Caseri, W. R. & Hahner, G. 1998. Orientation and electronic structure of ion exchanged dye molecules on mica: An x-ray absorption study. *Journal of Colloid and Interface Science*, 198 (2), 337–346.

Freyssinet, P., Butt, C. R. M., Morris, R. C. & Piantone, P. 2005. Ore-forming processes related to lateritic weathering. In: Hedenquist, J. W., Thompson, J. F. H., Goldfarb, R. J. & Richards, J. P. (eds) *Economic Geology 100th Anniversary, Volume 1905–2005.* Littleton, CO: Society of Economic Geologists.

Fripiat, J. J. & Toussaint, F. 1960. Predehydroxylation state of kaolinite. *Nature*, 186 (4725), 627–628.

Frost, R. L. & Vassallo, A. M. 1996. The dehydroxylation of the kaolinite clay minerals using infrared emission spectroscopy. *Clays and Clay Minerals*, 44 (5), 635–651.

Gaite, J. M., Ermakoff, P., Allard, T. & Muller, J. P. 1997. Paramagnetic Fe^{3+}: A sensitive probe for disorder in kaolinite. *Clays and Clay Minerals*, 45 (4), 496–505.

Galán, E. 2006. Genesis of clay minerals. In: Bergaya, F., Theng, B. K. G. & Lagaly, G. (eds) *Developments in Clay Science: Volume 1. Handbook of Clay Science.* New York: Elsevier.

Galán, E. & Ferrell, R. E. 2013. Genesis of clay minerals. In: Bergaya, F. & Lagaly, G. (eds) *Developments in Clay Science: Volume 5. Handbook of Clay Science*, 2nd edition. New York: Elsevier.

Gehring, A. U., Fry, I. V., Luster, J. & Sposito, G. 1993. Vanadium(IV) in a multimineral lateritic saprolite: A thermoanalytical and spectroscopic study. *Soil Science Society of America Journal*, 57 (3), 868–873.

Gil, A., Korili, S. A. & Vicente, M. A. 2008. Recent advances in the control and characterization of the porous structure of pillared clay catalysts. *Catalysis Reviews-Science and Engineering*, 50 (2), 153–221.

Gualtieri, A. F., Ferrari, S., Leoni, M., *et al.* 2008. Structural characterization of the clay mineral illite-1M. *Journal of Applied Crystallography*, 41 (2), 402–415.

Guthrie George, D. & Veblen David, R. 1990. High-resolution transmission electron microscopy applied to clay minerals. In: Coyne, L. M., McKeever, S. W. S. & Blake, D. F. (eds) *Spectroscopic Characterization of Minerals and their Surfaces.* Washington, DC: American Chemical Society.

Guven, N. 2009. Bentonites: Clays for molecular engineering. *Elements*, 5 (2), 89–92.

Habashi, F. & van Deventer, J. S. J. 2001. Correspondence. *Minerals Engineering*, 14 (9), 1121–1122.

Haest, M., Cudahy, T., Laukamp, C. & Gregory, S. 2012. Quantitative mineralogy from infrared spectroscopic data: II. Three-dimensional mineralogical characterization of the Rocklea channel iron deposit, Western Australia. *Economic Geology*, 107 (2), 229–249.

Hähner, G., Marti, A., Spencer, N. D. & Caseri, W. R. 1996. Orientation and electronic structure of methylene blue on mica: A near edge x-ray absorption fine structure spectroscopy study. *Journal of Chemical Physics*, 104 (19), 7749–7757.

Hatherly, P., Medhurst, T. & Zhou, B. 2013. Geotechnical modelling based on geophysical logging data. In: *13th Coal Operators Conference*. University of Wollongong, Australian Institute of Mining and Metallurgy, 21–26.

Heilman, M. D., Carter, D. L. & Gonzalez, C. L. 1965. Ethylene glycol monoethyl ether (EGME) technique for determining soil-surface area. *Soil Science*, 100 (6), 409–413.

Helfferich, F. G. 1962. *Ion Exchange*. New York: McGraw-Hill.

Holuszko, M. E. & Laskowski, J. S. 2004. Use of pelletization to assess the effect of particle–particle interactions on coal handleability. *Physicochemical Problems of Mineral Processing*, 38 (1), 23–35.

Inoue, A. 1995. Formation of clay minerals in hydrothermal environments. In: Velde, B. (ed.) *Origin and Mineralogy of Clays*. Berlin: Springer.

Jagiello, J., Bandosz, T. J., Putyera, K. & Schwarz, J. A. 1995. Determination of proton affinity distributions for chemical systems in aqueous environments using a stable numerical solution of the adsorption integral equation. *Journal of Colloid and Interface Science*, 172 (2), 341–346.

Kahr, G. & Madsen, F. T. 1995. Determination of the cation exchange capacity and the surface area of bentonite, illite and kaolinite by methylene blue adsorption. *Applied Clay Science*, 9 (5), 327–336.

Karathanasis, A. D. & Hajek, B. F. 1982. Revised methods for rapid quantitative determination of minerals in soil clays. *Soil Science Society of America Journal*, 46 (2), 419–425.

Kasperski, K. L. & Mikula, R. J. 2011. Waste streams of oil sands: Characteristics and remediation. *Elements*, 7 (6), 387–392.

Kaufhold, S., Ufer, K., Kaufhold, A., *et al.* 2010. Quantification of allophane from Ecuador. *Clays and Clay Minerals*, 58 (5), 707–716.

Kelly, S. D., Hesterberg, D. & Ravel, B. 2008. Analysis of soils and minerals using x-ray absorption spectroscopy. In: Ulery, A. L. & Drees, L. R. (eds) *Methods of Soil Analysis. Part 5. Mineralogical Methods.* Madison, WI: Soil Science Society of America.

Kelm, U., Helle, S., Jerez, O. & Pincheira, M. 2013. What are copper clays? Geometallurgical implications. Copper Cobre, 2013. Santiago, Chile.

Kim, W., Jeon, H., Shin, H., Rim, J. & Kim, S. 2004. Dry and wet grinding effect on kaolinite and its zeolite formation in NaOH solution. *Geosystem Engineering*, 7 (1), 27–32.

Kloprogge, J. T. 2005. *The Application of Vibrational Spectroscopy to Clay Minerals and Layered Double Hydroxides*. Aurora, CO: Clay Minerals Society.

Kloprogge, J. T. & Frost, R. L. 2005. Infrared emission spectroscopy of clay minerals. In: Kloprogge, J. T. (ed.) *The Application of Vibrational Spectroscopy to Clay Minerals and Layered Double Hydroxides*. Aurora, CO: Clay Minerals Society.

Kohyama, N. 1992. Development of multi-functional transmission electron microscope and its application to clay science. *Nendo Kagaku*, 32 (4), 247–258.

Komadel, P. & Madejová, J. 2006. Acid activation of clay minerals. In: Bergaya, F., Theng, B. K. G. & Lagaly, G. (eds) *Developments in Clay Science: Volume 1. Handbook of Clay Science.* New York: Elsevier.

Kostka, J. E., Wu, J., Nealson, K. H. & Stucki, J. W. 1999. The impact of structural Fe(III) reduction by bacteria on the surface chemistry of smectite clay minerals. *Geochimica et Cosmochimica Acta*, 63 (22), 3705–3713.

Lagaly, G. 1994. Bentonites: Adsorbents of toxic substances. In: Schwuger, M. J. & Haegel, F. H. (eds) *Surfactants and Colloids in the Environment.* Darmstadt: Steinkopff.

Lee, S. Y. & Kim, S. J. 2002. Transmission electron microscopy of hexadecyltrimethylammonium-exchanged smectite. *Clay Minerals*, 37 (3), 465–471.

Lim, J., de Kretser, R. G. & Scales, P. J. 2009. Investigating the influence of total electrolyte concentration and sodium–calcium ion competition on controlled dispersion of swelling clays. *International Journal of Mineral Processing*, 93 (2), 95–102.

Liu, C., Shi, B., Zhou, J. & Tang, C. 2011. Quantification and characterization of microporosity by image processing, geometric measurement and statistical methods: Application on SEM images of clay materials. *Applied Clay Science*, 54 (1), 97–106.

Ma, C. & Eggleton, R. A. 1999. Cation exchange capacity of kaolinite. *Clays and Clay Minerals*, 47 (2), 174–180.

Macht, F., Eusterhues, K., Pronk, G. J. & Totsche, K. U. 2011. Specific surface area of clay minerals: Comparison between atomic force microscopy measurements and bulk-gas (N-2) and -liquid (EGME) adsorption methods. *Applied Clay Science*, 53 (1), 20–26.

MacKenzie, K. J. D., Okada, K. & Temuujin, J. 2004. Nanoporous inorganic materials from mineral templates. *Current Applied Physics*, 4 (2), 167–170.

McBride, M. B. 1994. *Environmental Chemistry of Soils*. New York: Oxford University Press.

McFarlane, A. J., Bremmell, K. E. & Addai-Mensah, J. 2005. Optimising the dewatering behaviour of clay tailings through interfacial chemistry, orthokinetic flocculation and controlled shear. *Powder Technology*, 160 (1), 27–34.

Mermut, A. R. & Lagaly, G. 2001. Baseline studies of the Clay Minerals Society source clays: Layer-charge determination and characteristics of those minerals containing 2:1 layers. *Clays and Clay Minerals*, 49 (5), 393–397.

Mestdagh, M. M., Vielvoye, L. & Herbillon, J. 1980. Iron in kaolinite: II. The relationship between kaolinite crystallinity and iron content. *Clay Minerals*, 15 (1), 1–13.

Meunier, A. 2006. Why are clay minerals small? *Clay Minerals*, 41 (2), 551–566.

Morris, H. D., Bank, S. & Ellis, P. D. 1990. Aluminum-27 NMR spectroscopy of iron-bearing montmorillonite clays. *Journal of Physical Chemistry*, 94 (7), 3121–3129.

Murad, E. 2006. Mössbauer spectroscopy of clays and clay minerals. In: Bergaya, F., Theng, B. K. G. & Lagaly, G. (eds) *Developments in Clay Science: Volume 1. Handbook of Clay Science.* New York: Elsevier.

Murray, H. H. 2006. Structure and composition of the clay minerals and their physical and chemical properties. In: *Developments in Clay Science. Volume 2. Applied Clay Mineralogy: Occurrences, Processing and Applications of Kaolins, Bentonites, Palygorskitesepiolite, and Common Clays.* Amsterdam: Elsevier.

Murray, H. H. & Lyons, S. C. 1955. Correlation of paper-coating quality with degree of crystal perfection of kaolinite. *Clays and Clay Minerals*, 4(1), 31–40.

Murray, H. H. & Lyons, S. C. 1960. Further correlations of kaolinite crystallinity with chemical and physical properties. *Clays and Clay Minerals*, 8 (1), 11–17.

Omotoso, O., McCarty, D. K., Hillier, S. & Kleeberg, R. 2006. Some successful approaches to quantitative mineral analysis as revealed by the 3rd Reynolds Cup contest. *Clays and Clay Minerals*, 54 (6), 748–760.

Parfitt, R. L. 2009. Allophane and imogolite: Role in soil biogeochemical processes. *Clay Minerals*, 44 (1), 135–155.

Pevear, D. R. 1999. Illite and hydrocarbon exploration. *Proceedings of the National*

Academy of Science USA, 96, 3440–3446.
Quigley, R. M., Inculet, I. I. & Beisser, E. M. J. 1982. Clay–coal charge transfer and beneficiation by dry mineral removal. *Canadian Geotechnical Journal*, 19 (4), 508–511.
Quirk, J. P. 1955. Significance of surface areas calculated from water vapor sorption isotherms by use of the BET equation. *Soil Science*, 80 (6), 423–430.
Quirk, J. P. & Murray, R. S. 1999. Appraisal of the ethylene glycol monoethyl ether method for measuring hydratable surface area of clays and soils. *Soil Science Society of America Journal*, 63 (4), 839–849.
Rábago, K. R., Lovins, A. B. & Feiler, T. E. 2001. *Energy and Sustainable Development in the Mining and Minerals Industries*. London: International Institute for Environment and Development.
San Miguel, G., Lambert, S. D. & Graham, N. J. D. 2006. A practical review of the performance of organic and inorganic adsorbents for the treatment of contaminated waters. *Journal of Chemical Technology and Biotechnology*, 81 (10), 1685–1696.
Scherer, H. W. & Zhang, Y. 2002. Mechanisms of fixation and release of ammonium in paddy soils after flooding: III. Effect of the oxidation state of octahedral Fe on ammonium fixation. *Journal of Plant Nutrition and Soil Science*, 165 (2), 185–189.
Schoonheydt, R. A. 1994. Organization and spectroscopy of dyes on submicron-sized crystalline solids. In: Masuhara, H., De Shryver, F. C., Kitamura, N. & Tamai, N. (eds) *Microchemistry: Spectroscopy and Chemistry in Small Domains*. Amsterdam: North Holland, Elsevier Science B.V.
Schoonheydt, R. A. & Johnston, C. T. 2006. Surface and interface chemistry of clay minerals. In: Bergaya, F., Theng, B. K. G. & Lagaly, G. (eds) *Developments in Clay Science: Volume 1. Handbook of Clay Science*. New York: Elsevier.
Schroeder, P. A. & Erickson, G. 2014. Kaolin: From ancient porcelains to nanocomposites. *Elements*, 10 (3), 177–182.
Sing, K. S. W., Everett, D. H., Haul, R. A. W., *et al.* 1985. Reporting physisorption data for gas/solid systems with special reference to the determination of surface area and porosity (Recommendations 1984). *Pure and Applied Chemistry*, 57 (4), 603–619.
Sparks, D. L. 2003. *Environmental Soil Chemistry*. San Diego, CA: Academic Press.
Stumm, W. 1992. *Chemistry of the Solid–Water Interface: Processes at the Mineral–Water and Particle–Water Interface in Natural Systems*. New York: Wiley.
Su, C.-C. & Shen, Y.-H. 2005. Preparation and dispersive behaviors of reduced charge smectite. *Colloids and Surfaces A: Physical and Engineering Aspects*, 259 (1–3), 173–179.
Teh, E.-J., Leong, Y. K., Liu, Y., Fourie, A. B. & Fahey, M. 2009. Differences in the rheology and surface chemistry of kaolin clay slurries: The source of the variations. *Chemical Engineering Science*, 64 (17), 3817–3825.
van der Gaast, S. J., Wada, K., Wada, S. I. & Kakuto, Y. 1985. Small-angle x-ray powder diffraction, morphology, and structure of allophane and imogolite. *Clays and Clay Minerals*, 33 (3), 237–243.
Velde, B. 1995. *Origin and Mineralogy of Clays: Clays and the Environment*. Berlin: Springer-Verlag.
Verburg, K. & Baveye, P. 1994. Hysteresis in the binary exchange of cations on 2:1 clay minerals: A critical review. *Clays and Clay Minerals*, 42 (2), 207–220.
Ward, C. R., Nunt-Jaruwong, S. & Swanson, J. 2005. Use of mineralogical analysis in geotechnical assessment of rock strata for coal mining. *International Journal of Coal Geology*, 64 (1–2), 156–171.
Wolters, F., Lagaly, G., Kahr, G., Nueesch, R. & Emmerich, K. 2009. A comprehensive

characterization of dioctahedral smectites. *Clays and Clay Minerals*, 57 (1), 115–133.

Yermiyahu, Z., Lapides, I. & Yariv, S. 2002. Thermo-XRD analysis of the adsorption of Congo-red by montmorillonite saturated with different cations. *Journal of Thermal Analysis and Calorimetry*, 69 (1), 317–332.

Yu, C.-H., Newton, S. Q., Norman, M. A., *et al.* 2000. Molecular dynamics simulations of the adsorption of methylene blue at clay mineral surfaces. *Clays and Clay Minerals*, 48 (6), 665–681.

Zbik, M. 2006. Micro-structural explanation for differences in gelation properties of kaolinites from Birdwood (S. Australia) and Georgia (U.S.A.). *Clay Science*, 12 (Supplement 2), P31–P36.

Zbik, M. & Frost, R. L. 2009. Micro-structural differences in kaolinite suspensions. *Journal of Colloid and Interface Science*, 339 (1), 110–116.

Appendix: Surface Adsorption Parameters of Commodity Metals on Common Clays

Clay	Metal	Ox. state	Conc. range	Counter-ion	T (°C)	pH range	Ionic strength (*I*) range	Max sorbed or max adsorp. capacity	% of total	~pH of max adsorp	Observations/ comments	Ref.
Acid-activated kaolinite	Co	(II)	170–4244 μM	NO_3^-	30 °C	5.8	NR	205 μmol g^{-1}§		5.8	Increase in the no. of surface functional groups	1
Acid-activated kaolinite	Co	(II)	849 μM	NO_3^-	30–40°C	5.8	NR	~104–118 μmol g^{-1}		5.8	Tendency for adsorp. to increase over T range	1
Acid-activated kaolinite	Co	(II)	849 μM	NO_3^-	30 °C	1.0 10.0	NR	238 μmol g^{-1}		9.0		1
Acid-activated montmorillonite	Co	(II)	170–4244 μM	NO_3^-	30 °C	5.8	NR	504 μmol g^{-1}§		5.8	Increase in the no. of surface functional groups	1
Acid-activated montmorillonite	Co	(II)	849 μM	NO_3^-	30–40°C	5.8	NR	~255–289 μmol g^{-1}		5.8	Tendency for adsorp. to decrease with increasing T	1
Acid-activated montmorillonite	Co	(II)	849 μM	NO_3^-	30 °C	1.0–10.0	NR	~424 μmol g^{-1}		~9.0		1
Ca-montmorillonite	Co	(II)	10–150 μM	ClO_4^-	1 °C	6.5	NR	270 μmol g^{-1}	NR			2
Ca-montmorillonite	Co	(II)	10–150 μM	ClO_4^-	14 °C	6.5	NR	390 μmol g^{-1}	NR			2
Ca-montmorillonite	Co	(II)	10–150 μM	ClO_4^-	33 °C	6.5	NR	480 μmol g^{-1}	NR			2
Ca-montmorillonite	Co	(II)	10–150 μM	ClO_4^-	40 °C	6.5	NR	520 μmol g^{-1}	NR			2
Chlorite	Co	(II)	1698 μM	NO_3^-	25 °C	3.0	NR	not determinable	7	n/a	Very slow uptake at pH 3	3

Table (*cont.*)

Clay	Metal	Ox. state	Conc. range	Counter-ion	T (°C)	pH range	Ionic strength (*I*) range	Max sorbed or max adsorp. capacity	% of total	~pH of max adsorp	Observations/ comments	Ref.
Chlorite	Co	(II)	1698 μM	NO_3^-	25 °C	7	NR	not determinable	24	n/a	Increasingly negative edge charge promotes Co(II) uptake	3
K-kaolinite	Co	(II)	50 μM	NO_3^-	10 °C	3.0–10.0	0.005	7.8 μmol g^{-1}	100	~9.5	T dependent adsorp. edges for Co on kaolinite	4
K-kaolinite	Co	(II)	50 μM	NO_3^-	25 °C	3.0–10.0	0.005	7.8 μmol g^{-1}	100	~9.0	T dependent adsorp. edges for Co on kaolinite	4
K-kaolinite	Co	(II)	50 μM	NO_3^-	40 °C	3.0–10.0	0.005	7.8 μmol g^{-1}	100	~9.0	T dependent adsorp. edges for Co on kaolinite	4
K-kaolinite	Co	(II)	50 μM	NO_3^-	55 °C	3.0–10.0	0.005	7.8 μmol g^{-1}	100	~8.0	T dependent adsorp. edges for Co on kaolinite	4
K-kaolinite	Co	(II)	50 μM	NO_3^-	70 °C	3.0–10.0	0.005	7.8 μmol g^{-1}	100	~7.5	T dependent adsorp. edges for Co on kaolinite	4
K-kaolinite	Co	(II)	≤80 μM	NO_3^-	10–70°C	5.5	0.005	16.2 μmol g^{-1}		5.5	Max surface coverage possible at 10 °C, but max adsorp. within [Co] was at 70 °C ~8.8 μmol g^{-1}	4

K-kaolinite	Co	(II)	≤80 μM	NO_3^-	10–70 °C	7.0	0.005	32.4 μmol g^{-1}	7.0	K, the partitioning coefficient for Co increases with T at pH 7.0 and 5.5	4
Kaolinite	Co	(II)	up to 17 000 μM	SO_4^{2-}	25 °C	NR	NR	15.6 μmol g^{-1}§	NR		5
Kaolinite	Co	(II)	up to 17 000 μM	SO_4^{2-}	40 °C	NR	NR	25.0 μmol g^{-1}§	NR		5
Kaolinite	Co	(II)	100 μM	NO_3^-	NR	3.0–10.0	0.005	~6.4 μmol g^{-1}	~9.0		6
Kaolinite	Co	(II)	100 μM	NO_3^-, 100 μM Asp	NR	3.0–10.0	0.005	~6.4 μmol g^{-1}	~10.0		6
Kaolinite	Co	(II)	100 μM	NO_3^-, 500 μM Asp	NR	3.0–10.0	0.005	~4.1 μmol g^{-1}	~10.0		6
Kaolinite	Co	(II)	100 μM	NO_3^-, 1000 μM Asp	NR	3.0–10.0	0.005	~2.7 μmol g^{-1}	~9.0		6
Kaolinite	Co	(II)	0–50 μM	NO_3^-, 100 μM Asp	NR	5.5	0.005	~2.6 μmol g^{-1}	5.5		6
Kaolinite	Co	(II)	0–50 μM	NO_3^-, 100 μM Asp	NR	7.5	0.005	~4.1 μmol g^{-1}	7.5		6
Kaolinite	Co	(II)	170–4244 μM	NO_3^-	30 °C	5.8	NR	190 μmol g^{-1}§	5.8		1
Kaolinite	Co	(II)	849 μM	NO_3^-	30–40 °C	5.8	NR	~98–104 μmol g^{-1}	5.8	Tendency for adsorp. to increase over T range	1
Kaolinite	Co	(II)	849 μM	NO_3^-	30 °C	1.0–10.0	NR	~186 μmol g^{-1}	~9.0		1
Montmorillonite	Co	(II)	170–4244 μM	NO_3^-	30 °C	5.8	NR	486 μmol g^{-1}§	5.8		1

Table (*cont.*)

Clay	Metal	Ox. state	Conc. range	Counter-ion	T (°C)	pH range	Ionic strength (*I*) range	Max sorbed or max adsorp. capacity	% of total	~pH of max adsorp	Observations/ comments	Ref.
Montmorillonite	Co	(II)	849 μM	NO_3^-	30–40 °C	5.8	NR	~238–272 μmol g^{-1}		5.8	Tendency for adsorp. to decrease with increasing T	1
Montmorillonite	Co	(II)	849 μM	NO_3^-	30 °C	1.0–10.0	NR	~390 μmol g^{-1}		~9.0		1
Na-bentonite	Co	(II)	1698–42 445 μM	NO_3^-	25 °C	4.2–6.4	NR	2365 μmol g^{-1}		NR		7
Na-bentonite	Co	(II)	12 733–42 445 μM	NO_3^-	55 °C	4.2–6.4	NR	3032 μmol g^{-1}		NR		7
Na-bentonite	Co	(II)	340 μM	NO_3^-, Ca^{2+}	25 °C	2.0–13.0	0.03	1700 μmol g^{-1}	100	~9.0	Adsorp. edge shift to lower pH when fulvic acid is present, no effect from I,	8
Na-bentonite	Co	(II)	0–450 μM	NO_3^-, Ca^{2+}	25 °C	6.3	0.03	1400 μmol g^{-1}			No effect from 7.1 mg L^{-1} fulvic acid on Co(II) adsorp.	8
Na-bentonite	Co	(II)	0–450 μM	NO_3^-, Ca^{2+}	25 °C	7.3	0.03	1600 μmol g^{-1}			Small depressing effect from 7.1 mg L^{-1} fulvic adsorp.	8
Na-hectorite	Co	(II)	10 000 μM	Cl^-	NR	6.2	NR	265 μmol g^{-1}	NR	6.2		9
Sepiolite	Co	(II)	0.1–10 000 μM	NO_3^-	20 °C	<7.8	NR	80 μmol g^{-1}§		NR	T pH <7.8 assuring adsorp. and not ppt reactions dominate, endo.	10

Sepiolite	Co	(II)	0.1–10 000 μM	NO_3^-	30 °C	<7.8	NR	128 μmol g^{-1}§		NR	T pH <7.8 assuring adsorp. and not ppt reactions dominate, endo.	10
Sepiolite	Co	(II)	0.1–10 000 μM	NO_3^-	40 °C	<7.8	NR	158 μmol g^{-1}§		NR	T pH <7.8 assuring adsorp. and not ppt reactions dominate, endo.	10
Vermiculite	Co	(II)	0–15 μM	NR	25 °C	NR	NR	760–1237 μmol g^{-1}§		NR		11
Acid-activated kaolinite	Ni	(II)	170–852 μM	NO_3^-	30 °C	5.7	NR	169 μmol g^{-1}**		5.7	Creation of additional edge sites	12
Acid-activated montomorillonite	Ni	(II)	170–852 μM	NO_3^-	30 °C	5.7	NR	363 μmol g^{-1}**		5.7		12
Bentonite	Ni	(II)	1703 μM	Cl^-	30 °C	1.0–8.0	NR	65 μmol g^{-1}	95	6.0	pH	13
Bentonite	Ni	(II)	681–3407 μM	Cl^-	30 °C	6.0	NR	1577 μmol g^{-1}§		6.0	Very high 'monolayer' coverage, probability of $NiOHCO_3$ ppt	13
Ca-montmorillonite	Ni	(II)	170 μM	NO_3^-	20 °C	4.0–10.0	0.003	306 μmol g^{-1}	90	9	Background e'lyte is 0.001M $Ca(NO_3)_2$, adsorp. at pH 4 ~20%	14
Ca-montmorillonite	Ni	(II)	170 μM	NO_3^-	40 °C	4.0–10.0	0.003	323 μmol g^{-1}	95	9	Background e'lyte is 0.001M $Ca(NO_3)_2$, adsorp. at pH 4 ~40%	14

Table (*cont.*)

Clay	Metal	Ox. state	Conc. range	Counter-ion	T (°C)	pH range	Ionic strength (*I*) range	Max sorbed or max adsorp. capacity	% of total	~pH of max adsorp	Observations/ comments	Ref.
Ca-montmorillonite	Ni	(II)	170 μM	NO_3^-	60 °C	4.0–10.0	0.003	323 μmol g^{-1}	95	8.5	Background e'lyte is 0.001M $Ca(NO_3)_2$, adsorp. at pH 4 ~45%	14
Ca-montmorillonite	Ni	(II)	1–306 μM	NO_3^-	20 °C	6.4	0.003	153 μmol g^{-1}		6.4	Background e'lyte is 0.001M $Ca(NO_3)_2$, T	14
Ca-montmorillonite	Ni	(II)	1–306 μM	NO_3^--	40 °C	6.4	0.003	290 μmol g^{-1}		6.4	Background e'lyte is 0.001M $Ca(NO_3)_2$, T	14
Ca-montmorillonite	Ni	(II)	1–306 μM	NO_3^-	60 °C	6.4	0.003	375 μmol g^{-1}		6.4	Background e'lyte is 0.001M $Ca(NO_3)_2$, T	14
Kaolinite	Ni	(II)	170–852 μM	NO_3^-	30 °C	5.7	NR	121 μmol g^{-1}**		5.7		12
Kaolinite	Ni	(II)	136–647 μM	NO_3^-	25 °C	6.0	NR	906 μmol g^{-1}‡		6.0		15
Kaolinite	Ni	(II)	136–647 μM	NO_3^-	35 °C	6.0	NR	792 μmol g^{-1}‡		6.0		15
Kaolinite	Ni	(II)	136–647 μM	NO_3^-	45 °C	6.0	NR	728 μmol g^{-1}‡		6.0		15
Kaolinite	Ni	(II)	136–647 μM	NO_3^-	52 °C	6.0	NR	922 μmol g^{-1}‡		6.0		15
Kaolinite	Ni	(II)	1703 μM	Cl^-	30 °C	1.0–8.0	NR	61 μmol g^{-1}	90	6.0	pH	13
Kaolinite	Ni	(II)	681–3407 μM	Cl^-	30 °C	6.0	NR	2385 μmol g^{-1}§		6.0	Very high 'monolayer' coverage, probability of $NiOHCO_3$ ppt	13
Kaolinite	Ni	(II)	1 μM	$Ca(ClO_4)_2$	25 °C	5.5–7.5	0.03		50	5.8*	pH values are pH_{50}	16
Kaolinite	Ni	(II)	1 μM	$CaCl_2$	25 °C	5.5–7.5	0.03		50	5.82*	pH values are pH_{50}	16

Kaolinite	Ni	(II)	1 µM	$CaSO_4$	25 °C	5.5–7.5	0.03		50	6.02*	pH values are pH_{50}	16
Montmorillonite	Ni	(II)	852–5111 µM	NO_3^-	25 °C	~6	NR	220 µmol g^{-1}			pH/edge charge	17
Montmorillonite	Ni	(II)	1704 µM	NO_3^-	25 °C	2.5–9	NR	85 µmol g^{-1}	>99	>4	pzc, hydrolysis of Ni	17
Montmorillonite	Ni	(II)	170–852 µM	NO_3^-	30 °C	5.7	NR	359 µmol g^{-1}**		5.7		12
Montmorillonite	Ni	(II)	1 µM	$Ca(ClO_4)_2$	25 °C	5.5–7.5	0.03		50	5.28*	pH values are pH_{50}	16
Montmorillonite	Ni	(II)	1 µM	$CaCl_2$	25 °C	5.5–7.5	0.03		50	5.28*	pH values are pH_{50}	16
Montmorillonite	Ni	(II)	1 µM	$CaSO_4$	25 °C	5.5–7.5	0.03		50	5.81*	pH values are pH_{50}	16
Na-bentonite	Ni	(II)	0.26–10.4 µM	NR	NR	8	0.001	12 µmol g^{-1}	100		Linear adsorp. isotherm, no saturation obtained, also potentially due to $Ni(OH)_2$ ppt at pH 8	18
Na-illite	Ni	(II)	48.8 µM	$NaNO_3$	25 °C	3.0–9.0	0.001	1627 µmol g^{-1}	100	5.5	*I*, pH – edge charge	19
Na-illite	Ni	(II)	48.8 µM	$NaNO_3$	25 °C	3.0–9.0	0.01	1627 µmol g^{-1}	100	7.5	*I*, pH – edge charge	19
Na-illite	Ni	(II)	48.8 µM	$NaNO_3$	25 °C	3.0–9.0	0.1	1627 µmol g^{-1}	100	8.0	*I*, pH – edge charge	19
Na-montmorillonite	Ni	(II)	8.5 µM	NO_3^-	20 °C	4–7.5	0.01	4.2 µmol g^{-1}		7.5		20
Na-montmorillonite	Ni	(II)	8.5 µM	NO_3^-, 50 µM citr.	20 °C	4–7.5	0.01	0.8 µmol g^{-1}		5.0	Citr. complx.	20
Na-montmorillonite	Ni	(II)	0–4 µM†	NO_3^-	20 °C	6.0	0.01	1.8 µmol g^{-1}		6.0		20
Na-montmorillonite	Ni	(II)	0–4 µM†	NO_3^-, 10 µM citr.	20 °C	6.0	0.01	1.6 µmol g^{-1}		6.0	Citr. complx.	20
Na-montmorillonite	Ni	(II)	0–4 µM†	NO_3^-, 100 µM citr.	20 °C	6.0	0.01	1.0 µmol g^{-1}		6.0	Citr. complx.	20

Table (*cont.*)

Clay	Metal	Ox. state	Conc. range	Counter-ion	T (°C)	pH range	Ionic strength (*I*) range	Max sorbed or max adsorp. capacity	% of total	~pH of max adsorp	Observations/ comments	Ref.
Na-montmorillonite	Ni	(II)	0.3 μM	$NaClO_4$	NR	3.0–10.0	0.01	K_d ~4.0		~9.0	K_d ~10^4, pH ~4.8	21
Na-montmorillonite	Ni	(II)	0.3 μM	$NaClO_4$	NR	3.0–10.0	0.03	K_d ~4.0		~9.0	K_d ~10^3, pH ~4.8	21
Na-montmorillonite	Ni	(II)	0.3 μM	$NaClO_4$	NR	3.0–10.0	0.1	K_d ~4.0		~9.0	K_d ~10^2, pH ~3	21
Na-montmorillonite	Ni	(II)	1 nM–100 μM	$NaClO_4$	NR	5.9	0.1	6.3 μmol g^{-1}			Trace conc. of Ni were used to ensure only adsorp. and not ppt reactions were measured	21
Na-montmorillonite	Ni	(II)	1 nM–100 μM	$NaClO_4$	NR	7.0	0.1	6.3 μmol g^{-1}			Trace conc. of Ni were used to ensure only adsorp. and not ppt reactions were measured	21
Na-montmorillonite	Ni	(II)	1 nM–100 μM	$NaClO_4$	NR	8.2	0.1	25.1 μmol g^{-1}			Trace conc. of Ni were used to ensure only adsorp. and not ppt reactions were measured	21
Vermiculite	Ni	(II)	0–15 μM	NR	25 °C	NR	NR	930–1320** μmol g^{-1}		NR		11
Vermiculite	Ni	(II)	100 μM	Acet.	NR	2.5–8.0	0.01	80 μmol g^{-1}	80	8.0		22

Vermiculite	Ni	(II)	1000 μM	Acet.	NR	5.5	0.01	432 μmol g^{-1}		5.5		22
Vermiculite	Ni	(II)	100 μM	Acet.	NR	5.5	0.01			5.5		22
Vermiculite	Ni	(II)	100 μM	Acet. 1 mM EDTA	NR	5.5	0.01		6.3 of 100 μM	5.5	Ligand complx. repulsion	22
Vermiculite	Ni	(II)	100 μM	Acet. 1 mM NTA	NR	5.5	0.01		6.8 of 100 μM	5.5	Ligand complx. repulsion	22
Vermiculite	Ni	(II)	100 μM	Acet. 1 mM DTPA	NR	5.5	0.01		5.2 of 100 μM	5.5	Ligand complx. repulsion	22
Vermiculite	Ni	(II)	100 μM	Acet. 1 mM OxA	NR	5.5	0.01		64.4 of 100 μM	5.5	Ligand complx. repulsion	22
Vermiculite	Ni	(II)	100 μM	Acet. 1 mM CitA	NR	5.5	0.01		80.1 of 100 μM	5.5	Ligand complx. repulsion	22
Acid-activated green clay	Cu	(II)	800–10 000 μM	SO_4^{2-}	20 °C	5.5	NR	219 μmol g^{-1}§			Acid activation → increase in edge sites	23
Acid-activated red clay	Cu	(II)	800–10 000 μM	SO_4^{2-}	20 °C	5.5	NR	401 μmol g^{-1}§			Acid activation → increase in edge sites	23
Acid-activated bentonite	Cu	(II)	NR	NO_3^-	25 °C	NR		668 μmol g^{-1}, 6.1 μmol m^{-2}**			*I*, pH	24
Acid-activated kaolinite	Cu	(II)	157–787 μM	SO_4^{2-}	30–40 °C	1–6	NR	159 μmol g^{-1}§	NR	5.7	Acid-activation, edge charge, H^+ comp.	25
Acid-activated montmorillonite	Cu	(II)	157–787 μM	SO_4^{2-}	30–40 °C	1–6	NR	509 μmol g^{-1}§	NR	5.7	Edge-charge, H^+ comp.	25
Bentonite	Cu	(II)	NR	NO_3^-	25 °C	NR		507 μmol g^{-1}, 13.8 μmol m^{-2}**			*I*, pH	24
Ca-rectorite	Cu	(II)	1000 μM	Cl-	25 °C	2.0–8.0	NR	~158 μmol g^{-1}	~95	7.0	pH	26
Ca-rectorite	Cu	(II)	1000 – 12000 μM	Cl-	25 °C	5.2	NR	~210 μmol g^{-1}		5.2		26

Table (*cont.*)

Clay	Metal	Ox. state	Conc. range	Counter-ion	T (°C)	pH range	Ionic strength (*I*) range	Max sorbed or max adsorp. capacity	% of total	~pH of max adsorp	Observations/ comments	Ref.
Feldspar	Cu	(II)	Up to 4 µM	NO_3^-	25 °C	5.4	0.01–0.1	33 µmol g^{-1}, 1.8 µMol m^{-2}		5.4	Strong *I* depend. max adsorp. at I = 0.01, strong K–Cu exchange, adsorp. observed	27
Illite	Cu	(II)	Up to 1500 µM	NO_3^-	25 °C	5.45–6.05	0.1	~40 µmol g^{-1}		6.05	pH	28
Illite	Cu	(II)	1050–5440 µM	NO_3^-	25 °C	3.0–8.0	0.1		100	7.0	Adsorp. edge shifts to lower pH as the initial [Cu] decreases	28
Illite	Cu	(II)	160 µM	NO_3^-	25 °C	3.5–10.0		~500 µmol g^{-1}		7.0	Complx. ligands, pH and surface charge density	29
Kaolinite	Cu	(II)	157–787 µM	SO_4^{2-}	30–40 °C	1–6	NR	143 µmol g^{-1}§	NR	5.7	Edge-charge, H^+ comp.	25
Kaolinite	Cu	(II)	787 µM	SO_4^{2-}	30 °C	1.0–6.0	NR	61 µmol g^{-1}		6.0	pH, SSA	30
Kaolinite	Cu	(II)	157–787 µM	SO_4^{2-}	30 °C	5.7	NR	69 µmol g^{-1}**		5.7	pH, SSA	30
Kaolinite	Cu	(II)	160 µM	NO_3^-	25 °C	4.0–9.0	NR	~300 µmol g^{-1}		7.0	complx. ligands, pH and surface charge density	31
Kaolinite/illite mixture (green clay)	Cu	(II)	800–10 000 µM	SO_4^{2-}	20 °C	5.5	NR	120 µmol g^{-1}§			no. of edge sites, less structural Mg	23
Kaolinite/illite mixture (red clay)	Cu	(II)	800–10 000 µM	SO_4^{2-}	20 °C	5.5	NR	283 µmol g^{-1}§			no. of edge sites, more structural Mg	23

Montmorillonite	Cu	(II)	787–4724 µM	NO_3^-	25 °C	~6	NR	119 µmol g^{-1}§			pH/edge charge	17
Montmorillonite	Cu	(II)	1575 µM	NO_3^-	25 °C	2.5–9	NR	78 µmol g^{-1}§	>99	>7	pzc, hydrolysis of Cu	17
Montmorillonite	Cu	(II)	787 µM	SO_4^{2-}	30 °C	1.0–6.0	NR	331 µmol g^{-1}		6.0	pH, SSA	30
Montmorillonite	Cu	(II)	157–787 µM	SO_4^{2-}	30 °C	5.7	NR	474 µmol g^{-1}**		5.7	pH, SSA	30
Montmorillonite	Cu	(II)	160 µM	NO_3^-	25 °C	3.5–10.0		~700 µmol g^{-1}		7.0	Complx. ligands, pH and surface charge density	29
Montmorillonite	Cu	(II)	157–787 µM	SO_4^{2-}	30–40 °C	1–6	NR	500 µmol g^{-1}§	NR	5.7	Edge-charge, H^+ comp.	25
Muscovite	Cu	(II)	up to 4 µM	NO_3^-	25 °C	5.4	0.01–0.1	179 µmol g^{-1}, ~4.5 µmol m^{-2}		5.4	No *I* depend. strong K–Cu exchange, possible Cu–Cu dimerization and adsorp. to 001 basal planes and edge-sites	27
Sepiolite	Cu	(II)	1500 µM	ClO_4^-	25 °C	3.5–7.0	0.1	~160 µmol g^{-1}		7.0	At pH 4.9, max adsorp. ~10 µmol g^{-1}	32
Sepiolite	Cu	(II)	1500 µM	ClO_4^-	25 °C	3.5–7.0	0.01	~160 µmol g^{-1}		7.0	At pH 4.9, max adsorp. ~70 µmol g^{-1}	32
Sepiolite	Cu	(II)	1500 µM	ClO_4^-	25 °C	5.9	0.01	109 µmol g^{-1}		5.9	*I* depend. of Cu adsorp. on sepiolite	32
Sepiolite	Cu	(II)	1500 µM	ClO_4^-	25 °C	5.4	0.1	71 µmol g^{-1}		5.4	*I* depend. of Cu adsorp. on sepiolite	32
Vermiculite	Cu	(II)	0–15 µM	NR	25 °C	NR	NR	840–859 µmol g^{-1}§		NR		11

Table (*cont.*)

Clay	Metal	Ox. state	Conc. range	Counter-ion	T (°C)	pH range	Ionic strength (*I*) range	Max sorbed or max adsorp. capacity	% of total	~pH of max adsorp	Observations/ comments	Ref.
Vermiculite	Cu	(II)	3150 µM	Cl-	25 °C	~5.0	NR	688 µmol g^{-1}§				33
Vermiculite	Cu	(II)	3150 µM	Cl^-, Na^+	25 °C	~5.0	0.50	K_d ~95			Comp. from and nature of other cations	33
Vermiculite	Cu	(II)	3150 µM	Cl^-, Na^+	25 °C	~5.0	1	K_d ~60			Comp. from and nature of other cations	33
Vermiculite	Cu	(II)	3150 µM	Cl^-, Na^+	25 °C	~5.0	3	K_d ~40			Comp. from and nature of other cations	33
Vermiculite	Cu	(II)	3150 µM	Cl^-, Na^+	25 °C	~5.0	5	K_d ~25			Comp. from and nature of other cations	33
Vermiculite	Cu	(II)	3150 µM	Cl^-, K^+	25 °C	~5.0	0.50	K_d ~40			Comp. from and nature of other cations	33
Vermiculite	Cu	(II)	3150 µM	Cl^-, K^+	25 °C	~5.0	1	K_d ~35			Comp. from and nature of other cations	33
Vermiculite	Cu	(II)	3150 µM	Cl^-, K^+	25 °C	~5.0	3	K_d ~20			Comp. from and nature of other cations	33
Vermiculite	Cu	(II)	3150 µM	Cl^-, K^+	25 °C	~5.0	5	K_d ~15			Comp. from and nature of other cations	33
Vermiculite	Cu	(II)	3150 µM	Cl^-, NH_4^+	25 °C	~5.0	0.50	K_d ~25			Comp. from and nature of other cations	33

Vermiculite	Cu	(II)	3150 μM	Cl^-, NH_4^+	25 °C	~5.0	1	K_d ~15			Comp. from and nature of other cations	33
Vermiculite	Cu	(II)	3150 μM	Cl^-, NH_4^+	25 °C	~5.0	3	K_d ~12			Comp. from and nature of other cations	33
Vermiculite	Cu	(II)	3150 μM	Cl^-, NH_4^+	25 °C	~5.0	5	K_d ~5				33
Vermiculite	Cu	(II)	1000 μM	acet.	25 °C	5.5	0.01		>95	5.5	Ligands: lactic, succinic, GluA, tiron. cysteine	34
Vermiculite	Cu	(II)	100 μM	acet.	25 °C	2.5–8.5	0.01		100	2.5–8.5	*I*, pH	34
Vermiculite	Cu	(II)	100 μM	acet.	25 °C	2.5–8.5	0.10		100	7–8.5	*I*, pH	34
Acid-activated kaolinite	Zn	(II)	76.5–382.4 μM	NO_3^-	25 °C	NR	NR	253 μmol $g^{-1‡}$			Increased adsorp. due to increased no. of edge sites from acid activation	35
Acid-activated kaolinite	Zn	(II)	76.5–382.4 μM	NO_3^-	30 °C	NR	NR	265 μmol $g^{-1‡}$			Increased adsorp. due to increased no. of edge sites from acid activation	35
Acid-activated kaolinite	Zn	(II)	76.5–382.4 μM	NO_3^-	40 °C	NR	NR	284 μmol $g^{-1‡}$			Increased adsorp. due to increased no. of edge sites from acid activation	35
Bentonite	Zn	(II)	1042–8333 μM	NR	NR	4.9–8.8	NR	114.9 μmol g^{-1}		8.8	Adsorp. strongly pH dependent	36

Table (*cont.*)

Clay	Metal	Ox. state	Conc. range	Counter-ion	T (°C)	pH range	Ionic strength (*I*) range	Max sorbed or max adsorp. capacity	% of total	~pH of max adsorp	Observations/ comments	Ref.
Ca-kaolinite	Zn	(II)	0-62 500 µM	NO_3^-	NR	3.64	NR	~20 µmol g^{-1}			Ca is preferred over Zn on the exchange complex	37
Ca-kaolinite	Zn	(II)	0-62 500 µM	NO_3^-	NR	4.75	NR	~35 µmol g^{-1}			Ca is preferred over Zn on the exchange complex	37
Illite	Zn	(II)	167–2667 µM	NR	NR	4.9–8.8	NR	71 µmol g^{-1}		8.8	Adsorp. strongly pH dependent	36
Kaolinite	Zn	(II)	41.6–1333µM	NR	NR	4.9–8.8	NR	9.6 µmol g^{-1}		8.8	Adsorp. strongly pH dependent, no Zn adsorp. below pH 7.6	36
Kaolinite	Zn	(II)	1 µM	$Ca(ClO_4)_2$	25 °C	5.5–7.5	0.03		50	5.37*	pH values refer to 50/50 distribution between solid and liquid phase according to Kurbatov plots	16
Kaolinite	Zn	(II)	1 µM	$CaCl_2$	25 °C	5.5–7.5	0.03		50	5.12*	pH values refer to 50/50 distribution between solid and liquid phase according to Kurbatov plots	16

Kaolinite	Zn	(II)	1 µM	$CaSO_4$	25 °C	5.5–7.5	0.03		50	5.39*	pH values refer to 50/50 distribution between solid and liquid phase according to Kurbatov plots	16
Kaolinite	Zn	(II)	100 µM	NO_3^-	25 °C	3.0–11.0	0.001	24.2 µmol g^{-1}, 2.22 µmol m^{-2}	100	~8.9, ~6.5*	* Refers to $pH_{50/50}$ similar to ref 27	38
Kaolinite	Zn	(II)	100 µM	Cl^-	25 °C	3.0–11.0	0.001	24.2 µmol g^{-1}, 2.22 µmol m^{-2}	100	~8.9, ~6.0*	* Refers to $pH_{50/50}$ similar to ref 27	38
Kaolinite	Zn	(II)	100 µM	NO_3^-	25 °C	3.0–11.0	0.01	24.2 µmol g^{-1}, 2.22 µmol m^{-2}	100	8.5–9.0, ~7.0*	* Refers to $pH_{50/50}$ similar to ref 27	38
Kaolinite	Zn	(II)	100 µM	Cl^-	25 °C	3.0–11.0	0.01	24.2 µmol g^{-1}, 2.22 µmol m^{-2}	100	8.5–9.0, ~6.5*	* Refers to $pH_{50/50}$ similar to ref 27	38
Kaolinite	Zn	(II)	100 µM	NO_3^-	25 °C	3.0–11.0	0.1	24.2 µmol g^{-1}, 2.22 µmol m^{-2}	100	8.5–9.0, ~7.25*	* Refers to $pH_{50/50}$ similar to ref 27	38
Kaolinite	Zn	(II)	100 µM	Cl^-	25 °C	3.0–11.0	0.1	24.2 µmol g^{-1}, 2.22 µmol m^{-2}	100	8.5–9.0, ~7.25*	* Refers to $pH_{50/50}$ similar to ref 27	38
Kaolinite	Zn	(II)	28–432 µM	KNO_3	25 °C	4.0	0.01	3.7 µmol g^{-1}§		4.0		39
Kaolinite	Zn	(II)	28–432 µM	KNO_3	25 °C	5.0	0.01	6.1 µmol g^{-1}§		5.0		39
Kaolinite	Zn	(II)	28–432 µM	KNO_3	25 °C	6.0	0.01	7.8 µmol g^{-1}§		6.0		39
Kaolinite	Zn	(II)	28–432 µM	KNO_3	25 °C	7.0	0.01	9.7 µmol g^{-1}§		7.0		39
Kaolinite	Zn	(II)	28–432 µM	KNO_3	25 °C	8.0	0.01	8.1 µmol g^{-1}§		8.0		39
Kaolinite	Zn	(II)	28–432 µM	KNO_3	25 °C	4.0	0.001	3.9 µmol g^{-1}§		4.0		39
Kaolinite	Zn	(II)	28–432 µM	KNO_3	25 °C	5.0	0.001	7.8 µmol g^{-1}§		5.0		39

Table (*cont.*)

Clay	Metal	Ox. state	Conc. range	Counter-ion	T (°C)	pH range	Ionic strength (*I*) range	Max sorbed or max adsorp. capacity	% of total	~pH of max adsorp	Observations/ comments	Ref.
Kaolinite	Zn	(II)	28–432 μM	KNO_3	25 °C	6.0	0.001	10.1 μmol g^{-1}§		6.0		39
Kaolinite	Zn	(II)	28–432 μM	KNO_3	25 °C	7.0	0.001	21.6 μmol g^{-1}§		7.0		39
Kaolinite	Zn	(II)	28–432 μM	KNO_3	25 °C	8.0	0.001	16.4 μmol g^{-1}§		8.0		39
Kaolinite	Zn	(II)	76.5–382.4 μM	NO_3^-	25 °C	NR	NR	211 μmol g^{-1}‡			Endo. reaction surface complx. mechanism	35
Kaolinite	Zn	(II)	76.5–382.4 μM	NO_3^-	30 °C	NR	NR	212 μmol g^{-1}‡			Endo. reaction surface complx. mechanism	35
Kaolinite	Zn	(II)	76.5–382.4 μM	NO_3^-	40 °C	NR	NR	231 μmol g^{-1}‡			Endo. reaction surface complx. mechanism	35
Kaolinite	Zn	(II)	48.8 μM	NO_3^-	25 °C	3.0–9.0	0.1	6.3 μmol g^{-1}	100	7.0	pH_{50} ~6.0	40
Kaolinite	Zn	(II)	48.8 μM	NO_3^-	25 °C	3.0–9.0	0.01	6.3 μmol g^{-1}	100	7.0	pH_{50} ~5.2	40
Kaolinite	Zn	(II)	48.8 μM	NO_3^-	25 °C	3.0–9.0	0.001	6.3 μmol g^{-1}	100	7.0	pH_{50} ~4.5	40
Montmorillonite	Zn	(II)	1 μM	$Ca(ClO_4)_2$	25 °C	5.5–7.5	0.03		50	4.76*	pH values refer to 50/50 distribution between solid and liquid phase according to Kurbatov plots	16
Montmorillonite	Zn	(II)	1 μM	$CaCl_2$	25 °C	5.5–7.5	0.03		50	4.13*	pH values refer to 50/50 distribution between solid and liquid phase according to Kurbatov plots	16

Montmorillonite	Zn	(II)	1 μM	$CaSO_4$	25 °C	5.5–7.5	0.03		50	4.00*	pH values refer to 50/50 distribution between solid and liquid phase according to Kurbatov plots	16
Na-bentonite	Zn	(II)	191–3059 μM	Cl^-	NR	4.0	NR	~765 μmol g^{-1}				41
Na-bentonite	Zn	(II)	1530 μM	Cl^-	NR	3.0–8.0	NR	~887 μmol g^{-1}	89	7.0	No pH effect on adsorp. at pH >7, ppt reactions appear to occur for 1530 μM Zn	41
Na-kaolinite	Zn	(II)	up to 50 μM	NO_3^-	NR	NR	NR	~25 μmol g^{-1}				42
Na-montmorillonite	Zn	(II)	NR	NO_3^-	25 °C	3.55–5.10	0.01	340 μmol g^{-1}		5.1	Level of adsorp. in the pH range studied varied by 6 μmol g^{-1}	43
Na-montmorillonite	Zn	(II)	up to 50 μM	NO_3^-	NR	NR	NR	~260 μmol g^{-1}				42
Na-montmorillonite	Zn	(II)	0.001 μM	$NaClO_4$	NR	2.0–10.0	0.1	K_d ~100–$10^{5.6}$		2.0 to 8.5	Surface conc. of Zn NR	21
Na-montmorillonite	Zn	(II)	1.6 nM ~1000 μM	$NaClO_4$	NR	5.6	0.1	100 μmol g^{-1}		5.6		21
Na-montmorillonite	Zn	(II)	1.6 nM – ~1000 μM	$NaClO_4$	NR	7.0	0.1	39 μmol g^{-1}		7.0		21
Na-montmorillonite	Zn	(II)	trace to 14 000 μM	Cl^-	20 °C	5.0	NR	419 μmol g^{-1}		5.0		44
Na-montmorillonite	Zn	(II)	100 μM	Acet.	NR	2.0–8.5	0.1	100 μmol g^{-1}		2.0–8.5	No pH effect for an initial conc. of 100 μM Zn	45

Table (*cont.*)

Clay	Metal	Ox. state	Conc. range	Counter-ion	T (°C)	pH range	Ionic strength (*I*) range	Max sorbed or max adsorp. capacity	% of total	~pH of max adsorp	Observations/ comments	Ref.
Na-montmorillonite	Zn	(II)	100 µM	Acet. 10^{-3} M EDTA	NR	2.0–8.5	0.1	8.5 µmol g^{-1}		5.5	Chelation of Zn	45
Na-montmorillonite	Zn	(II)	100 µM	Acet. 10^{-3} M NTA	NR	2.0–8.5	0.1	5.9 µmol g^{-1}		5.5	Chelation of Zn	45
Na-montmorillonite	Zn	(II)	100 µM	Acet. 10^{-3} TarA	NR	2.0–8.5	0.1	65.3 µmol g^{-1}		5.5	Partial repulsive effect due to neg charged Zn-tartrate complexes	45
Na-montmorillonite	Zn	(II)	100 µM	Acet. 10^{-3} M OxA	NR	2.0–8.5	0.1	93.9 µmol g^{-1}		5.5	Possibility of ternary complex formation at the montmorillonite surface	45
Na-montmorillonite	Zn	(II)	100 µM	Acet. 10^{-3} M MalA	NR	2.0–8.5	0.1	100 µmol g^{-1}		5.5	Possibility of ternary complex formation at the montmorillonite surface	45
Na-montmorillonite	Zn	(II)	100 µM	Acet. 10^{-3} M SucA	NR	2.0–8.5	0.1	96.3 µmol g^{-1}		5.5	Possibility of ternary complex formation at the montmorillonite surface	45
Na-montmorillonite	Zn	(II)	100 µM	Acet. 10^{-3} M GluA	NR	2.0–8.5	0.1	98.6 µmol g^{-1}		5.5	Possibility of ternary complex formation at the montmorillonite surface	45

Na-montmorillonite	Zn	(II)	100 µM	Acet. 10^{-3} M CitA	NR	2.0–8.5	0.1	95.3 µmol g^{-1}		5.5	Possibility of ternary complex formation at the montmorillonite surface	45
Natural bentonite	Zn	(II)	191–3059 µM	Cl^-	NR	4.0	NR	~320 µmol g^{-1}				41
Natural bentonite	Zn	(II)	1530 µM	Cl^-	NR	3.0–8.0	NR	~320 µmol g^{-1}	32	2.0–7.0	No pH effect on adsorp. at pH >7, ppt reactions appear to occur for 1530 µM Zn	41
Sepiolite	Zn	(II)	1500 µM	ClO_4^-	25 °C	3.5–7.0	0.1	~100 µmol g^{-1}		7.0	At pH 3.5, max adsorp. is ~10 µmol g^{-1}	32
Illite (Scalla red-type 392)	Au	(I)	64 µM	CN^-	18	10.5	0.004		2.7		Mixed clay mineralogy***	46
Kaolinite	Au	(III)	388 µM	Cl^-	RT	3–10	1	90 µmol g^{-1}		6	pH	47
Kaolinite	Au	(III)	0.5 µM	Cl^-	RT	3–11	0.1	0.0475 µmol g^{-1}	95	3–8.5	Reaction time, colloid formation	48
Kaolinite	Au	(III)	0.25 µM	Cl^-	RT	3–9	NR	0.02488 µmol g^{-1}	98	3–6	pH-edge charge, crystallinity/no. of edge sites	49
Kaolinite	Au	(III)	2.54 µM	Cl^-	RT	3–9	NR	0.252 µmol g^{-1}	99	3–6	pH-edge charge, crystallinity/no. of edge sites	49
Kaolinite	Au	(III)	50.8 µM	Cl^-	RT	3–9	NR	4.92 µmol g^{-1}	97	3–4	pH-edge charge, crystallinity/no. of edge sites	49
Kaolinite	Au	(III)	0.25 µM	Cl^-	120	3–9	NR	0.02488 µmol g^{-1}	98	3–5	pH-edge charge, crystallinity/no. of edge sites	49

Table (*cont.*)

Clay	Metal	Ox. state	Conc. range	Counter-ion	T (°C)	pH range	Ionic strength (*I*) range	Max sorbed or max adsorp. capacity	% of total	~pH of max adsorp	Observations/ comments	Ref.
Kaolinite	Au	(III)	2.54 μM	Cl^-	120	3–9	NR	0.253 μmol g^{-1}	99.6	3–5	pH-edge charge, crystallinity/no. of edge sites	49
Kaolinite	Au	(III)	50.8 μM	Cl^-	120	3–9	NR	5.07 μmol g^{-1}	99.8	3–4	pH-edge charge, crystallinity/no. of edge sites	49
Kaolinite	Au	(III)	0.025 μM	Cl^-	RT	2.5–8	NR	0.51 μmol g^{-1}	100	2.5–3	pH-edge charge, crystallinity/no. of edge sites	50
Montmorillonite-brown (Scalla 392)	Au	(I)	65 μM	CN^-	18	10.5	0.004		1.9		mixed clay mineralogy***	46
Montmorillonite-white (Scalla 432)	Au	(I)	67 μM	CN^-	18	10.5	0.004		2.7		mixed clay mineralogy***	46
Nontronite-montmorillonite (Scalla 392)	Au	(I)	60 μM	CN^-	18	10.5	0.004		7.6		mixed clay mineralogy***	46
Pyrophyllite	Au	(III)	0.025 μM	Cl^-	RT	2.5–8	NR	0.51 μmol g^{-1}	100	2.5–3	pH-edge charge, crystallinity/no. of edge sites	50
Smectite	Au	(III)	0.5 μM	Cl^-	RT	3–11	0.1	0.0490 μmol g^{-1}	95	3–8.5	reaction time, colloid formation	48
Acid-activated kaolinite	Pb	(II)	241 μM	NO_3^-	30 °C	1.0–6.0	NR	~33 μmol g^{-1}		6.0	~15 μmol g^{-1} adsorp. at pH 1	51
Acid-activated kaolinite	Pb	(II)	48–241 μM	NO_3^-	30 °C	5.7	NR	149.9 μmol g^{-1}§				51

Acid-activated kaolinite	Pb	(II)	48 µM	NO_3^-	30 °C	5.7	NR	~8.7 µmol g^{-1}			Exo. adsorp.	51
Acid-activated kaolinite	Pb	(II)	48 µM	NO_3^-	35 °C	5.7	NR	~7.7 µmol g^{-1}			Exo. adsorp.	51
Acid-activated kaolinite	Pb	(II)	48 µM	NO_3^-	40 °C	5.7	NR	~6.7 µmol g^{-1}			Exo. adsorp.	51
Acid-activated montmorillonite	Pb	(II)	241 µM	NO_3^-	30 °C	1.0–6.0	NR	~110 µmol g^{-1}		6.0	~95 µmol g^{-1} adsorp. at pH 1	51
Acid-activated montmorillonite	Pb	(II)	48–241 µM	NO_3^-	30 °C	5.7	NR	151.3 µmol g^{-1}§				51
Acid-activated montmorillonite	Pb	(II)	48 µM	NO_3^-	30 °C	5.7	NR	~24.1 µmol g^{-1}			Exo. adsorp.	51
Acid-activated montmorillonite	Pb	(II)	48 µM	NO_3^-	35 °C	5.7	NR	~22.2 µmol g^{-1}			Exo. adsorp.	51
Acid-activated montmorillonite	Pb	(II)	48 µM	NO_3^-	40 °C	5.7	NR	~21.2 µmol g^{-1}			Exo. adsorp.	51
Bentonite	Pb	(II)	Up to ~1000 µM	NO_3^-	30 °C	3.0	NR	70 µmol g^{-1}		3.0	Dubinin max adsorp. for Pb was calculated as 240 µmol g^{-1}	52
Bentonite	Pb	(II)	10 000 µM	NO_3^-	27 °C	~5.0	NR	1.33 µmol g^{-1}		<5.0		53
Bentonite	Pb	(II)	48.3 µM	$NaNO_3$	18 °C	1.0–12.0	0.01	96.6 µmol g^{-1}	80	~9.0	pH_{50} for all 3 e'lyte systems is ~5.0	54
Bentonite	Pb	(II)	48.3 µM	$LiNO_3$	18 °C	1.0–12.0	0.01	96.6 µmol g^{-1}	80	~7.0	pH_{50} for all 3 e'lyte systems is ~5.0	54
Bentonite	Pb	(II)	48.3 µM	KNO_3	18 °C	1.0–12.0	0.01	96.6 µmol g^{-1}	80	~7.5	pH_{50} for all 3 e'lyte systems is ~5.0	54
Illite	Pb	(II)	Up to 2000 µM	$NaNO_3$	25 °C	2.5	0.01	37 µmol g^{-1}		2.5		55
Illite	Pb	(II)	Up to 2000 µM	$NaNO_3$	25 °C	3.5	0.01	78 µmol g^{-1}		3.5		55

Table (*cont.*)

Clay	Metal	Ox. state	Conc. range	Counter-ion	T (°C)	pH range	Ionic strength (*I*) range	Max sorbed or max adsorp. capacity	% of total	~pH of max adsorp	Observations/ comments	Ref.
Illite	Pb	(II)	Up to 2000 μM	$NaNO_3$	25 °C	5.0	0.01	124 μmol g^{-1}		5.0		55
Illite	Pb	(II)	up to 2000 μM	$NaNO_3$	25 °C	6.0	0.01	248 μmol g^{-1}		6.0		55
Kaolinite	Pb	(II)	241 μM	NO_3^-	30 °C	1.0–6.0	NR	~26 μmol g^{-1}		6.0	~11 μmol g^{-1} adsorp. at pH 1	51
Kaolinite	Pb	(II)	48–241 μM	NO_3^-	30 °C	5.7	NR	55.6 μmol g^{-1}§				51
Kaolinite	Pb	(II)	48 μM	NO_3^-	30 °C	5.7	NR	~7.2 μmol g^{-1}			exo. adsorp.	51
Kaolinite	Pb	(II)	48 μM	NO_3^-	35 °C	5.7	NR	~5.5 μmol g^{-1}			exo. adsorp.	51
Kaolinite	Pb	(II)	48 μM	NO_3^-	40 °C	5.7	NR	~4.5 μmol g^{-1}			exo. adsorp.	51
Kaolinite	Pb	(II)	48.8 μM	$NaNO_3$	25 °C	3.0–9.0	0.1	6.3 μmol g^{-1}	100	~5.5	pH_{50}~4.5	40
Kaolinite	Pb	(II)	48.8 μM	$NaNO_3$	25 °C	3.0–9.0	0.01	6.3 μmol g^{-1}	100	~5.5	pH_{50}~3.5	40
Kaolinite	Pb	(II)	48.8 μM	$NaNO_3$	25 °C	3.0–9.0	0.001	6.3 μmol g^{-1}	100	~5.5	pH_{50}~3.5	40
Kaolinite	Pb	(II)	0.1–31.6 μM	$NaNO_3$	20 °C	4.0	0.01	~3.2 μmol g^{-1}		4.0		56
Kaolinite	Pb	(II)	0.1–31.6 μM	$NaNO_3$	20 °C	5.0	0.01	~5.6 μmol g^{-1}		5.0		56
Kaolinite	Pb	(II)	0.1–31.6 μM	$NaNO_3$	20 °C	6.0	0.01	~17.8 μmol g^{-1}		6.0		56
Kaolinite	Pb	(II)	0–50 μM	KNO_3, 100 μM Asp	25 °C	5.5	0.005	~5.9 μmol g^{-1}		5.5		57
Kaolinite	Pb	(II)	100 μM	KNO_3	25 °C	3.0–10.0	0.005	~14.7 μmol g^{-1}		7.0	At pH 5.5, Pb surface conc. is ~8.8 μmol g^{-1}	57
Kaolinite	Pb	(II)	0-1930 μM	NO_3^-	20 °C	5.0	NR	153.2 μmol g^{-1}		5.0		58
Montmorillonite	Pb	(II)	241 μM	NO_3^-	30 °C	1.0–6.0	NR	~106 μmol g^{-1}		6.0	~89 μmol g^{-1} adsorp. at pH 1	51
Montmorillonite	Pb	(II)	48–241 μM	NO_3^-	30 °C	5.7	NR	59.5 μmol g^{-1}§				51
Montmorillonite	Pb	(II)	48 μM	NO_3^-	30 °C	5.7	NR	~23.1 μmol g^{-1}			Exo. adsorp.	51
Montmorillonite	Pb	(II)	48 μM	NO_3^-	35 °C	5.7	NR	~21.2 μmol g^{-1}			Exo. adsorp.	51

Montmorillonite	Pb	(II)	48 µM	NO_3^-	40 °C	5.7	NR	~20.5 µmol g^{-1}			Exo. adsorp.	51
Montomorillonite	Pb	(II)	50 µM	$NaNO_3$	NR	3.0–10.0	0.001	100 µmol g^{-1}	100	8.0	pH_{50}~3.0	59
Montomorillonite	Pb	(II)	50 µM	$NaNO_3$	NR	3.0–10.0	0.02	100 µmol g^{-1}	100	8.0	pH_{50}~5.5	59
Montomorillonite	Pb	(II)	50 µM	$NaNO_3$	NR	3.0–10.0	0.1	100 µmol g^{-1}	100	~8.8	pH_{50}~6.8	59
Montomorillonite	Pb	(II)	5 µM	$NaNO_3$	NR	3.0–10.0	0.02	10 µmol g^{-1}	100	7.0	pH_{50}~5.0	59
Montomorillonite	Pb	(II)	5 µM	$NaNO_3$	NR	3.0–10.0	0.1	10 µmol g^{-1}	100	>7.0	pH_{50}~6.0	59
Montomorillonite	Pb	(II)	0.5 µM	$NaNO_3$	NR	3.0–10.0	0.1	1 µmol g^{-1}	100	>9.0	pH_{50}~5.0–6.0	59
Na-illite	Pb	(II)	48.8 µM	$NaNO_3$	25 °C	3.0–9.0	0.1	16.6 µmol g^{-1}	99	~6.5	pH_{50}~4.5, 3 µmol g^{-1} Pb adsorp. at pH 3	19
Na-illite	Pb	(II)	48.8 µM	$NaNO_3$	25 °C	3.0–9.0	0.01	16.6 µmol g^{-1}	100	~6.0	pH_{50}~3 → ~8.4 µmol g^{-1}	19
Na-illite	Pb	(II)	48.8 µM	$NaNO_3$	25 °C	3.0–9.0	0.001	16.6 µmol g^{-1}	100	~5.0	pH_{50}~3 → ~8.4 µmol g^{-1}	19
Na-montmorillonite	Pb	(II)	100 µM	acet.	NR	2.0–8.5	0.1	100 µmol g^{-1}		5.5	40 µmol g^{-1} adsorp. at pH 2.5	45
Na-montmorillonite	Pb	(II)	100 µM	Acet. 10^{-3} M EDTA	NR	2.0–8.5	0.1	9.8 µmol g^{-1}	9.80	5.5	Chelation of Pb	45
Na-montmorillonite	Pb	(II)	100 µM	Acet. 10^{-3} M NTA	NR	2.0–8.5	0.1	10 µmol g^{-1}	10	5.5	Chelation of Pb	45
Na-montmorillonite	Pb	(II)	100 µM	Acet. 10^{-3} M TarA	NR	2.0–8.5	0.1	99.1 µmol g^{-1}	99.10	5.5	Possibility of ternary complex formation at the montmorillonite surface	45
Na-montmorillonite	Pb	(II)	100 µM	Acet. 10^{-3} M OxA	NR	2.0–8.5	0.1	97.7 µmol g^{-1}	97.70	5.5	Possibility of ternary complex formation at the montmorillonite surface	45

Table (*cont.*)

Clay	Metal	Ox. state	Conc. range	Counter-ion	T (°C)	pH range	Ionic strength (*I*) range	Max sorbed or max adsorp. capacity	% of total	~pH of max adsorp	Observations/ comments	Ref.
Na-montmorillonite	Pb	(II)	100 µM	Acet. 10^{-3} M MalA	NR	2.0–8.5	0.1	100 µmol g^{-1}	100	5.5	Possibility of ternary complex formation at the montmorillonite surface	45
Na-montmorillonite	Pb	(II)	100 µM	Acet. 10^{-3} M SucA	NR	2.0–8.5	0.1	85.5 µmol g^{-1}	85.50	5.5	Partial repulsive effect due to –ve Zn-tartrate complexes	45
Na-montmorillonite	Pb	(II)	100 µM	Acet. 10^{-3} M GluA	NR	2.0–8.5	0.1	89.3 µmol g^{-1}	89.30	5.5	Partial repulsive effect due to –ve Zn-tartrate complexes	45
Na-montmorillonite	Pb	(II)	100 µM	Acet. 10^{-3} M CitA	NR	2.0–8.5	0.1	96.5 µmol g^{-1}	96.50	5.5	Possibility of ternary complex formation at the montmorillonite surface	45
Na-montmorillonite	Pb	(II)	48.8 µM	$NaNO_3$	25 °C	3.0–9.0	0.1	31.8 µmol g^{-1}	~95	~6.2	pH_{50}~5.0	60
Na-montmorillonite	Pb	(II)	48.8 µM	$NaNO_3$	25 °C	3.0–9.0	0.01	33.4 µmol g^{-1}	100	~6.5	pH_{50}<3, at pH 3: 21.7 µmol g^{-1} → ~65	60
Na-montmorillonite	Pb	(II)	48.8 µM	$NaNO_3$	25 °C	3.0–9.0	0.001	33.4 µmol g^{-1}	100	~5.5	pH_{50}<3, at pH 3: 28.4 µmol g^{-1} → ~85	60

Na-montmorillonite	Pb	(II)	NR	NO_3^-	25 °C	5.0		442.5 μmol g^{-1}		5.0	Isotherm constructed from variable masses of montmorillonite: 0.05–1.0 g in a constant vol.	61
Na-montmorillonite	Pb	(II)	NR	$CaClO_4$	25 °C	5.0	0.1	445.5 μmol g^{-1}		5.0	isotherm constructed from variable masses of montmorillonite: 0.05–1.0 g in a constant vol.	61
Vermiculite	Pb	(II)	0–15 μM	NR	25 °C	NR	NR	590–593 μmol g^{-1}§		NR		11
Vermiculite	Pb	(II)	100 μM	Acet.	NR	2.0–8.5	0.1	100 μmol g^{-1}	100	7.0		22
Vermiculite	Pb	(II)	100 μM	Acet.	NR	5.5	0.01	309 μmol g^{-1}		5.5		22
Vermiculite	Pb	(II)	100 μM	Acet. 10^{-3} M EDTA	NR	5.5	0.01	5.5 μmol g^{-1}	5.50	5.5	Strong chelation effect	22
Vermiculite	Pb	(II)	100 μM	Acet. 10^{-3} M NTA	NR	5.5	0.01	5.5 μmol g^{-1}	5.5	5.5	Strong chelation effect	22
Vermiculite	Pb	(II)	100 μM	Acet. 10^{-3} M DTPA	NR	5.5	0.01	7.0 μmol g^{-1}	7.00	5.5	Strong chelation effect	22
Vermiculite	Pb	(II)	100 μM	Acet. 10^{-3} M LacA	NR	5.5	0.01	99.9 μmol g^{-1}	99.9	5.5		22
Vermiculite	Pb	(II)	100 μM	Acet. 10^{-3} M OxA	NR	5.5	0.01	90.4 μmol g^{-1}	90.40	5.5		22
Vermiculite	Pb	(II)	100 μM	Acet. 10^{-3} M MalA	NR	5.5	0.01	99.7 μmol g^{-1}	99.7	5.5		22
Vermiculite	Pb	(II)	100 μM	Acet. 10^{-3} M SucA	NR	5.5	0.01	99.9 μmol g^{-1}	99.90	5.5		22

Table (*cont.*)

Clay	Metal	Ox. state	Conc. range	Counter-ion	T (°C)	pH range	Ionic strength (*I*) range	Max sorbed or max adsorp. capacity	% of total	~pH of max adsorp	Observations/ comments	Ref.
Vermiculite	Pb	(II)	100 µM	Acet. 10^{-3} M GluA	NR	5.5	0.01	99.8 µmol g^{-1}	99.8	5.5		22
Vermiculite	Pb	(II)	100 µM	Acet. 10^{-3} M TarA	NR	5.5	0.01	99.9 µmol g^{-1}	99.90	5.5		22
Vermiculite	Pb	(II)	100 µM	Acet. 10^{-3} M CitA	NR	5.5	0.01	98.7 µmol g^{-1}	98.7	5.5		22
Vermiculite	Pb	(II)	100 µM	Acet. 10^{-3} M IDA	NR	5.5	0.01	99.8 µmol g^{-1}	99.80	5.5		22
Vermiculite	Pb	(II)	100 µM	Acet. 10^{-3} M tiron	NR	5.5	0.01	99.5 µmol g^{-1}	99.5	5.5		22
Vermiculite	Pb	(II)	100 µM	Acet. 10^{-3} M cysteine	NR	5.5	0.01	99.9 µmol g^{-1}	99.90	5.5		22
Vermiculite	Pb	(II)	50 µM	Cl^-	NR	4.0–8.0	NR	~7.5 µmol g^{-1}		8.0	$pH_{50} = 5.07$	62
Vermiculite	Pb	(II)	50 µM	NaCl	NR	4.0–8.0	0.003	~3.7 µmol g^{-1}		8.0	$pH_{50} = 5.60$	62
Vermiculite	Pb	(II)	10 000 µM	NO_3^-	27 °C	~5.0	NR	1.0 µmol g^{-1}		<5.0		53
Ca-smectite, SAz-1	U	(VI)	46-100 400 µM *	0.099 M KNO_3	22 °C	2.89–6.48	0.099	2060 µmol g^{-1}	0.153	2.89	No pH control, greater adsorp. efficiency (> 90%) is achieved with lower (<100 µM) initial U conc.	63

Ca-smectite, SAz-1	U	(VI)	506–46 600 μM *	$Ca(NO_3)_2$	NR	3.4–5.6	NR	208 μmol g^{-1}	0.212	3.4	No pH control, greated adsorp. efficiency (~100%) is achieved with lower (506 μM) initial U conc.	64
Ca-smectite, SAz-1	U	(VI)	9870 μM *	NaCl	NR	3.24	0	380 μmol g^{-1}	0.0385	3.24	Strong H^+ supp. of U sorp. (pH 3.85 vs 4.11) for similar initial U conc. with additional *I* (0 vs 0.01) effects	65
Ca-smectite, SAz-1	U	(VI)	100 μM *	NaCl	NR	4.11	0	99.8 μmol g^{-1}	0.998	4.11	Strong ionic depend. (0 vs 0.1) on extent of U sorp. for similar pH (4.06–4.11)	65
Ca-smectite, SAz-1	U	(VI)	9790 μM *	NaCl	NR	4.11	0.01	78.3 μmol g^{-1}	0.8	4.11	Strong H^+ supp. of U sorp. (pH 3.85 vs 4.11) for similar initial U conc. with additional *I* (0 vs 0.01) effects	65

Table (*cont.*)

Clay	Metal	Ox. state	Conc. range	Counter-ion	T (°C)	pH range	Ionic strength (*I*) range	Max sorbed or max adsorp. capacity	% of total	~pH of max adsorp	Observations/ comments	Ref.
Ca-smectite, SAz-1	U	(VI)	100 μM *	NaCl	NR	4.06	0.1	10.9 μmol g^{-1}	0.109	4.06	Strong pH depend. of UO_2^{2+} adsorp. favoured at higher pH for high (0.1) *I*, strong ionic depend. (0 vs 0.1) on extent of U sorp. for similar pH (4.06–4.11)	65
Ca-smectite, SAz-1	U	(VI)	102 μM *	NaCl	NR	6.41	0.1	74.4 μmol g^{-1}	0.729	6.41	Strong pH depend. of UO_2^{2+} adsorp. favoured at higher pH for high (0.1) I	65
chlorite CCa-2 (unaltered)	U	(VI)	1 μM **	$NaClO_4$	NR	5.0–8.5	0.1	~0.070 μmol g^{-1}	0.879	6.3–7.5	Potential contribution to U sorp. from weathering products, specifically ferrihydrite	66
Chlorite Grimsel (hydrotherm. altered)	U	(VI)	1 μM **	$NaClO_4$	NR	5.0–8.5	0.1	~0.056 μmol g^{-1}	0.7	5.8–6.9	μmol U g^{-1} based on 12.5 g L^{-1} suspension densities as reported by authors	66

Ferrogenous beidellite	U	(VI)	8.4 μM *	$NaClO_4$	25 °C	4.0–9.0	0.001	~5.4 μmol g^{-1}	> 95%	> 5.5	*I* depend. is observed during pH dependent sorp. of uranyl with high (0.1) *I* suppressing U sorp.	67
Ferrogenous beidellite	U	(VI)	8.4 μM *	$NaClO_4$	25 °C	4.0–9.0	0.01	~5.4 μmol g^{-1}	> 95%	> 6.0	*I* depend. is observed during pH dependent sorp. of uranyl with high (0.1) *I* suppressing U sorp.	67
Ferrogenous beidellite	U	(VI)	8.4 μM *	$NaClO_4$	25 °C	4.0–9.0	0.1	~5.4 μmol g^{-1}	> 95%	> 6.5	*I* depend. is observed during pH dependent sorp. of uranyl with high (0.1) *I* suppressing U sorp.	67
Ferrogenous beidellite	U	(VI)	8.4 μM *	$Ca(ClO_4)_2$	25 °C	4.0–9.0	0.015	~5.4 μmol g^{-1}	> 95%	> 6.5	*I* depend. is lost when the background e'lyte contains appreciable Ca ions	67
Ferrogenous beidellite	U	(VI)	8.4 μM *	$Ca(ClO_4)_2$	25 °C	4.0–9.0	0.15	~5.4 μmol g^{-1}	> 95%	> 6.5	*I* depend. is lost when the background e'lyte contains appreciable Ca ions	67

Table (*cont.*)

Clay	Metal	Ox. state	Conc. range	Counter-ion	T (°C)	pH range	Ionic strength (*I*) range	Max sorbed or max adsorp. capacity	% of total	~pH of max adsorp	Observations/ comments	Ref.
Glauconite	U	(VI)	0.3508–100 μM *	0.01 M NaCl	5 °C	~7	0.01	1.062 $\mu mol\ g^{-1}$	0.97			68
Glauconite	U	(VI)	0.3508–100 μM *	0.01 M NaCl	25 °C	~7	0.01	1.057 $\mu mol\ g^{-1}$	0.97			68
Glauconite	U	(VI)	0.3508–100 μM *	0.01 M NaCl	65 °C	~7	0.01	1.057 $\mu mol\ g^{-1}$	0.97			68
Glauconite	U	(VI)	0.3508–100 μM *	0.01 M $NaHCO_3$	5 °C	8–9	0.01	0.6401 $\mu mol\ g^{-1}$	0.59		U sorp. in $NaHCO_3$ increases with T due to the heat sensitivity of uranyl–carb. complx.	68
Glauconite	U	(VI)	0.3508–100 μM *	0.01 M $NaHCO_3$	25 °C	8–9	0.01	0.6452 $\mu mol\ g^{-1}$	0.59		U sorp. in $NaHCO_3$ increases with T due to the heat sensitivity of urany–carb. complx.	68
Glauconite	U	(VI)	0.3508–100 μM *	0.01 M $NaHCO_3$	65 °C	8–9	0.01	1.041 $\mu mol\ g^{-1}$	0.96		U sorp. in $NaHCO_3$ increases with T due to the heat sensitivity of uranyl–carb. complx.	68
Hectorite	U	(VI)	20 μM	NaCl, N_2 atmosphere	22 °C	3.5–9.5	0.5	~10 $\mu mol\ g^{-1}$	~100%	5.5–8.8	pH_{50} ~4.0	69
Hydrobiotite	U	(VI)	920 μM *	NR	NR	1.8–2.0	NR	524 $\mu mol\ g^{-1}$	0.095	1.8–2.0	No pH control	70

Hydrobiotite	U	(VI)	1100 μM *	NR	NR	2.5–2.6	NR	14.9 μmol g^{-1}	0.024	2.5–2.6	No pH control	70
Hydrobiotite	U	(VI)	3.2 μM *	0.0005 M $MgCl_2$	NR	6.5–6.9	0.0015	8.6 μmol g^{-1}	0.539	6.5–6.9	no pH control, the exchange-site cation depend. (K vs. Mg) is not observed For hydrobiotite	70
Hydrobiotite	U	(VI)	2.8 μM *	0.001 M KCl	NR	6.5–7.5	0.001	8.8 μmol g^{-1}	0.628	6.5–7.5	No pH control, the exchange-site cation depend. (K vs. Mg) is not observed for hydrobiotite	70
Illite	U	(VI)	0.3508–100 μM *	0.01 M NaCl	5 °C	~7	0.01	1.097 μmol g^{-1}	> 99%		Sample has minor Qz (SiO_2) contam.	68
Illite	U	(VI)	0.3508–100 μM *	0.01 M NaCl	25 °C	~7	0.01	1.078 μmol g^{-1}	0.98		Sample has minor Qz (SiO_2) contam.	68
Illite	U	(VI)	0.3508–100 μM *	0.01 M NaCl	65 °C	~7	0.01	0.9307 μmol g^{-1}	0.84		Sample has minor Qz (SiO_2) contam.	68
Illite	U	(VI)	0.3508–100 μM *	0.01 M $NaHCO_3$	5 °C	8–9	0.01	0.4251 μmol g^{-1}	0.38		Sample has minor Qz (SiO_2) contam.; U sorp. in $NaHCO_3$ increases with T due to the heat sensitivity of uranyl-carb. complx.	68

Table (*cont.*)

Clay	Metal	Ox. state	Conc. range	Counter-ion	T (°C)	pH range	Ionic strength (*I*) range	Max sorbed or max adsorp. capacity	% of total	~pH of max adsorp	Observations/ comments	Ref.
Illite	U	(VI)	0.3508–100 μM *	0.01 M $NaHCO_3$	25 °C	8–9	0.01	0.9069 μmol g^{-1}	0.82		Sample has minor Qz (SiO_2) contam.; U sorp. in $NaHCO_3$ increases with T due to the heat sensitivity of uranyl-carb. complx.	68
Illite	U	(VI)	0.3508–100 μM *	0.01 M $NaHCO_3$	65 °C	8–9	0.01	1.059 μmol g^{-1}	0.96		Sample has minor Qz (SiO_2) contam.; U sorp. in $NaHCO_3$ increases with T due to the heat sensitivity of uranyl-carb. complx.	68
Kaolinite	U	(VI)	0.3508–100 μM *	0.01 M NaCl	5 °C	~7	0.01	1.401 μmol g^{-1}	0.92			68
Kaolinite	U	(VI)	0.3508–100 μM *	0.01 M NaCl	25 °C	~7	0.01	1.095 μmol g^{-1}	0.98			68
Kaolinite	U	(VI)	0.3508–100 μM *	0.01 M NaCl	65 °C	~7	0.01	1.083 μmol g^{-1}	0.96			68
Kaolinite	U	(VI)	0.3508–100 μM *	0.01 M $NaHCO_3$	5 °C	8–9	0.01	0.2200 μmol g^{-1}	0.2		U sorp. in $NaHCO_3$ increases with T due to the heat sensitivity of uranyl–carb. complx.	68

Kaolinite	U	(VI)	0.3508–100 μM *	0.01 M $NaHCO_3$	25 °C	8–9	0.01	0.1875 μmol g^{-1}	0.17		U sorp. in $NaHCO_3$ increases with T due to the heat sensitivity of uranyl–carb. complx.	68
Kaolinite	U	(VI)	0.3508–100 μM *	0.01 M $NaHCO_3$	65 °C	8–9	0.01	0.2447 μmol g^{-1}	0.29		U sorp. in $NaHCO_3$ increases with T due to the heat sensitivity of uranyl–carb. complx.	68
Kaolinite	U	(VI)	10 μM **	$NaClO_4$, 10 mg L^{-1} HumA	NR	5	0.1	2.26 μmol g^{-1}	0.905	5	Open to atm.	66
Kaolinite	U	(VI)	10 μM **	$NaClO_4$, 10 mg L^{-1} HumA	NR	6.1	0.1	2.44 μmol g^{-1}	0.974	6.1	Open to atm.	66
Kaolinite	U	(VI)	10 μM **	$NaClO_4$, 10 mg L^{-1} HumA	NR	6.9	0.1	2.42 μmol g^{-1}	0.969	6.9	Open to atm.	66
Kaolinite	U	(VI)	10 μM **	$NaClO_4$, 10 mg L^{-1} HumA	NR	8.4	0.1	1.67 μmol g^{-1}	0.666	8.4	Open to atm.	66
Kaolinite	U	(VI)	10 μM **	$NaClO_4$, 10 mg L^{-1} HumA	NR	8.6	0.1	2.45 μmol g^{-1}	0.978	8.6	Open atm. replaced by N_2 atm. – the increase in U sorp. can be attributed to a reduction in carb. complx. of U at this pH (8.4)	66
Kaolinite	U	(VI)	10 μM **	$NaClO_4$, no HumA	NR	7	0.1	2.30 μmol g^{-1}	0.921	7	Open to atm.	66

Table (*cont.*)

Clay	Metal	Ox. state	Conc. range	Counter-ion	T (°C)	pH range	Ionic strength (*I*) range	Max sorbed or max adsorp. capacity	% of total	~pH of max adsorp	Observations/ comments	Ref.
Montmorillonite	U	(VI)	1 µM *	NR	NR	4	NR	1.982 µmol g^{-1}	0.99			71
Montmorillonite Swy-1	U	(VI)	8.4 µM *	$NaClO_4$	25 °C	4.0–8.5	0.001	~16 µmol g^{-1}	0.95	4	*I* depend. is observed during pH dependent sorp. of uranyl with high (0.1) *I* suppressing U sorp.	72
Montmorillonite Swy-1	U	(VI)	8.4 µM *	$NaClO_4$	25 °C	4.0–8.5	0.01	~15 µmol g^{-1}	0.9	8	*I* depend. is observed during pH dependent sorp. of uranyl with high (0.1) *I* suppressing U sorp.	72
Montmorillonite Swy-1	U	(VI)	8.4 µM *	$NaClO_4$	25 °C	4.0–8.5	0.1	~15 µmol g^{-1}	0.9	7.5	*I* depend. is observed during pH dependent sorp. of uranyl with high (0.1) *I* suppressing U sorp.	72
Montmorillonite	U	(VI)	0.3508–100 µM *	0.01 M NaCl	5 °C	~7	0.01	2.043 µmol g^{-1}	0.9			68
Montmorillonite	U	(VI)	0.3508–100 µM *	0.01 M NaCl	25 °C	~7	0.01	1.097 µmol g^{-1}	0.97			68
Montmorillonite	U	(VI)	0.3508–100 µM *	0.01 M NaCl	65 °C	~7	0.01	2.129 µmol g^{-1}	0.94			68

Montmorillonite	U	(VI)	0.3508–100 μM *	0.01 M $NaHCO_3$	5 °C	8–9	0.01	0.1218 μmol g^{-1}	0.05		U sorp. in $NaHCO_3$ increases with T due to the heat sensitivity of uranyl–carb. complx.	68
Montmorillonite	U	(VI)	0.3508–100 μM *	0.01 M $NaHCO_3$	25 °C	8–9	0.01	0.08329 μmol g^{-1}	0.07		U sorp. in $NaHCO_3$ increases with T due to the heat sensitivity of uranyl–carb. complx.	68
Montmorillonite	U	(VI)	0.3508–100 μM *	0.01 M $NaHCO_3$	65 °C	8–9	0.01	1.790 μmol g^{-1}	0.79		U sorp. in $NaHCO_3$ increases with T due to the heat sensitivity of uranyl–carb. complx.	68
Montmorillonite	U	(VI)	840–8400 μM *	NR	25 °C	3.1–4.4	NR	395 μmol g^{-1}	~75%		No pH control, max sorp. efficiency achieved at lowest initial U conc. (200 ppm U)	75
Montmorillonite	U	(VI)	20 μM	NaCl, N_2 atmosphere	22 °C	3.5–9.5	0.5	~10 μmol g^{-1}	~100%	7.2–8.8	pH_{50} ~5.2	69
Nontronite	U	(VI)	0.3508–100 μM *	0.01 M NaCl	5 °C	~7	0.01	2.086 μmol g^{-1}	0.94			68
Nontronite	U	(VI)	0.3508–100 μM *	0.01 M NaCl	25 °C	~7	0.01	0.8060 μmol g^{-1}	0.73			68
Nontronite	U	(VI)	0.3508–100 μM *	0.01 M NaCl	65 °C	~7	0.01	2.137 μmol g^{-1}	0.96			68

Table (*cont.*)

Clay	Metal	Ox. state	Conc. range	Counter-ion	T (°C)	pH range	Ionic strength (*I*) range	Max sorbed or max adsorp. capacity	% of total	~pH of max adsorp	Observations/ comments	Ref.
Nontronite	U	(VI)	0.3508–100 μM *	0.01 M $NaHCO_3$	5 °C	8–9	0.01	0.2051 μmol g^{-1}	0.09		U sorp. in $NaHCO_3$ increases with T due to the heat sensitivity of uranyl–carb. complx.	68
Nontronite	U	(VI)	0.3508–100 μM *	0.01 M $NaHCO_3$	25 °C	8–9	0.01	0.1490 μmol g^{-1}	0.13		U sorp. in $NaHCO_3$ increases with T due to the heat sensitivity of uranyl–carb. complx.	68
Nontronite	U	(VI)	0.3508–100 μM *	0.01 M $NaHCO_3$	65 °C	8–9	0.01	0.2159 μmol g^{-1}	0.1		U sorp. in $NaHCO_3$ increases with T due to the heat sensitivity of uranyl–carb. complx.	68
Saponite	U	(VI)	100 μM	$NaClO_4$	NR	2.8–7.8	0.01	~33 μmol g^{-1}	~66%	6.0–7.0	U sorp. (ion exchange) at $I = 0.01$ was ~10–20 μmol g^{-1} (20–40% of total U) over a pH range of 2.8–4.8	73

Saponite	U	(VI)	100 μM	$NaClO_4$	NR	2.8–7.8	0.1	~33 μmol g^{-1}	~66%	6.0–7.0	U sorp. (ion exchange) at $I = 0.1$ was ~0 over a pH range of 2.8–4.8	73
Saponite	U	(VI)	100 μM	$NaClO_4$, 100 μM EDTA	NR	2.0–8.0	0.01	~28 μmol g^{-1}	~56%	~7.0	EDTA suppressed U sorp. at 10 μmol g^{-1} (~20% of total U) up to pH ~5.8	73
Saponite	U	(VI)	100 μM	$NaClO_4$, 100 mg L^{-1} FulA	NR	2.0–8.0	0.01	~12 μmol g^{-1}	~24%	~4.0	FulA most significantly reduced U sorp. on saponite	73
Saponite	U	(VI)	100 μM	$NaClO_4$, 5000 μM Na_2CO_3	NR	2.0–8.0	0.01	~32 μmol g^{-1}	~64%	~7.0	Little or no effect of the pH range studied	73
Sepiolite	U	(VI)	1050 μM	NR	30 °C	3	NR	100.8 μmol g^{-1}	~96%	3		74
Sepiolite	U	(VI)	420–2100 μM	NR	30 °C	3	NR	147 μmol g^{-1}	~98%	3		74
Vermiculite	U	(VI)	840–8400 μM*	NR	25 °C	3.1–4.4	NR	336 μmol g^{-1}	~40%		No pH control, max sorp. efficiency achieved at lowest initial U conc. (200 ppm U)	75
Vermiculite	U	(VI)	920 μM *	NR	NR	1.8–2.0	NR	437 μmol g^{-1}	0.095	1.8–2.0	No pH control	70
Vermiculite	U	(VI)	1100 μM *	NR	NR	2.4–2.5	NR	19.8 μmol g^{-1}	0.036	2.4–2.5	No pH control	70
Vermiculite	U	(VI)	3.2 μM *	0.0005 M $MgCl_2$	NR	6.5–6.9	0.0015	12.4 μmol g^{-1}	0.776	6.5–6.9	No pH control, strong depend. on exchange-site cation, K^+ suppressing U adsorp. more than Mg^{2+}	70

Table (*cont.*)

Clay	Metal	Ox. state	Conc. range	Counter-ion	T (°C)	pH range	Ionic strength (*I*) range	Max sorbed or max adsorp. capacity	% of total	~pH of max adsorp	Observations/ comments	Ref.
Vermiculite	U	(VI)	2.8 μM *	0.001 M KCl	NR	6.5–7.5	0.001	4.62 μmol g^{-1}	0.33	6.5–7.5	No pH control, strong depend. on exchange-site cation, K^+ suppressing U adsorp. more than Mg^{2+}	70

acet. – acetate; adsorp. – adsorption; Asp – aspartic acid; atm. – atmosphere; carb. – carbonate; citr. – citrate; CitA – citric acid; comp. – competition; complx. – complexation; conc. – concentration; contam. – contamination; depend. – dependence; DTPA – diethylene triamino-penta-acetic acid; e'lyte – electrolyte; EDTA – ethylene diamine tetra-acetic acid; endo. – endothermic; exo. – exothermic; FulA – fulvic acid; GluA – glutaric acid; HumA – humic acid; *I* – ionic strength and is defined: SUM ($C^* Z^2$)i; IDA – iminodiacetic acid; K_d is the distribution coefficient: $(q_e/C_e)^{1/n}$; LacA – lactic acid; M – molar; MalA – malonic acid; no. – number; NR – not reported; NTA – nitrilotriacetic acid; OxA. – oxalic acid; pH_{50} – values refer to 50/50 distribution between solid and liquid phase according to Kurbatov plots; ppt – precipitation; Qz – quartz; sorp. – sorption; SSA – specific surface area; SucA – succinic acid; supp. – suppression; TarA – tartaric acid; T – temperature; tiron – 1,2-dihydroxybenzene-3,5-disulphonic acid; -ve – negative; vol. – volume; § Langmuir max qe; ‡ calculated from the Dubinin–Radushkevich isotherm; † concentrations refer to Ni in solution after sorption; * U(VI) species as the uranyl UO_2^{2+} and added as a nitrate NO_3^-; ** U(VI) species as the uranyl UO_2^{2+} and added as a perchlorate ClO_4^-; Ref 5 – individual contributions from clays cannot be ascertained. In some samples, iron oxides (goethite) were present

References

1. Bhattacharyya, K. G. & Sen Gupta, S. 2007. Adsorption of Co(II) from aqueous medium on natural and acid activated Kaolinite and montmorillonite. *Separation Science and Technology*, 42, 3391–3418.
2. Nagy, N. M., Konya, J. & Urbin, Z. 1997. The competitive exchange of hydrogen and cobalt ions on calcium- montmorillonite. *Colloids and Surfaces A- Physicochemical and Engineering Aspects*, 121, 117–124.
3. Koppelman, M. H. & Dillard, J. G. 1978. X-ray photoelectron spectroscopic (XPS) study of cobalt adsorbed on clay mineral chlorite. *Journal of Colloid and Interface Science*, 66, 345–351.
4. Angove, M. J., Johnson, B. B. & Wells, J. D. 1998. The influence of temperature on the adsorption of cadmium(II) and cobalt(II) on kaolinite. *Journal of Colloid and Interface Science*, 204, 93–103.
5. Yavuz, O., Altunkaynak, Y. & Guzel, F. 2003. Removal of copper, nickel, cobalt and manganese from aqueous solution by kaolinite. *Water Research*, 37, 948–952.
6. Ikhsan, J., Wells, J. D., Johnson, B. B. & Angove, M. J. 2004. The effect of aspartic acid on the binding of transition metals to kaolinite. *Journal of Colloid and Interface Science*, 273, 6–13.
7. Shahwan, T., Erten, H. N. & Unugur, S. 2006. A characterization study of some aspects of the adsorption of aqueous Co^{2+} ions on a natural bentonite clay. *Journal of Colloid and Interface Science*, 300, 447–452.
8. Yu, S. M., Ren, A. P., Chen, C. L., Chen, Y. X. & Wang, X. 2006. Effect of pH, ionic strength and fulvic acid on the sorption and desorption of cobalt to bentonite. *Applied Radiation and Isotopes*, 64, 455–461.
9. Davison, N., Mcwhinnie, W. R. & Hooper, A. 1991. X- ray photoelectron spectroscopic study of cobalt (II) and nickel (II) sorbed on hectorite and montmorillonite. *Clays and Clay Minerals*, 39, 22–27.
10. Kara, M., Yuzer, H., Sabah, E. & Celik, M. S. 2003. Adsorption of cobalt from aqueous solutions onto sepiolite. *Water Research*, 37, 224–232.
11. Da Fonseca, M. G., De Oliveira, M. M., Arakaki, L. N. H., Espinola, J. G. P. & Airoldi, C. 2005. Natural vermiculite as an exchanger support for heavy cations in aqueous solution. *Journal of Colloid and Interface Science*, 285, 50–55.
12. Bhattacharyya, K. G. & Sen Gupta, S. 2008. Influence of acid activation on adsorption of Ni(II) and Cu(II) on kaolinite and montmorillonite: Kinetic and thermodynamic study. *Chemical Engineering Journal*, 136, 1–13.
13. Sathyanarayana, B. & Seshaiah, K. 2011. Kinetics and equilibrium studies on the sorption of manganese(II) and nickel(II) onto kaolinite and bentonite. *E-Journal of Chemistry*, 8, 373–385.
14. Tan, X. L., Hu, J., Montavon, G. & Wang, X. K. 2011. Sorption speciation of Nickel(II) onto Ca-montmorillonite: Batch, EXAFS techniques and modelling. *Dalton Transactions*, 40, 10953–10960.
15. Riza, K. A., Tolga, D., Ihsan, A., Salih, A. & Yunus, O. 2011. Equilibrium, kinetic and thermodynamic studies of nickel adsorption onto natural and modified kaolinites. *Fresenius Environmental Bulletin*, 20, 1155–1166.
16. Puls, R. W. & Bohn, H. L. 1988. Sorption of cadmium, nickel, and zinc by kaolinite and montmorillonite suspensions. *Soil Science Society of America Journal*, 52, 1289–1292.
17. Ijagbemi, C. O., Baek, M. H. & Kim, D. S. 2009. Montmorillonite surface properties and sorption characteristics for heavy metal removal from aqueous solutions. *Journal of Hazardous Materials*, 166, 538–546.
18. Greenpedersen, H., Jensen, B. T. & Pind, N. 1997. Nickel adsorption on MnO_2, $Fe(OH)_3$, montmorillonite, humic acid and calcite: A comparative study. *Environmental Technology*, 18, 807–815.

19. Gu, X. & Evans, L. J. 2007. Modelling the adsorption of Cd(II), Cu(II), Ni(II), Pb(II), and Zn(II) onto Fithian illite. *Journal of Colloid and Interface Science*, 307, 317–325.
20. Marcussen, H., Holm, P. E., Strobel, B. W. & Hansen, H. C. B. 2009. Nickel sorption to goethite and montmorillonite in presence of citrate. *Environmental Science & Technology*, 43, 1122–1127.
21. Baeyens, B. & Bradbury, M. H. 1997. A mechanistic description of Ni and Zn sorption on Na-montmorillonite:1. Titration and sorption measurements. *Journal of Contaminant Hydrology*, 27, 199–222.
22. Abollino, O., Giacomino, A., Malandrino, M. & Mentasti, E. 2008. Interaction of metal ions with montmorillonite and vermiculite. *Applied Clay Science*, 38, 227–236.
23. Eloussaief, M., Jarraya, I. & Benzina, M. 2009. Adsorption of copper ions on two clays from Tunisia: pH and temperature effects. *Applied Clay Science*, 46, 409–413.
24. Eren, E. & Afsin, B. 2008. An investigation of Cu(II) adsorption by raw and acidactivated bentonite: A combined potentiometric, thermodynamic, XRD, IR, DTA study. *Journal of Hazardous Materials*, 151, 682–691.
25. Bhattacharyya, K. G. & Sen Gupta, S. 2011. Removal of Cu(II) by natural and acid activated clays: An insight of adsorption isotherm, kinetic and thermodynamics. *Desalination*, 272, 66–75.
26. Zhang, G. K., Xia, Y., Ying, L., Jia, Y. Y., Yu, G. W. & Ouyang, S. X. 2004. Copper(II) adsorption on Ca-rectorite, and effect of static magnetic field on the adsorption. *Journal of Colloid and Interface Science*, 278, 265–269.
27. Farquhar, M. L., Vaughan, D. J., Hughes, C. R., Charnock, J. M. & England, K. E. R. 1997. Experimental studies of the interaction of aqueous metal cations with mineral substrates: Lead, cadmium, and copper with perthitic feldspar, muscovite, and biotite. *Geochimica et Cosmochimica Acta*, 61, 3051–3064.
28. Du, Q., Sun, Z. X., Forsling, W. & Tang, H. X. 1997. Adsorption of copper at aqueous illite surfaces. *Journal of Colloid and Interface Science*, 187, 232–242.
29. Farrah, H. & Pickering, W. F. 1976a. Sorption of copper species by clays: 2. Illite and montmorillonite. *Australian Journal of Chemistry*, 29, 1177–1184.
30. Bhattacharyya, K. G. & Sen Gupta, S. 2006. Kaolinite, montmorillonite, and their modified derivatives as adsorbents for removal of Cu(II) from aqueous solution. *Separation and Purification Technology*, 50, 388–397.
31. Farrah, H. & Pickering, W. F. 1976b. Sorption of copper species by clays:1. Kaolinite. *Australian Journal of Chemistry*, 29, 1167–1176.
32. Vico, L. I. 2003. Acid- base behaviour and Cu^{2+} and Zn^{2+} complexation properties of the sepiolite/water interface. *Chemical Geology*, 198, 213–222.
33. El- Bayaa, A. A., Badawy, N. A. & Alkhalik, E. A. 2009. Effect of ionic strength on the adsorption of copper and chromium ions by vermiculite pure clay mineral. *Journal of Hazardous Materials*, 170, 1204–1209.
34. Malandrino, M., Abollino, O., Giacomino, A., Aceto, M. & Mentasti, E. 2006. Adsorption of heavy metals on vermiculite: Influence of pH and organic ligands. *Journal of Colloid and Interface Science*, 299, 537–546.
35. Kul, A. R. & Caliskan, N. 2009. Equilibrium and kinetic studies of the adsorption of Zn(II) ions onto natural and activated kaolinites. *Adsorption Science & Technology*, 27, 85–105.
36. Reddy, M. R. & Perkins, H. F. 1974. Fixation of zinc by clay-minerals. *Soil Science Society of America Journal*, 38, 229–231.
37. Garcia-Miragaya, J. & Davalos, M. 1986. Sorption and desorption of Zn on Ca-kaolinite. *Water Air Soil Pollution*, 27, 217–224.
38. Spark, K. M., Wells, J. D. & Johnson, B. B. 1995. Characterizing trace metal adsorption on kaolinite. *European Journal of Soil Science*, 46, 633–640.
39. Dimirkou, A., Ioannou, A., Papadopoulos, P. & Paschalidou, C. 2002. Zinc sorption by kaolinite: Influence of pH, electrolyte, and initial Zn concentrations with

simultaneous release of Mg, Ca, Mn, and Cu ions. *Communications in Soil Science and Plant Analysis*, 33, 2917–2934.
40. Gu, X. & Evans, L. J. 2008. Surface complexation modelling of Cd(II), Cu(II), Ni(II), Pb(II) and Zn(II) adsorption onto kaolinite. *Geochimica Et Cosmochimica Acta*, 72, 267–276.
41. Kaya, A. & Oren, A. H. 2005. Adsorption of zinc from aqueous solutions to bentonite. *Journal of Hazardous Materials*, 125, 183–189.
42. Garcia-Miragaya, J., Cardenas, R. & Page, A. L. 1983. Sorption of Cd and Zn on kaolinite and montmorillonite. *Heavy Metals in the Environment*, 2, 1244–1248.
43. Maes, A. & Cremers, A. 1975. Cation-exchange hysteresis in montmorillonite: pHdependent effect. *Soil Science*, 119, 198–202.
44. Weiss, Z., Klika, Z., Capkova, P., Janeba, D. & Kozubova, S. 1998. Sodium-cadmium and sodium-zinc exchangeability in montmorillonite. *Physics and Chemistry of Minerals*, 25, 534–540.
45. Abollino, O., Aceto, M., Malandrino, M., Sarzanini, C. & Mentasti, E. 2003. Adsorption of heavy metals on Na- montmorillonite: Effect of pH and organic substances. *Water Research*, 37, 1619–1627.
46. Tremolada, J., Dzioba, R., Bernardo- Sanchez, A. & Menendez- Aguado, J. M. 2010. The preg- robbing of gold and silver by clays during cyanidation under agitation and heap leaching conditions. *International Journal of Mineral Processing*, 94, 67–71.
47. Nechaev, Y. A. & Nikolenko, N. 1986. Adsorption of chloride complexes of gold(III)-oxide compounds of iron. *Kolloidn. Zhurnal USSR*, 48, 1196–1201.
48. Cohen, D. R. & Waite, T. D. 2004. Interaction of aqueous Au species with goethite, smectite and kaolinite. *Geochemistry: Exploration Environment Analysis*, 4, 279–287.
49. Hong, H. L., Sun, Z. Y., Fu, Z. Y. & Min, X. M. 2003. Adsorption of $AuCl_4^-$ by kaolinites: Effect of pH, temperature and kaolinite crystallinity. *Clays and Clay Minerals*, 51, 493–501.
50. Mohammadnejad, S., Provis, J. L. & Van Deventer, J. S. J. 2011. Gold sorption by silicates in acidic and alkaline chloride media. *International Journal of Mineral Processing*, 100, 149–156.
51. Bhattacharyya, K. G. & Sen Gupta, S. 2006. Pb(II) uptake by kaolinite and montmorillonite in aqueous medium: Influence of acid activation of the clays. *Colloids and Surfaces A: Physicochemical and Engineering Aspects*, 277, 191–200.
52. Donat, R., Akdogan, A., Erdem, E. & Cetisli, H. 2005. Thermodynamics of Pb^{2+} and Ni^{2+} adsorption onto natural bentonite from aqueous solutions. *Journal of Colloid and Interface Science*, 286, 43–52.
53. Inglezakis, V. J., Stylianou, M. & Loizidou, M. 2010. Ion exchange and adsorption equilibrium studies on clinoptilolite, bentonite and vermiculite. *Journal of Physics and Chemistry of Solids*, 71, 279–284.
54. Xu, D., Tan, X. L., Chen, C. L. & Wang, X. K. 2008. Adsorption of Pb(II) from aqueous solution to MX- 80 bentonite: Effect of pH, ionic strength, foreign ions and temperature. *Applied Clay Science*, 41, 37–46.
55. Echeverria, J. C., Zarranz, I., Estella, J. & Garrido, J. J. 2005. Simultaneous effect of pH, temperature, ionic strength, and initial concentration on the retention of lead on illite. *Applied Clay Science*, 30, 103–115.
56. Heidmann, I., Christl, I., Leu, C. & Kretzschmar, R. 2005. Competitive sorption of protons and metal cations onto kaolinite: Experiments and modeling. *Journal of Colloid and Interface Science*, 282, 270–282.
57. Ikhsan, J., Wells, J. D., Johnson, B. B. & Angove, M. J. 2004. The effect of aspartic acid on the binding of transition metals to kaolinite. *Journal of Colloid and Interface Science*, 273, 6–13.
58. Sari, A., Tuzen, M., Citak, D. & Soylak, M. 2007. Equilibrium, kinetic and thermodynamic studies of adsorption of Pb(II) from aqueous solution onto Turkish kaolinite clay. *Journal of Hazardous Materials*, 149, 283–291.

59. Akafia, M. M., Reich, T. J. & Koretsky, C. M. 2011. Assessing Cd, Co, Cu, Ni, and Pb sorption on montmorillonite using surface complexation models. *Applied Geochemistry*, 26, S154–S157.
60. Gu, X., Evans, L. J. & Barabash, S. J. 2007. Modeling the adsorption of Cd (II), Cu (II), Ni (II), Pb (II) and Zn (II) onto montmorillonite. *Geochimica Et Cosmochimica Acta*, 74, 5718–5728.
61. Griffin, R. A. & Au, A. K. 1977. Lead adsorption by montmorillonite using a competitive Langmuir equation. *Soil Science Society of America Journal*, 41, 880–882.
62. Das, N. C. & Bandyopadhyay, M. 1991. Selectivity sequence of adsorption of heavy metals by vermiculite. *Asian Environment*, 13, 13–20.
63. Morris, D. E., Chisholm-Brause, C. J., Barr, M. E., Conradson, S. D. & Eller, P. G. 1994. Optical spectroscopic studies of the sorption of UO_2^{2+} species on a reference smectite. *Geochimica et Cosmochimica Acta*, 58, 3613–3623.
64. Chisholm-Brause, C. J., Conradson, S. D., Buscher, C. T., Eller, P. G. & Morris, D. E. 1994. Speciation of uranyl sorbed at multiple binding sites on montmorillonite. *Geochimica et Cosmochimica Acta*, 58, 3625–3631.
65. Sylvester, E. R., Hudson, E. A. & Allen, P. G. 2000. The structure of uranium (VI) sorption complexes on silica, alumina, and montmorillonite. *Geochimica et Cosmochimica Acta*, 64, 2431–2438.
66. Krawcyk-Barsch, E., Arnold, T., Reuther, H., Brandt, F., Bosbach, D. & Bernhard, G. 2004. Formation of secondary Fe-oxyhydroxide phases during the dissolution of chlorite: Effects on uranium sorption. *Applied Geochemistry*, 19, 1403–1412.
67. Turner, G. D., Zachara, J. M., Mckinley, J. P. & Smith, S. C. 1996. Surface-charge properties of UO_2^{2+} adsorption of a subsurface smectite. *Geochimica et Cosmochimica Acta*, 60, 3399–3414.
68. Ames, L. L., McGarrah, J. E. & Walker, B. A. 1983. Sorption of trace constituents from aqueous solutions onto secondary minerals: I. Uranium. *Clays and Clay Minerals*, 31, 321–334.
69. Schlegel, M. L. & Descostes, M. 2009. Uranium uptake by hectorite and montmorillonite: A solution chemistry and polarized EXAFS study. *Environmental Science and Technology*, 43, 8593–8598.
70. Hudson, E. A., Terminello, L. J., Viani, B. E., Denecke, M., Reich, T., Allen, P. G., Bucher, J. J., Shuh, D. K. & Edelstein, N. M. 1999. The structure of U^{6+} sorption complexes on vermiculite and hydrobiotite. *Clays and Clay Minerals*, 47, 439–457.
71. Dent, A. J., Ramsay, J. D. F. & Swanton, S. W. 1992. An EXAFS study of uranyl ion in solution and sorbed onto silica and montmorillonite clay colloids. *Journal of Colloid and Interface Science*, 150, 45–60.
72. McKinley, J. P., Zachara, J. M., Smith, S. C. & Turner, G. D. 1995. The influence of uranyl hydrolysis and multiple site-binding reactions on adsorption of U(VI) to montomorillonite. *Clays and Clay Minerals*, 43, 586–598.
73. Pshinko, G. N., Kobets, S. A., Bogolepov, A. A. & Goncharuk, V. V. 2010. Treatment of waters containing uranium with saponite clay. *Journal of Water Chemistry and Technology*, 32, 10–16.
74. Donat, R. 2009. The removal of uranium (VI) from aqueous solution onto natural sepiolite. *Journal of Chemistry and Thermodynamics*, 41, 829–835.
75. Sikalidis, C. A. & Alexiades, C. 1989. Adsorption of uranium and thorium from aqueous solutions by the clay minerals montmorillonite and vermiculite. *Toxicology and Environmental Chemistry*, 20–21, 175–180.

2

Fundamentals of Clays

Surface and Colloid Science, and Rheology

Elizaveta Forbes and Andrew Chryss

2.1 Introduction

Relating bulk properties of clays and clay suspensions to their particle morphology or surface chemistry is a complex process due to the multiple possible types of interaction and the large number of bodies over which these forces act. The work that has been done on simple, idealized suspensions can be extended to complex multi-component suspensions, such as clay in water, with caution. This allows the observation of trends in behaviour to be given some meaning and predictions to be made. This chapter combines the surface and colloid science with macroscopic rheological behaviour due to the strong causal relationships involved. For example, what affects clay swelling behaviour also affects the rheology, and the swelling itself leads to complex rheological behaviour.

2.2 Mineral Structure

2.2.1 Silicate Minerals

Silicate minerals are by far the most commonly occurring minerals in the world. All silicate minerals are based on tetrahedral building blocks. These tetrahedral units are composed of a central Si^{4+} atom bonded to four surrounding O^{2-} atoms, more commonly known as the silica tetrahedron.

Different families of silicate minerals are based upon the numerous ways in which these tetrahedra may be combined (Klein and Hurlbut, 1993), either as isolated tetrahedra (e.g. olivine), single or double chains of tetrahedra (e.g. orthopyroxene, hornblende), as an interlocking tetrahedral framework (e.g. feldspar, quartz), or as tetrahedral sheets (e.g. talc, kaolinite). The different combinations of tetrahedral blocks are illustrated in Fig. 2.1. This review focuses on the phyllosilicate mineral family, as all clay minerals fall within this family (Deer *et al.*, 1992).

2.2.2 Phyllosilicate Minerals

Phyllosilicate minerals are so named because the sheets of silica tetrahedral tend to give the minerals a platy/leafy habit, as illustrated in an SEM image in Fig. 2.2.

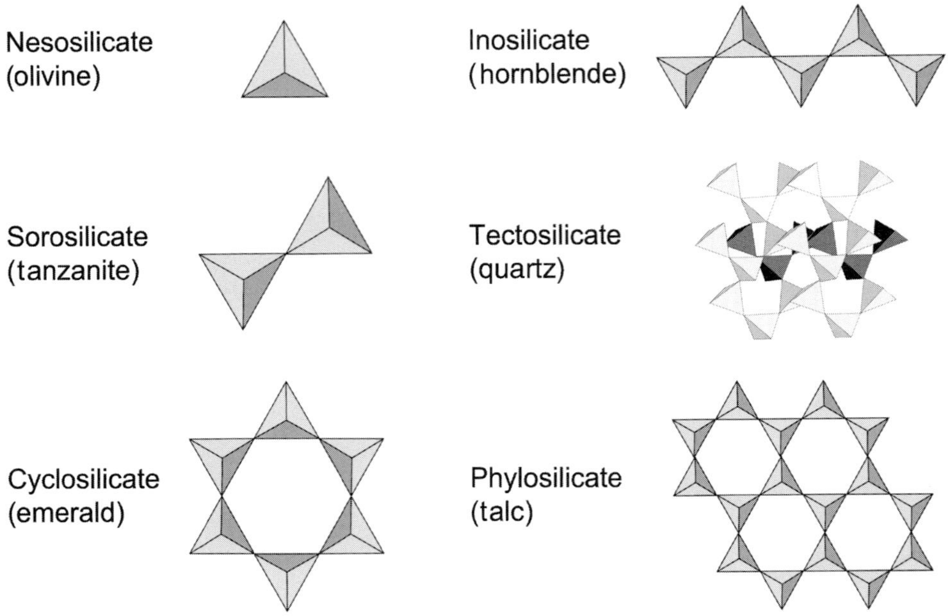

Figure 2.1. Types of silicates (Deer *et al.*, 1992).

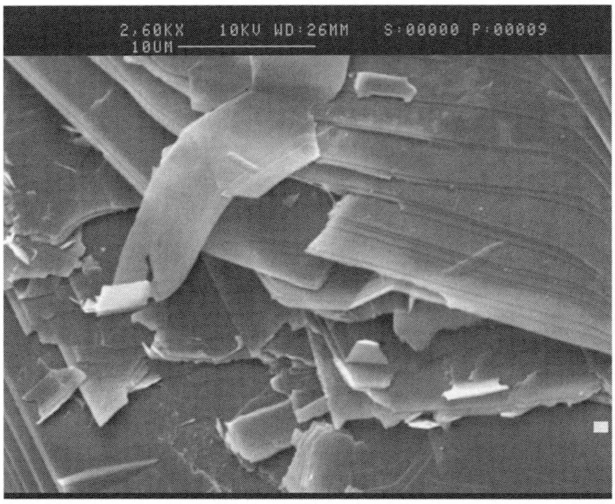

Figure 2.2. Plate-like smectite particles. Image courtesy of Angus McFarlane.

There are a large number of different types of phyllosilicates in existence with a variety of different properties. The detailed structure of phyllosilicates is composed of a combination of tetrahedral and octahedral sheets. Individual octahedra are made up of two layers of four oxygen/hydroxyl atoms, with embedded magnesium or aluminium atoms arranged in an octahedral coordination, as illustrated in Fig. 1.4 and already discussed in Chapter 1.

The variation of the stacking of the tetrahedral (T) and octahedral (O) sheets makes up the different types of phyllosilicate minerals. The way in which these minerals are sub-classified is not uniform across the literature. For the purposes of this review the taxonomic approach taken is that as outlined in Table 1.1, i.e. (1) type, (2) group, (3) subgroup and (4) species. While in practical consideration of the mineral processing value chain (MPVC) most of the review concentrates on the kaolinite and smectite groups as illustrative 1:1 and 2:1 clays, it would be erroneous to assume that the deleterious effects often seen in industrial practice are limited to the minerals within these two groups. In fact, minerals such as talc and serpentine often get clustered together with kaolinite (and other swelling and non-swelling clays), and are generically referred to as 'clay minerals'. The disparity in the terminology used to describe deleterious minerals often leads to confusion with regards to what constitutes a 'clay mineral' or a 'clay'.

It would be more correct to say that the common use of the term 'clay minerals' refers to phyllosilicates in general. Within the phyllosilicates, different groups contain minerals that have different degrees to which they affect the behaviour of ore slurries, ranging from mild to extreme. The clay mineral that arguably exhibits the most extreme behaviour of all (chrysotile) belongs to the serpentine group, not the 'clay' group as normally perceived.

2.2.2.1 Talc and Chlorite Groups

Talc and chlorite are largely similar in structure, both consisting of a three-layered T–O–T structure. In the case of talc, the sheets are held together by weak van der Waals forces and easily come apart when ground. The resulting particles tend to be smooth platelets, commonly used in paints, coatings and cosmetics. In the case of chlorite, the T–O–T sheets are similar to that of talc, but separated by an extra octahedral brucite layer (Deer *et al.*, 1992).

Talc and chlorite minerals occur as gangue components of many base metal sulphide ore deposits around the world (Schouwstra *et al.*, 2000). These ores are generally beneficiated by means of the flotation process. Due to natural floatability, talc and chlorite readily enter the flotation concentrates, reducing their grade.

2.2.2.2 Mica Group

The continuous stacking of consecutive T–O–T units results in the thin, platy morphology of minerals such as muscovite. The space between the T–O–T layers is occupied by interlayer cations. The two most prevalent mica types, biotite and muscovite, are characterized by K^+ ions in their interlayers (Deer *et al.*, 1992). In this, micas are more similar to swelling clays than to talc. In fact, micas serve as precursors to swelling clays such as vermiculites and smectites. The transformation of micas to swelling clays is well documented (Bracke *et al.*, 1995; Churchman, 1980) and occurs as a result of the replacement of the non-exchangeable interlayer cations by exchangeable hydrated cations. This exchange of cations occurs as a result of weathering within the mineral deposits.

2.2.2.3 Kaolinite and Smectite Groups

This group is the more commonly perceived 'clays'; kaolinite in particular is commonly known as 'pottery clay'. These minerals are commonly used in ceramics, as well as paints and coatings applications.

Minerals in this group can be further classified into the swelling (smectite) and non-swelling (kaolinite) varieties. The capacity to swell originates from the interlayer cations found between the T–O–T sheets of these minerals. The interlayer cations are typically strongly hydrated ions (Mg^{2+}, Ca^{2+}). These ions attract water molecules that accumulate between the sheet layers of minerals such as montmorillonite or smectite, forcing the adjoining sheets apart and causing the mineral to swell. The degree of swelling is determined by the hydration capacity of the interlayer cations (Klein and Hurlbut, 1993; Luckham and Rossi, 1999). The swelling leads to an effective increase in the volumetric solids fraction of the suspensions, which in turn leads to an increase in particle–particle interactions and complex rheological behaviour.

Suspensions containing swelling clay minerals such as montmorillonite present substantial dewatering and impoundment challenges in tailings treatment due to their low settling and poor compaction rates (McFarlane *et al.*, 2005). Problems associated with swelling clays are commonly encountered in the processing of coal, phosphate and bauxite industries (de Kretser *et al.*, 1997; Power *et al.*, 2011; Sofrá and Boger, 2002).

2.2.2.4 Serpentine Group

The serpentine group requires a closer inspection, as the minerals within it can deviate from the more common platy habit and form fibrous or corrugated structures. This deviation is caused by the mismatch of bond lengths between the octahedral and tetrahedral units. The T–O structure of serpentine minerals is very similar to that of kaolinite (see Fig. 1.4). In the case of kaolinite, the octahedral unit contains aluminium atoms, and the bond length between the octahedral and tetrahedral sheets is well matched. For this reason, kaolinite particles form distinct platelets. However, when the octahedral unit contains magnesium ions, the bond length in the octahedral unit is slightly shorter than that of the tetrahedral unit. This is the case with minerals such as antigorite and chrysotile. In order to compensate for the bond length mismatch, the entire structure begins to curve (Mellini and Zanazzi, 1987; Yada, 1971).

In the case of antigorite, this results in a corrugated configuration. In the case of chrysotile (often termed 'asbestos'), the plates roll up into tight tubes. These tubes then take the form of long, thin fibres, which result in extremely problematic behaviour when present in mineral processing streams (Burdukova *et al.*, 2008; Ndlovu *et al.*, 2011; Ralston and Fornaserio, 2006; Senior and Thomas, 2005). The morphology of chrysotile is illustrated in Fig. 2.3. This bond length mismatch does not cause similar curving behaviour in other magnesium-based phyllosilicates, as the presence of the second tetrahedral layer (T–O–T configuration) tends to balance out the structure.

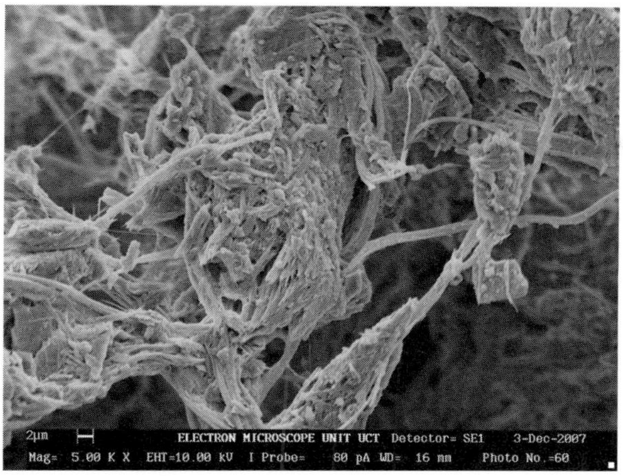

Figure 2.3. SEM of chrysotile fibres.

2.3 The Electrical Double Layer and Inter-Particle Forces

When discussing the impact of clay minerals on mineral processing operations, one generally refers to their behaviour in aqueous suspension, i.e. as a mineral slurry. When minerals are present in an aqueous environment, they become subject to different inter-particle forces, which then have a strong impact on the macroscopic flow properties of these same suspensions. The types of forces acting between mineral particles and the impact this has on particle–particle interactions are summarized below.

2.3.1 DLVO Theory

DLVO theory was independently developed by Derjaguin and Landau (1941) and Verwey and Overbeek (1948). Although Derjaguin's paper was published first, both sets of authors are given equal credit for the theory, as the intervening years cover the duration of the Second World War.

The theory states that the overall total force between two particles in aqueous suspension is determined by the balance of attractive and repulsive forces, as outlined in Eq. (2.1).

$$F_{Tot} = F_{attractive} - F_{repulsive} \tag{2.1}$$

In the classic DLVO theory, the forces responsible for the attraction are the attractive van der Waals force and heterocoagulation, while the repulsive force is attributed to double-layer repulsion (Hogg *et al.*, 1966). However, the theory was later revised to include the attractive hydrophobic force and the repulsive steric force. The nature of these surface forces has been described in great detail by a number of authors (Adamson, 1985; Dukhin and Derjaguin, 1976; Gregory, 2006; Hiemenz and Rajagopolan, 1997; Hunter, 1988; Israelachvili, 1985), and only a brief overview will be presented here.

2.3.1.1 Van der Waals Force

The van der Waals force arises due to the interaction between polarizable molecules or molecules and a surface. For two spherical particles, the force can be expressed in terms of Eq. (2.2), where a is the particle radius, H is the distance between particles, and A_H is the Hamaker constant. The van der Waals attraction has a short range and comes into play if the distance H is significantly smaller than the particle radius (typically 1–10 nm) (Israelachvili and Pashley, 1983).

$$F_{VDW} = -\frac{aA_H}{12H^{12}} \tag{2.2}$$

2.3.1.2 Double-Layer Repulsion

When mineral particles are placed in aqueous solution, they typically acquire a strong electrical charge. The presence of the surface charge causes the ions in solution to become re-distributed around the mineral particle. The oppositely charged counter-ions form an inner layer close to the particle surface, called the adsorbed layer. Another, less concentrated layer of co- and counter-ions forms the outer shell, known as the diffuse layer. The concentration of counter-ions decreases with an increasing distance away from the particle surface, until it reaches that of the bulk solution. This double-layer model is illustrated in Fig. 2.4 (Dukhin and Derjaguin, 1976).

The charge potential of the particle surface is commonly approximated by the zeta potential (ζ), which is the charge potential on the outer plane of the adsorbed layer, also known as the Stern plane. The thickness of the double layer is termed the Debye length (κ^{-1}) and is a strong function of the ionic strength of the aqueous medium (I), as shown in Eq. (2.3). Hence the range of the repulsion can vary significantly depending on the electrolyte concentration in the bulk solution.

When two similarly charged particles are brought together, the double layers around the particles begin to overlap, causing the concentration of ions between the two surfaces to be greater than of the bulk aqueous medium. The decrease in the entropy and the increase in the osmotic pressure result in the formation of a repulsive force between the two particles, as described by Eq. (2.4), where H is the distance from the surface, ζ is the zeta potential, ε is the dielectric constant of the bulk solution and ε_0 is the permittivity of free space.

$$\kappa^{-1} \propto \frac{1}{\sqrt{I}} \tag{2.3}$$

$$F_{ELD} = \frac{2\pi a \varepsilon_0 \varepsilon \kappa \zeta^2 e^{-\kappa H}}{1 + e^{-\kappa H}} \tag{2.4}$$

The zeta potential of the particles can be altered by the varying concentrations of potential determining ions in the bulk solution. For oxide minerals, potential is determined by H^+ and OH^- ions, i.e. the solution pH.

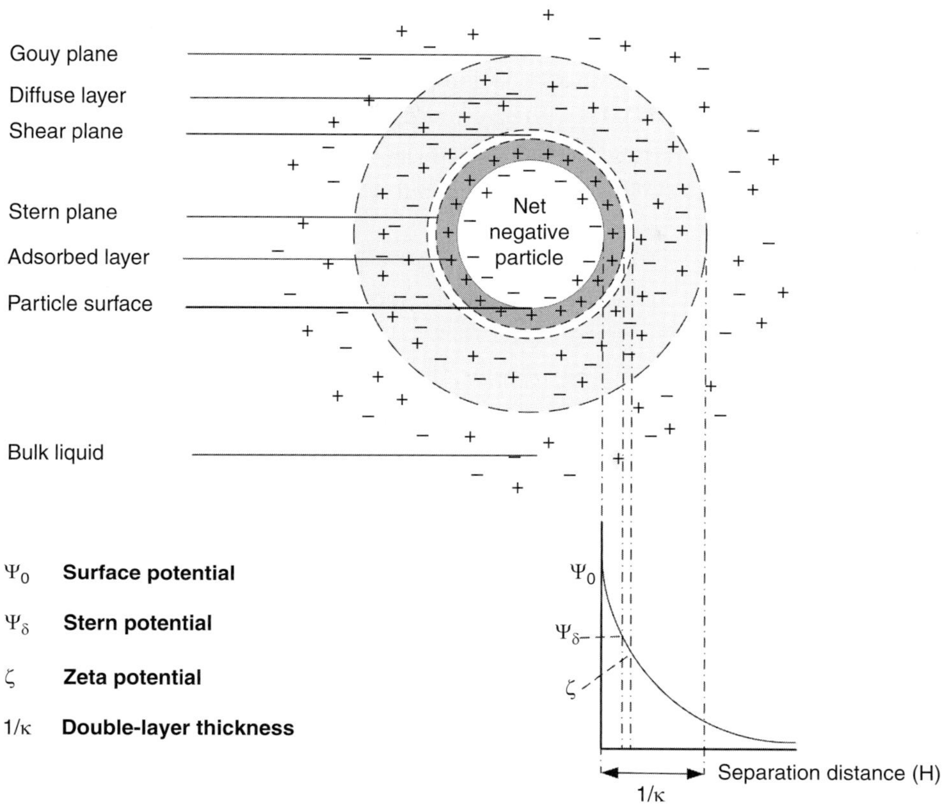

Figure 2.4. Theoretical structure of the electrical double layer.

2.3.1.3 Hydrophobic Attraction

The nature and the existence of the hydrophobic force are still debated. The concept of hydrophobic attraction gained prominence when it was discovered that in the case of non-polar solids, the magnitude of the long-range attractive force could not be explained by means of classic DLVO theory. The early work of Israelachvili and Pashley (1983) showed that when two such surfaces are in close proximity in water, there is a strong attraction between them. This and other early work mainly found it to be a short-range attraction, similar to that of the van der Waals force (in the order of 1–10 nm) (Israelachvili and Pashley, 1983; Meagher and Craig, 1994; Pashley *et al.*, 1985). Later investigations found the attraction had a significantly longer range, with estimates varying from 20 to 300 nm (Craig *et al.*, 1998, 1999; Ishida *et al.*, 2000a; Kurihara *et al.*, 1990; Mantel *et al.*, 1995; Rabinovich and Derjaguin, 1988).

Many researchers now believe that the long-range component of hydrophobic attraction is mainly due to nano-bubbles present on the hydrophobic surfaces (Ishida *et al.*, 2000b), which can bridge a gap of a few hundred nanometres between the surfaces and create capillary adhesion (Ishida *et al.*, 2000a; Meyer *et al.*, 2006). Nano-bubbles would only be expected when there is a 'supply' of air to the surface, such as when the surface is inserted into the liquid from air or when solvent conditions are changed so that bubbles nucleate on the surface (Zhang *et al.*, 2008). The current estimate of the range of the 'true' hydrophobic attraction is between 1 and 10 nm, and the exact nature of this attraction is still not fully understood (Meyer *et al.*, 2006).

2.3.1.4 Steric Repulsion

In mineral processing operations, steric repulsion occurs as a result of the presence of molecular species on the mineral surfaces. These molecules are typically water-soluble polymers of a short to medium chain length. While longer chain polymers (such as typical flocculants) cause particles to aggregate through bridging, shorter chain polymers can instead cause the suspension to become dispersed.

This dispersion can take place through two mechanisms. When two polymer-coated surfaces are brought together, the overall polymer concentration between the two surfaces increases, creating an increase in the osmotic pressure existing between the two surfaces. In addition, the compression of the extended polymer molecules results in an elastic force, creating additional resistance to contact (Laskowski and Pugh, 1992). In the case where the polymers themselves carry an electrical charge, an additional component of the repulsive force can be present due to the enhancement of double-layer repulsion with the additional polymer charge (Marra and Hair, 1988).

2.4 Clay Particles in Aqueous Suspension

2.4.1 Anisotropic Nature of Phyllosilicate Particles

Because of the platy habits of phyllosilicate minerals, phyllosilicate particles possess distinct basal faces and edges. The different particle planes have different surface charge characteristics, causing the particles to become charge anisotropic. As a result, these particles display distinctly different behaviour from the more uniformly charged isotropic minerals (such as quartz or galena), and have complex modes of particle–particle association, which change depending on the surface charge distribution of the different particle facets, which in turn is dependent upon solution pH. The exact nature of these modes of association is still debated in the literature, with no uniform consensus. The edge versus basal interaction is critically important in the coagulation and flocculation behaviour of clays, and this is addressed in detail in Chapter 8.

Early research into clay minerals proposed that the tetrahedral basal planes of phyllosilicates carry a permanent negative charge due to the isomorphous substitutions of Si^{4+} ions in the tetrahedral layer with metal ions (such as Al^{3+} and Ti^{3+}). The presence of lower valence metal ions causes a charge imbalance in the tetrahedral plane, leaving the faces with a permanent negative charge (Schofield and Samson, 1954; Swartzen-Allen and Matijevic, 1974; van Olphen, 1951).

On the other hand, other researchers argue that the magnitude of ion substitution was insufficient to cause the negative charge behaviour of tetrahedral basal plane surfaces (Nishimura *et al.*, 1995; Scales *et al.*, 1990). They propose that basal plane (face) charge is not constantly negative, but is instead strongly pH dependent, with an isoelectric point at pH 2. These assertions are made based on the use of zeta potential and streaming potential techniques. The authors used their findings to conclude that the pH-dependant nature of the basal plane surfaces was a result of the hydrolysis of the Si–O–Si sites, making the surface charge of tetrahedral planes similar to that of quartz. The simplest way to view the basal faces are as areas dominated by silanol groups with a negative charge over most of the pH range.

The edge faces of phyllosilicates are more complex because of the role of aluminol as well as silanol groups, and are thought to carry a charge that changes from positive (low pH) to negative (high pH) as a function of pH. This charge arises from the presence of amphoteric hydroxyl sites on the edge surface: SiO sites in the tetrahedral portion and the AlO/MgO sites in the octahedral portion of the particle edge (Johnson *et al.*, 1998; Tombácz and Szekeres, 2006; Williams and Williams, 1977).

These surface sites are amenable to protonation/deprotonation by H^+ and OH^- ions in a manner illustrated in Eq. (2.5). For this reason H^+ and OH^- ions are often referred to as 'potential determining ions' and this mechanism is responsible for the alteration of mineral surface charge with pH (see also Fig. 8.5).

$$H_2O + MO^- \underset{OH^-}{\overset{H^+}{\rightleftharpoons}} MOH \underset{OH^-}{\overset{H^+}{\rightleftharpoons}} MOH_2^+ \qquad (2.5)$$

2.4.2 Charge Measurement for Anisotropic Particles

The exact measurement of the surface charge characteristics of both the basal faces and edge faces of clay particles is exceedingly difficult. Before surface (force) specific microscopy techniques (such as atomic force microscopy (AFM)) became available, attempts to quantify the surface charge of anisotropic minerals relied on bulk (or average) measurements. Such techniques cannot measure the charge on isolated surface planes and therefore require extensive interpretation, which can often lead to misleading results.

2.4.2.1 Zeta Potential

One of the more controversial techniques for charge magnitude estimation is the use of zeta potential measurements. Zeta potential is commonly calculated by means of measuring the mobility of a particle in suspension within an electric field (electrophoretic mobility). The resultant zeta potential estimate is then treated as a close approximation of surface potential.

The use of zeta potential measurements with regards to clay minerals can be problematic. Since the minerals are anisotropic, the mobility of the particles in the aqueous medium represents an 'apparent' value that is (at best) an average zeta potential of all the particle planes. Furthermore, the value of zeta potential measurements is strongly dependent on the thickness of the electrical double layer, which in turn is dependent on the concentration of electrolyte in solution. Some researchers feel that this added complexity further invalidates zeta potential as an accurate proxy for surface potential (Hou *et al.*, 2009).

Several authors have attempted to 'deconstruct' the average zeta value using an estimate of the particle aspect ratio, i.e. the relative surface areas of differently charged planes (Johnson *et al.*, 1998; Williams and Williams, 1977). These studies indicate that the overall zeta potential of clay minerals strongly resembled that of quartz, which was used to conclude that the negatively charged tetrahedral (basal) faces dominate the overall charge of the particle surfaces. This approach relies heavily on the assumption that clay particles occur as perfect crystalline units, which has since been shown to be incorrect (Du *et al.*, 2010; Żbik *et al.*, 2008).

Conversely, a number of studies have looked at the surface potential of talc, with an incorrect underlying assumption that that talc basal planes are perfectly charge compensated and therefore neutral. The zeta potential measurements were then assumed to represent the surface potential of talc edges (Bremmell and Addai-Mensah, 2005; Fuerstenau and Huang, 2003; Fuerstenau *et al.*, 1988; Steenberg and Harris, 1984).

These instances highlight the fact that the use of zeta potential measurements to estimate the surface charge distribution of anisotropic minerals can be highly misleading and certain researchers advise to avoid it altogether when working with anisotropic clay minerals (Burdukova *et al.*, 2007; Hou *et al.*, 2009; Missana and Adell, 2000). Nonetheless, there are ways in which zeta potential measurements can be used to estimate the way in which clay minerals can interact with other minerals on the basis of charge. This is specifically pertinent in determining whether clay minerals are likely to cause 'slime coatings' in flotation applications (Arnold and Aplan, 1986; Attia and Deason, 1989; Fuerstenau *et al.*, 1958; Iwasaki *et al.*, 1962).

One such method makes use of the distribution of zeta potential measurements in a binary mineral system (Liu *et al.*, 2002; Xu *et al.*, 2003). The measurement involves obtaining the zeta distribution of each individual mineral and then compares it to the zeta distribution of a binary mineral mixture. This technique can give an indication of whether or not one of the minerals had been effectively coated with another (a single distribution peak vs a bimodal distribution). This method does not estimate the exact nature of the clay surface charge, but it does provide a measure that can predict the behaviour of that mineral in an industrial setting.

2.4.2.2 Streaming Potential

A method similar to zeta potential measurements, though more restricted, is that of streaming potential. In this technique, a particle is not freely suspended in the aqueous medium, but rather a single mineral plate is exposed to the surrounding liquid (Nalaskowski *et al.*, 2007; Scales *et al.*, 1990). This approach typically requires a relatively large, isolated basal plane surface, which is very difficult to obtain in practice. Of all the phyllosilicate minerals, the only one that is amenable to obtaining perfectly cleaved flakes large enough to perform such measurements is mica. Surface streaming potential studies are thus confined to studying mica surfaces and not those of other minerals. Furthermore, streaming potential measurements are subject to the same scepticism as zeta potential measurements when it comes to using them as adequate proxies for surface potential measurements, and has been strongly questioned (Hou *et al.*, 2009; Missana and Adell, 2000).

2.4.2.3 Potentiometric Titrations

The use of titrations to study the surface charge distribution is one of the more widely accepted and implemented methodologies (Burdukova *et al.*, 2007; Duc *et al.*, 2005a, 2005b; Tombácz and Szekeres, 2004, 2006; Zhao *et al.*, 2008). The technique involves estimating the net surface charge density of minerals through the exchange between potential determining ions (H^+ and OH^-) and ions of an indifferent electrolyte (e.g. NaCl) on the mineral surface (Laskowski and Sobieraj, 1969; Mular and Roberts, 1966). The magnitude of the surface charge density can be calculated using Eq. (2.6), where σ_0 is the net surface charge density, F is the Faraday constant and γ_i is the adsorption density of potential determining ions (Hunter, 1988).

$$\sigma_0 = \mathrm{F} \cdot \left(\gamma_{H^+} - \gamma_{OH^-}\right) \tag{2.6}$$

When indifferent electrolyte is present in solution, the overall ionic strength of that solution increases. This causes the electrical double layer to become compressed, resulting in a shorter Debye length (see Eq. (2.3)). As a result, the value of the Stern Potential (Ψ_S) decreases (see Fig. 2.4). This in turn causes potential determining ions to be adsorbed onto the surface in order to compensate for the change. The adsorption of these ions from solution is then reflected in the overall solution pH. Therefore, the net surface potential can be measured by measuring the change in solution pH as a function of an increase in the concentration of background electrolyte, as illustrated in Fig. 2.5. Unlike zeta potential and streaming potential measurements, this technique gives a direct estimate of surface charge that is not dependent on the thickness of the electrical double layer.

The obtained point of zero charge (pzc) is thought to represent the net point of zero charge (NPZC) of all the pH dependent mineral facets (i.e. broken ion sites subject to adsorption and desorption of H^+ and OH^- ions). Several researchers believe that this NPZC represents the NPZC of the edge surfaces only (Janek and Lagaly, 2001; Penner and Lagaly, 2001; Tombácz and Szekeres, 2004, 2006). This stance rests on the assumption that the tetrahedral basal planes of clay minerals carry a permanent/constant negative charge that is present entirely due to the isomorphic ion substitution into the tetrahedra and is therefore pH independent (van Olphen, 1951). This means that the potentiometric titration results can be used to estimate the surface charge density of clay particle edges as a function of pH.

However, as was mentioned in the previous sections, not all researchers believe that the surface charge of the tetrahedral basal planes is pH independent (Nishimura *et al.*, 1995; Scales *et al.*, 1990). If this is the case, then potentiometric titration measurements are subject to the same uncertainty as zeta potential measurements, where only an average or 'apparent' charge distribution can be obtained using the technique. On the other hand, the very

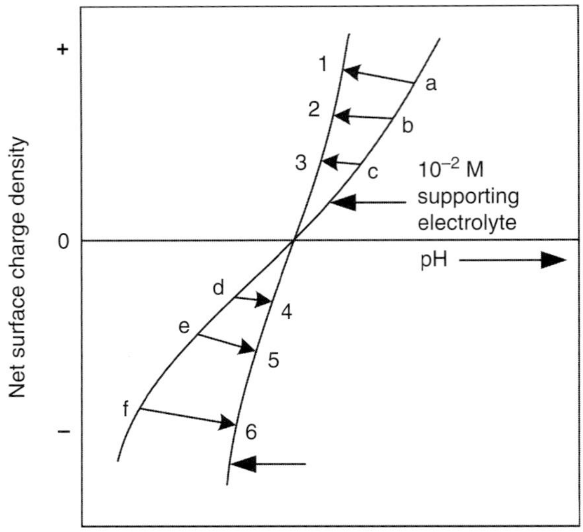

Figure 2.5. Schematic representation of a potentiometric titration (Mular and Roberts, 1966).

assertion that the basal plane surfaces are pH dependant is based on zeta potential measurements, which many researchers believe are inappropriate. This means that many of the arguments with regards to the validity of various techniques can rapidly become circular.

2.4.2.4 Microscopic Methods

Recent advances in the field of AFM technology have enabled researchers to look directly at the specific particle planes and directly measure the surface charge distribution (Nalaskowski *et al.*, 2007; Zhao *et al.*, 2008). The results of the studies have largely confirmed the existing notions of the surface charge distributions – that the basal planes are not pH dependent and the edge charges agreed with estimates generated by titration measurements. The difference is dramatically illustrated in Fig. 2.6.

Exact magnitudes of the basal face and edge face charges are still difficult to measure, in part as it is exceedingly difficult to obtain a sufficiently smooth surface on which to perform the measurements. However, in more recent years improved estimates of surface-specific surface potentials have begun to emerge (Yan *et al.*, 2011; Yin *et al.*, 2012). These measurements once more confirm the presence of the consistently negative charge of the basal plane surfaces, which corroborates the hypothesis of the isomorphous substitution as the source of the negative charge.

2.5 Models of Particle Interactions

The charge disparity between particle basal and edge faces means that the simple model of an electrical double layer that holds for a largely symmetric, uniformly charged mineral particle cannot be applied to an anisotropic particle. In the case of simple isotropic

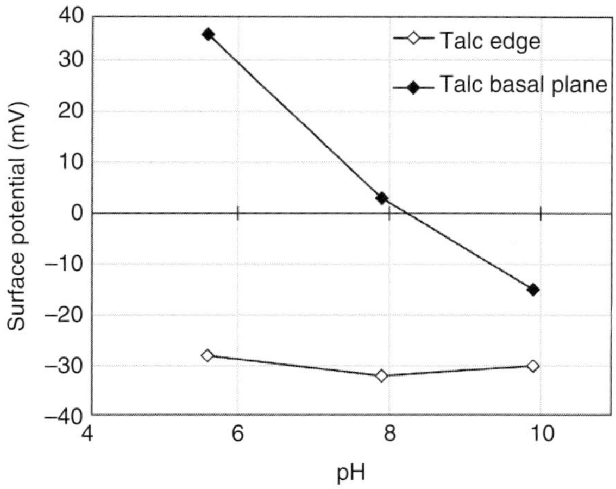

Figure 2.6. Surface potentials of basal planes and edges of talc (from reported data in Yin *et al.*, 2012)

particles, the double-layer forces are always repulsive, since the particles are similarly charged. The net inter-particle forces usually become attractive at a point where the overall surface charge of the particle is zero, and the attractive van der Waals forces become dominant.

This simple model does not hold for anisotropic particles, where different particle faces carry a different electrical charge. These differently charged faces will interact via electrical double-layer forces in such a way that the oppositely charged surfaces result in an attractive force (Hogg *et al.*, 1966). As a result, interactions between particles of the same clay mineral are often described in the form of complex particle–particle coordinations. Van Olphen (1951) first proposed that there are three main types of particle–particle interactions between anisotropic clay particles: face–face (FF), face–edge (EF) and edge–edge (EE), as illustrated in Fig. 2.7. These interactions are thought to arise from the differences in the electrical charge of the basal plane and edge surfaces at a given pH.

Van Olphen (1951) postulated that at a pH when both the basal planes and edges of particles carry a negative charge, the overall inter-particle forces would be repulsive, resulting in a randomly dispersed particle orientation (Fig. 2.7a). This would typically take place at high pH values. However, as pH decreases, the charge on the particle edges would begin to decrease, approaching a point of zero charge. At this point the dominant force between particle edges would be the attractive van der Waals force, while the force between negatively charged particle planes would remain a strongly repulsive double-layer force. This would cause the particles to coagulate in the EE configuration (Fig. 2.7d).

As the pH decreases further, and the charge on the particle edge faces becomes positive, an attractive force develops between negatively charged basal planes and positively charged edges, engendering a 'house of cards' type particle interaction (Fig. 2.7c).

Face–face coagulation is slightly more complex, as it could take place in two ways depending on the structure of the mineral. The T–O–T structured minerals (e.g. montmorillonite)

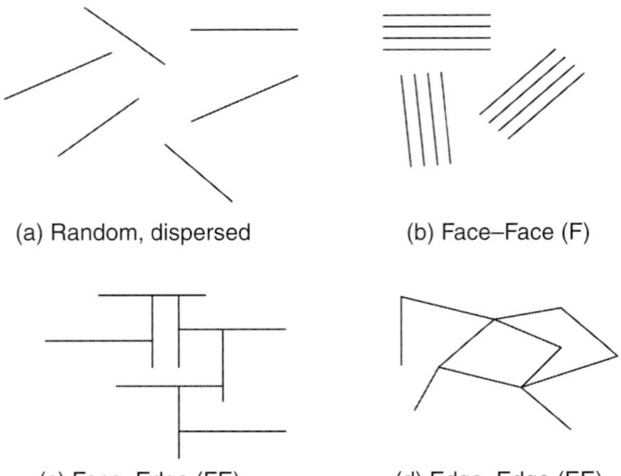

Figure 2.7. Possible orientation of clay particles (van Olphen, 1951).

have two identical basal planes. This means that in order for FF coagulation to take place, the negative charge on the basal planes needs to approach zero, in order for attractive van der Waals forces to become dominant. On the other hand, minerals like kaolinite, which have a T–O structure, have two types of basal plane surface. One is a permanently negative tetrahedral face, while the other is an octahedral face with a surface charge that alters as a function of pH. The coagulation between these surfaces can therefore occur as a result of an attractive force between two oppositely charged basal plane surfaces.

Although a large number of authors adhere to van Olphen's original propositions (Du *et al.*, 2010; Kitchener, 1969; Rand and Melton, 1975; Schofield and Samson, 1954; Swartzen-Allen and Matijevic, 1974; Williams and Williams, 1977; Zhao *et al.*, 2008), others refute them. Several researchers claim that due to the aspect ratio of the platelet particles, the edge face surface area was insufficiently large to provide a sufficient charge density to engender heterocoagulation with the basal face surface (Callaghan and Ottewill, 1974; Lagaly, 1989; Tateyama *et al.*, 1997). Instead, it is claimed that the predominant model of particle–particle interaction is that of FF coagulation.

Despite claims that the 'house of cards' or 'honeycomb' EF and EE particle arrangements are unlikely, their existence has been well demonstrated by the use of cryo-SEM techniques (Du *et al.*, 2010; McFarlane *et al.*, 2006). The source of the disagreement may lay in the assumptions with regards to the shape and aspect ratio of clay particles. A number of researchers assume a perfect thin-plate shape when making predictions about clay particle behaviour (Johnson *et al.*, 1998; Williams and Williams, 1977). However, studies have also clearly demonstrated that phyllosilicate particles do not often occur as thin, perfect flakes, but rather as poorly ordered structures, with sub-hedral facets and islands, resulting in a much higher proportion of edge surfaces being present (Żbik *et al.*, 2008, 2010).

Overall, different researchers have put forward a number of different potential mechanisms for predicting the particle–particle associations of clays in solution. In all likelihood, there is no single correct way to describe such associations, and the overall clay behaviour is determined by a combination of all the described factors. To date, there is no firm understanding of the degree to which these different factors influence clay behaviour and how they vary with different clay types and clays from different sources.

2.6 Rheological Behaviour of Clay Suspensions

The rheology of a clay suspension, i.e. how it deforms in response to an applied stress, τ, has a significant effect on unit operations in mineral processing, such as pumping and pipeline transport, mixing and separation behaviour, and how tailings slurries behave on deposition in a tailings storage facility. Clay suspensions often behave in similar ways to other fine suspensions, as will be shown below, but their swelling behaviour in water makes them problematic for fluid-based mineral processing. The behaviour of viscous suspensions of clay in water will primarily be considered as these are most relevant to unit operations, water recovery, or disposal of waste streams (i.e. non-plastic systems). For clay as a semi-solid in a geotechnical context, see Murray (2006); for clay as a nano-composite filler in polymers, see Bhattacharya *et al.* (2008). A rudimentary knowledge of viscosity is assumed – otherwise, refer to Barnes *et al.* (1993).

2.6.1 Characterizing Suspensions

Viscosity at its most simple is a measure of a fluid's resistance to flow. For many rheologically complex materials the traditional definition of a fluid does not apply and an apparent viscosity is defined as the ratio between the applied shear stress and the resultant shear rate. The behaviour of the apparent viscosity as a function of shear rate is how a material is categorized rheologically. The broad characterization of time-independent, inelastic fluids that describe clay suspension is shown in the rheogram in Fig. 2.8. Newtonian behaviour is constant viscosity (at constant temperature) and is typical of simple fluids such as water, light hydrocarbons, and some low-concentration suspensions. Pseudoplastic behaviour can be described as 'shear thinning' with an apparent viscosity that decreases with increasing shear rate. This is seen in many aqueous polymers and molten thermoplastics. A yield pseudoplastic is also shear thinning but includes a yield stress, that is, a minimum applied shear stress that must be exceeded before flow commences. This is the most common behaviour observed for suspensions, including clay suspensions, and will be further discussed below. Shear thickening behaviour is an increase in the apparent viscosity with increasing shear rate. It was formerly known as 'dilatancy' but has been re-titled to distinguish it from volume dilatancy, which may occur simultaneously. Shear thickening occurs only in moderately to highly concentrated multi-phase mixtures, which are usually observed to have a yield stress as well (as is shown in Fig. 2.8). Shear thickening does not occur in typical clay suspensions, but the presence of an additional coarse solid phase at high enough concentration may produce this effect. Pseudoplastic, yield pseudoplastic and shear thickening behaviour are grouped as non-Newtonian material behaviours.

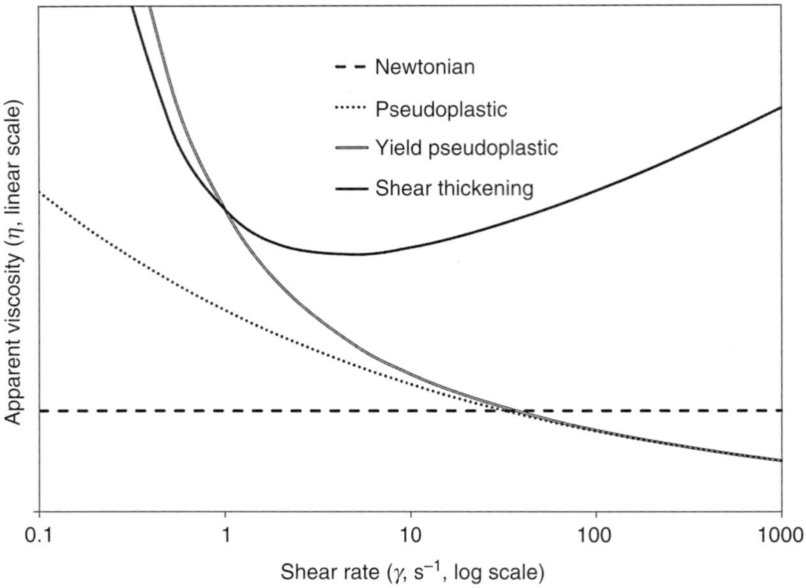

Figure 2.8. Representation of typical rheological behaviour of suspensions.

Suspension viscosity may also show time of shear dependency. If viscosity decreases with time at a constant shear stress or shear rate, and the effect is reversible by ceasing shear, then it is classified as thixotropic behaviour (see Fig. 2.9). If the reverse occurs it is anti-thixotropy (in earlier literature this was referred to as rheopexy). These terms are often confused with shear thinning or thickening, which are time independent, or rheomalaxis, which is the irreversible viscosity decrease with time. Thixotropic behaviour occurs in some more concentrated clay suspensions and mixtures; a well-studied industrial example is 'red mud' from bauxite processing (Nguyen and Boger, 1987). Thixotropic effects may occur due to shearing as a consequence of mixing, pumping or pipeline transport.

2.6.2 Factors Influencing Suspension Rheology

The rheological response of a suspension is a result of particle–fluid and particle–particle interactions. Suspension rheology can also be roughly categorized as resulting from mainly hydrodynamic forces or mainly from the influence of surface forces; this can vary with shear rate for a given suspension. Yield pseudoplastic behaviour can arise from both types of interaction, but usually begins at lower solids concentrations for suspensions with strong surface forces. The rheology of a suspension is therefore strongly dependant on the viscosity of the suspending medium and the inter-particle forces (see Section 2.4). The presence of electric double layers produces effects on the flow of the suspending medium that are referred to as electroviscous effects. The primary electroviscous effect is an increase in viscosity due to distortion of the electric double layer adding to energy dissipation from

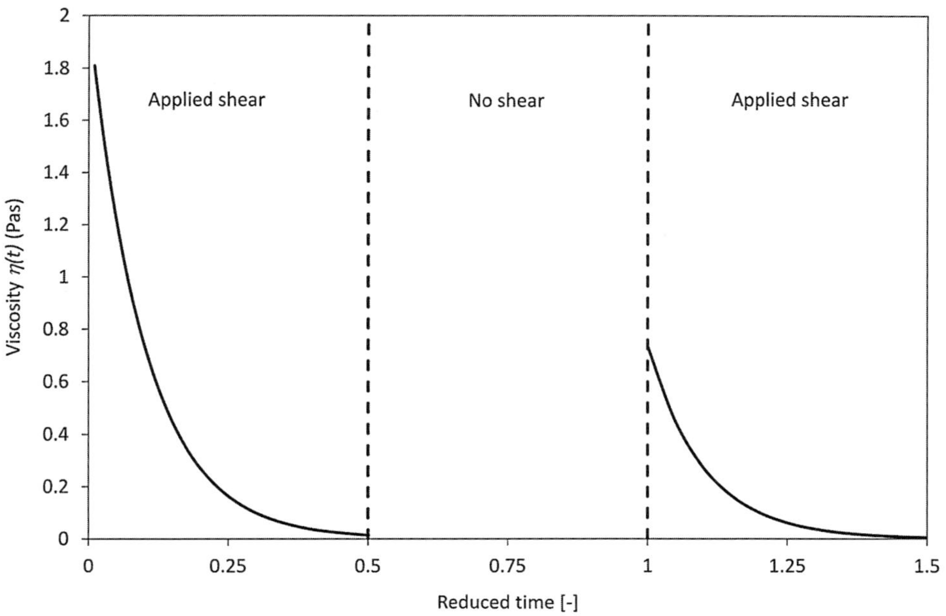

Figure 2.9. Thixotropic behaviour showing viscosity reduction during shearing, partial structure rebuild during rest and continued viscosity reduction with reapplication of shear.

the flow of the suspending medium. Secondary electroviscous effect is due to repulsion between double layers and the tertiary viscous effect includes all other inter- and intra-particle interactions that alter the geometry of the suspended medium. The suspending medium for the majority of the clay suspensions of interest is water, its viscosity showing only minor changes from varying salt content, but decreasing with increasing temperature. The number of interactions increase with concentration of the suspended phase, a dilute suspension having close to Newtonian behaviour (viscosity invariant with rate of shearing, $\dot{\gamma}$) and a higher concentration developing non-Newtonian effects (viscosity varying with shear rate and shear history). Types of non-Newtonian rheology displayed by clay suspensions are shown in Fig. 2.8.

Although it is not possible to predict a suspension's rheology a priori, a useful guide to how it will vary can be gained from the available semi-empirical models. The dimensional requirements of a model of suspension rheology are given by Stickel and Powell (2005), and they consider suspension viscosity η to be a function of particle and suspending medium properties:

$$\eta = f\left(a, \rho_p, \rho_n, \eta_m, \rho_f, kT, \dot{\gamma}.or.\tau, t\right) \tag{2.7}$$

where a is particle radius, ρ_p is particle density, ρ_n is particle number density, η_m is suspending medium viscosity, ρ_f is fluid density and kT is thermal energy. It should be noted that even this function is not exhaustive as a assumes a single particle size, not a size distribution, and does not consider particle morphology, both of which have significant effects on rheological behaviour. These variables can be formed into six dimensionless groups:

$$\eta_r = f\left(C_v, \rho_r, N_{Pe}, N_{Re\dot{\gamma}}, t_r\right) \tag{2.8}$$

where $\eta_r = \eta / \eta_m$ (relative viscosity), $C_v = 4\pi\rho_n a^3/3$ (volume fraction), $\rho_r = \rho_p / \rho_f$ (reduced density), $N_{Pe} = 6\pi\eta a^3\dot{\gamma}/kT$ (Peclet number), $N_{Re\dot{\gamma}} = \rho_f a^2\dot{\gamma}/\eta$ (sheared particle Reynolds number) and $t_r = tkT/\eta a^3$ (reduced time).

For the special condition of a neutrally buoyant system, ρ_r can be ignored, and similarly at steady-state t_r may be neglected, leaving:

$$\eta_r = f\left(C_v, N_{Pe}, N_{Re\dot{\gamma}}\right) \tag{2.9}$$

Under conditions where $1/N_{Pe} \rightarrow 0$ and the Schmidt number ($N_{Sc} = N_{Pe} / N_{Re}$, a measure of shear-driven particle diffusivity) is large the suspension is likely to be Newtonian. This is the case when C_v is low and consequently suspension viscosity is a function of C_v only. This was derived theoretically by Einstein (1956):

$$\eta_r = 1 + C_v[\eta] \tag{2.10}$$

where $[\eta] = \lim_{C_v \to 0}\left(\frac{\eta_r - 1}{C_v}\right)$, the intrinsic viscosity which is largely a function of shape, with a value of 2.5 for hard spheres. The equation is effective for $C_v < 0.1$ and was improved upon by Batchelor (1977):

$$\eta_r = 1 + C_v[\eta] + k[\eta]^2 \tag{2.11}$$

where k is a function of suspension type, and has a value of 6.2 for Brownian suspensions. The parameter variations are due to changes in the primary electroviscous effects. A better approximation is obtained by including the maximum packing fraction C_{vm}, the geometrically highest achievable concentration where:

$$\lim_{C_v \to C_{vm}} \eta_r = \infty \tag{2.12}$$

C_{vm} is strongly dependant on particle size distribution and increases with polydispersity (Farris, 1968). One such model employing a reduced volume fraction is that of Krieger and Dougherty (1959):

$$\eta_r = \left(1 - \frac{C_v}{C_{vm}}\right)^{-[\eta]C_{vm}} \tag{2.13}$$

A comparison of these models is shown in Fig. 2.10. An extension to this model was introduced by Clarke (1967). By changing $[\eta]$ values to reflect particle shape, non-spherical particles could be modelled; for a constant concentration, spheres yielded the lowest viscosity, followed by plates. The highest viscosity was for rods or high-aspect-ratio objects. These three models all show the necessary increase in viscosity with increasing concentration but do not capture the non-Newtonian variation of viscosity with shear rate. This can be incorporated by using the technique of Wildemuth and Williams (1984, 1985) that was developed to describe the shear thinning behaviour of colloidal spheres and irregularly shaped coal suspensions (<100 μm; both are flocculating suspensions). This method considers that the maximum packing fraction C_{vm} is a function of shear stress:

$$\frac{1}{C_{vm}} = \frac{1}{C_{vm0}} - \left(\frac{1}{C_{vm0}} - \frac{1}{C_{vm\infty}}\right) \cdot C_{vd} \tag{2.14}$$

where C_{vm0} is a minimum value of packing fraction (at zero stress), $C_{vm\infty}$ is the highest value of packing fraction and C_{vd} is the fraction of particles that remain unflocculated and is related to the shear stress by:

$$C_{vd} = \frac{1}{1 + A\tau^{-m}} \tag{2.15}$$

where A and m are fitted parameters. A useful aspect of the Wildemuth and Williams method is that a yield stress, τ_y, is predicted for the condition $C_v < C_{vm}(\tau)$. As the imposed shear stress on a suspension decreases, so too does $C_{vm}(\tau)$ (from Eqs (2.14) and (2.15) until it is less than C_v). The yield stress will be discussed further below. Meaning can be assigned to the various parameters in Eqs (2.14) and (2.15) in terms of strength and type of interaction; however, they are still largely experimentally derived or fitted values and cannot be estimated from knowledge of suspended phase conditions. As concentration of the dispersed phase increases, the multiple body interactions increase rapidly and make prediction of bulk properties overly complex.

Useful models of the rheological behaviour of suspensions have been developed for describing the transport properties of suspensions such as clay in water. Most models suitable for yield pseudoplastics are mathematical fits with limited physical meaning in their parameters. The Herschel–Bulkley model (see Eqs (2.16) and (2.17)) is one such instance (Coussot *et al.*, 1996; Pignon *et al.*, 1997):

$$\tau = \tau_y + k\dot{\gamma}^n \tag{2.16}$$

$$\eta = \frac{\tau_y}{\dot{\gamma}} + \dot{\gamma}^{(n-1)} \tag{2.17}$$

where k is a measure of the consistency of the fluid; and n the flow behaviour index. Real, non-dilute clay suspensions cannot be described effectively by simple two-parameter models such as the power-law equation, Bingham plastic equation or Casson equation (Casson, 1959). This is because the structures and interactions that give rise to shear thinning will also give rise to yield stresses. The Herschel–Bulkley model imperfectly captures the viscosity behaviour of many suspensions, but is adequate over a limited range of shear rates.

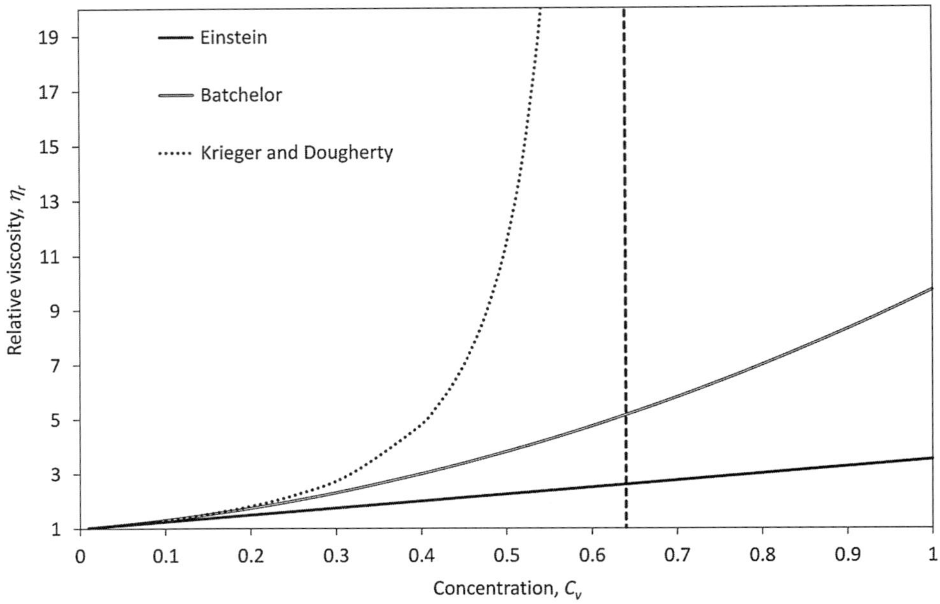

Figure 2.10. Three models of relative viscosity for suspensions.

The yield stress has been mentioned already, but its role in suspension rheology requires some further discussion. A yield stress can arise in a suspension of interacting or non-interacting particles, but will be seen more readily at lower concentrations when strong surface forces are present. This is typically the case for clay-based slurries. Due to the relative ease of taking repeatable yield stress measurements compared to more complete shear stress rheograms (see Section 2.6.3) and the strong dependence on network structures, the yield stress is a common means of monitoring changes in clay suspension behaviour. There are a limited number of studies that attempt to relate the physical properties of particles to the yield stress response of suspensions (e.g. Gay *et al.*, 1969; Thomas, 1961) without incorporating the mechanisms responsible. The effects of surface chemistry on rheology, such as van der Waals and electrical double-layer forces, has been related to the yield stress of suspensions of spheres by Scales *et al.* (1998). The yield stress is found to be a strong function of the zeta potential, and at a maximum when the zeta potential is zero (i.e. at the iso-electric point). The relationship is more complicated in mineral suspensions and does not simply rely on a single parameter (Franks *et al.*, 1999). In the studies cited above, where suspensions are colloidal and inter-particle forces are significant, the yield stress is often related to the breaking of bonds or the breaking up of agglomerates. Experimentally the presence of a static yield stress in dense suspensions is familiar to those working with them regularly and their existence has been established convincingly by Husband *et al.* (1993) and by Huang *et al.* (2005), among others.

2.6.3 Rheometry

Measurement of rheological properties (rheometry) is the characterization of changes to the momentum transport properties of a material. Modelling rheological data is useful for engineering design in pipelines, mixing and thickening. Rheology can also be used as a surrogate for other less readily testable phenomena due to the relative ease of testing bulk properties in place of phenomena at the scale of particle dimensions. Rheometry, however, does not provide a priori insights into material structure as the measured property changes can arise from many causes. It is only a useful indicator once it has been 'calibrated' against the phenomena of interest by other, non-rheological, tests. Many of the studies cited in Section 2.6.4 are of this type.

As many of the clay suspensions of interest are non-Newtonian, a suitable rheometer needs to measure a range of defined shear rates and shear stresses for a given sample. Older types of viscometer, such as the ubiquitous Brookfield, Fann or Hercules, do not measure true rheology as they lack a defined shear field. Important distinctions between sample behaviour are therefore missed. Rotational rheometers using cone and plate, parallel plate or Couette geometries are suitable for clay–water suspensions (Whorlow, 1980), but the cone and plate geometries are not suitable if larger particles or structures are present. Pressure-driven capillary instruments are useful for higher shear rates if the non-Newtonian effects are allowed for in the interpretation of the data using the Rabinowitsch–Mooney equation (Barnes *et al.*, 1993). Apart from unidirectional rotational rheometry, it is possible to apply oscillatory shear to deform a sample without causing unrecoverable strain or disrupting any possible network structure. This technique is not useful for determining transport properties, but may provide some insight into the condition of particle arrangement, notwithstanding the caveat regarding a priori insights from rheometry.

Thixotropy is often detected in concentrated clay suspensions and can manifest as a hysteresis loop on a rheogram. Although commonly taken as a means of comparison between samples (e.g. Brandenburg and Lagaly, 1988; Christidis *et al.*, 2006), a hysteresis loop is actually not a fundamental property but rather an artefact of how the test is conducted. Thixotropic behaviour is best characterized at constant shear rate (or shear stress) and the decay in viscosity recorded as in Fig. 2.9. This allows for analysis using simpler models of $\eta(t)$.

Rheometry is further complicated by experimental artefacts such as wall slippage, which occurs when the suspension being sheared moves relative to the surfaces of the measurement geometry. Slip is more likely to occur and increase in severity as concentration of the dispersed phase increases. It is possible to minimize (but not eliminate) wall slip through the roughening or profiling of measurement surfaces, although it is preferable to use one of many available techniques to de-convolute the rheological and slip behaviour of a suspension (see Chryss *et al.*, 2005 or Mooney, 1931). These techniques involve tests in multiple measuring geometries of different dimensions and consequently increase the number of tests required per sample. Unsurprisingly, the majority of experimental data presented for shear rheometry of clay suspensions, or suspensions generally, are not proofed against slip effects. A simple work-around to avoid slip artefacts is to use a vane rheometer to provide a value of the yield stress (Nguyen and Boger, 1987). This is a highly repeatable technique that provides valuable data for comparison between samples and an indication of a materials slump characteristics (see Schowalter and Christensen, 1998). The main drawback to the vane method is that it only provides data at a single point, i.e. at a zero shear rate. Several of the studies cited in Section 2.6.4 utilize this method.

2.6.4 Chemistry Changes and Rheology

The rheology of clay–water suspensions has been studied extensively in relation to the ceramics industry (Worrall, 1986) and to drilling muds for petroleum extraction (Kassab *et al.*, 2011). Consequently, bentonite and kaolinite tend to dominate the literature. The effect of pH, electrolyte concentration and Na:Ca:Mg ratio on rheology, in particular yield stress, has been widely reported. A maximum in the yield stress of bentonite at its 'natural' (i.e. acidic) pH is reported by Kelessidis *et al.* (2007), Benna *et al.* (1999) and Brandenburg and Lagaly (1988). Similar results were reported for sepiolite (Çınar *et al.*, 2009), but with a basic natural pH. Attapulgite is largely unaffected by pH or NaCl content (Chang *et al.*, 1993), hence its efficacy as a drilling fluid in salty conditions. Yong and Ohtsubo (1987) observed that kaolinite has a monotonically decreasing yield stress with pH, but the addition of iron hydroxide as ferrihydrate can produce maxima similar to those of bentonites.

The effect of NaCl concentration on bentonite has been covered by many workers, with similar outcomes. Abend and Lagaly (2000) found small minima in rheological properties at low NaCl content. They also found similar minima with other chlorides and with Na_2SO_4. Penner and Lagaly (2001) found viscosity minima for chloride and nitrate salt addition, again at low concentration (~1 mmol). De Kretser *et al.* (1998) found a large maximum in yield stress with NaCl content at concentrations two orders of magnitude higher. Güler and Balci (1998) saw an increase in viscosity for kaolinitic clays with the addition of salts, the effect being stronger for chlorides over sulphides and no minima were present. Alther (1986) determined a strong correlation between the ratios of exchangeable cations in bentonite and the viscosity of a clay suspension.

Brandenburg and Lagaly (1988) state that the dependence of flow behaviour on salt concentration and pH can be explained by different types of particle aggregation. Similarly, de Kretser *et al.* (1998), in examining thixotropic behaviour in sodium-montmorillonite-based suspensions, explains the effect of higher NaCl concentrations as due to FF aggregation, which reduces the number of interacting flow units and consequently the rheological parameters such as yield stress.

Clay rheology deviates from simple concentration dependency such as Eq. (2.10) due to particle anisotropy (Packter, 1962). Yuan and Murray (1997) tested the viscosity of high-solids kaolin suspensions with differing particle morphology. The lowest viscosity was shown by a spherical halloysite, a platy kaolinite was higher and a tubular halloysite was highest. This corresponds to the modelling of suspensions of differing shaped particles by Clarke (1967).

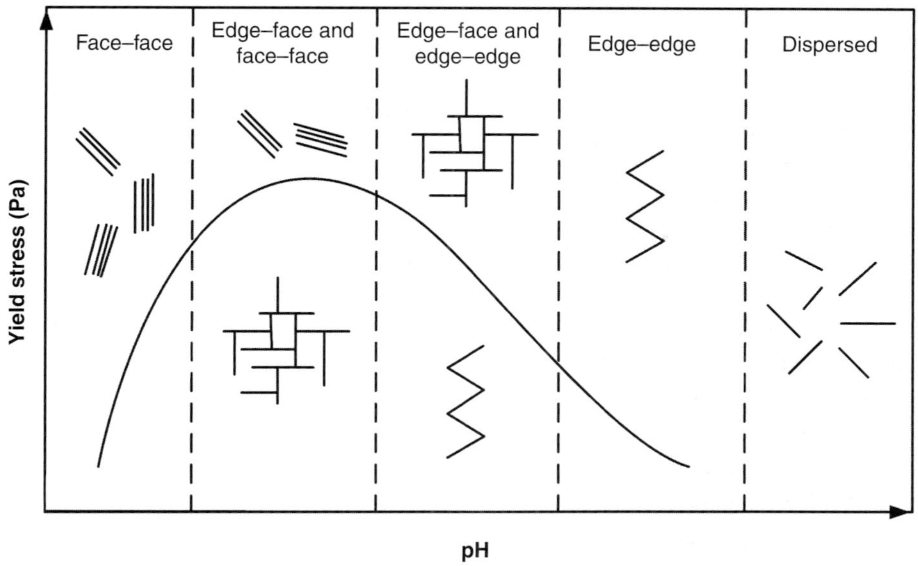

Figure 2.11. Schematic representation of rheological behaviour resulting from different particle orientations (based on observed Bingham yield values; Rand and Melton, 1976).

2.6.5 Effect of Particle Orientation on Suspension Rheology

The different orientations of the clay particles can directly influence the rheological behaviour of mineral suspension through factors like effective volume fraction and the presence of strong inter-particle forces (both attractive and repulsive).

The effect of particle orientations on suspension yield stress was first proposed by Rand and Melton (1976), as described in Fig. 2.11. It must be noted that in their interpretation Rand and Melton focus on lower pH regions and have left out the mention of FF coagulation. Figure 2.11 was adapted from the original publication and altered to reflect the presence of FF coagulation in the acidic pH region.

Rand and Melton postulate that at high pH values, particles will be completely dispersed, thus resulting in low (or zero) Bingham yield stresses. As the pH decreases, the particles begin to aggregate into different particle–particle orientation, described in the previous section.

At relatively high pH values, the predominant orientation is thought to be that of EE, forming loosely bonded, open structures. This causes an elevation in the yield stress of the suspension. As the pH decreases further, the charge on the particle edges approaches neutrality, and EF (house of cards) structures begin to form. Such open structures are thought to be bonded by strong heterocoagulation forces and trap water within them. A combination of these two factors is thought to further increase the suspension yield stress.

As the pH continues to decrease, the predominant orientation becomes that of EF structures. However more compact FF aggregates also begin to form. FF aggregates have a lower effective volume than EF aggregates, which halts the continuous increase in the suspension yield stress, resulting in maximum peak. A further decrease in suspension pH results in FF aggregates becoming dominant, driving the suspension yield stress further down.

It is important to stress that the particle coordination/rheology relationship described in Fig. 2.11 is highly generalized and speculative. It can be further modified by including the effect of electrolytes (Lagaly and Ziesmer, 2003), which will alter the type of coordination that occurs at a given pH, and consequently the rheology. The yield stress/pH relationship for anisotropic clays generally adheres to the shape of the described curve (i.e. exhibits a peak), but neither the location of the peak with regards to the pH, nor the magnitude of the peak can be predicted with any degree of accuracy for a specific mineral under a specified set of chemical conditions.

2.7 Summary

Overall, the lack of agreement as to the exact nature of the mechanisms that drive the interparticle interactions of phyllosilicate suspensions only serves to emphasize the complexity of the process. As a result, there is no fundamental or mechanistic model in existence that adequately describes or predicts the behaviour of clay minerals in suspension. Some authors have even claimed that DLVO theory as a whole is not sufficient to describe and predict colloidal stability of anisotropic clay particles (Missana and Adell, 2000). This assertion is based on the persistent incongruity between model predictions and experimental results.

However, these inconsistencies appear to be based on inadequate assumptions and estimates of the mineral properties that act as input parameters into the DLVO equations (such as Hamaker constants and surface potential values), rather than the theory itself. Hopefully, the latest developments in cryo-SEM, AFM and X-ray diffraction techniques will pave the way for future work that will finally elucidate consistent mechanisms for clay particle interactions and allow for the development of accurate predictive models.

Similarly, rheology suspension theory rapidly fails as moderate concentrations are examined. Prediction of rheological properties is difficult due to the confounding of multiple factors; a combination of particle–particle and fluid–particle interactions (which are affected by pH and electrolyte type and strength) with the particle morphology and size distribution produce rheological behaviour that is not possible to predict well from non-rheological measurements. This does not mean that understanding the rheology of clay suspensions is not useful. When appropriate measurement techniques are applied information that leads to engineering prediction and design can be derived, and means to rheology modification are indicated.

References

Abend, S. & Lagaly, G. 2000. Sol–gel transitions of sodium montmorillonite dispersions. *Applied Clay Science*, 16 (3–4), 201–227.

Adamson, A. W. 1985. *Physical Chemistry of Surfaces*. New York: Wiley & Sons.

Alther, G. R. 1986. The effect of the exchangeable cations on the physico-chemical properties of Wyoming bentonites. *Applied Clay Science*, 1 (3), 273–284.

Arnold, B. J. & Aplan, F. F. 1986. The effect of clay slimes on coal flotation, part I: The nature of the clay. *International Journal of Mineral Processing*, 17 (3–4), 225–242.

Attia, Y. A. & Deason, D. M. 1989. Control of slimes coating in mineral suspensions. *Colloids and Surfaces*, 39 (1), 227–238.

Barnes, H. A., Hutton, J. F. & Walters, K. 1993. *An Introduction to Rheology*. Amsterdam: Elsevier.

Batchelor, G. K. 1977. The effect of Brownian motion on the bulk stress in a suspension of spherical particles. *Journal of Fluid Mechanics*, 83 (1), 97–117.

Benna, M., Kbir-Ariguib, N., Magnin, A. & Bergaya, F. 1999. Effect of pH on rheological properties of purified sodium bentonite suspensions. *Journal of Colloid and Interface Science*, 218 (2), 442–455.

Bhattacharya, S. N., Kamal, M. R. & Gupta, R. K. 2008. *Polymeric Nanocomposites: Theory and Practice*. Berlin: Hanser Publishers.

Bracke, G., Satir, M. & Krauss, P. 1995. The cryptand [222] for exchanging cations of micas. *Clays and Clay Minerals*, 43 (6), 732–737.

Brandenburg, U. & Lagaly, G. 1988. Rheological properties of sodium montmorillonite dispersions. *Applied Clay Science*, 3 (3), 263–279.

Bremmell, K. E. & Addai-Mensah, J. 2005. Interfacial chemistry mediated behaviour of colloidal talc dispersions. *Journal of Colloid and Interface Science*, 283, 385–391.

Burdukova, E., Bradshaw, D. J. & Laskowski, J. S. 2007. Effect of CMC and pH on the rheology of suspensions of isotropic and anisotropic minerals. *Canadian Metallurgical Quarterly*, 46, 273–278.

Burdukova, E., Becker, M., Ndlovu, B., Mokgethi, B. & Deglon, D. A. 2008. Relationship between slurry rheology and its mineralogical content. In: Wang, D. D., Xao, S. C., Wang, F. L., Cheng, Z. U. & Long, H., (eds) *24th Int. Minerals Processing Congress*. Beijing: China Scientific Book Service Co. Ltd, 2169–2178.

Callaghan, I. C. & Ottewill, R. H. 1974. Interparticle forces in montmorillonite gels. *Faraday Discussions of the Chemical Society*, 57, 110–118.

Casson, N. 1959. A flow equation for pigment-oil suspensions of the printing ink type. In: Mill, C. C. (ed.) *Rheology of Disperse Systems*. London: Pergamon Press.

Chang, S. H., Ryan, M. H. & Gupta, R. K. 1993. The effect of pH, ionic strength, and temperature on the rheology and stability of aqueous clay suspensions. *Rheologica Acta*, 32 (3), 263–269.

Christidis, G. E., Blum, A. E. & Eberl, D. D. 2006. Influence of layer charge and charge distribution of smectites on the flow behaviour and swelling of bentonites. *Applied Clay Science*, 34 (1–4), 125–138.

Chryss, A., Bhattacharya, S. & Pullum, L. 2005. Rheology of shear thickening suspensions and the effects of wall slip in torsional flow. *Rheologica Acta*, 45 (2), 124–131.

Churchman, G. J. 1980. Clay minerals from micas and chlorites in some New Zealand soils. *Clay Minerals*, 15, 59–76.

Çınar, M., Can, M. F., Sabah, E., Karagüzel, C. & Çelik, M. S. 2009. Rheological properties of sepiolite ground in acid and alkaline media. *Applied Clay Science*, 42 (3–4), 422–426.

Clarke, B. 1967. Rheology of coarse settling suspensions. *Transactions of the Institution of Chemical Engineers*, 45, 251–256.

Coussot, P., Proust, S. & Ancey, C. 1996. Rheological interpretation of deposits of yield stress fluids. *Journal of Non-Newtonian Fluid Mechanics*, 66(1), 55–70.

Craig, V. S. J., Ninham, B. W. & Pashley, R. M. 1998. Study of the long-range hydrophobic attraction in concentrated salt solutions and its implications for electrostatic models. *Langmuir*, 14 (12), 3326–3332.

Craig, V. S. J., Ninham, B. W. & Pashley, R. M. 1999. Direct measurement of hydrophobic forces: A study of dissolved gas, approach rate, and neutron irradiation. *Langmuir*, 15 (4), 1562–1569.

de Kretser, R., Scales, P. J. & Boger, D. V. 1997. Improving clay-based tailings disposal: Case study on coal tailings. *American Institute of Chemical Engineers Journal*, 43 (7), 1894–1903.

de Kretser, R. G., Scales, P. J. & Boger, D. V. 1998. Surface chemistry–rheology interrelationships in clay suspensions. *Colloids and Surfaces A: Physicochemical and Engineering Aspects*, 137 (1), 307–318.

Deer, W. A., Howie, R. A. & Zussman, J. 1992. *Introduction to Rock-Forming Minerals*. New York: Prentice Hall.

Derjaguin, B. V. & Landau, L. 1941. Theory of stability of strongly charged lyophobic sols and the adhesion of strongly charged particles in solution of electrolytes. *Acta Physiochimica*, 14, 633–662.

Du, J., Morris, G., Pushkarova, R. A. & St. C. Smart, R. 2010. Effect of surface structure of kaolinite on aggregation, settling rate, and bed density. *Langmuir*, 26 (16), 13227–13235.

Duc, M., Gaboriaud, F. & Thomas, F. 2005a. Sensitivity of the acid–base properties of clays to the methods of preparation and measurement: 1. Literature review. *Journal of Colloid and Interface Science*, 289 (1), 139–147.

Duc, M., Gaboriaud, F. & Thomas, F. 2005b. Sensitivity of the acid–base properties of clays to the methods of preparation and measurement: 2. Evidence from continuous potentiometric titrations. *Journal of Colloid and Interface Science*, 289 (1), 148–156.

Dukhin, S. S. & Derjaguin, B. V. 1976. *Electrophoresis*. Moscow: Academy of Sciences of the USSR.

Einstein, A. 1956. *Investigations on the Theory of the Brownian Movement*. Mineola, NY: Dover Publications, Inc.

Farris, R. J. 1968. Prediction of the viscosity of multimodal suspensions from unimodal viscosity data. *Transactions of the Society of Rheology*, 12 (2), 281–302.

Franks, G. V., Johnson, S. B., Scales, P. J., Boger, D. V. & Healy, T. W. 1999. Ion-specific strength of attractive particle networks. *Langmuir*, 15 (13), 4411–4420.

Fuerstenau, D. W. & Huang, P. 2003. Interfacial phenomena involved in talc flotation. In: D. J. Bradshaw (ed.) *22nd Int. Minerals Processing Congress*. Cape Town: Document Transformation Technologies, 1034–1043.

Fuerstenau, D. W., Gaudin, A. M. & Miaw, H. L. 1958. Iron oxide slime coatings in flotation. *Transactions of the American Institute of Mining, Metallurgy, and Petroleum Engineers*, 211, 792–793.

Fuerstenau, M. C., Valdivieso, A. & Fuerstenau, D. W. 1988. Role of hydrolyzed cations in the natural hydrophobicity of talc. *International Journal of Minerals Processing*, 23, 161–170.

Gay, E. C., Nelson, P. A. & Armstrong, W. P. 1969. Flow properties of suspensions with high solids concentration. *AIChE Journal*, 15 (6), 815–822.

Gregory, J. 2006. *Particles in Water, Properties and Processes*. Boca Raton, FL: Taylor & Francis.

Güler, Ç. & Balci, E. 1998. Effect of some salts on the viscosity of slip casting. *Applied Clay Science*, 13 (3), 213–218.

Hiemenz, P. C. & Rajagopolan, R. 1997. *Principles of Colloid Chemistry.* New York: Dekker.

Hogg, R., Healy, T. W. & Fuerstenau, D. W. 1966. Mutual coagulation of colloidal dispersions. *Transactions of the Faraday Society*, 62, 1638–1651.

Hou, J., Li, H., Zhu, H. & Wu, L. 2009. Determination of clay surface potential: A more reliable approach. *Soil Science Society of America Journal*, 73 (5), 1658–1663.

Huang, N., Ovarlez, G., Bertrand, F., *et al.* 2005. Flow of wet granular materials. *Physical Review Letters*, 94 (2), 028301.

Hunter, R. J. 1988. *Zeta Potential in Colloid Science: Principles and Applications.* San Diego, CA: Academic Press.

Husband, D. M., Aksel, N. & Gleissle, W. 1993. The existence of static yield stresses in suspensions containing noncolloidal particles. *Journal of Rheology*, 37 (2), 215–235.

Ishida, N., Sakamoto, M., Miyahara, M. & Higashitani, K. 2000a. Attraction between hydrophobic surfaces with and without gas phase. *Langmuir*, 16, 5681–5687.

Ishida, N., Inoue, T., Miyahara, M. & Higashitani, K. 2000b. Nano bubbles in a hydrophobic surface in water observed by tapping-mode atomic force microscopy. *Langmuir*, 16, 6377–6380.

Israelachvili, J. 1985. *Intermolecular & Surface Forces*. London: Academic Press.

Israelachvili, J. N. & Pashley, R. M. 1983. Measurement of the hydrophobic interaction between two hydrophobic surfaces in aqueous electrolyte solution. *Journal of Colloid and Interface Science*, 98 (2), 500–514.

Iwasaki, I., Cooke, S. R. B., Haraway, D. H. & Choi, H. S. 1962. Iron wash slimes: Some mineralogical characteristics. *Transactions AIME*, 223, 97–108.

Janek, M. & Lagaly, G. 2001. Proton saturation and rheological properties of smectite dispersions. *Applied Clay Science*, 19 (1–6), 121–130.

Johnson, S. B., Russell, A. S. & Scales, P. J. 1998. Volume fraction effects in shear rheology and electroacoustic studies of concentrated alumina and kaolin suspensions. *Colloids and Surfaces*, 141, 119–130.

Kassab, S. Z., Ismail, A. S. & Elessawi, M. M. 2011. Drilling fluid rheology and hydraulics for oil fields. *European Journal of Scientific Research*, 57 (1), 68–86.

Kelessidis, V. C., Tsamantaki, C. & Dalamarinis, P. 2007. Effect of pH and electrolyte on the rheology of aqueous Wyoming bentonite dispersions. *Applied Clay Science*, 38 (1–2), 86–96.

Kitchener, J. A. 1969. Colloid minerals: Chemical aspects of their dispersion, flocculation and filtration. *Filtration and Separation*, 6 (5), 553–558.

Klein, C. & Hurlbut, C. S. 1993. *Manual of Mineralogy.* New York: Wiley & Sons.

Krieger, I. M. & Dougherty, T. J. 1959. A mechanism for non-Newtonian flow in suspensions of rigid spheres. *Transactions of the Society of Rheology*, 3 (1), 137–152.

Kurihara, K., Kato, S. & Kunitake, T. 1990. Very strong long range attractive forces between stable hydrophobic monolayers of a polymerized ammonium surfactant. *Chemistry Letters*, 19 (9), 1555.

Lagaly, G. 1989. Principles of flow of kaolin and bentonite dispersions. *Applied Clay Science*, 4 (2), 105–123.

Lagaly, G. & Ziesmer, S. 2003. Colloid chemistry of clay minerals: The coagulation of montmorillonite dispersions. *Advances in Colloid and Interface Science*, 100–102, 105–128.

Laskowski, J. S. & Pugh, R. J. 1992. Dispersion stability and dispersing agents. In: Laskowski, J. S. & Ralston, J. (eds) *Colloid Chemistry in Mineral Processing.* New York: Elsevier.

Laskowski, J. S. & Sobieraj, S. 1969. Zero points of charge of spinel minerals. *Transactions of the Canadian Institute of Mining and Metallurgy*, 78, C161–C162.

Liu, J., Zhou, Z., Xu, Z. & Masliyah, J. 2002. Bitumen–clay interactions in aqueous media studied by zeta potential distribution measurement. *Journal of Colloid and Interface Science*, 252 (2), 409–418.

Luckham, P. S. & Rossi, S. 1999. The colloidal and rheological properties of bentonite suspensions. *Journal of Colloid and Interface Science*, 82, 43–92.

Mantel, M., Rabinovich, Y. I., Wightman, J. P. & Yoon, R. H. 1995. A study of hydrophobic interactions between stainless steel and salinated glass surfaces using atomic force microscopy. *Journal of Colloid and Interface Science*, 170, 203–214.

Marra, J. & Hair, M. L. 1988. Forces between two poly(2-vinylpyridine)-covered surfaces as a function of ionic strength and polymer charge. *The Journal of Physical Chemistry*, 92 (21), 6044–6051.

McFarlane, A., Bremmell, K. & Addai-Mensah, J. 2005. Microstructure, rheology and dewatering behaviour of smectite dispersions during orthokinetic flocculation. *Minerals Engineering*, 18 (12), 1173–1182.

McFarlane, A., Bremmell, K. & Addai-Mensah, J. 2006. Improved dewatering behavior of clay minerals dispersions via interfacial chemistry and particle interactions optimization. *Journal of Colloid and Interface Science*, 293 (1), 116–127.

Meagher, L. & Craig, V. S. J. 1994. Effect of dissolved gas and salt on the hydrophobic force between polypropylene surfaces. *Langmuir*, 10 (8), 2736–2742.

Mellini, M. & Zanazzi, P. R. 1987. Crystal structures of lizardite-IT and lizardite-2Hl from Coli, Italy. *American Mineralogist*, 72, 941–948.

Meyer, E. E., Rosenberg, K. J. & Israelachvili, J. 2006. Recent progress in understanding hydrophobic interactions. *Proceedings of the National Academy of Sciences of United States of America*, 103 (43), 15739–15746.

Missana, T. & Adell, A. 2000. On the applicability of DLVO theory to the prediction of clay colloids stability. *Journal of Colloid and Interface Science*, 230 (1), 150–156.

Mooney, M. 1931. Explicit formulas for slip and fluidity. *Journal of Rheology*, 2 (2), 210–222.

Mular, A. L. & Roberts, R. B. 1966. A simplified method to determine isoelectric points of oxides. *Transactions of the Canadian Institute of Mining and Metallurgy*, 69, 438–439.

Murray, H. H. 2006. Exploration, mining, and processing. In: Haydn, H. M. (ed.) *Developments in Clay Science.* New York: Elsevier.

Nalaskowski, J., Abdul, B., Du, H. & Miller, J. D. 2007. Anisotropic character of talc surfaces as revealed by streaming potential measurements, atomic force microscopy, molecular dynamics simulations and contact angle measurements. *Canadian Metallurgical Quarterly*, 46 (3), 227–235.

Ndlovu, B. N., Forbes, E., Becker, M., *et al.* 2011. The effects of chrysotile mineralogical properties on the rheology of chrysotile suspensions. *Minerals Engineering*, 24 (9), 1004–1009.

Nguyen, Q. D. & Boger, D. V. 1987. The rheology of concentrated bauxite residue suspensions: a complete story. In: Wagh, A. S. & Desai, P. (eds) *Bauxite Tailings 'Red Mud'*. Kingston: The Jamaica Bauxite Institute.

Nishimura, S., Scales, P. J., Tateyama, H., Tsunematsu, K. & Healy, T. W. 1995. Cationic modification of muscovite mica: An electrokinetic study. *Langmuir*, 11 (1), 291–295.

Packter, A. 1962. Studies in the rheology of clay–water systems. *Rheologica Acta*, 2 (1), 44–50.

Pashley, R. M., McGuiggan, P. M. & Ninham, B. W. 1985. Attractive forces between uncharged hydrophobic surfaces: Direct measurement in aqueous solution. *Science*, 229, 1088–1089.

Penner, D. & Lagaly, G. 2001. Influence of anions on the rheological properties of clay mineral dispersions. *Applied Clay Science*, 19 (1–6), 131–142.

Pignon, F., Magnin, A., Piau, J.-M., *et al.* 1997. Yield stress thixotropic clay suspension: Investigations of structure by light, neutron, and x-ray scattering. *Physical Review E*, 56 (3), 3281–3289.

Power, G., Gräfe, M. & Klauber, C. 2011. Bauxite residue issues: I. Current management, disposal and storage practices. *Hydrometallurgy*, 108 (1–2), 33–45.

Rabinovich, Y. I. & Derjaguin, B. V. 1988. Interaction of hydrophobized filaments in aqueous electrolyte solutions. *Colloids and Surfaces*, 30 (3–4), 243–251.

Ralston, J. & Fornaserio, D. 2006. Effect of MgO minerals on pentlandite flotation. In: Onal, G., Acarkan, N., Celik, M. S., *et al.* (eds) *23rd International Minerals Processing Congress*. Istanbul:Promed Advertising Ltd, 750–755.

Rand, B. & Melton, I. E. 1975. Isoelectric point of edge surface of kaolinite. *Nature*, 257, 214–216.

Rand, B. & Melton, I. E. 1976. Particle interactions in aqueous kaolinite suspensions. *Journal of Colloid and Interface Science*, 60 (2), 308–320.

Scales, P. J., Grieser, F. & Healy, T. W. 1990. Electrokinetics of the muscovite mica–aqueous solution interface. *Langmuir*, 6 (3), 582–589.

Scales, P. J., Johnson, S. B., Healy, T. W. & Kapur, P. C. 1998. Shear yield stress of partially flocculated colloidal suspensions. *American Institute of Chemical Engineers Journal*, 44 (3), 538–544.

Schofield, R. K. & Samson, H. R. 1954. Flocculation of kaolinite due to the attraction of oppositely charged crystal faces. *Discussions of the Faraday Society*, 18, 135–145.

Schouwstra, R. P., Kinloch, E. D. & Lee, C. A. 2000. A short geological review of the Bushvelt Complex. *Platinum Metals Review*, 44, 33–39.

Schowalter, W. R. & Christensen, G. 1998. Toward a rationalization of the slump test for fresh concrete: Comparisons of calculations and experiments. *Journal of Rheology*, 42 (4), 865–870.

Senior, G. D. & Thomas, S. A. 2005. Development and implementation of a new flowsheet for the flotation of a low grade nickel ore. *International Journal of Mineral Processing*, 78 (1), 49–61.

Sofrá, F. & Boger, D. V. 2002. Environmental rheology for waste minimisation in the minerals industry. *Chemical Engineering Journal*, 86 (3), 319–330.

Steenberg, E. & Harris, P. J. 1984. Adsorption of carboxymethyl cellulose, guar gum and starch onto talc, sulphides, oxides and salt type minerals. *South African Journal of Chemistry*, 37, 85–90.

Stickel, J. J. & Powell, R. L. 2005. Fluid mechanics and rheology of dense suspensions. *Annual Review of Fluid Mechanics*. 37, 129–149.

Swartzen-Allen, S. L. & Matijevic, E. 1974. Surface and colloid chemistry of clays. *Chemical Reviews*, 74 (3), 385–400.

Tateyama, H., Scales, P. J., Ooi, M., *et al.* 1997. X-ray diffraction and rheology study of highly ordered clay platelet alignment in aqueous solutions of sodium tripolyphosphate. *Langmuir*, 13 (9), 2440–2446.

Thomas, D. G. 1961. Transport characteristics of suspensions: 3. Laminar-flow properties of flocculated suspensions. *AIChE Journal*, 7 (3), 431–437.

Tombácz, E. & Szekeres, M. 2004. Colloidal behavior of aqueous montmorillonite suspensions: The specific role of pH in the presence of indifferent electrolytes. *Applied Clay Science*, 27 (1–2), 75–94.

Tombácz, E. & Szekeres, M. 2006. Surface charge heterogeneity of kaolinite in aqueous suspension in comparison with montmorillonite. *Applied Clay Science*, 34, 105–124.

van Olphen, H. 1951. Rheological phenomena of clay sols in connection with the charge distribution on the micelles. *Discussions of the Faraday Society*, 11, 83–96.

Verwey, E. J. W. & Overbeek, J. T. G. 1948. *Theory of Stability of Lyophobic Solids*. Amsterdam: Elsevier.

Whorlow, R. W. 1980. *Rheological Techniques*. Sussex: Ellis Horwood.

Wildemuth, C. R. & Williams, M. C. 1984. Viscosity of suspensions modeled with a shear-dependent maximum packing fraction. *Rheologica Acta*, 23 (6), 627–635.

Wildemuth, C. R. & Williams, M. C. 1985. A new interpretation of viscosity and yield stress in dense slurries: Coal and other irregular particles. *Rheologica Acta*, 24 (1), 75–91.

Williams, D. J. A. & Williams, K. P. 1977. Electrophoresis and zeta potential of kaolinite. *Journal of Colloid and Interface Science*, 65 (1), 79–87.

Worrall, D. M. 1986. *Clays and Ceramic Raw Materials*. Dordrecht: Springer.

Xu, Z., Liu, J., Choung, J. W. & Zhou, Z. 2003. Electrokinetic study of clay interactions with coal in flotation. *International Journal of Mineral Processing*, 68 (1–4), 183–196.

Yada, K. 1971. Study of microstructure of chrysotile asbestos by high resolution electron microscopy. *Acta Chrystallographica A*, A27 (6), 659–664.

Yan, L., Englert, A. H., Masliyah, J. H. & Xu, Z. 2011. Determination of anisotropic surface characteristics of different phyllosilicates by direct force measurements. *Langmuir*, 27 (21), 12996–13007.

Yin, X., Gupta, V., Du, H., Wang, X. & Miller, J. D. 2012. Surface charge and wetting characteristics of layered silicate minerals. *Advances in Colloid and Interface Science*, 179–182, 43–50.

Yong, R. N. & Ohtsubo, M. 1987. Interparticle action and rheology of kaolinite–amorphous iron hydroxide (ferrihydrite) complexes. *Applied Clay Science*, 2 (1), 63–81.

Yuan, J. & Murray, H. H. 1997. The importance of crystal morphology on the viscosity of concentrated suspensions of kaolins. *Applied Clay Science*, 12 (3), 209–219.

Żbik, M. S., Smart, R. S. C. & Morris, G. E. 2008. Kaolinite flocculation structure. *Journal of Colloid and Interface Science*, 328 (1), 73–80.

Żbik, M. S., Martens, W. N., Frost, R. L., *et al.* 2010. Smectite flocculation structure modified by Al13 macro-molecules – As revealed by the transmission X-ray microscopy (TXM). *Journal of Colloid and Interface Science*, 345 (1), 34–40.

Zhang, X. H., Quinn, A. & Ducker, W. A. 2008. Nanobubbles at the interface between water and a hydrophobic solid. *Langmuir*, 24, 4756–4764.

Zhao, H., Bhattacharjee, S., Chow, R., *et al.* 2008. Probing surface charge potentials of clay basal planes and edges by direct force measurements. *Langmuir*, 24 (22), 12899–12910.

3

Leaching and Solution Chemistry

Jian Li, Angus McFarlane, Craig Klauber and Peter Smith

3.1 Fundamental Dissolution Mechanisms of Clays

The mechanisms of dissolution and precipitation at the mineral–water interface have been reasonably well understood in terms of natural mineral weathering and metal cycling processes (Aldushin *et al.*, 2006; Hering and Stumm, 1990; Kalinowski and Schweda, 1996). The same mechanisms may be applicable to an understanding of mineral dissolution under hydrometallurgical conditions where non-equilibrium conditions with faster reaction kinetics prevail, due to higher lixiviant concentrations, temperature and pressure.

Surface complexation models, within the framework of the transition state theory (TST) (Fig. 3.1) are often applied to explain the dissolution and precipitation of major rock-forming minerals in natural weathering processes (Schott *et al.*, 2009 and references therein). The interaction between reactants A and B need to overcome an energy barrier (E_a) to form an activated complex species ($AB^{\ddagger}$) at the top of the barrier, which subsequently rearranges to yield products (C and D) at a lower final (Gibbs) free energy state. A simplified conceptual mineral dissolution model can be modified from that idea; the rate-determining step involves irreversible decomposition of the activated complex to form product species (Walther and Wood, 1986):

$$\text{Mineral (A)} + \text{fluid (B)} \rightleftharpoons \text{activated complex } (AB^{\ddagger}) \rightarrow \text{product species } (C + D) \text{ (dissolution)}$$

Transition state theory treats the activated complex as a true chemical species. The surface chemistry concept of mineral dissolution developed suggests that oxides and oxide minerals in aqueous environments are covered with surface hydroxyl groups (S–OH) (Schindler and Stumm, 1987). Adsorption of H^+ and/or OH^- ions causes protonation or deprotonation of the surface hydroxyl groups, forming a high-energy, activated complex or a combination of activated complexes:

$$S\text{–}OH + H^+ = S\text{–}OH_2^+ \quad (3.1)$$

$$S\text{–}OH + (OH^-) = S\text{–}O^- + H^+ (+ H_2O) \quad (3.2)$$

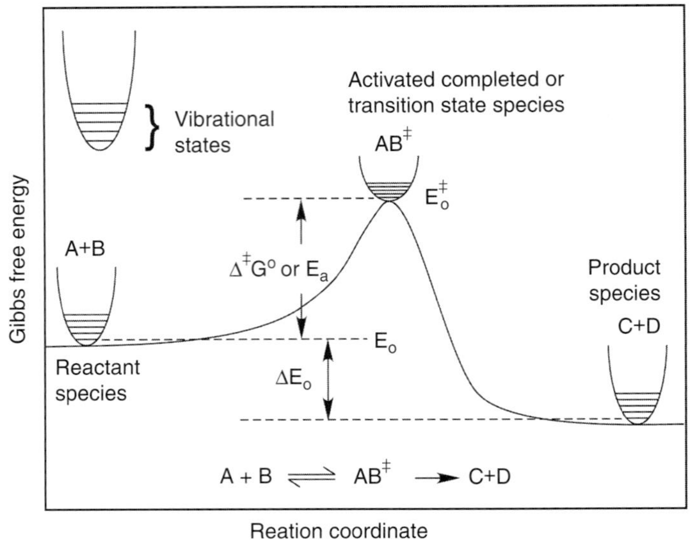

Figure 3.1. Illustration of the transition state or activated complex theory.

Proton adsorption to the mineral surface weakens the metal–oxygen bond, probably by depolarizing bonding electrons and therefore promoting the detachment of the metal ion from the bulk mineral (Cornell and Schwertmann, 2003).

Adsorption of metal ions onto an oxide surface can be viewed as a competitive reaction involving one or more hydroxyl groups:

$$S\text{–}OH + M^{z+} = S\text{–}OM^{(z-1)+} + H^{+} \quad (3.3)$$

Phyllosilicates can simply be viewed as consisting of various metal–oxygen bonds. The dissolution rate of any silicate mineral is primarily governed by the breakage of the slowest metal–oxygen bond essential for maintaining the given mineral structure. To understand the differences in metal–oxygen bonding strength, it is necessary to first explain the link between surface and aqueous chemistry (Schott *et al.*, 2009). Metal ions in aqueous solution are coordinated by water molecules and are thus complexes of water as ligands. Ligand water molecules can be replaced (exchanged) by negatively charged or other polar inorganic and organic ligands (like OH^-, Cl^-, $C_2O_4^{2-}$ etc.). The rate constant of water exchange and mean residence time for water molecules in the primary hydration shell varies between different cations (Fig. 3.2a) and depends on the charge, radius and crystal-field effects of the cation (Schott *et al.*, 2009).

The rate of surface complex formation at any oxide surface is therefore approximately proportional to the rate of exchange of water molecules (or any other coordinating ligand) from the original metal coordination sphere. This is supported by findings that the dissolution rate of single metal oxide minerals in acidic conditions correlates to the rate constant for water exchange (Fig. 3.2b). Based on these correlations, Oelkers (2001) proposed

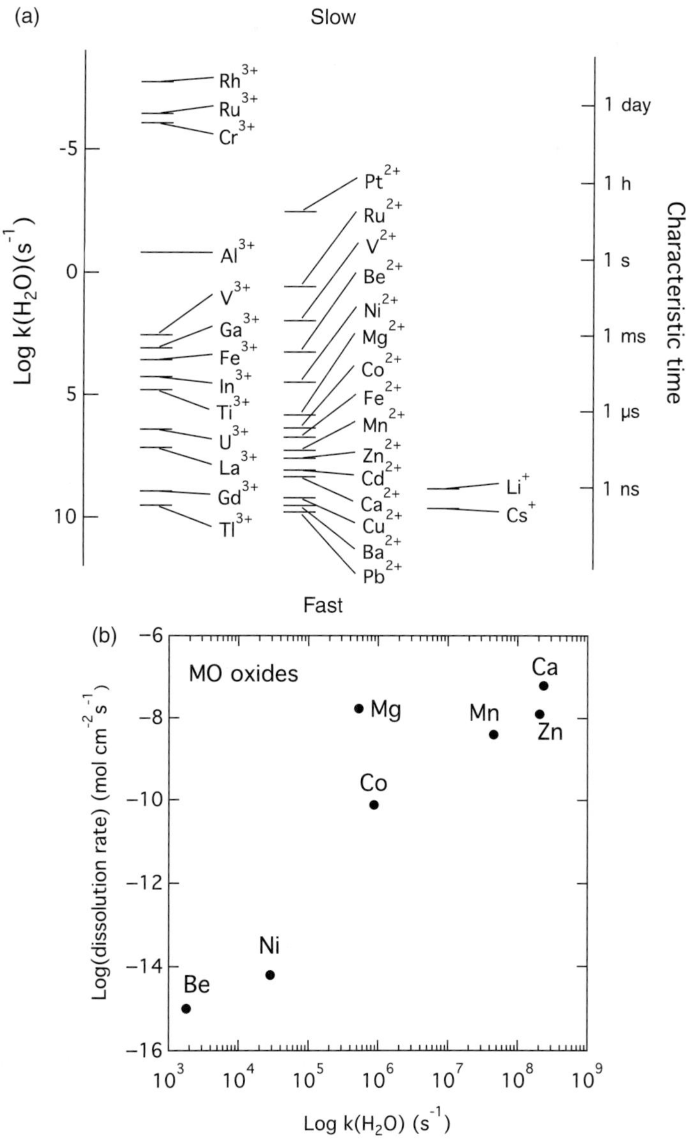

Figure 3.2. (a) The rate of exchange and residence time of water molecules in metals' inner coordination spheres (redrawn from Schott *et al.* (2009) with permission); (b) rate of dissolution at ~25 °C and pH = 2 of metal oxide minerals versus the rate constant for exchange of water molecules in the coordination sphere of the corresponding metal in solution (Casey, 1991 and Neely and Connick, 1970 for the corrected value for Mg).

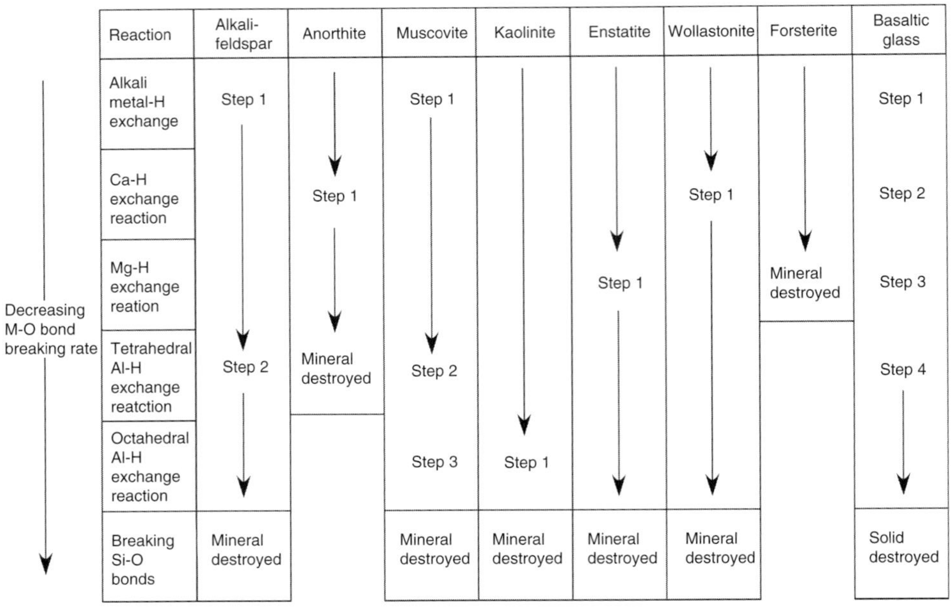

Figure 3.3. Summary of the dissolution mechanism of a variety of systems involving M–O bonds (redrawn from Schott *et al.* (2009) after Oelkers (2001), with permission).

that the relative rate of breaking each type of metal–oxygen bond would be identical in multi-oxide minerals. For instance, kaolinite can be seen as silica and aluminium oxides. A dissolution mechanism for major multi-oxide silicates including aluminosilicates, basic silicates and natural glasses was proposed by Oelkers (2001) based on the structure of the multi-oxide silicates and the relative rates for breaking the metal–oxygen bond, as shown in Fig. 3.3. In general, under acidic solution conditions, the scission rates of metal–oxygen bonds reduce with increasing metal valency, i.e. rates decrease in the order of monovalent > divalent > trivalent > tetravalent (Oelkers, 2001), because the higher the metal valency the stronger the metal–oxygen bonding strength. The dissolution of a multi-oxide silicate will proceed by breaking the weakest to the strongest metal–oxygen bonds, releasing various metals from the mineral surface. The removal of a metal ion from the mineral structure is coupled with the addition of protons to the mineral surface (Oelkers *et al.*, 2009), so the dissolution process is effectively one of metal–proton exchange reactions. Dissolution continues by successive removal of metals in their order of relative rates for breaking their corresponding metal–oxygen bonds until the original mineral structure is destroyed. The removal of the metal that forms the strongest metal–oxygen bond essential to maintain the mineral structure is the slowest step, and therefore is the rate-limiting step.

3.1.1 Dissolution of Kaolinite

The pH-dependent dissolution of kaolinite $Al_2Si_2O_5(OH)_4$ has been explained using a two-site complexation model (Carroll and Walther, 1990; Carroll-Webb and Walther, 1988).

The dissolution rate is controlled by adsorption of H^+ and OH^- ions to aluminol (>AlOH) and silanol (>SiOH) reaction sites on the kaolinite surface. At a certain temperature, the detachment of Si from the surface is the rate-limiting factor in very acidic or alkaline conditions, while at near-neutral to basic pH, the detachment of Al is the rate-limiting factor.

Subsequently a three-site model was proposed to describe the acid–base characteristics of kaolinite dissolution (Wieland and Stumm, 1992):

Site 1: Aluminol groups at the edge surface (=AlOH)
Site 2: Aluminol groups at the 'gibbsite' surface (Al–OH–Al)
Site 3: Negatively charged =XO sites in siloxane surface; X can be either Si or Al.

Ion exchange reactions may occur at Site 3 due to isomorphic substitution of Si by Al in kaolinite's tetrahedral sheet, leading to the formation of negatively charged $(Si\text{–}O\text{–}Al)^-$ groups at the surface of the siloxane layer. In addition to the protonation/deprotonation reactions mentioned in the two-site model, electrostatic ion exchange reactions involving these terminal groups are considered:

$$=XOH + Na^+ \rightleftharpoons =XO^- \text{----} Na^+ + H^+ \tag{3.4}$$

where =XO sites are the terminal groups of the siloxane layer available for ion exchange reactions with H^+, Na^+, and Al^{3+}.

It is hypothesized that the dissolution process at the aluminol edge surface (Site 1) dominates in the pH range 5.0–7.5. Dissolution reactions at aluminol Sites 1 and 2 account for the observed overall dissolution reaction of kaolinite at $pH \leq 4$. The detachment of an Al centre is the rate-determining elementary step of the dissolution reaction at Sites 1 and 2. In a subsequent step, Si is hydrolysed and rapidly released from the surface. The protonation of Site 3 in acidic conditions will produce =SiOH, but not $=SiOH_2^+$. The protonation of Sites 1 and 2 will lead to the breakdown of the metal–oxygen bond, directly contributing to the dissolution of kaolinite structure. The protonation of Site 3 indirectly affects the dissolution of kaolinite by the formation of =XOH, especially at $pH < 5$, therefore contributing to the changed activity of the proton.

3.1.2 Dissolution of Smectite

Smectites, illites and micas are 2:1 phyllosilicates in which the crystallography is based on T–O–T layers as opposed to T–O layers, but with similar a and b unit cell parameters. The major differences are caused by isomorphic substitutions in tetrahedral and/or octahedral sheets and by cations in the interlayer space (Newman and Brown, 1987).

Bickmore *et al.* (2001) used atomic force microscopy (AFM) to monitor the change of nontronite (an iron-rich di-octahedral mineral in the montmorillonite group) crystal morphology over time in acidic solution (pH 1.5–2 HCl for less than 5 hours). They observed almost exclusive edge dissolution without significant change in particle height, i.e. no significant basal plane attack occurred. The authors postulated that the oxygen sites connecting octahedral and tetrahedral sheets (nontronite's (010), (110) and $\left(110\right)$ edge faces) are all fully charge-satisfied and their scission becomes the rate-limiting step in dissolution. Further, the basal plane of the stable siloxane bonds do not appear to participate in the dissolution. These observations obtained with AFM are generally well accepted (Aldushin *et al.*, 2006; Köhler *et al.*, 2005; Kuwahara, 2006; Oelkers *et al.*, 2008).

Rozalen *et al.* (2009) compared the dissolution rates of montmorillonite in their study with the values to previous studies (see references therein), and other 2:1 phyllosilicates including chlorite and illite on the same mass and temperature basis. The authors demonstrated that the change of the dissolution rates versus pH at various temperatures was very similar and almost independent of mineral species (Fig. 3.4). The authors gave no preference to the theories of Bickmore *et al.* (2001) or Oelkers (2001); however, the overall phyllosilicate dissolution rate was best explained by the scission of bridging oxygen bonds in the T–O–T layer (Rozalen *et al.*, 2009).

The hypothesis of Bickmore *et al.* (2001) can be seen as a special case of the silicate mineral dissolution model proposed by Oelkers (2001), as discussed in Section 3.1. The latter suggested that the rate-limiting step for acidic dissolution of Si–O–Si networks is the dissolution rate of high valence metal–oxygen bonds – this is often the Si^{4+}–O bonds. Support for Oelkers' theory can be found in acid leaching of clay minerals. For instance, Okada *et al.* (2003) used 2–4 *M* H_2SO_4 to leach talc (2:1 layer Mg silicate) at 150 °C, and monitor the change of residue mineralogy by X-ray powder diffraction (XRD). They observed a continuous reduction of Mg levels and a concomitant increase of Si levels in the residue with leach time, along with a reduction of the talc XRD reflections. They proposed that the acid leaching of talc commenced with removal of the octahedral Mg–O layer from the structure, with the residual tetrahedral Si–O layer initially remaining and gradually (re-)polymerizing to amorphous silica ($SiO_2 \cdot xH_2O$); this leaching sequence agrees well with the leaching model of Oelkers (2001). Further discussions of the leaching of silicate minerals can be found in Section 3.2.2.

Temperature directly contributes to the reaction rate. The Arrhenius equation has been applied frequently to explain the temperature dependence of the dissolution rate constant (k_r) at fixed pH (Carroll and Walther, 1990):

$$k_r = A\left(-Ea/RT\right) \tag{3.5}$$

where A is the pre-exponential factor, and E_a the activation energy, i.e. the energy with which the reactants must collide in order for reaction to occur.

From the above discussions it is apparent that a good understanding of the clay mineral structure and its relationship to reactivity in acidic (to a lesser extend in alkaline) media has been reached. While the dissolution of all silicate minerals can be understood using generalized models such as that of Oelkers (2001), clay mineralogy and its reactivity need to be brought into the context specific to hydrometallurgical leaching conditions.

So far the dissolution has centred on a molecular basis, emphasizing differences between metal–oxygen bond strengths and reactivities in different crystal orientations (i.e. mineral edge versus basal plane). Hereby the entire crystal structure is considered as one molecule

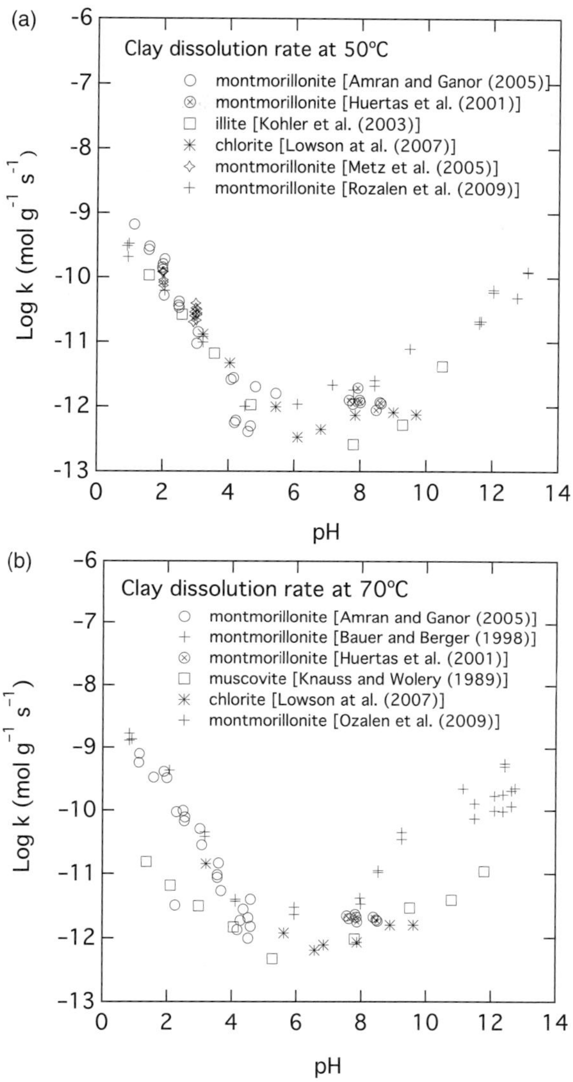

Figure 3.4. Dissolution rates at (a) 50 °C and (b) 70 °C for montmorillonite and other 2:1 clay minerals (redrawn from Rozalen *et al.* (2009), with permission).

and all other crystal surface features are disregarded (Lasaga and Lüttge, 2003). Lasaga and Lüttge (2001, 2003) proposed a completely new concept, the Stepwave Model, that has also been used to explain the dissolution of clays and other phyllosilicate minerals (Arvidson and Lüttge, 2010; Beig and Lüttge, 2006; Kurganskaya *et al.*, 2012; Stopar *et al.*, 2006).

The Stepwave Model considers two aspects: (1) localized dissolution and (2) the overall dissolution of minerals. The localized dissolution concept stipulates that dissolution initially starts at defects or etch pits on the surface of mineral particles; each etch pit

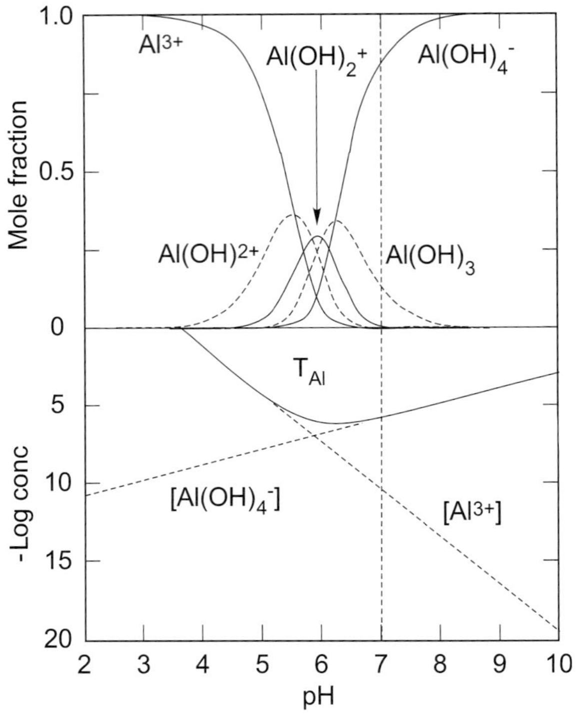

Figure 3.5. Speciation of Al versus pH in aqueous solution with no presence of complexing species, shown as mole fraction of the species and concentration of the two limiting species. T_{Al} is total dissolved Al (redrawn from Martin (1991), with permission).

will then generate a dissolution stepwave that spreads across the whole mineral surface. Simultaneously, the entire crystal surface is retreating via global dissolution. Both aspects contribute to net mineral dissolution. Global dissolution is fairly uniform and systematic over the entire crystal surface and therefore governs the bulk dissolution rate (Lasaga and Lüttge, 2001, 2003). The model fits well with the idea that the same bonds need to be broken regardless of the direction of attack and changes in solution composition (e.g. lower or higher pH; Beig and Lüttge, 2006).

3.1.3 Chemistry of Aluminium and Silicon

The speciation of Al in aqueous solution over a broad pH range can be summarized in Fig. 3.5 according to Martin (1991) and Swaddle (2001). Only a narrow pH range exists over which the presence of Al species other than Al^{3+}, $AlOH^{2+}$ and $Al(OH)_4^-$ are significant. The only species that can be detected at pH < 3 is Al^{3+}, while in the alkaline range the main Al species is $Al(OH)_4^-$ (also see Section 3.2.1.1).

Compared to Al, the speciation and chemistry of Si is more complex, especially in the alkaline range. In dilute neutral and acidic pH range, silicic acid is present as monomeric $Si(OH)_4$ (Swaddle, 2001). When present in high concentration, silicic acid undergoes

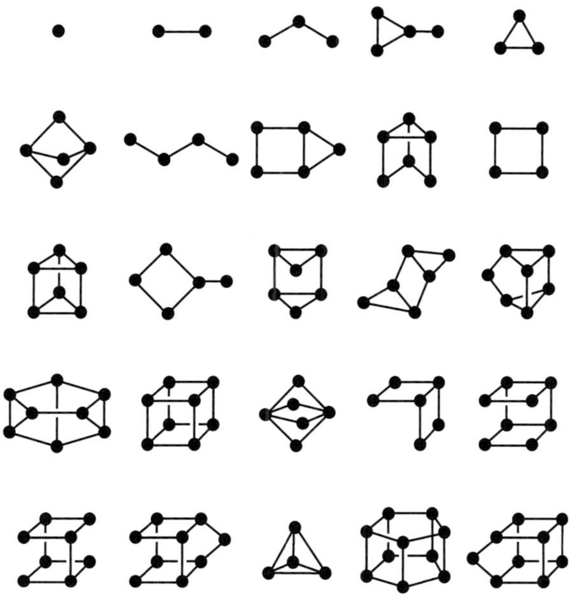

Figure 3.6. Silicate anion species detected in alkaline solution by 29Si-NMR (redrawn from Swaddle *et al.* (1994), with permission).

various degrees of oligomerization or polymerization. In alkaline conditions, however, $Si(OH)_4$ is protonated to various degrees. A variety of oligomeric silicate anions have been shown to exist (Fig. 3.6). Larger silicate oligomers are favoured by high silica concentrations, low alkalinity, and lower temperature.

Silica chemistry in the hydrometallurgical context has been summarized by Queneau and Berthold (1986). Amorphous silica, the most soluble form of silica, forms when silicic acid precipitates from solution. Near room temperature, the solubility of amorphous silica remains constant with increasing pH until ~9, with a sharp increase of solubility above pH 9 (Fig. 3.7a). Solubility of amorphous silica increases with increasing temperature and pH, as illustrated in Fig. 3.7b.

Below pH 7, soluble silica is present as monomeric $Si(OH)_4$. At pH >9, silicic acid reacts with hydroxyl ions to form various silicate ions such as $(OH)_3SiO^-$ and $(OH)_2SiO^{2-}$ (Queneau and Berthold, 1986). High levels of silica can be encountered during acid leaching of silicate minerals as a result of polymerization in silica supersaturated solutions. Ageing of such high levels of soluble silica can lead to the formation of amorphous silica as either a colloid (i.e. a sol), a precipitate, or a gel. Colloid particles are small in size and therefore do not settle well. A silica gel forms when these colloid particles aggregate with the incorporation of water. In contrast, when a silica gel forms as discontinuous agglomerates, they are termed precipitates. The formation chemistry for anhydrous silica agglomerates, the polymerization process, can be simply expressed as:

$$nSi(OH)_4 = (SiO_2)_n + 2nH_2O \tag{3.6}$$

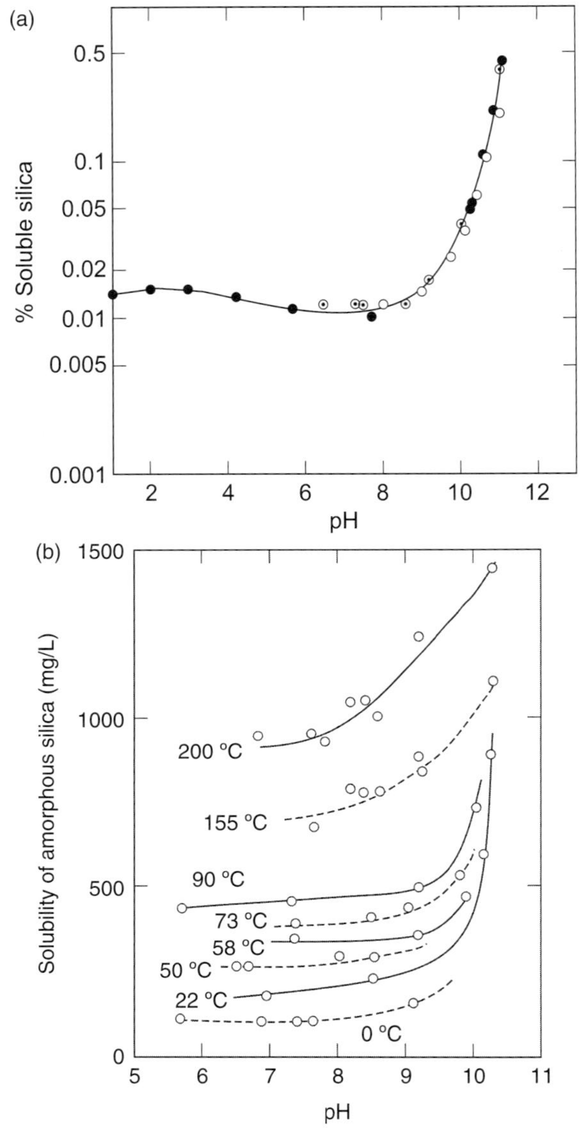

Figure 3.7. (a) Solubility of amorphous silica in water at 25 °C; ⊙ and • represent two types of sols, ○ represents calculated solubility (redrawn from Alexander *et al.*, 1954). (b) The effect of temperature and pH on the solubility of amorphous silica (redrawn from Okamoto *et al.* (1957), with permission).

Formation of a gel or a precipitate largely depends on (1) the degree of silica supersaturation, (2) the availability of silica surface to deposit fresh silica, (3) temperature, (4) pH and (5) ionic strength. Continuous dense layers of anhydrous silica tend to form at a slow rate from hot solutions of low supersaturation with an active surface for seeding in a pH range of 5–10 in high ionic strength. These conditions favour formation of silica agglomeration with minimum entrainment of water. The opposite conditions favour the formation of hydrated silica gel.

The rate at which silica polymerizes to form colloid particles, and the rate of subsequent agglomeration of these particles to form precipitates, is important to hydrometallurgical processes (Queneau and Berthold, 1986). Polymerization is favoured by increasing silica supersaturation, temperature, and concentration of active seed. pH has a large impact on the rate of polymerization. At 25 °C, polymerization occurs slowly in the acidic pH range, most rapidly at approximately pH 7, and slowing down again in more alkaline pH conditions. An induction period often proceeds polymerization, which can be eliminated by seeding, elevated supersaturation and/or temperature.

pH also affects the rate of colloid agglomeration (Fig. 3.8). Silica gel is most stable between pH 1 and 3, where the polymerization rate is proportional to H^+ concentration for pH <2. At pH >2, the agglomeration rate is proportional to the concentration of OH^-, with maximum agglomeration or gelling rate occurring between pH 5 and 8. In more alkaline conditions, the negatively charged colloid particles are prevented from coagulating by their mutual repulsion.

The occurrence of amorphous silica can potentially result in plant shutdown (Queneau *et al.*, 1983). In this specific case, silica gelation occurred during atmospheric leaching of zinc silicates using sulphuric acid. The situation was avoided by applying basic silica chemistry, where pH was controlled in the range of pH 4.0–5.5 so that rapid silica precipitation and coagulation occurred.

3.2 Clay Dissolution in Hydrometallurgical Processes

3.2.1 Caustic Dissolution

3.2.1.1 Bayer Process (Bauxite)

Clays, and predominately kaolinite, represents the majority of 'reactive silica' in bauxite that dissolves in the strongly caustic Bayer liquor at the process temperatures of ~100–250 °C. Before looking at the mechanism and characteristics of aluminosilicate dissolution in highly caustic environments, it is instructive to review what is known about aluminium and silicon speciation at high pH – especially when both metals are present.

Aluminium speciation in strongly caustic solutions has been studied widely because of its importance in the mechanism of gibbsite precipitation during the Bayer process. Various poly-ions of monomeric aluminate ($Al(OH)_4^-$) have been proposed using evidence from the measurement of colligative properties (mainly from Russian studies, see e.g. Zambo (1986). However, reviews of aluminate speciation (Chen *et al.*, 2006; Sipos, 2009) have concluded that monomeric aluminate ($Al(OH)_4^-$) and a small concentration of the corresponding de-protonated dimer are the only likely aluminium-containing species. By contrast, silicates have long been known to form various polymeric ions in dynamic equilibrium in caustic solutions (Iler, 1979) – aluminosilicate ions behave similarly (Kinrade

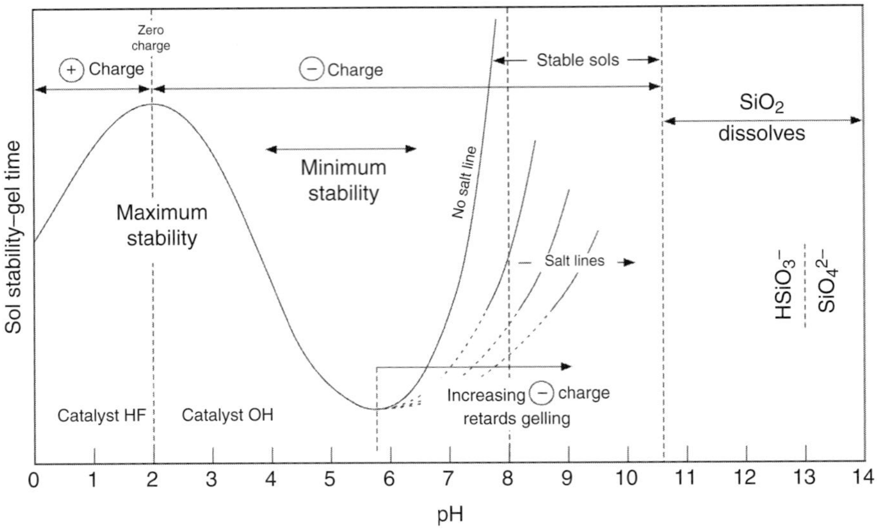

Figure 3.8. Silica gel stability in relation to solution pH in the absence of salts, after Iler (1979) and Queneau and Berthold (1986).

and Swaddle, 1989). Laws (2005) concluded, by comparing calculated equilibrium constants for various aluminosilicate ions to the very successful empirical solubility correlations of Oku and Yamada (1971), that the most likely aluminosilicate ion in Bayer-type liquors was the tetra-aluminosilicate ion – $Si[OAl(OH)_3]_4^{4-}$.

In the Bayer process (and in particular in the unit operation called pre-desilication), kaolinite in the bauxite dissolves in the Bayer liquor, and at a critical concentration of dissolved silica in liquor the precipitation of a sodium aluminium silicate (desilication product or DSP) is triggered. The desilication product is predominately a caged structured zeolitic material similar to sodalite. This process is schematically illustrated in Fig. 3.9. During the pre-desilication process, both kaolinite dissolution and DSP precipitation occur together. Because of this, many studies (both theoretical and experimental) have studied the combined process (Cousineau and Fulford, 1987; Cresswell, 1984; Tizon *et al.*, 2004). However, there have been some studies that have isolated the process of kaolinite dissolution (Carroll and Walther, 1990; Eremin *et al.*, 1978; Johnston *et al.*, 1985; Roach and White, 1988).

These studies are mainly concerned with determining the kinetics of kaolinite dissolution, including rate law dependence (generally first order seems to be the consensus), rate constants and activation energies (the process is usually described as being surface controlled). Rates are increased by temperature, free caustic (i.e. sodium hydroxide) concentration, decreased by aluminate and silica in liquor concentrations, and unaffected by particle size and sodium carbonate concentration.

In terms of the mechanism, Roach and White (1988) noted from SEM images that kaolinite effectively 'delaminates' – i.e. the platy structure of the kaolinite opens up during the dissolution process. Subsequently Cardile *et al.* (1994) and later Laws (2005) both

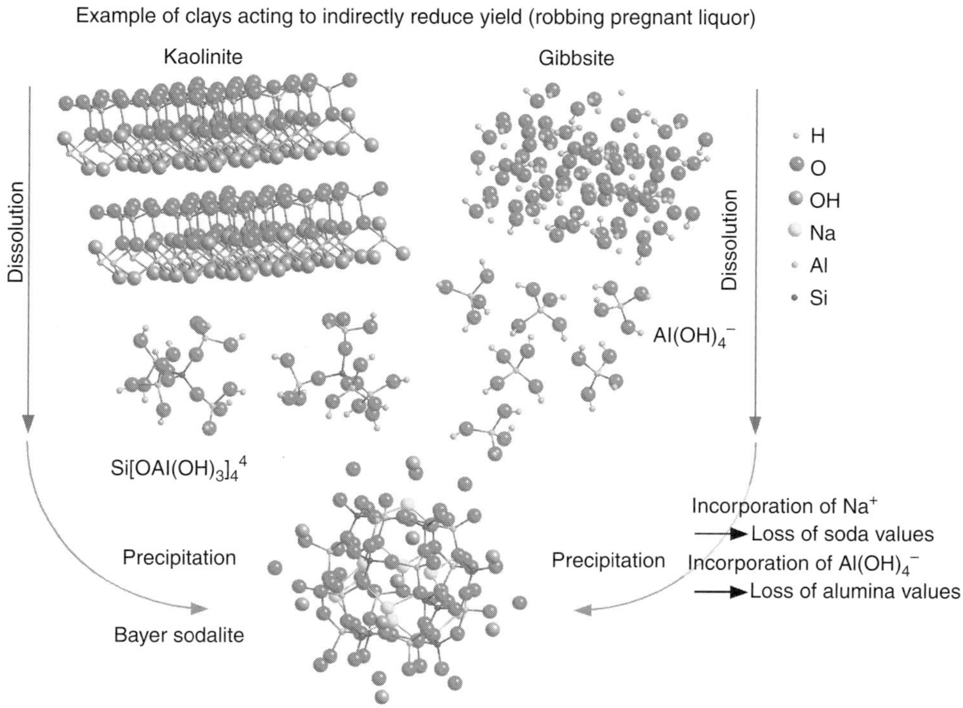

Figure 3.9. An example of clays acting to indirectly reduce gibbsite yield in the Bayer process by providing a source of solubilized aluminosilicate (most likely the tetra-aluminosilicate ion) to allow formation of unwanted sodalite. A black and white version of this figure will appear in some formats. For the colour version, please refer to the plate section.

used ^{23}Na NMR to show that sodium was incorporated into the delaminated structure of the kaolin (or possibly was the cause of the delamination) prior to the dissolution of aluminosilicate framework into solution. Interestingly, dissolution of kaolinite in potassium hydroxide rather than sodium hydroxide is substantially slower (Laws, 2005). This may be related to the larger ionic diameter of K^+ (2.8 Å vs 2.0 Å for sodium) and the associated difficulty in penetrating/delaminating the aluminosilicate layers. At present it is unknown whether the penetration of the aluminosilicate layers by the cation is the cause (or an effect of) the delamination.

More evidence regarding the mechanism of kaolinite dissolution can be found in the way DSP precipitates. Laws (2005) added isotope-enriched silica to solution before dissolving kaolinite to show that the silica in precipitated DSP was in direct proportion to solution concentrations, i.e. the mechanism was predominately one of kaolinite dissolution and DSP re-precipitation. Although this essentially excludes the possibility of solid-state transformation, it also suggests that the building block for DSP crystallization is in fact monomeric silicate (or aluminosilicate) ions rather than aluminosilicate oligomers from kaolinite fragments.

The state of impurities like silica in Bayer liquor are also important from the standpoint of gibbsite (product) quality. Sodium, silica, iron and a number of other impurities report to product not as components of included Bayer liquor, but rather by atomic incorporations into the gibbsite structure.

3.2.1.2 Cyanide Leach (Gold Ores)

Gold ores can be classified based on their gold recovery using the conventional cyanidation process: 'free milling' ores recover >90 per cent gold (La Brooy *et al.*, 1994); 'refractory' ores recover <80 per cent gold after fine grinding (Iglesias and Carranza, 1994); while 'complex' ores are defined to give acceptable economic gold recovery only with the use of significantly higher cyanide or oxygen requirements (La Brooy *et al.*, 1994). Gold leaching with alkaline cyanide solutions is assumed to proceed through two steps according to these overall reactions:

$$2Au + 4NaCN + O_2 + 2H_2O \rightarrow 2NaAu(CN)_2 + 2NaOH + H_2O_2 \quad (3.7)$$

$$2Au + 4NaCN + H_2O_2 \rightarrow 2NaAu(CN)_2 + 2NaOH \quad (3.8)$$

in which hydrogen peroxide (H_2O_2), formed as an intermediate product in the first step, becomes the oxidizing agent in the second (Guzman *et al.*, 1999).

Three methods were developed to recover leached gold from solution: carbon in pulp (CIP), carbon in leach (CIL) and carbon in column or carbon in clear solution (CIC) (Gönen *et al.*, 2007). The CIP process uses activated carbon granules of ~3–5 mm added to the leach slurry after completion of the extraction to adsorb aurocyanide species from leach solution. In CIL, activated carbon is added already during leaching to prevent/minimize leached gold from re-adsorbing to residual gangue surfaces such as carbonaceous materials or clay minerals in the ore. CIC is used in heap leaching, with the leachate solutions passing through adsorption columns containing activated carbon to remove gold from solution. The main disadvantage of the cyanidation process is the cyanide toxicity and disposal of the waste to the environment (Gönen *et al.*, 2007). Alternative leaching methods, including thiourea, thiosulphate, chlorine and acid leaching, have been developed in recent years, but none of them have been applied in commercial production (see Dai *et al.*, 2013 and Gönen *et al.*, 2007 and references therein) other than the thiosulphate process. The thiosulphate leach process was developed and first used in commercial production by Barrick Gold Corporation to treat refractory sulphidic and carbonaceous ores (Choi *et al.*, 2013). Any clay-related issues would in theory be similar to those encountered in the traditional cyanide gold leach process.

The presence of clays in feed ore tends to reduce the cyanide leaching efficiency of the gold; this is thought to be by two factors: (1) forming impermeable coatings on the surface of gold after grinding; and (2) gold being present as fine colloid particles, which are difficult to separate from equally fine clay minerals (Gasparini, 1989; Karimi *et al.*, 2010). Examples of 4–5 per cent micaceous material (sericite) reducing gold cyanide recovery by 10 per cent and more are presented in the study by Gasparini (1989). It is best to remove the barren clay fraction by flotation of the sulphide minerals before the cyanidation process. However, such beneficiation might not always be possible if the gold occurs in the same size fraction as clays. For example Wei *et al.* (1997) found clay fines to contain 38 per cent of total gold in a refractory gold ore in China. Only 76 per cent gold recovery was achieved from cyanide leaching after flotation and bioleach pre-treatment, and most of the unrecovered gold was thought to be associated with the clay minerals. As noted in Section 1.8.1 and Table 1.2, clays have the ability to adsorb a wide variety of commodity metals, so it is equally valid to consider a low recovery of gold due to the presence of clays in a 'preg-robbing' context.

Carbonaceous material present in gold ores can adsorb gold from alkaline cyanide solutions, and such ores are most often referred to as 'preg-robbing' ores (Miller *et al.*, 2005). Carbonaceous materials in refractory gold-bearing ores refer to (1) the native carbon component acting like activated charcoal in adsorbing gold cyanide from solution; (2) high molecular weight hydrocarbon mixture which appears not to interact directly with gold; and (3) 'humic acid'-like components containing functional groups capable of complexing soluble gold (Tan *et al.*, 2005). To a much lesser extent, however, significant non-carbonaceous materials including clays and silicates have been observed to contribute to preg-robbing by adsorbing $Au(CN)_2^-$ (see La Brooy *et al.*, 1994; Miller *et al.*, 2005 and references therein; also refer to Table 1.2 in Chapter 1).

A study of Au and Ag cyanide adsorption to nontronite, montmorillonite, illite and muscovite (among others) indicated that clays can adsorb up to 7.5 per cent of total Au from cyanide solution, while much greater values of up to 88 per cent were found with silver adsorption (Tremolada *et al.*, 2010). Gold adsorption on clays maximized before settling at an apparent, lower final surface concentration. The authors concluded that preg-robbing may be one of the main causes of reduced gold and silver recovery from clay ores when cyanide solutions are used in heap leaching extraction procedures. The mechanism of gold cyanide adsorption to clay minerals is not well understood. Although transmission electron microscopy (TEM) investigations provided evidence of submicrometre gold appearing on the edges of kaolinite and illite particles in a natural ore sample (Stumm, 1992), the study did not explain or provide a mechanism as to how gold cyanide was adsorbing onto the clay minerals. It is speculated that the negatively charged Au complex ($Au(CN)_2^-$) would most likely adsorb to positively charged clay surface though electrostatic forces and/or chemical complexation (Hong *et al.*, 2001 and references therein).

There are also reports of clay adsorption of gold in leach media other than cyanide. For example, it is indicated that the cationic thiourea complex may be even more susceptible to preg-robbing by clays than the anionic aurocyanide complex (La Brooy *et al.*, 1994). Clays, including kaolinite and pyrophyllite, along with quartz and feldspars all appear to adsorb $AuCl_4^-$, particularly in acidic conditions (pH 2.5–3.0), but more weakly in alkaline conditions (pH 7.5–8.0) (Mohammadnejad *et al.*, 2011). The authors propose chemisorption as the main adsorption mechanism, occurring between gold chloro-hydroxy anions and protonated silanol groups on the surface of feldspar and quartz, or positively charged edge surfaces of clays. Gel formation as a mechanism for gold loss is observed mostly for quartz and feldspar, but appears to be effective only in acidic conditions. A quantum chemical calculation has suggested that the edge surfaces of kaolinite are more likely to adsorb $Au(HS)_2^-$ relative to its basal plane, especially at aluminol groups (Hong *et al.*, 2001). It should be noted that CIL or resin-in-leach (RIL) can reduce mild preg-robbing behaviour, as gold adsorption to activated carbon or resin has faster kinetics (La Brooy *et al.*, 1994; Miller *et al.*, 2005; Tan *et al.*, 2005).

3.2.1.3 Carbonate Leaching (Calcrete Uranium Ores)

The leaching of uranium (as U(IV)) from its ores can be considered to be an oxidation process in which the less soluble tetravalent U(IV) is transformed to its soluble hexavalent form (Mattus and Torma, 1980) which then assumes uranyl geometry (UO_2^{2+}) in aqueous solution. An alkaline leach is normally applied to carbonate rich UO_2^{2+} ores in nature as acid leaching is uneconomic (Seidel, 1979). The most common alkaline leaching solutions are mixtures of sodium carbonate and sodium bicarbonate to produce a well-buffered lixiviant between pH 10 and 11. The carbonate leach chemistry can be summarized as follows (Du Preez *et al.*, 1981; Mattus and Torma, 1980). Uranium is oxidized by oxygen based on the following reaction:

$$UO_2 + \tfrac{1}{2}O_2 \rightarrow UO_2^{2+} + O^{2-} \tag{3.9}$$

Uranyl (UO_2^{2+}) will then react with excess carbonate ions to form a stable anionic complex, with a net reaction:

$$2UO_2 + 6CO_3^{2-} + O_2 + 2H_2O \rightarrow 2UO_2(CO_3)_3^{4-} + 4OH^- \tag{3.10}$$

Sodium bicarbonate is often added to the solution to neutralize any free hydroxyl concentration (Eq. 3.11) to prevent the precipitation of diuranate species (Eq. 3.12) that would otherwise occur:

$$NaOH + NaHCO_3 \rightarrow Na_2CO_3 + H_2O \tag{3.11}$$

$$UO_2(CO_3)_3^{4-} + 6OH^- + 2Na^+ \rightarrow Na_2U_2O_7\downarrow + 6CO_3^{2-} + 3H_2O \tag{3.12}$$

The uranyl carbonate system is much more complex than the above equations suggest. Uranium speciation occurs as UO_2^{2+} for pH <3.7 and with increasing pH as a series of hydroxyl complexes, eventually changing into carbonated species in aqueous solution and in the presence of CO_2 (Korichi and Bensmaili, 2009; see Fig. 3.10). Under acidic conditions the uranyl (UO_2^{2+}) cation is prone to adsorption both by iron oxides (Yusan and Erenturk, 2011 and references therein) and clay minerals (Aytas *et al.*, 2009; Korichi and Bensmaili, 2009; McKinley *et al.*, 1995 and references therein). However, at pH >5 uranyl carbonate complexes are increasingly formed, and become anionic at pH >8, and therefore can no longer be efficiently adsorbed onto iron hydroxides (Gruber *et al.*, 2013). The adsorption of uranium carbonate complexes in carbonate media is not well documented. Catalano and Brown Jr. (2005) indicated that uranyl-carbonato forms surface complex on edge sites of montmorillonite, and uranyl preferentially binds to $[Fe(O,OH)_6]$ sites over

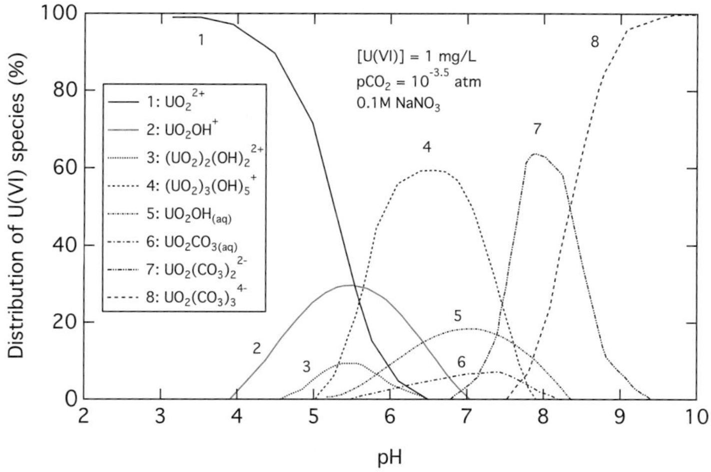

Figure 3.10. Uranium species in aqueous solution in the presence of CO_2 over a range of pH values (redrawn from Korichi and Bensmaili (2009), with permission).

$[Al(O,OH)_6]$ sites on the edges of Wyoming montmorillonite. Canterford and Sparrow (1984) reported that adsorption of uranium onto Wyoming 'bentonite' (standard montmorillonite No. 25) over the pH range of 8–10 was significant in carbonate media, and believed to occur by the anionic uranyl triscarbonate complex adsorbing rapidly onto edge sites of the clay, followed by a slow reaction of adsorption of uranium hydroxyl complexes to clay surfaces, eventually leading to uranium precipitates as hydroxides. The authors recommended that pH would need to be kept above 10 to prevent significant loss of uranium through adsorption by the clay mineral (montmorillonite). This applied to the counter-current decantation washing circuit of a conventional leaching process, as well as in the leaching and restoration phases of any carbonate solution mining.

3.2.2 Acid Dissolution

3.2.2.1 High-Pressure Acid Leach (Ni Laterite Ores)

High-pressure acid leaching (HPAL) has been applied to extract nickel (Ni) and cobalt (Co) from low-grade nickel laterite ores since 1959 in the Cuba Moa Bay operation (Dalvi *et al.*, 2004). It has the advantages of high (>90 per cent) nickel and cobalt extraction (Dalvi *et al.*, 2004; Whittington and Muir, 2000) and lower acid consumption compared to other low-temperature processes (McDonald and Whittington, 2008), but the disadvantage of high capital cost. The chemistry of HPAL leaching for nickel laterite ores has been summarized in the review of Whittington and Muir (2000). Ore or mixed ores of desired chemical composition as slurry is heated to 250–255 °C in a titanium-clad autoclave; sulphuric acid is then injected to the autoclave to react with the ore. Complete dissolution of most minerals, including clay minerals, present in the laterite ores can be achieved under those conditions with the exception of quartz and other minor refractory minerals (Kaya and Topkaya, 2011; Whittington *et al.*, 2003). Nickel and cobalt dissolve into the acidic solution along with iron and aluminium. Under HPAL conditions, the high concentrations of iron and aluminium normally hydrolyse and precipitate as hematite and a range of mixed alunite/jarosite phases, depending on the leaching conditions.

Based on the earlier work of Terry (1983), two mechanisms as to how acid might act on the silicate minerals including kaolinite and smectite under HPAL conditions were considered: (1) complete breakdown of the silicate structure with the dissolution of metal cations and silica. A silica gel may then form due to ready polymerization of silica in aqueous solutions. (2) Partial decomposition of the silicate structure leading to dissolution of cations and leaving a siliceous residue. It was speculated that both processes proceed quickly during the dissolution of silicate minerals in HPAL conditions with process (1) dominating the behaviour of resulting amorphous silica. A better understanding of the HPAL mechanism may be achieved from applying both the Oelkers (2001) and the Lasaga and Lüttge (2001) Stepwave Model concepts, as elaborated earlier.

The chemistry of silica under highly acidic and high-temperature conditions is unique to the HPAL process. As discussed earlier (Section 3.1.3), as well as in the work of Queneau *et al.* (1983), the solubility of silica is consistently low in acidic conditions, and also insensitive to the change in pressure levels required for laterite leaching (i.e. 1–10 000 KPa), but it increases with temperature up to 300 °C. Furthermore, high temperature and agitation will also accelerate silica polymerization. Slurry containing high silica precipitation can cause scaling and poor settling in the HPAL process during the cooling down period from above 200 °C to ambient temperature (Queneau *et al.*, 1983).

The specific interactions between silica and soluble metals present in the HPAL post-leach liquor and their subsequent extent depend on the metal species. Metals like sodium and potassium in the presence of sulphate have little effect on silica polymerization in acidic conditions. Interaction between divalent cations with silica is also rare. For pH <3.5, ferric iron can form a soluble complex with $Si(OH)_4$. In neutral solution, a molar ratio of 3:1 $Si(OH)_4$ to Fe^{3+} will inhibit Fe and Al hydroxide precipitation. The presence of chromate (Cr(VI)) can retard silica gel formation between pH 0 and 3 by forming a loose complex with silica of unknown nature, thereby usefully reducing the concentration of polysilicic acid (Queneau and Berthold, 1986 and references therein). For example, Sobol (1972) added elemental sulphur to the leach liquors of HPAL of nickel laterite ores to reduce Cr(VI) to Cr(III). This resulted in faster and higher recovery of nickel and cobalt, but also led to more rapid silica gel precipitation. This then appeared in the thickeners, causing poor settling of the solids.

Silica build-up or scale can impact on HPAL pressure relief systems, so routine silica removal during plant operation is required. Since the silica deposits are often soft, they can be easily removed by on-stream hydroblasting (Queneau and Berthold, 1986). Reducing the supersaturation, providing more active surfaces in the form of recycled amorphous silica seeding, and pH control are often applied to minimize the quantity of silica scaling.

Even though the recovery of nickel in the HPAL process is normally high, any loss of nickel is economically significant. As the nickel level in post-leach residue is normally <0.1 wt%, an understanding of nickel association among residue phases, and hence some possible understanding of the loss mechanism, is scientifically challenging. Incorporation of nickel during the precipitation of amorphous silica has previously been cited as the nickel loss mechanism; however, recent research has placed some doubt on that mechanism (McDonald and Li, unpublished data).

3.2.2.2 Ambient Pressure and Elevated Temperature Acid Leach

Atmospheric acid leaching (AL) of nickel laterite ores has the advantages of lower capital expenditure and maintenance costs, lower energy consumption and easier operations compared to HPAL (McDonald and Whittington, 2008). Various concentrations of nickel are believed to be present in clay minerals typical to nickel laterite ores, including iron-rich smectite (nontronite), serpentine and chlorite (McDonald and Whittington, 2008; Whittington and Muir, 2000 and references therein).

The dissolution kinetics of smectite ores under AL conditions are proposed to follow the sequence of (1) replacement of exchangeable cations with H^+; (2) release of octahedral Al, Mg and Fe; and (3) removal of tetrahedral Si and Al (McDonald and Whittington, 2008). This sequence is more or less in agreement with the basic mechanism proposed earlier.

As the AL leach operates below 100 °C, the resulting post-leach liquors can contain significant concentrations of iron and aluminium, which will not undergo further hydrolysis and precipitate as hematite and jarosite/alunite as observed under HPAL conditions. Therefore, acid consumption is normally high in AL conditions, especially AL of serpentine type ores. For example, Agatzini-Leonardou and Zafiratos (2004) reported stir tank leaching of a serpentinitic 'Kastoria' ore at 40–80 °C using 1.5–2.5 M sulphuric acid recovered 55 kg H_2SO_4 per kilogram of Ni, compared to a recovery of 25 kg H_2SO_4 per kilogram of Ni observed for the HPAL process (Crundwell *et al.*, 2011).

For simple agitation leaching of nontronite and limonite ores (goethite-dominated ores) between 25 and 90 °C, significantly higher nickel and iron recoveries were observed in the nontronite ore compared to the limonite ore under the same leaching conditions (Büyükakinci and Topkaya, 2009). The formation of amorphous silica from clay minerals is often encountered in AL of nickel laterite ores (McDonald and Whittington, 2008). Dissolved silicon will further polymerize and form fine precipitates that are difficult to settle and filter, as noted earlier in the case of acid dissolution of zinc silicate (Queneau and Berthold, 1986).

Acid attack on silicate minerals can result in complete breakdown of the silicate structure, as well as partial decomposition of silicate structure leading to dissolution of cations and leaving siliceous residues (Terry, 1983). This mode of dissolution is more pronounced in sheet silicate minerals like clays (compared to general silicates) as they are composed of infinite Si–O tetrahedral sheets (Terry, 1983). Temuujin *et al.* (2006) indicated that the formation of amorphous silica phases does not require the complete destruction of the montmorillonite structure (0.5–2 M sulphuric acid in 90 °C). The framework of the silicon tetrahedral sheet probably persists after the dissolution of the octahedral cations. An example can be seen in the TEM images of montmorillonite particles before and after leaching (Fig. 3.11). Chen *et al.* (2000) monitored a large specimen of serpentine (16 mm) with SEM, and noticed that after leaching at 95 °C in 7 M HCl for up to 16 hours, most of the specimen became amorphous despite the morphology of the original serpentine being retained. No significant degree of silica dissolution was detected chemically and silica gelation was not observed. Das and his group (Das and de Lange, 2011; Senanayake *et al.*, 2015) used AL to leach a smectite ore with and without sulphur dioxide. With significant recovery of iron and nickel in solution, the level of silicon dissolution was well below 1 g/L from the ore containing 19.4 per cent Si (Das, personal communication).

Controlling ore particle size to minimize or avoid the formation of fine silica gel can be a useful approach. In leaching fine and coarse fractions of silicate minerals for metal recovery using 6 M HCl at 105 °C, Gjelsvik and Torgersen (1983) found a similar metal recovery for both the fine and coarse fractions, but the leachate of the fine fraction was harder to filter and wash and had poor settling properties, while the opposite was obtained for the coarse fraction. O'Neill (1978) also indicated that although lower nickel recovery was obtained from coarser magnesium ore in AL conditions, the leach residues had much better settling properties compared to finer fractions.

3.2.2.3 Heap Leach

Extraction of metal values via percolation of lixiviant solutions through a stacked pile of coarse ore has been extensively utilized for various ore types (e.g. low-grade copper sulphide ores, low-grade nickel laterite ores) over the last few decades. Heap leaching generally returns lower metal recoveries over longer time periods compared to most processing routes; however, its attraction is due to lower capital and/or operating costs for deposits where grades are low (e.g. gold and copper sulphide) or plant equipment is capital intensive (e.g. nickel laterites and copper oxides). Ore preparation prior to percolation leaching may include one or more pre-leaching optimization strategies before and during heap construction (Bouffard and West-Sells, 2009; Dhawan *et al.*, 2013). These decision factors are commodity dependent (Table 3.1) but may variously include the following:

(1) Blast management in order to minimize fines generation in clayey ores, or increasing fractures in harder ores.
(2) Crusher control to select a coarser crush size for high-clay ores.
(3) Agglomeration of fines by the addition of optimum moisture and pre-addition of lixiviant can both accelerate initial leaching and promote permeability by ensuring an even distribution of fines and coarse particles in the final heap.
(4) Conveyor/stacker control may require an optimum moisture point for conveyance and stacking following drum agglomeration. This is likely to be a trade-off – excess moisture can block transfer point, while insufficient moisture may not enable 'ideal' agglomerate formation.

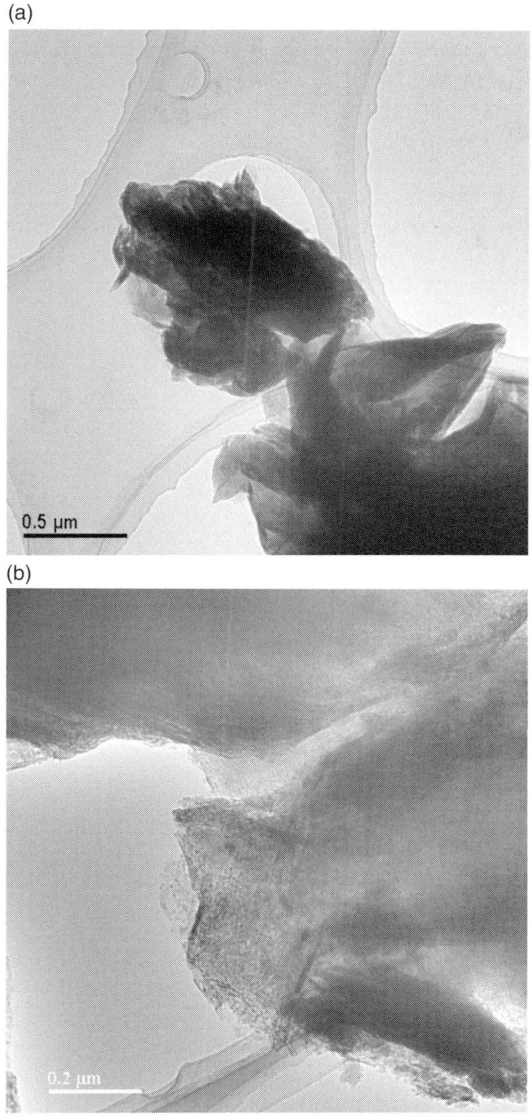

Figure 3.11. TEM of (a) raw montmorillonite and (b) after acid treatment 2 mol/L H_2SO_4 for 4 hours at 90 °C (reproduced from Temuujin *et al.* (2006), with permission).

The presence of clay minerals affects heap leach performance detrimentally in two interrelated ways: permeability and reagent consumption. The effects of clay minerals on heap hydrology and some of the steps that can be taken to mitigate their negative influence are illustrated in Fig. 3.12. The presence of fine particles impedes solution percolation through the stacked ore pile, slowing the rate at which lixiviant can reach valuable mineral particles and/or the rate pregnant leach solution (PLS) is eluted from the heap for further refining.

Table 3.1. *Decision factors in whether heap leaching is used as the extraction route.*

Commodity	Lixiviant	Decision factor
Copper oxide	Acid	Grade/mineralogy (poor oxide flotability)
Copper sulphide	Fe^{3+}/acid/bacteria	Low grade
Gold	Cyanide	Low grade
Uranium	Acid or carbonate	Mineralogy (*i.e.* amounts of carbonate minerals)
Nickel (laterite)	Acid	Capital cost

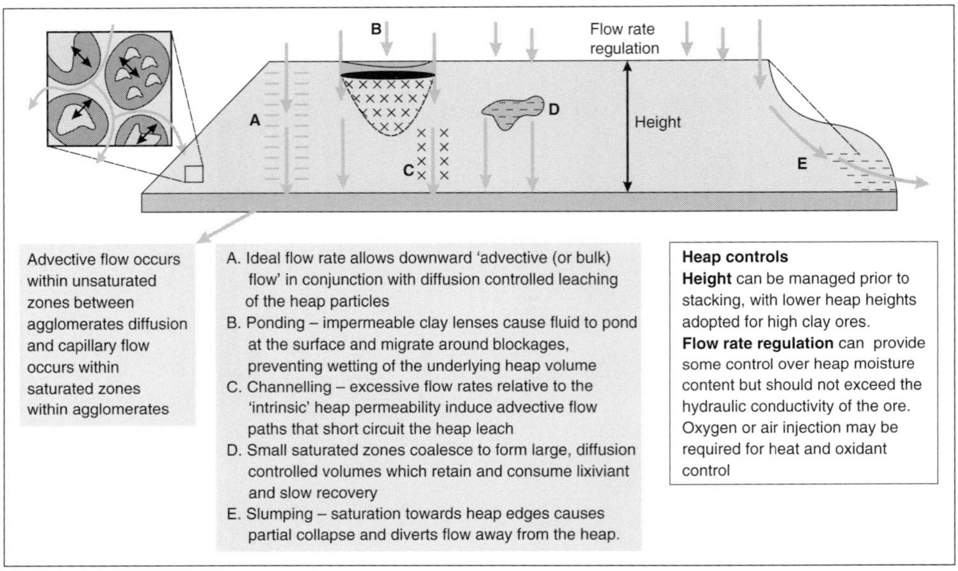

Figure 3.12. An overview of the effect of clay minerals on heap leach hydrology and issues regarding detrimental effects. Unlike many other unit processes, optimization of heap conditions after initiation is limited, making careful characterization and preparation of feed materials, especially those with high clay content, essential. A black and white version of this figure will appear in some formats. For the colour version, please refer to the plate section.

Heap leaching of sulphides and gold are also critically dependent on O_2, thus both solution and air percolation must be considered. Meanwhile, slower percolation increases the solution residence time and can impart additional reagent consumption either through direct contact with soluble gangue, soluble clay particles (which have a slow specific dissolution rate but high surface area) or other means.

There are a number of options for managing clays in heap leach operations. Since clay minerals tend to deport predominantly to the fines fraction of crushed ore, removal of the fine material from heap leach feed enables some mitigation. If disposal of fines is the only option, some loss of valuable metal is also likely to occur, thus it is preferable to direct the fine material through an alternative processing route. Herkenhoff and Dean (1987) suggested that heap leach operators should consider leaching the 'slimes' of gold ores. While avoiding permeability problems, preferential association of gold with the fine fraction may also be recovered more quickly in tank leaches. The success of the Ni laterite heap leach at Murrin Murrin in Australia is also likely due to the fact that the heap only receives the coarse fraction of the ore, although in this instance the HPAL process is the primary Ni recovery process (Lottering *et al.*, 2008; Readett and Fox, 2009).

Whether fines are removed from heap leach feeds or not, agglomeration of the crushed ore is generally regarded as an important pre-treatment step. Initially, agglomeration was thought to be a requirement for ores where the <75 μm material exceeded 10 wt% of the crushed ore with greater than 25 wt% regarded as unfit for heap leaching due to permeability issues. Clay minerals are not necessarily the only component of the fines fraction, nor are they necessarily absent from the coarse fractions; thus, their impact on processing is best inferred through a combination of both mineralogical and (physical) testing (Helle *et al.*, 2010). Regardless of clay content, agglomeration of clay-bearing ore feeds is thought to provide at least one or more of the following benefits: (1) increasing the effective particle size of the ore feed (Bartlett, 1997); (2) improving the distribution of fines within the heap to prevent channelling or stagnation (Herkenhoff and Dean, 1987; O'Kane *et al.*, 1999;Pautler *et al.*, 1990); (3) pre-dispersion of lixiviant and particle wetting, and increased geotechnical strength to allow greater stacking height (Padilla *et al.*, 2008); and (4) improved tailings management and water consumption (Moats and Janwong, 2008;Pautler *et al.*, 1990). More recently, agglomeration has been demonstrated to (5) favourably influence the hydrology of heaped ores, such that operators can ensure that the relative volume fraction of moisture and (where required) air is able to sustain economically viable flow rates (Guzman *et al.*, 2008; Readett and Fox, 2011).

Smectite-rich bentonites are added to iron ores to increase the binding strength of fine particles to improve their resistance in the sintering process (Forsmo *et al.*, 2008). While it is plausible that similar strength benefits might be possible for heap leach ores prior to heap irrigation, the high surface area available to confer strength benefits, as might be the case in the study of Quaicoe *et al.* (2014), is also capable of significant moisture retention that weakens particle–particle forces and induces other negative effects. There may be other means by which clay minerals contribute to agglomerate binding strength, such as a source of dissolved silica that precipitates to form silicic cements (Lu *et al.*, 2007), and while mineralogical considerations appear to be a factor in determining the curing time for agglomerated ores (Baum, 1999), there is insufficient information on the topic to draw firm conclusions about the role of clays.

In terms of operating costs, a heap is profitable when the value of the metal in solution exceeds the cost of the acid needed to extract it (once mining and refining costs have been accounted for). The individual acid consumption of the various clay minerals can be calculated from stoichiometry. Unlike the more intensive leaching situations of finer grinding and stronger dissolution conditions, complete dissolution of clay minerals is not inevitable in heap leaching environments, and stoichiometric estimates are likely to be overestimated.

The low-intensity reaction conditions (ambient to low temperature and low acid concentration) along with inconsistent solution distribution (due to particle coarseness and uneven percolation within the heap) tend to produce partial dissolution that is estimated through empirical means from pilot or operational trials. As an example of this, a study of copper oxide heap leach operations at Manto Verde, Chile, concluded that reduction in acid consumption can be achieved by applying most of the acid in the curing stage during the agglomeration step (Salgado *et al.*, 2009).

An extensive description of the negative effects of clays on heap leach viability is given by Miller (2003). In his analysis of a series of heap leach operations and column trials, the detrimental effects of clay minerals are: (1) increasing flow-path tortuosity via particle shape and size effects; (2) increasing moisture retention; (3) decreasing geotechnical strength; and (4) decreasing heap porosity. As a result, acid consumption increases due to longer residence times at particle surfaces while the lixiviant addition rate and the elution rate of dissolved metal from the heap are reduced leading to a significant reduction of the economic viability of high-clay ores.

Knowledge of the clay mineral variability in the ore deposits of heap leach operations using methods such as those referenced in Chapter 4 of this work is therefore critical for anticipating processing problems and optimizing their economic benefit (Baum, 1999; Bouffard and West-Sells, 2009).

3.3 Summary

This chapter reviewed fundamental aspects of clay dissolution based on surface complexation models within the framework of TST. Here, dissolution is considered as a process of overcoming an energy barrier from reactants to dissolution products. Clay dissolution has been viewed as a sequence of breaking low to high valency metal–oxygen bonds. The dissolution of kaolinite and smectite have been discussed in detail to demonstrate the actual metal–oxygen dissolution processes in 1:1 and 2:1 layer clays. A new Stepwave Model has been proposed recently. It consists of an initial, localized dissolution reaction at defects or etch pits on the surface of mineral particles and spreads across the whole mineral surface like a dissolution stepwave. Simultaneous global dissolution of the entire crystal surface is also occurring.

Leaching and reactivity of clay as valuable as well as gangue minerals in a few typical hydrometallurgical processes have been discussed. The problems due to the presence and the dissolution of clays were explained in various leaching environments. The fundamental chemistry of aluminium and silicon in aqueous solution, and in relevant hydrometallurgical processes, were carefully examined in this chapter. Although clay dissolution in acidic environments has been extensively studied, a lack of in-depth knowledge is perceived for clays with different elemental substitutions. The influence of dissolution products from clays (notably Al and Si) on the polymerization/gelation of silicon and/or aluminium and the concomitant loss of valuable metals to the post-leach residues are not well understood. More studies in these areas will potentially help to increase metal recovery and therefore decrease the metal loading to residue areas and their environmental risk. With the increasing failure of HPAL in nickel laterite processing, heap leaching, although gaining momentum, often faces problems of low and slow nickel recovery and high acid consumption. Increasing efforts are undertaken to find more efficient and selective lixiviants as alternatives to sulphuric acid. For instance, some success has been achieved using a number of organic acids (e.g. citric acid) or less common acids to leach more selectively the targeted valuable metals without extensively destroying gangue minerals. Although the state of knowledge is a long way from becoming economically viable at this point in time, such research indicates future industrial needs (e.g. cost savings and lower environmental impact) in hydrometallurgical extraction of valuable metals from increasingly lower-grade ores. Future research into this area is thus highly warranted.

References

Agatzini-Leonardou, S. & Zafiratos, I. G. 2004. Beneficiation of a Greek serpentinic nickeliferous ore Part II: Sulphuric acid heap and agitation leaching. *Hydrometallurgy*, 74 (3–4), 267–275.

Aldushin, K., Jordan, G. & Schmahl, W. W. 2006. Basal plane reactivity of phyllosilicates studied in situ by hydrothermal atomic force microscopy (HAFM). *Geochimica et Cosmochimica Acta*, 70 (17), 4380–4391.

Alexander, G. B., Heston, W. M. & Iler, R. K. 1954. The solubility of amorphous silica in water. *The Journal of Physical Chemistry*, 58 (6), 453–455.

Arvidson, R. S. & Lüttge, A. 2010. Mineral dissolution kinetics as a function of distance from equilibrium: New experimental results. *Chemical Geology*, 269 (1–2), 79–88.

Aytas, S., Yurtlu, M. & Donat, R. 2009. Adsorption characteristic of U(VI) ion onto thermally activated bentonite. *Journal of Hazardous Materials*, 172 (2–3), 667–674.

Bartlett, R. W. 1997. Metal extraction from ores by heap leaching. *Metallurgical and Materials Transactions B: Process Metallurgy and Materials Processing Science*, 28 (4), 529–545.

Baum, W. 1999. The use of a mineralogical data base for production forecasting and troubleshooting in copper leach operations. In: Young, S. K., Dreisinger, D. B., Hackl, R. P. & Dixon, D. G. (eds) *Copper 99: Cobre 99 International Conference*. Phoenix, AZ: The Minerals, Metals and Materials Society, 393–408.

Beig, M. S. & Lüttge, A. 2006. Albite dissolution kinetics as a function of distance from equilibrium: Implications for natural feldspar weathering. *Geochimica et Cosmochimica Acta*, 70 (6), 1402–1420.

Bickmore, B. R., Bosbach, D., Hochella Jr., M. F., Charlet, L. & Rufe, E. 2001. In situ atomic force microscopy study of hectorite and nontronite dissolution: Implications for phyllosilicate edge surface structures and dissolution mechanisms. *American Mineralogist*, 86, 411–423.

Bouffard, S. C. & West-Sells, P. G. 2009. Hydrodynamic behavior of heap leach piles: Influence of testing scale and material properties. *Hydrometallurgy*, 98 (1–2), 136–142.

Büyükakinci, E. & Topkaya, Y. A. 2009. Extraction of nickel from lateritic ores at atmospheric pressure with agitation leaching. *Hydrometallurgy*, 97 (1–2), 33–38.

Canterford, J. H. & Sparrow, G. J. 1984. Adsorption of uranium from carbonate media. *Minerals and Metallurgical Processing*, 1 (3), 241–246.

Cardile, C. M., Willing, M. J. & Hughes, C. A. 1994. Conversion of kaolin to DSP in the Bayer process. *Sixth AusIMM Extractive Metallurgy Conference July 3–6, 1994*. Brisbane, QLD: Australasian Institute of Mining and Metallurgy, 321–324.

Carroll, S. A. & Walther, J. V. 1990. Kaolinite dissolution at 25°, 60°, and 80°C. *American Journal of Sciences*, 290 (7), 797–810.

Carroll-Webb, S. A. & Walther, J. V. 1988. A surface complex reaction model for the pH-dependence of corundum and kaolinite dissolution rates. *Geochimica et Cosmochimica Acta*, 52 (11), 2609–2623.

Casey, W. H. 1991. On the relative dissolution rates of some oxide and orthosilicate minerals. *Journal of Colloid and Interface Science*, 146 (2), 586–589.

Catalano, J. G. & Brown Jr, G. E. 2005. Uranyl adsorption onto montmorillonite: Evaluation of binding sites and carbonate complexation. *Geochimica et Cosmochimica Acta*, 69 (12), 2995–3005.

Chen, T. T., Dutrizac, J. E. & White, C. W. 2000. Serpentine ore microtextures occurring in the Magnola magnesium process. *Journal of the Minerals, Metals & Materials Society*, 52 (4), 20–22.

Chen, Y., Feng, Q., Liu, K., Chen, Y. & Zhang, G. 2006. Study on the structure of Bayer liquor with spectroscopy and MD simulation. *Chemical Physics Letters*, 422 (4–6), 406–411.

Choi, Y., Baron, J. Y., Wang, Q., Langhans, J. & Kondos, P. 2013. Thiosulfate processing: From lab curiosity to commercial application. In: Lorenzen, L. (ed.) *World Gold 2013, 26–29 September 2013*. Brisbane, QLD: The Australasian Institute of Mining and Metallurgy, 45–50.

Cornell, R. M. & Schwertmann, U. 2003. Dissolution. *The Iron Oxides: Structure, Properties, Reactions, Occurrences and Uses*. 2nd ed. Weinheim: Wiley-VCH.

Cousineau, P. G. & Fulford, G. D. 1987. Aspects of the desilication of Bayer liquors. In: Zabreznik, R. (ed.) *116th Annual Meeting of Light Metals*. Denver, CO: TMS, 11–17.

Cresswell, P. J. 1984. Factors affecting desilication of Bayer process liquors. In: Barton, A. C. T. (ed.) *Chemeca 84: The 12th Australian Chemical Engineering Conference; Winning in the Competitive World 1984*. Melbourne, VIC: Institution of Engineers Australia, 285–292.

Crundwell, F. K., Moats, M. S., Ramachandran, V., Robinson, T. G. & Davenport, W. G. 2011. High-temperature sulfuric acid leaching of laterite ores. *Extractive Metallurgy of Nickel, Cobalt and Platinum Group Metals*. Oxford: Elsevier.

Dai, X., Breuer, P., Hewitt, D. & Bergamin, A. 2013. Thiosulfate process for treating gold concentrates. In: Lorenzen, L. (ed.) *World Gold Conference 2013, 26–29 September 2013*. Brisbane, QLD: The Australasian Institute of Mining and Metallurgy, 61–69.

Dalvi, A. D., Bacon, W. G. & Osborne, R. C. 2004. The past and the future of nickel laterites. PDAC 2004 International Convention, Trade Show and Investors Exchange, 7–10 March 2004. Canada. PDAC.

Das, G. K. & de Lange, J. A. B. 2011. Reductive atmospheric acid leaching of West Australian smectitic nickel laterite in the presence of sulphur dioxide and copper(II). *Hydrometallurgy*, 105 (3–4), 264–269.

Dhawan, N., Safarzadeh, M. S., Miller, J. D., Moats, M. S. & Rajamani, R. K. 2013. Crushed ore agglomeration and its control for heap leach operations. *Minerals Engineering*, 41, 53–70.

Du Preez, J. G. H., Morris, D. C., Van Vuuren, C. P. J., Hendriks, P. & Oertell, M. 1981. The chemistry of uranium: Part XXVIII. The development of a combined gold and uranium leach of randfontein ore. *Hydrometallurgy*, 6 (3–4), 203–218.

Eremin, N. I., Tkacheva, L. V. & Makarenkov, V. N. 1978. Investigation of the kinetics of the decomposition of kaolinite in alkaline and aluminate solutions. *Soviet Non-ferrous Metals Research*, 6, 197–199.

Forsmo, S. P. E., Samskog, P. O. & Bjorkman, B. M. T. 2008. A study on plasticity and compression strength in wet iron ore green pellets related to real process variations in raw material fineness. *Powder Technology*, 181 (3), 321–330.

Gasparini, C. 1989. Gold. In: *Gold and Other Precious Metals: The Lure and the Trap.* Toronto, ON: Space Eagle Publishing Co.

Gjelsvik, N. & Torgersen, J. H. 1983. Method of acid leaching of silicates. US Patent 4367215.

Gönen, N., Körpe, E., Yıldırım, M. E. & Selengil, U. 2007. Leaching and CIL processes in gold recovery from refractory ore with thiourea solutions. *Minerals Engineering*, 20 (6), 559–565.

Gruber, C., Harpaz, L., Zhu, C., Bullen, T. D. & Ganor, J. 2013. A new approach for measuring dissolution rates of silicate minerals by using silicon isotopes. *Geochimica et Cosmochimica Acta*, 104, 261–280.

Guzman, L., Segarra, M., Chimenos, J. M., Fernandez, M. A. & Espiell, F. 1999. Gold cyanidation using hydrogen peroxide. *Hydrometallurgy*, 52 (1), 21–35.

Guzman, A., Scheffel, R. & Flaherty, S. 2008. The fundamentals of physical characterisation of ore for heap leach. In: Young, C. A., Taylor, P. R., Anderson, C. G. & Choi, Y. (eds) *Hydrometallurgy 2008: Proceedings of the Sixth International Symposium.* Littleton, CO: Society for Mining, Metallurgy and Exploration, 937–954.

Helle, S., Jerez, O., Kelm, U., Pincheira, M. & Varela, B. 2010. The influence of rock characteristics on acid leach extraction and re-extraction of Cu-oxide and sulfide minerals. *Minerals Engineering*, 23 (1), 45–50.

Hering, J. G. & Stumm, W. 1990. Oxidative and reductive dissolution of minerals. *Reviews in Mineralogy*, 23 (1), 427–465.

Herkenhoff, E. C. & Dean, J. G. 1987. Heap leaching: Agglomerate or deslime? A critique of current wisdom and the case for splitting fines to a mini-agitator leach. *Engineering and Mining Journal*, 188 (6), 32–39.

Hong, H., Fu, Z. & Min, X. 2001. The adsorption of $[Au(HS)2]^{-}$ on kaolinite surfaces: Quantum chemistry calculations. *The Canadian Mineralogist*, 39 (6), 1591–1596.

Iglesias, N. & Carranza, F. 1994. Refractory gold-bearing ores: A review of treatment methods and recent advances in biotechnological techniques. *Hydrometallurgy*, 34 (3), 383–395.

Iler, R. K. 1979. *The Chemistry of Silica: Solubility, Polymerization, Colloid and Surface Properties and Biochemistry of Silica.* New York: Wiley & Sons.

Johnston, C. T., Sposito, G. & Birge, R. R. 1985. Raman spectroscopic study of kaolinite in aqueous suspension. *Clays and Clay Minerals*, 33 (6), 483–489.

Kalinowski, B. & Schweda, P. 1996. Kinetics of muscovite, phlogopite, and biotite dissolution and alteration at pH 1–4, room temperature. *Geochimica et Cosmochimica Acta*, 60 (3), 367–385.

Karimi, P., Abdollahi, H., Amini, A., *et al.* 2010. Cyanidation of gold ores containing copper, silver, lead, arsenic and antimony. *International Journal of Mineral Processing*, 95 (1–4), 68–77.

Kaya, Ş. & Topkaya, Y. A. 2011. High pressure acid leaching of a refractory lateritic nickel ore. *Minerals Engineering*, 24 (11), 1188–1197.

Kinrade, S. D. & Swaddle, T. W. 1989. Direct detection of aluminosilicate species in aqueous-solution by Si-29 and Al-27 NMR-spectroscopy. *Inorganic Chemistry*, 28 (10), 1952–1954.

Köhler, S. J., Bosbach, D. & Oelkers, E. H. 2005. Do clay mineral dissolution rates reach steady state? *Geochimica et Cosmochimica Acta*, 69 (8), 1997–2006.

Korichi, S. & Bensmaili, A. 2009. Sorption of uranium (VI) on homoionic sodium smectite experimental study and surface complexation modeling. *Journal of Hazardous Materials*, 169 (1–3), 780–793.

Kurganskaya, I., Arvidson, R. S., Fischer, C. & Lüttge, A. 2012. Does the stepwave model predict mica dissolution kinetics? *Geochimica et Cosmochimica Acta*, 97, 120–130.

Kuwahara, Y. 2006. In-situ AFM study of smectite dissolution under alkaline conditions at room temperature. *American Mineralogist*, 91, 1142–1149.

La Brooy, S. R., Linge, H. G. & Walker, G. S. 1994. Review of gold extraction from ores. *Minerals Engineering*, 7 (10), 1213–1241.

Lasaga, A. C. & Lüttge, A. 2001. Variation of crystal dissolution rate based on dissolution stepwave model. *Science*, 291 (5512), 2400–2404.

Lasaga, A. C. & Lüttge, A. 2003. A model for crystal dissolution. *European Journal of Mineralogy*, 15 (4), 603–615.

Laws, M. 2005. *Reactions of kaolinite in the predesilication processing of bauxite ores.* PhD, University of Melbourne.

Lottering, M. J., Lorenzen, L., Phala, N. S., Smit, J. T. & Schalkwyk, G. A. C. 2008. Mineralogy and uranium leaching response of low grade South African ores. *Minerals Engineering*, 21 (1), 16–22.

Lu, J., Dreisinger, D. & West-Sells, P. 2007. Acid curing and agglomeration. In: Riveros, P. A., Dixon, D. G., Dreisinger, D. B. & Collins, M. J. (eds) *The John E. Dutrizac International Symposium on Copper Hydrometallurgy*. Toronto: The Canadian Institute of Mining, Metallurgy and Petroleum, 453–464.

Martin, R. B. 1991. Fe^{3+} and Al^{3+} hydrolysis equilibria: Cooperativity in Al^{3+} hydrolysis reactions. *Journal of Inorganic Biochemistry*, 44 (2), 141–147.

Mattus, A. J. & Torma, A. E. 1980. A comparison of carbonate leaching of a low-grade uranium ore at atmospheric and increased pressures. *Hydrometallurgy*, 5 (2–3), 179–190.

McDonald, R. G. & Whittington, B. I. 2008. Atmospheric acid leaching of nickel laterites review Part I: Sulphuric acid technologies. *Hydrometallurgy*, 91 (1–2), 35–55.

McKinley, J. P., Zachara, J. M., Smith, S. C. & Tuner, G. D. 1995. The influence of uranyl hydrolysis and multiple site-binding reactions on adsorption of U(VI) to montmorillonite. *Clays and Clay Minerals*, 43 (5), 586–598.

Miller, G. 2003. Ore geotechnical effects on copper heap leach kinetics. In: Young, C., Alfantazi, A., Anderson, C., James, A., Dreisinger, D. & Harris, B. (eds) *Hydrometallurgy 2003: Proceedings of the Fifth International Symposium Honoring Professor Ian Ritchie*. Vancouver, BC: The Minerals, Metals and Materials Society, 329–342.

Miller, J. D., Wan, R. Y. & Díaz, X. 2005. Preg-robbing gold ores. In: Adams, M. D. & Wills, B. A. (eds) *Developments in Mineral Processing*. New York: Elsevier.

Moats, M. S. & Janwong, A. 2008. The art and science of crushed ore agglomeration for heap leaching. In: Young, C. A., Taylor, P. R., Anderson, C. G. & Choi, Y. (eds) *Proceedings of the Sixth International Symposium on Hydrometallurgy*. Phoenix, AZ: Society for Mining, Metallurgy, and Exploration, Inc., 912–917.

Mohammadnejad, S., Provis, J. L. & van Deventer, J. S. J. 2011. Gold sorption by silicates in acidic and alkaline chloride media. *International Journal of Mineral Processing*, 100 (3–4), 149–156.

Neely, J. & Connick, R. 1970. Rate of water exchange from hydrated magnesium ion. *Journal of the American Chemical Society*, 92 (11), 3476–3478.

Newman, A. C. D. & Brown, G. 1987. The chemical constitution of clays. In: Newman, A. C. D. (ed.) *Chemistry of Clays and Clay Minerals*. Chantilly, VA: Mineralogical Society Monographs.

O'Kane, M., Barbour, S. L. & Haug, M. D. 1999. A framework for improving the ability to understand and predict the performance of heap leach piles. In: Young, S. K., Dreisinger, D. B., Hackl, R. P. & Dixon, D. G. (eds) *Hydrometallurgy of Copper*. Warrendale, PA: The Minerals, Metals and Materials Society, 409–419.

O'Neill, C. E. 1978. Leaching nickeliferous silicate ores with hydrochloric acid. Canadian Patent 1024353.

Oelkers, E. H. 2001. General kinetic description of multioxide silicate mineral and glass dissolution. *Geochimica et Cosmochimica Acta*, 65 (21), 3703–3719.

Oelkers, E. H., Schott, J., Gauthier, J.-M. & Herrero-Roncal, T. 2008. An experimental study of the dissolution mechanism and rates of muscovite. *Geochimica et Cosmochimica Acta*, 72 (20), 4948–4961.

Oelkers, E. H., Golubev, S. V., Chairat, C., Pokrovsky, O. S. & Schott, J. 2009. The surface chemistry of multi-oxide silicates. *Geochimica et Cosmochimica Acta*, 73 (16), 4617–4634.

Okada, K., Temuujin, J., Kameshima, Y. & Mackenzie, K. J. D. 2003. Selective acid leaching of talc. *Clay Science*, 12 (3), 159–165.

Okamoto, G., Okura, T. & Goto, K. 1957. Properties of silica in water. *Geochimica et Cosmochimica Acta*, 12 (1–2), 123–132.

Oku, T. & Yamada, K. 1971. The dissolution rate of quartz and the rate of desilication in the Bayer liquor. In: Edgeworth, T. G. (ed.) *100th AIME Annual Meeting of Light Metals*. New York: TMS, 31–45.

Padilla, G. A., Cisternas, L. A. & Cueto, J. Y. 2008. On the optimization of heap leaching. *Minerals Engineering*, 21 (9), 673–678.

Pautler, J. B., Strominger, M. G. & Gross, A. E. 1990. *New Polymeric Agglomeration Aid Improves Heap Leach Efficiency at Brewer Gold*. Naperville, IL: Nalco Company.

Quaicoe, I., Nosrati, A., Skinner, W. & Addai-Mensah, J. 2014. Single and mixed oxide and clay particle agglomeration: Influence of feed mineralogy and percent drum volume loading. *Powder Technology*, 253, 568–579.

Queneau, P. B. & Berthold, C. E. 1986. Silica in hydrometallurgy: An overview *Canadian Metallurgical Quarterly*, 25 (3), 201–209.

Queneau, P. B., Berggren, M. H., Cooperrider, M. W. & Doane, R. E. 1983. Control of silica deposition during pressure let-down of acidic leach slurries. In: Osseo-Asare, K. & Millers, J. D. (eds) *Hydrometallurgy Research, Development and Plant Practice, March 6–10. 1983*. Atlanta, GA: The Metallurgical Society of AIME.

Readett, D. J. & Fox, J. 2009. *Development of Heap Leaching and its Integration into the Murrin Murrin Operations*. Perth, WA: ALTA Metallurgical Services.

Readett, D. J. & Fox, J. 2011. Agglomeration: The key to success for the Murrin Murrin Ni laterite heap leach. In: Australasian Institute of Mining and Metallurgy (ed.) *MetPlant 2011, 8–9 August, 2011*. Perth, WA: Australasian Institute of Mining and Metallurgy.

Roach, G. I. D. & White, A. J. 1988. Dissolution kinetics of kaolin in caustic liquors. In: Boxall, L. G. (ed.) *117th TMS Annual Meeting of Light Metals. January 25–28, 1988*. Phoenix, AZ: TMS, 41–47.

Rozalen, M., Huertas, F. J. & Brady, P. V. 2009. Experimental study of the effect of pH and temperature on the kinetics of montmorillonite dissolution. *Geochimica et Cosmochimica Acta*, 73 (13), 3752–3766.

Salgado, C., Perez, C., Alvayai, C. & Zarate, G. 2009. Acid management in heap leaching: Are we doing right? In: Domic, E. & Casas, J. (eds) *Hydrocopper*. Santiago: GECAMIN, 47–56.

Schindler, P. W. & Stumm, W. 1987. The surface chemistry of oxides, hydroxides, and oxide minerals. In: Stumm, W. (ed.) *Aquatic Surface Chemistry*. New York: Wiley & Sons.

Schott, J., Pokrovsky, O. S. & Oelkers, E. H. 2009. The link between mineral dissolution/precipitation kinetics and solution chemistry. *Reviews in Mineralogy & Geochemistry*, 70 (1), 207–258.

Seidel, D. C. 1979. Extracting uranium from its ores. *IAEA Bulletin*, 23 (2), 25–28.

Senanayake, G., Das, G. K., de Lange, A., Li, J. & Robinson, D. J. 2015. Reductive atmospheric acid leaching of lateritic smectite/nontronite ores in H2SO4/Cu(II)/SO2 solutions. *Hydrometallurgy*, 152, 44–54.

Sipos, P. 2009. The structure of Al(III) in strongly alkaline aluminate solutions: A review. *Journal of Molecular Liquids*, 146 (1–2), 1–14.

Sobol, S. I. 1972. Secondary losses of nickel and cobalt during the autoclave leaching oxidized nickel ores. *Tsvetnye Metally*, 45 (5), 25–28.

Stopar, J. D., Jeffrey Taylor, G., Hamilton, V. E. & Browning, L. 2006. Kinetic model of olivine dissolution and extent of aqueous alteration on Mars. *Geochimica et Cosmochimica Acta*, 70 (24), 6136–6152.

Stumm, W. 1992. *Chemistry of the Solid–Water Interface: Processes at the Mineral–Water and Particle–Water Interface in Natural Systems*. New York: Wiley.

Swaddle, T. W. 2001. Silicate complexes of aluminum(III) in aqueous systems. *Coordination Chemistry Reviews*, 219–221, 665–686.

Swaddle, T. W., Salerno, J. & Tregloan, P. A. 1994. Aqueous aluminates, silicates, and aluminosilicates. *Chemical Society Reviews*, 23 (5), 319–325.

Tan, H., Feng, D., Lukey, G. C. & van Deventer, J. S. J. 2005. The behaviour of carbonaceous matter in cyanide leaching of gold. *Hydrometallurgy*, 78 (3–4), 226–235.

Temuujin, J., Senna, M., Jadambaa, T., *et al.* 2006. Characterization and bleaching properties of acid-leached montmorillonite. *Journal of Chemical Technology & Biotechnology*, 81 (4), 688–693.

Terry, B. 1983. The acid decomposition of silicate minerals part II: Hydrometallurgical applications. *Hydrometallurgy*, 10 (2), 151–171.

Tizon, E., Clerin, P. & Cristol, B. 2004. Effect of predesilication and digestion conditions on silica level in Bayer liquor. In: Tabereaux, A. T. (ed.) *133rd TMS Annual Meeting of Light Metals*. Charlotte, NC: TMS, 9–14.

Tremolada, J., Dzioba, R., Bernardo-Sánchez, A. & Menéndez-Aguado, J. M. 2010. The preg-robbing of gold and silver by clays during cyanidation under agitation and heap leaching conditions. *International Journal of Mineral Processing*, 94 (1–2), 67–71.

Walther, J. V. & Wood, B. J. 1986. Mineral–fluid reaction rates. In: Walther, J. V. & Wood, B. J. (eds) *Fluid–Rock Interactions During Metamorphism*. New York: Springer-Verlag.

Wei, Y., Zhong, K., Adamov, E. V. & Smith, R. W. 1997. Semi-continuous biooxidation of the Chongyang refractory gold ore. *Minerals Engineering*, 10 (6), 577–583.

Whittington, B. I. & Muir, D. 2000. Pressure acid leaching of nickel laterites: A review. *Mineral Processing and Extractive Metallurgy Review*, 21 (6), 527–600.

Whittington, B. I., McDonald, R. G., Johnson, J. A. & Muir, D. M. 2003. Pressure acid leaching of arid-region nickel laterite ore: Part I. Effect of water quality. *Hydrometallurgy*, 70 (1–3), 31–46.

Wieland, E. & Stumm, W. 1992. Dissolution kinetics of kaolinite in acidic aqueous solutions at 25°C. *Geochimica et Cosmochimica Acta*, 56 (9), 3339–3355.

Yusan, S. & Erenturk, S. 2011. Sorption behaviors of uranium (VI) ions on α-FeOOH. *Desalination*, 269 (1–3), 58–66.

Zambo, J. 1986. Structure of sodium aluminate liquors: Molecular model of the mechanism of their decomposition. In: Miller, R. E. & Peterson, W. S. (eds) *115th AIME Annual Meeting of Light Metals, March 2–6, 1986*. New Orleans, LA: TMS Light Metals, 199–215.

4

Identification and Quantification of Clays

Mark Raven and Peter Self

4.1 Introduction

Clays are part of the group of sheet silicates that generally have a very fine particle size, typically less than 2 μm. They often have a platy morphology, but can also be found as tubes, laths and spheres. Elemental substitution in their structures is common, which may give rise to charge imbalance. This excess negative charge is balanced by cations positioned between the layers. The interlayer cations are fixed in some clays and exchangeable in others. Clays with exchangeable cations also incorporate variable amounts of water. The amount of water incorporated is dependent on the hydration energy of the interlayer cation and relative humidity of the environment. A consequence of the structural complexity of clays is that methods for identifying clays can be quite involved, requiring a sequence of tests (e.g. Poppe *et al.*, 2001) in order to uniquely identify the clays present. Furthermore, any discussion on the identification and quantification of clays will necessarily also encompass those groups, such as the micas, that are often associated with clays in mineral assemblages.

4.2 Clays

4.2.1 Crystal Structure

As described in Chapters 1 and 2, the backbone of clay structures (sheet silicates) is two atomic planes that consist of a mix of oxygen atoms and hydroxyl groups[1] in a close-packing arrangement. These two atomic planes define the basal plane of the clay structure and correspond to (001) in crystallographic Miller indices notation for the standard unit cell parameters used for sheet silicates. The octahedral sites between the oxygen/hydroxyl (O/OH) planes are occupied by metal centres (e.g. Al^{3+}, Fe^{3+}, Mg^{2+}) and the O/OH planes plus the octahedrally coordinated metals are described as an octahedral sheet. The valence of the metal centres determines how many of the octahedral sites are occupied. For divalent metals (e.g. Mg^{2+}) all octahedral sites would be expected to be occupied. For trivalent

[1] Fluorine-based analogues of sheet silicates where the hydroxyl groups are replaced by fluorine are found but are not common in most environments.

metals (e.g. Al^{3+} and Fe^{3+}) only two-thirds of the octahedral sites would be occupied. This, plus crystal structure considerations, gives rise to the notation 'tri-octahedral', meaning all octahedral sites are occupied and 'di-octahedral' meaning two out of every three octahedral sites are occupied. A single layer of a sheet silicate structure is formed by adding a further plane of oxygen atoms to one side (1:1 sheet silicates) or to both sides (2:1 sheet silicates) of the O/OH sheets containing the octahedrally coordinated metal centres. In these additional planes the oxygen atoms are approximately arranged in a hexagonal array with every second position of every second row unoccupied (see Chapter 1, Fig. 1.6b). The hexagonal net for an oxygen plane is smaller than the hexagonal net for the O/OH sheets making up the octahedral layer. Additionally, the oxygens in these planes are not positioned in a close-packing arrangement relative to the O/OH of the octahedral sheet. Rather, in an idealized clay structure, the oxygen atoms are positioned above the midpoint between two oxygen atoms of the octahedral sheet. This means that the additional oxygen sheets form only tetrahedral sites with the oxygen atoms of the octahedral sheet. The tetrahedral sites formed between the oxygen planes and the octahedral sheets are predominately occupied by Si atoms, but may also contain other atoms that have a sufficiently small atomic radius and reasonably high charge (~3+ to 4+). The most common substitution for Si is Al^{3+} with Fe^{3+} substituting for Si in sheet silicates with low Al content (e.g. Gates *et al.*, 2002). The form of the oxygen plane and the tetrahedral sites is shown schematically in the phyllosilicates diagram shown in Fig. 2.1 of Chapter 2. Thus, the individual layers in sheet silicates will consist of a plane of tetrahedrally coordinated metals plus one plane of octahedrally coordinated metals (1:1 sheet silicates) or two planes of tetrahedrally coordinated metals plus one plane of octahedrally coordinated metals (2:1 sheet silicates). This gives rise to the 'T–O' notation for the stacking of sheet silicates whereby an individual layer in a 1:1 sheet silicate would be described as T–O and an individual layer in a 2:1 sheet silicate would be described as T–O–T.

In terms of crystal structure, it is useful to describe the oxygen planes that make up the tetrahedral sheets as being part of an array of corner-sharing tetrahedra (see Chapter 2, Fig. 2.1). The existence of the relatively large empty spaces (cavities) in the tetrahedral oxygen plane has several consequences for the crystallography of sheet silicates. Mismatches in the spacing of the octahedral O/OHs and tetrahedral oxygen atoms can be accommodated by rotations of the tetrahedra about an axis roughly perpendicular to the sheets. These rotations can be quite large (13° in muscovite; Radoslovich, 1960) and they impose a trigonal symmetry on the cavities. Consequently, the cavities in a tetrahedral sheet are referred to as 'di-trigonal cavities'. Further consequences for crystallography are that in the octahedral sheet adjacent to the tetrahedral sheet the O/OH sites below the centre of a di-trigonal cavity will always be occupied by an OH group and the hydrogen atoms in the OH groups will point into the di-trigonal cavity. Correspondingly, the octahedral O/OH sites forming a corner with a tetrahedron will always be occupied by oxygen atoms.

The principles being discussed can be demonstrated by examining the structure of the 2:1 sheet silicates, talc and pyrophyllite. Talc [$Mg_3Si_4O_{10}(OH)_2$], a tri-octahedral sheet silicate, and pyrophyllite [$Al_2Si_4O_{10}(OH)_2$], a di-octahedral sheet silicate, consist of T–O–T layers that carry no net charge. The Mg atoms (talc) or the Al atoms (pyrophyllite) occupy octahedral sites in the structure while the Si atoms occupy all tetrahedral sites, with two Si atoms per formula unit in each of the tetrahedral sheets on either side of the octahedral sheet. The OH groups lie below the di-trigonal cavities of the tetrahedral sheet with one OH group per formula unit for each tetrahedral sheet. That is, the T–O–T layer will consist of the following stacking sequence:

(A) A plane of oxygen atoms containing di-trigonal cavities.
(B) Tetrahedrally coordinated metal centres (mainly Si).
(C) A plane consisting of oxygen atoms forming the apexes of tetrahedra surrounding the tetrahedrally coordinated metal centres and OH groups approximately centred on the di-trigonal cavities.
(D) Octahedrally coordinated metal centres.
(E) Another plane consisting of oxygen atoms forming the apexes of tetrahedra surrounding the tetrahedrally coordinated metal centres and OH groups approximately centred on the di-trigonal cavities.
(F) Tetrahedrally coordinated metal centres (mainly Si).
(G) A plane of oxygen atoms containing di-trigonal cavities.

In the overall talc and pyrophyllite structures, T–O–T layers combine by way of van der Waal's forces and stack together forming a …T–O–T–T–O–T… structure (see Fig. 1.4 in Chapter 1).

A similar arrangement of metal atoms is found in the 1:1 sheet silicates lizardite [$Mg_3Si_2O_5(OH)_4$, tri-octahedral] and kaolinite [$Al_2Si_2O_5(OH)_4$, di-octahedral] that consist of T–O layers that carry no net charge. One OH group per formula unit will sit below the di-trigonal cavity of the tetrahedral layer while the other three OH per formula unit will form a complete plane on the side of the octahedral layer not directly joined to the tetrahedral sheet. That is, the T–O layer will consist of the following stacking sequence:

(A) A plane of oxygen atoms containing di-trigonal cavities.
(B) Tetrahedrally coordinated metal centres (mainly Si).
(C) A plane consisting of oxygen atoms forming the apexes of tetrahedra surrounding the tetrahedrally coordinated metal centres and OH groups approximately centred on the di-trigonal cavities.
(D) Octahedrally coordinated metal centres.
(E) A plane consisting of OH groups.

In the overall lizardite and kaolinite structures, T–O layers combine by way of hydrogen bonding and stack together forming a …T–O–T–O… structure (see Fig. 1.4 in Chapter 1).

When elemental substitution occurs in the sheet silicate layers, the layers have a permanent charge. Extreme examples of charged layers are the brittle micas such as margarite [$CaAl_2(Si_2Al_2)O_{10}(OH)_2$] and clintonite [$Ca(Mg,Al)_3(Al_3Si)O_{10}(OH)_2$]. Brittle micas have a net layer charge of –2 per formula unit (Rieder *et al.*, 1998). In margarite the layer charge is the result of the substitution of two Al^{3+} for every two Si^{4+} in the tetrahedral sheet. The

Table 4.1. *Endmember compositions of some smectites.*

Mineral	Formula	Classification	Dominant charged sheet
Montmorillonite	$(Al_{2-x}Mg_x)Si_4O_{10}(OH)_2$. $(½Ca,½Mg,Na)_x(H_2O)_n$	Di-octahedral	Octahedral
Beidellite	$Al_2(Si_{4-x}Al_x)O_{10}(OH)_2$. $(½Ca,½Mg,Na)_x(H_2O)_n$	Di-octahedral	Tetrahedral
Saponite	$[Mg_{3-y}(Al,Fe)_y](Si_{4-x}Al_x)O_{10}(OH)_2$. $(½Ca,½Mg,Na)_{x-y}(H_2O)_n$	Tri-octahedral	Tetrahedral
Nontronite	$Fe^{3+}{}_2(Si_{4-x}Al_x)O_{10}(OH)_2$. $(½Ca,½Mg,Na)_x(H_2O)_n$	Di-octahedral	Tetrahedral

layer charge in clintonite arises from the replacement of three out of every four Si^{4+} atoms in the tetrahedral sheet with Al^{3+} atoms and the substitution of Al^{3+} atoms for Mg^{2+} atoms in the octahedral sheet. Correspondingly, the charge of the tetrahedral sheet in clintonite is –3 per formula unit that is partially balanced by a charge of +1 per formula unit for the octahedral sheet. The net layer charge is fully balanced by a divalent cation (Ca^{2+} in margarite and clintonite) that sits in the interlayer between the T–O–T layers, giving an …T–O–T M^{2+} T–O–T structure (Fig. 1.4 in Chapter 1).

In the more common micas (i.e. the true micas) such as muscovite $[KAl_2(Si_3Al)O_{10}(OH)_2]$ and phlogopite $[KMg_3(Si_3Al)O_{10}(OH)_2]$, the overall layer charge is –1 per formula unit (Rieder *et al.*, 1998). The layer charge is balanced by a non-hydrated, monovalent cation sitting in the interlayer between the T–O–T layers. Although the aforementioned chemical formula for muscovite and phlogopite show the layer charge being entirely from the tetrahedral sheet, as in the case of margarite, there can be some layer charge derived from atomic substitutions in the octahedral sheet. For example, the phengites, that form a mineral series between muscovite, celadonite, $K[(Mg,Fe^{2+})(Fe^{3+},Al)][Si_4]O_{10}(OH)_2$, and alumino-celadonite, $K[(Mg,Fe^{2+})Al][Si_4]O_{10}(OH)_2$, can carry significant octahedral charge as shown by the general chemical formula for the phengites, namely $K[Al_{2-x}(Mg,Fe^{2+})_x][Si_{3+x}(Al,Fe^{3+})_{1-x}]O_{10}(OH)_2$. In general terms, atomic substitutions will give rise to a negatively charged tetrahedral sheet, a positively charged octahedral sheet for the tri-octahedral sheet silicates (e.g. Al^{3+} substituting for Mg^{2+}), and a negatively charged octahedral sheet for the di-octahedral sheet silicates (e.g. Mg^{2+} substituting for Al^{3+}).

For the sheet silicate examples discussed above, the permanent layer charge has an integer value (0, –1, –2). This is not the case for all 2:1 clay minerals for which there is a progression in layer charge per formula unit (Guggenheim *et al.*, 2006):

(1) From –0.2 to –0.6 for the smectites. Smectites take a variety of forms, some of which are shown in Table 4.1. The smectites can be di- or tri-octahedral and the sheet containing the major component of the layer charge can be either the tetrahedral or the octahedral sheet. The amount of water in the interlayer (denoted as n in Table 4.1) is variable,

but the presence of water in the interlayer of vermiculites and smectites is crucial to the stability of these minerals. As water has a high dielectric constant (i.e. is a highly polar molecule) it acts to distribute the charge of the interlayer cation over the T–O–T layers that necessarily do not have a uniform distribution of charge.

(2) –0.6 to –0.9 for the vermiculites. Vermiculites are sheet silicates with cations and water in the interlayer between the T–O–T layers. The general chemical formula for vermiculites is $(Mg,Al,Fe)_3(Si,Al)_4O_{10}(OH)_2.(Mg,Ca)_x(H_2O)_4$ where x is half the layer charge per formula unit. The chemical formula lists x as the half-layer charge because the interlayer cations are given as divalent. For the vermiculites the layer charge arises predominantly from substitution of Al for Si in the tetrahedral sheet.

(3) –0.65 to –0.9 for the illites. Originally the illites were classified as clay-sized micas but, while being similar to muscovite, have more Si and Mg and less K than ideal muscovite and also contain some structural water (Grim *et al.*, 1937). With time this definition was modified and in 1984 the American Clay Minerals Society (CMS) Nomenclature Committee defined three requirements for illite (Bailey *et al.*, 1984). These requirements are: '(1) the micaceous layers are non-expansible; (2) the octahedral sheet is dioctahedral; and (3) the composition deviates from that of muscovite in two main ways: (a) a phengitic component is present in which substitution of R^{2+} cations for octahedral Al is balanced by addition of tetrahedral Si beyond the ideal Si:Al ratio of 3:1 for muscovite … (b) interlayer vacancies or water molecules amounting to about 0.2 to 0.4 atoms per formula unit are compensated by additional tetrahedral Si cations.' The CMS Nomenclature Committee revised this definition in 1998 (Rieder *et al.*, 1998) by stating that the interlayer water attributed to illite, as in the 1984 definition, was because of mixed layering (i.e. interstratification) of illite and smectite. Accordingly the term 'hydromicas' was replaced with 'interlayer-cation deficient micas' or 'interlayer-deficient micas'. The CMS Nomenclature Committee also gave a representative formula for illites as well as for other interlayer-deficient micas such as glauconite. Glauconite is a di-octahedral, interlayer-deficient mica with a high Fe content in the octahedral sites (as opposed to Al in illite) and a higher octahedral sheet charge, lower tetrahedral sheet charge and a lower overall layer charge than illite.

4.2.1.1 Polytypes

The way in which the T–O–T or T–O layers stack to form a complete structure is determined by the layer charge, the interlayer cations and the presence or absence of water in the interlayer. For 2:1 sheet silicates with a high layer charge and little or zero interlayer water (i.e. the micas and illite), the interlayer cations sit between the di-trigonal cavities of successive T–O–T layers and the di-trigonal cavities are aligned from one T–O–T layer to the next. Even with this structural constraint, because of the nature of the T–O–T layer, particularly the translational shift in the di-trigonal cavity on one side of the T–O–T layer relative to the di-trigonal cavity on the other side of the T–O–T layer, the approximately hexagonal symmetry of the T–O–T layer surfaces will allow 120° rotations. As there are

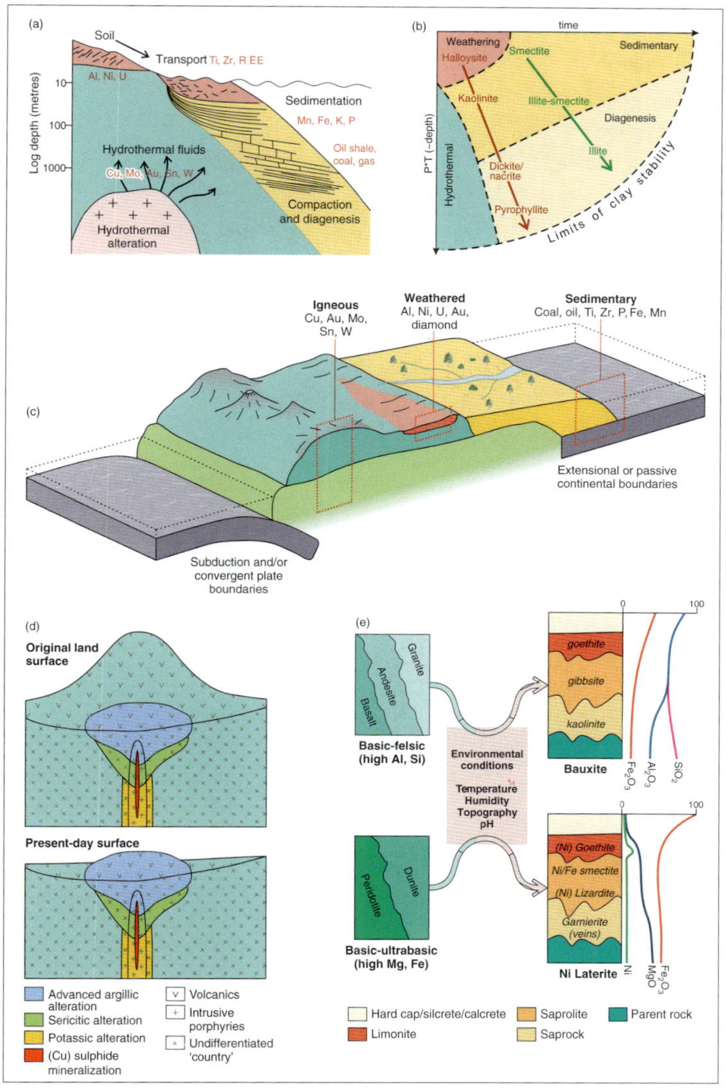

Figure 1.2. Clay minerals are hosted in a variety of geological settings and are likewise associated with numerous metallic ore deposits. In igneous-hosted deposits (such as Cu, Mo, Au, Sn and W), clays can form via both hydrothermal alteration and weathering of the volcanoclastic sediment pile. Lateritic ores can form in stable intracontinental settings, through the interaction of suitable host rocks and favourable environmental circumstances, resulting in clays associated with Au, Ni, Al and U concentrations. Clays are also common in sedimentary hosted ore deposits formed at extensional or passive continental boundaries through erosion/deposition (e.g. Ti, Zr, P) and diagenetic (e.g. coal, oil and gas) processes. Regardless of the formation conditions, ongoing tectonic evolution can modify the clay species through both prograde and retrograde modification. An overview of the geological context of clay mineral formation and ore types they may be associated with: (a) clays and ore types as a function of depth (based on Velde 1995); (b) clay stability as a function of pressure/temperature (depth and geothermal gradient) and time (based on Velde, 1995); (c) ore deposits in their tectonic context; (d) evolution of clay minerals in a porphyry copper deposit; (e) evolution of clay minerals in lateritic Al and Ni deposits.

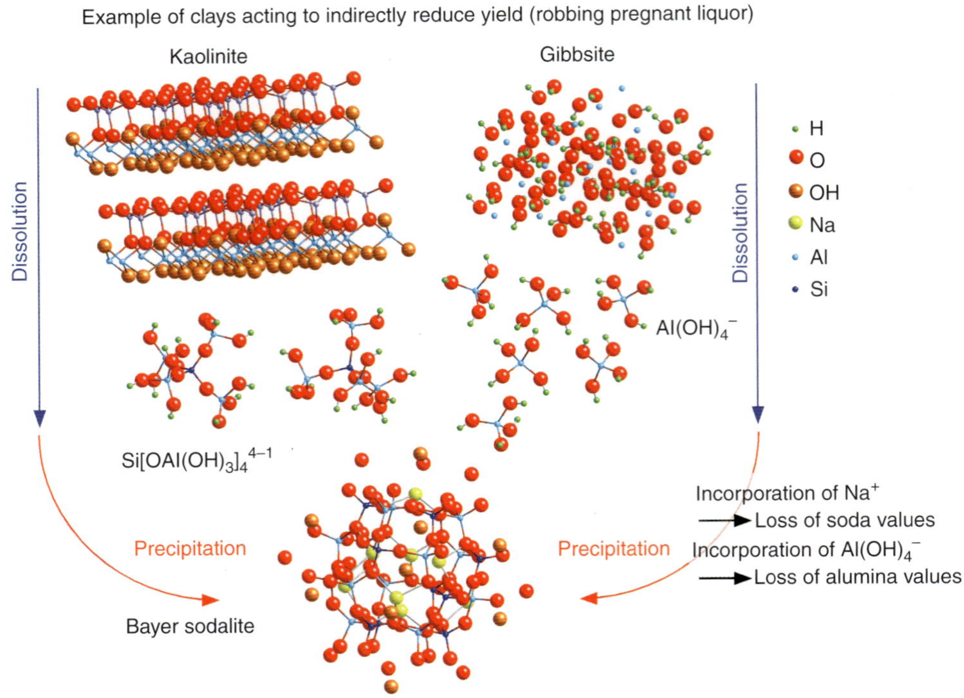

Figure 3.9. An example of clays acting to indirectly reduce gibbsite yield in the Bayer process by providing a source of solubilized aluminosilicate (most likely the tetra-aluminosilicate ion) to allow formation of unwanted sodalite.

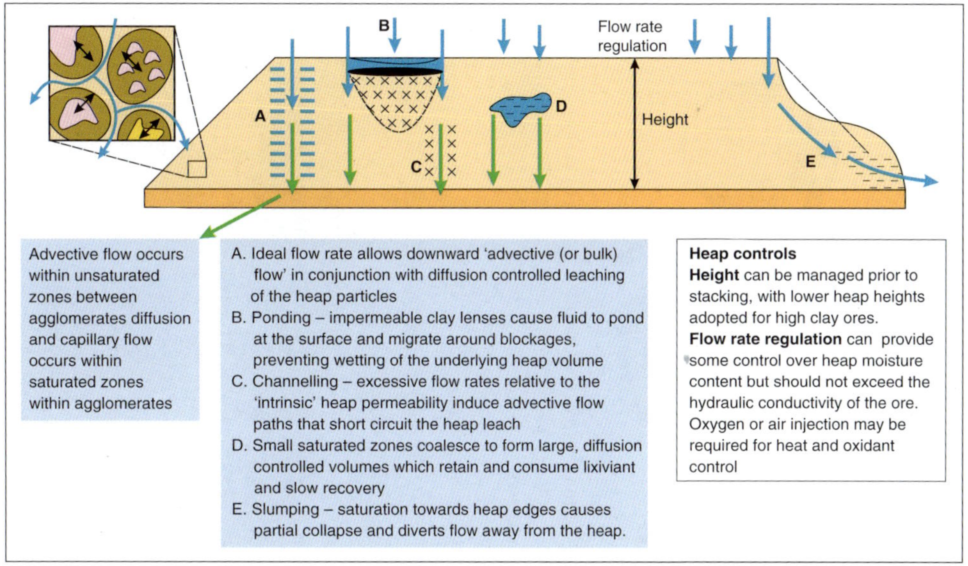

Figure 3.12. An overview of the effect of clay minerals on heap leach hydrology and issues regarding detrimental effects. Unlike many other unit processes, optimization of heap conditions after initiation is limited, making careful characterization and preparation of feed materials, especially those with high clay content, essential.

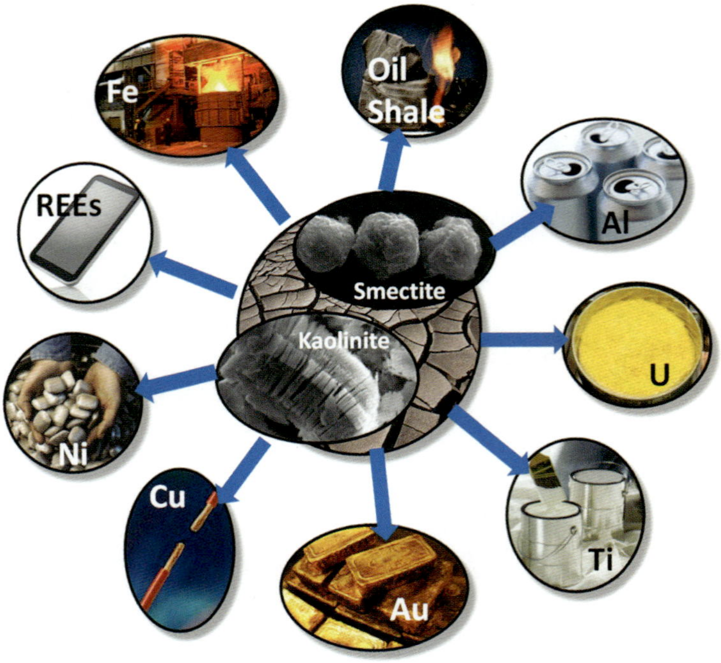

Figure 5.1. Kaolinite and smectite may be associated with a range of economically important commodities, typically as deleterious phases that can adversely affect the ore grade, processing and recovery of the commodity of interest.

Figure 5.5. Mine bench face of mainly vertical clay pipes (white) exposed in the Yandicoogina CID. Photograph by Erick Ramanaidou.

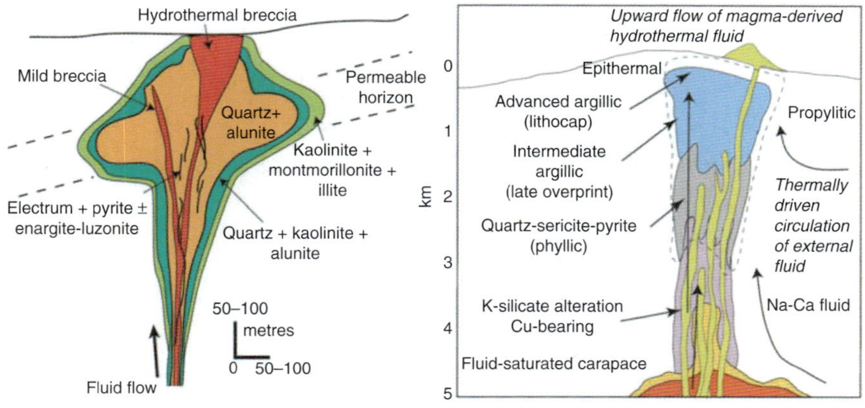

Figure 6.1. Schematic cross-section of a quartz–alunite epithermal (left) and porphyry Cu–Au (right) deposit (Tosdal *et al.*, 2009) showing argillic (i.e. kaolinite + montmorillonite + illite), advanced argillic alteration (i.e. quartz + kaolinite + montmorillonite) and phyllic alteration.

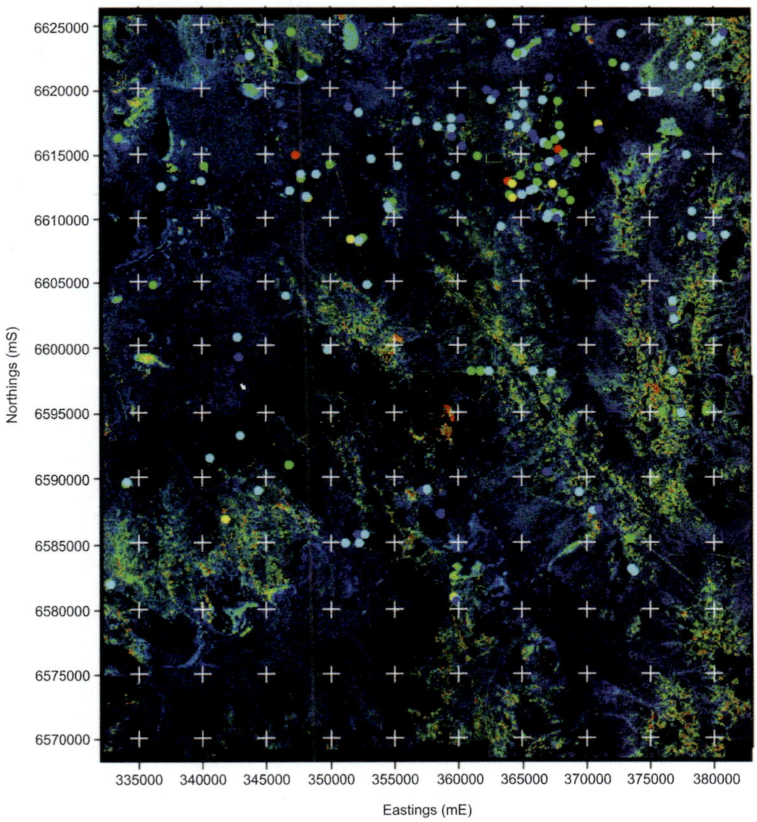

Figure 6.9. Kaolin disorder spectral index (i.e. kaolin crystallinity) derived from airborne hyperspectral (HyMap) data (Cudahy *et al.*, 2005). Coloured points indicate sample locations of field spectrometer and XRD measurements. In both datasets, the kaolin disorder is displayed using a rainbow stretch, with blue colours showing low crystalline order and red colours showing high crystalline order. Black pixels in the map are masked out due to absence of kaolinite or high influence of non-mineralogical material (e.g. vegetation).

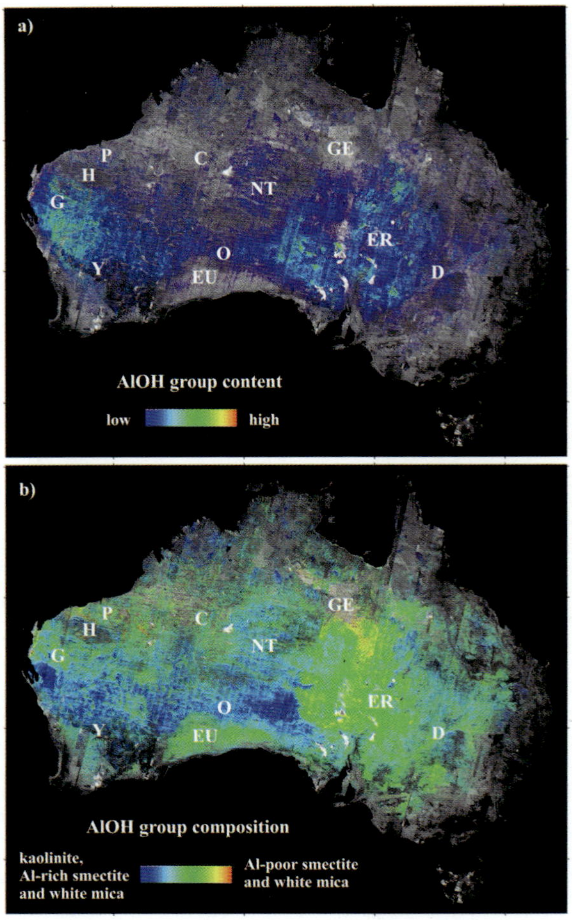

Figure 6.8. Continental-scale ASTER Geoscience Product showing geological provinces in the AlOH group content (a) and composition (b) maps (Cudahy *et al.*, 2012). Pixels containing no colour information are masked out due to absence of mapped material or significant non-mineral related influence (e.g. clouds). C – Canning Basin, D – Darling River; EU – Eucla Basin, ER – Eromanga Basin, G – Gascoyne Province, GE – Georginia Basin, H – Hamersley Basin, O – Officer Basin, NT – Northern Territory Provinces (e.g. Amadeus Basin, Arunta Inlier), P – Pilbara, Y – Yilgarn Craton.

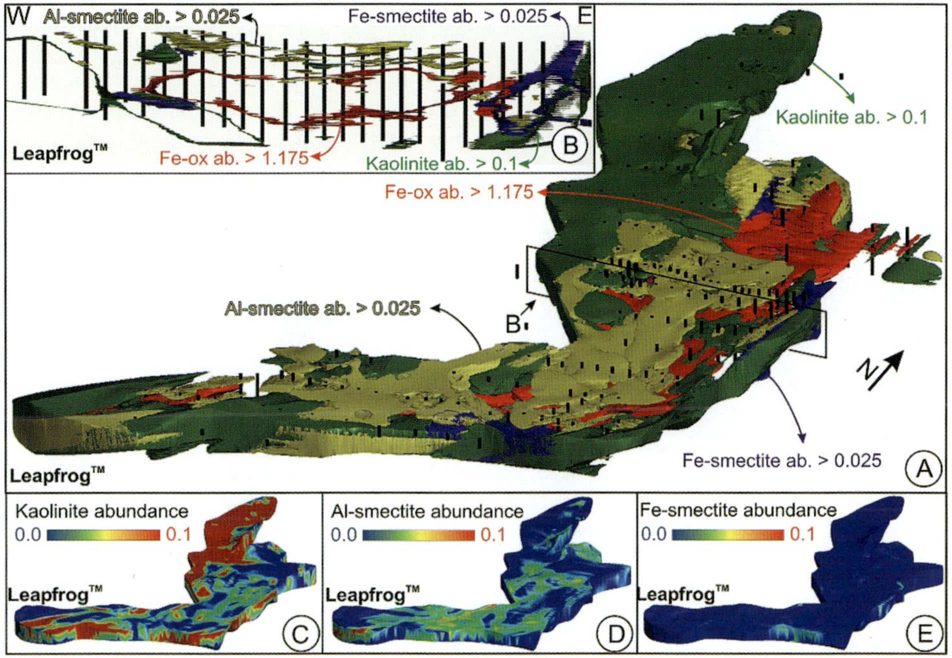

Figure 6.11. Three-dimensional representation of the Rocklea Dome CID (Western Australia) showing relative mineral abundances derived from HyLogging data using a multiple-feature extraction method modified after Haest *et al.* (2012a). (A) Shells showing the Fe (oxyhydr-)oxide (Kruse, 1988), kaolinite (green), Al-smectite (yellow) and Fe-smectite (purple) abundances. (B) East–west profile through the centre of the CID (view from south-west), showing the distribution of the clay shells, relative to the zone with spectrally derived Fe (oxyhydr-)oxide abundances. (C–E) Three-dimensional volume outlining all RC drill cores scanned at Rocklea Dome, evaluated as a function of the different spectrally defined clay types including kaolinite (C), Al-smectite (D) and Fe-smectite (E). Colour stretches showing increasing relative abundance of clay species from blue to red (i.e. 0.0 to 0.1). For setup of applied algorithms and thresholds see Haest *et al.* (2012a, 2012b).

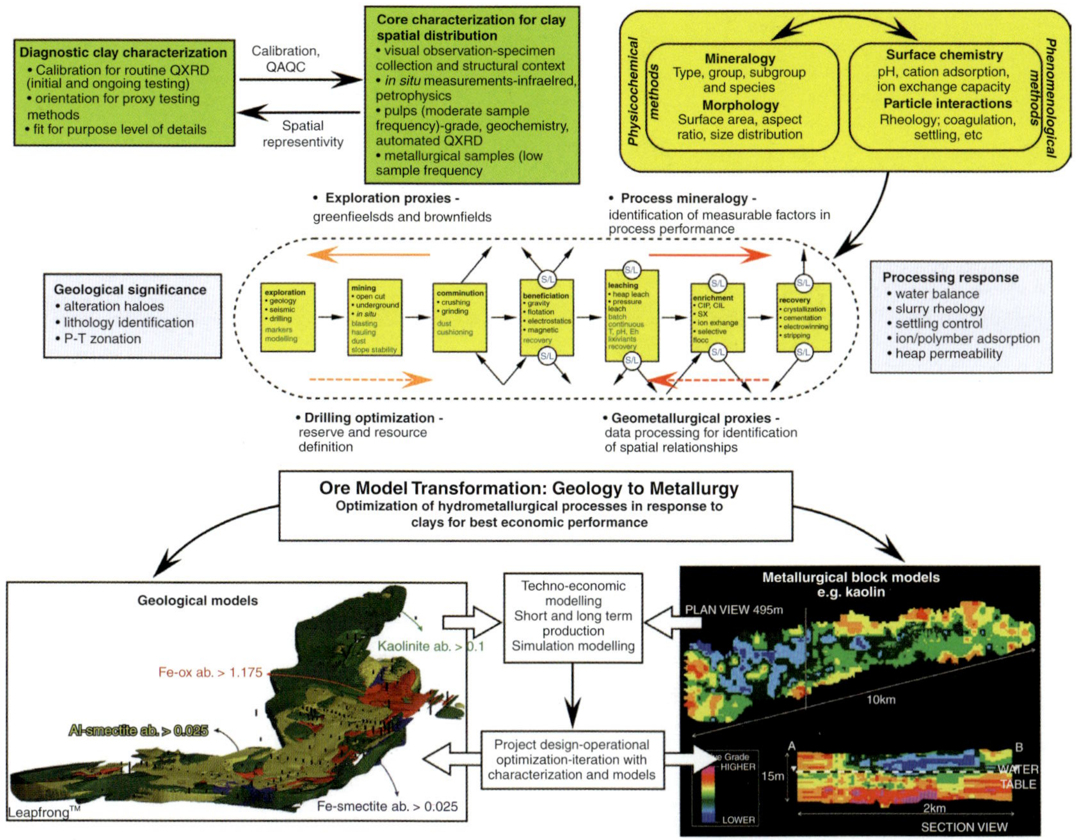

Figure 10.1. Integrated characterization of clays across the minerals value chain. Since clays will inevitably affect different unit processes, thorough diagnostic characterization of the clay mineral species present in the ore is an essential foundation for beginning and sustaining quantitative and qualitative analyses of ores and their processing behaviour. Analysis of drill core sections at various length scales provides the basis for modelling the spatial distribution of ore properties and their processing responses. Ongoing developments in clay-sensitive measurement methods such as XRD and infrared spectrometry can provide cost-effective data at high sampling frequency. Linking clay mineral content of ores to observed processing behaviour is aided by knowledge of the clay species present. As mining and processing operations for the deposit progress, the feedback of understanding gained leads to ongoing efficiency improvements.

also different possibilities for occupying octahedral sites in the octahedral layer, there are several ways in which the layer can stack to form a three-dimensional structure (Bailey, 1980). Six unique, standard structures (i.e. polytypes) are recognized for the micas (Smith and Yoder, 1955). The most common nomenclature used to distinguish the polytypes is the Ramsdell-style notation that consists of a three-part symbol (Guggenheim *et al.*, 2011). The first part is a number indicating the number of T–O–T layers in the repeating unit in the stacking direction; the second part is a symbol to denote the lattice symmetry; and the third part is number subscripts used to distinguish between polytypes with the same repeat and lattice symmetry. The six ideal mica polytypes are 1M (M = monoclinic), $2M_1$, $2M_2$, 3T (T = trigonal), 2O (O = orthorhombic) and 6H (H = hexagonal). In reality, a stack of T–O–T layers can contain many stacking faults where the structure switches from one polytype to another. The density of stacking faults can be so high that there is complete disorder in the stacking sequence. Such a stacking sequence is denoted $1M_d$. The identification of mica polytypes can be of considerable importance because the dominance of one polytype over another can be an indication of the genesis of the mica (e.g. Bailey *et al.*, 1962). X-ray diffraction provides one of the few methods for distinguishing polytypes.

The positioning of the interlayer cations relative to the di-trigonal cavities provides a rigid constraint on the stacking of the layers of high-layer-charge sheet silicates. The stacking of layers in low-layer-charge sheet silicates and even in no-layer-charge sheet silicates is also not without constraint. For the electrostatically neutral 2:1 layer silicates (e.g. talc and pyrophyllite), the successive layers position themselves to minimize the repulsion between Si atoms in the tetrahedral sites of successive T–O–T layers. This positioning means that there is a large number of possible polytypes for these structures (Bailey, 1980), but the weak bonding between layers gives rise to highly disordered stacking sequences so that only the 1A (A = anorthic = triclinic) structure for talc and 1A and 2M structures for pyrophyllite are found in nature. Similarly for the electrostatically neutral 1:1 sheet silicates (e.g. kaolinite and lizardite) each oxygen atom of the tetrahedral sheet of one T–O layer pairs with an OH group of the next T–O layer. This leads to 12 standard polytypes for tri-octahedral 1:1 sheet silicates and, because of the positioning of the vacant octahedral site in di-octahedral sheet silicates, numerous polytypes for di-octahedral sheet silicates (Bailey, 1988a). In reality only a few 1:1 sheet silicate polytypes are found in nature. For example, the only polytypes found in the kaolin group minerals are kaolinite-1A, dickite-$2M_1$ and nacrite-$2M_2$. Invariably, the sheet silicates with electrostatically neutral layers have a high density of stacking faults for which the notation $1M_d$ is used.

Smectites have a small layer charge (–0.2 to –0.6 per formula unit) so the required interlayer cations are less numerous and they are thinly spread throughout the interlayer. Furthermore, in smectites elemental substitutions may be in both the tetrahedral and octahedral sheets (while in the vermiculites elemental substitutions are mainly in the tetrahedral sheet). The low-charge, highly variable elemental composition and structural variations within T–O–T layers make generalizations about smectite structures difficult. Certainly, smectites that have a layer charge principally arising from substitutions in the tetrahedral layer (e.g. beidellite and saponite) can take 1M and $2M_1$ structures (Güven, 1988; Méring, 1975). This tends to be the exception rather than the rule for smectites and the structure of the majority of smectites can be classified in two ways: (1) semi-random stacking of T–O–T layers with 60° rotations between layers; and (2) completely random rotations and translations between layers (i.e. turbostratic stacking). One broad generalization about smectite structures is that the interlayer cations will try to position themselves close to charge centres in the T–O–T layer.

With vermiculites the higher layer charge derived from substitutions in the tetrahedral layer (–0.6 to –0.9 per formula unit) leads to more predictable structural behaviour. The interlayer cations are strongly attracted to the substituted tetrahedral sites and, in structurally ordered vermiculites, will sit directly atop these sites (Bailey, 1980; de la Calle and Suquet, 1988). When relative humidity conditions are such that divalent interlayer cations are octahedrally coordinated by water molecules, the hydrogen atoms of the water molecules pair with the oxygen atoms of the substituted tetrahedral sheet (Bailey, 1980; Skipper *et al.*, 1991; Wilson, 2013). This arrangement leads to an incomplete octahedral sheet of water molecules in the interlayer. Although it is possible to postulate a number of polytypes for vermiculites, only the 1M, $2M_1$ and $1M_d$ structures are found in nature (de la Calle and Suquet, 1988; Wilson, 2013). However, there is evidence of an ordering of the substituted tetrahedral sites in some vermiculites (Shirozu and Bailey, 1966) and short-range ordering of the interlayer cations (Alcover *et al.*, 1973). Vermiculites can be formed by the alteration of tri-octahedral micas (e.g. phlogopite and biotite) and chlorites (see following). When a vermiculite is derived from a mica, the form of the parent material can often be determined by exchanging the divalent interlayer cation of vermiculite with K^+, thereby reverting the structure to a form that closely mimics the structure of the original mica (Bailey, 1980; Barshard, 1948; de la Calle and Suquet, 1988).

The presence of water in the interlayer of a sheet silicate structure is not restricted to smectites and vermiculites. Other examples of sheet silicates that contain interlayer water are kerolite, a hydrated, Ni-bearing talc $[(Mg,Ni)_3Si_4O_{10}(OH)_2(H_2O)_n]$ and halloysite $[Al_2Si_2O_5(OH)_4.2H_2O]$ (Churchman *et al.*, 2016; Joussein *et al.*, 2005). Halloysite has the same electrostatically neutral T–O layer as kaolinite but with water molecules between successive T–O layers. The interlayer water molecules in halloysite take two forms: one located atop and slightly buried into the di-trigonal cavities of the tetrahedral layers and the other located atop the tetrahedra occupied by Si in the tetrahedral layer (Giese and Costanzo, 1987). The presence of water in the interlayer allows strain between the tetrahedral and octahedral sheets to be compensated by a curving of the T–O layers rather than by a rotation of the Si-occupied tetrahedra (Singh, 1996). This means that halloysite predominantly takes a tubular form, with other morphologies (spherical, blocky, etc.) also found (Churchman *et al.*, 2016; Joussein *et al.*, 2005).

Curved and tubular forms can also be found in the tri-octahedral 1:1 sheet silicates, namely the serpentines minerals (Wicks and O'Hanley, 1988). There is no interlayer water in the serpentines and the curvature is attributed to misfit between the octahedral and tetrahedral sheets. Serpentines take three forms: lizardite, chrysotile and antigorite. Lizardite has a platy morphology and has already been introduced as the tri-octahedral analogue of kaolinite. Chrysotile shows tubular morphology (Fig. 2.3 in Chapter 2) and is recognized

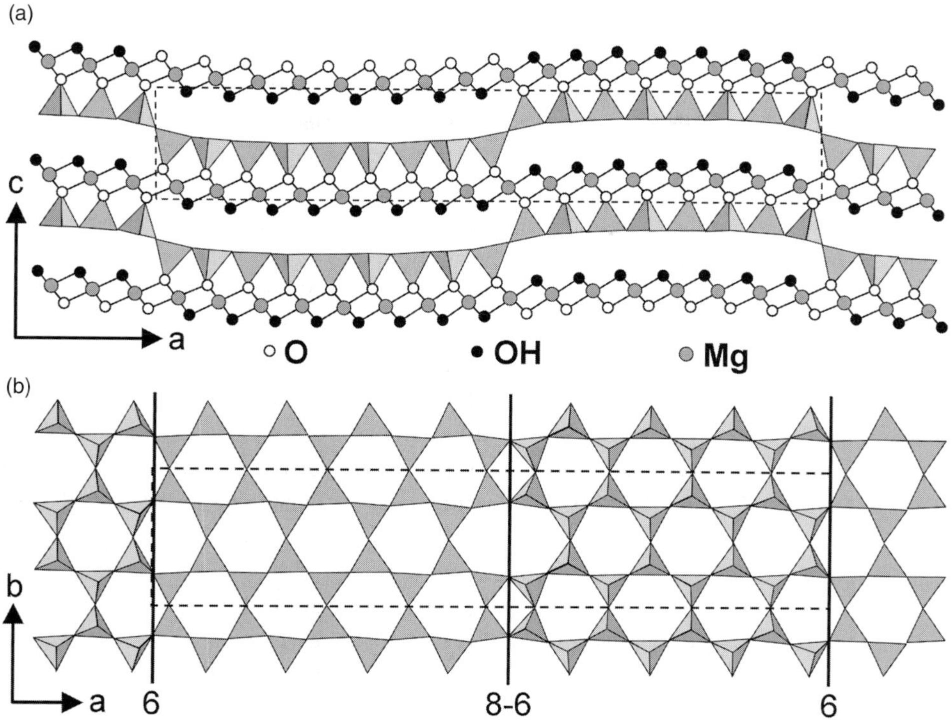

Figure 4.1. Schematic of the antigorite m = 17 polysome. (a) view along b axis and (b) view along c-axis. The view shown in (b) is the generally accepted form of the tetrahedral reversals, which are indicated by the solid lines. The O/OH entities at the bases of tetrahedra are not shown in (a). Dashed lines in the figures indicate the unit cell.

to have three structural forms: two clino ($2M_{c1}$ and $1M_{c1}$; subscript c indicates cylindrical) forms and an ortho ($2O_{c1}$) form. Description of a second orthorhombic form, parachrysotile, was disallowed in 2006 by the International Minerals Association (IMA). Instead of the use of ortho and clino for chrysotile, the IMA recommend using chrysotile followed by the polytype symbol (e.g. chrysotile-$2M_{c1}$) when describing chrysotile. Transmission electron microscopy (TEM) shows that chrysotile assumes either a ring or spiral structure (Wicks and O'Hanley, 1988). Antigorite has a repeating wave structure where the octahedral and tetrahedral sheets are continuous but the tetrahedral sheets reverse polarity every half-wave (Fig. 4.1). Figure 4.1 shows a schematic of the antigorite structure where the wave repeat distance contains 17 tetrahedral sites and 16 octahedral sites. This structure is described as the antigorite m = 17 polysome based on the number of tetrahedra making up the repeat distance. Other polysomes are observed in nature, with most having values between m = 10 and m = 20 (Bailey, 1980). Necessarily, the number of octahedra in the repeat distance for an m polysome is m-1. Thus, for the m = 17 polysome the ideal structural formula is $Mg_{48}Si_{34}O_{85}(OH)_{62}$. Transmission electron microscopy studies show that natural antigorites often consist of intergrowths of different polysomes.

The exact structure of antigorite at the tetrahedra reversals is the subject of debate. Figure 4.1b shows the generally accepted structure at the reversals. This structure involves alternating planes of what are described as '6-reversals' and '8-reversals' (Capitani and Mellini, 2007). The '6-reversals' consist of rings of six tetrahedra with four tetrahedra pointing in one direction and two pointing in the other direction. The '8-reversals' consist of rings of eight tetrahedra alternating with rings of four tetrahedra. In the rings of eight tetrahedra, four tetrahedra point in one direction and four tetrahedra point in the opposite direction; in the rings of four tetrahedra, two point in one direction and two point in the opposite direction. The occupied octahedral positions in the structure are shifted by b / 2 at each reversal. For the m = even polysomes, Capitani and Mellini (2007) show that the rings of tetrahedra in the '8-reversals' are shifted by half a unit cell (b / 2 shift) at every second '8-reversal'. This doubles the reversal distance for m = even polysomes and gives rise to a c-centred unit cell.

Dódony and co-workers (Dódony and Buseck, 2004; Dódony *et al.*, 2002) propose a different structure for the reversals in antigorite polysomes. The structure proposed by these workers has only alternating planes of '6-reversals' (i.e. no '8-reversals') with the octahedrally coordinated atoms having a b / 2 shift at every second '6-reversal'. Dódony *et al.* (2002) also recognize the existence of monoclinic and triclinic polytypes of antigorite. These polytypes are dependent on the number of tetrahedra in the antigorite wave repeat and the presence or absence of b / 3 shifts between successive 1:1 (T–O) layers.

Thus far, discussion on clay structure has concentrated on the 1:1 and 2:1 sheet silicates. Another important class of clay minerals is the 2:1:1 sheet silicates, the chlorites (see Figs 1.4 and 1.5 in Chapter 1). Chlorite structures consist of T–O–T layers joined by an octahedral interlayer. The octahedral interlayer consists of two planes of OH groups in a close-packed arrangement, with the octahedral sites occupied by divalent or trivalent metals. For example, the ideal formula for clinochlore is $Mg_5Al(Si_3Al)O_{10}(OH)_8$, signifying one tetrahedrally coordinated Al atom, one octahedrally coordinated Al atom and five octahedrally coordinated Mg atoms per formula unit. In the clinochlore structure, the octahedrally coordinated Al and Mg atoms are partitioned between the octahedral sites of the T–O–T layer and octahedral interlayer. Thus, in terms of crystal structure, the formula for clinochlore can be written as $(Mg_xAl_{3-x})(OH)_6$–$(Mg_{5-x}Al_{x-2})_3(Si_3Al)O_{10}(OH)_2$ where necessarily $x \geq 2$. Based on this formula, it is theoretically possible to devise a structure where both the T–O–T layer and octahedral interlayer are electrostatically neutral, that is $Mg_3(OH)_6$–$(Mg_2Al)(Si_3Al)O_{10}(OH)_2$. Because of the electrostatic neutrality of the layers, the term 'talc-like layer' is often used to describe the T–O–T layer and the term 'brucite-like layer' used to describe the octahedral interlayer. However, such a structure has not been found in nature, with all natural clinochlore specimens having charged T–O–T layers and octahedral interlayers. Indeed, from a review of experimental evidence from a wide range of sources, Bailey (1988b) suggests that the stable form of clinochlore has the formula $(Mg_2Al)(OH)_6$–$Mg_3(Si_3Al)O_{10}(OH)_2$. That is, the T–O–T layer carries a charge of –1 per formula unit and the octahedral interlayer carries a charge of +1 per formula unit. In such a case the T–O–T layer is equivalent to the T–O–T layer of the true mica phlogopite.

Consequently, the term 'talc-like' layer is somewhat misleading and the term 'mica-like' layer, as is used by some, may be more appropriate.

Within the chlorite group there is a wide range of elemental substitutions so that a general formula for chlorite minerals is $(R^{2+}_{6-x-3y}R_o^{3+}{}_{x+2y}\square_y)(Si_{4-x}R_t^{3+}{}_x)O_{10}(OH)_8$ where □ indicates vacant octahedral sites in either the T–O–T layer or the octahedral interlayer. R_o^{3+} indicates an octahedrally coordinated entity of which Al^{3+} and Fe^{3+} are the most common, but which can also be Cr, Ni and Mn. R_t^{3+} indicates a tetrahedrally coordinated entity that generally is Al^{3+} but occasionally is Fe^{3+}. The most common divalent octahedral entities (R^{2+}) are Mg and Fe with Mn, Ni and Zn also possible. Fe^{2+} can readily substitute for Mg with chamosite $[(Fe^{2+},Mg,Fe^{3+})_5Al(Si_3Al)O_{10}(OH)_8]$ being the Fe endmember equivalent of clinochlore. Although there is no universal evidence, and it certainly is possible to find chlorites where the T–O–T layer charge is not equal to –1 (and consequently the octahedral interlayer charge is not equal to +1) per formula unit, single crystal studies using X-ray diffraction, Mössbauer and infrared spectroscopies tend to confirm Bailey's hypothesis that the stable form for chlorites occurs when the T–O–T layer charge is –1 per formula unit (see, for example, the refined crystal structure for chlorites listed in the International Centre for Diffraction Data database: www.icdd.com). Even when the amount of Si and Al in the tetrahedral sites deviates from the ideal 3:1 ratio, the excess layer charge is at least partially compensated by an oppositely charged octahedral T–O–T layer.

In clinochlore and chamosite, both the T–O–T layer and the octahedral interlayer are tri-octahedral and so these chlorites are known as tri-octahedral chlorites. In nature it is possible to find chlorites where both the T–O–T layer and the octahedral interlayer are di-octahedral. For example, donbassite $[Al_{4+x/3}(Si_{4-x}Al_x)O_{10}(OH)_8]$ has Al in the octahedral sites of the T–O–T layer and the octahedral interlayer. These chlorites are described as di-octahedral chlorites or di, di-octahedral chlorites. Chlorites with a di-octahedral T–O–T layer and a tri-octahedral interlayer are described as di, tri-octahedral chlorites following the conventional nomenclature that the form of the T–O–T layer is placed first in the chlorite descriptor. Examples of di, tri-octahedral chlorites are sudoite $[Mg_2Al(OH)_6{\cdot}Al_2(Si_3Al)O_{10}(OH)_2]$ and cookeite $[LiAl_2(OH)_6{\cdot}Al_2(Si_3Al)O_{10}(OH)_2]$. The only known mineral described as a tri-di-octahedral chlorite is franklinfurnaceite $[Ca_2(Fe^{3+},Mn3^{+})_2(OH)_6{\cdot}Mn^{2+}{}_3(Zn_2Si_2)O_{10}(OH)_8]$. Although this mineral has a tri-octahedral T–O–T layer and a di-octahedral interlayer, there are also Ca atoms positioned between each T–O–T layer and octahedral interlayer.

In many ways the chlorite structures are similar to the vermiculite structure, but with a complete interlayer structure rather than the partially occupied interlayer found in vermiculites. Because both the T–O–T layer and the octahedral interlayer carry a charge and because of the OH entities of the octahedral interlayer, the bonding between chlorite layers is a combination of electrostatic and hydrogen bonding. Brown and Bailey (1962) demonstrate that there are two possible configurations for OH positions, plus two possible configurations for the positioning of the octahedral cations in the octahedral interlayer. In the one-layer polytypes, subsequent T–O–T layers can take any of six positions on the octahedral interlayer, leading to 24 possible one-layer structures. This number reduces to 14 after removal of structures related by a 180° rotation about the perpendicular axis to the direction of the stagger between the two tetrahedral sheets of the T–O–T layer. A further reduction to 12 unique polytypes occurs when enantiomorphic (mirror image isomer) structures are taken into account. Similar to the micas, a Ramsdell-style three-symbol notation for chlorites (Guggenheim *et al.*, 2011) is used to distinguish the one-layer chlorite polytypes. The first symbol (I or II) indicates the orientation of the interlayer sheet relative to an adjoining T–O–T layer. The second symbol (italic *a* or *b*) indicates the positioning of the cations in the octahedral interlayer relative to the hexagonal ring of the tetrahedral sheet of the adjoining T–O–T layer. The final symbol (1 to 6 separated by a dash from the second symbol) indicates the positioning of the subsequent T–O–T layer on the octahedral interlayer (Bailey, 1988b).

As is the case with most sheet silicates, the chlorites often have an irregular stacking of layers. In the chlorites the stacking will be such that optimum electrostatic and hydrogen bonding is maintained between the T–O–T layers and the octahedral interlayers. This limits the possible stacking arrangements for the disordered structures to six general types, of which the IIb-even type is the most abundant in nature (Brown and Bailey, 1962). The term 'even' in the Ramsdell-style notation is used to indicate that, although the stacking sequence is random, T–O–T layers occupy only the even sites of the six possible stacking sites (Bailey, 1988b).

The similar basic structural building blocks of the sheet silicates and the fact that many sheet silicates are the weathering or alteration products of other sheet silicates means that it is reasonably common to find sheet silicates where the stacking sequences are a composite of the basic layer structure of two or more individual sheet silicates. Such composite structures are termed interstratified. For example, rectorite is a regular interstratification of mica and montmorillonite, where the stacking sequence is such that mica and montmorillonite layers alternate to form the overall rectorite structure (Bradley, 1950; Brindley, 1956). Reynolds (1980) gives a list of common, naturally occurring interstratified materials (see also Table 4.6). Interstratified layer silicates are also reviewed by Sawhney (1989).

Two minerals that, strictly speaking, are not clays (i.e. sheet silicates) but that fall within the more general grouping of clay minerals are palygorskite and sepiolite. These minerals have continuous tetrahedral oxygen sheets but the octahedral O/OH sheets are discontinuous so that the T–O–T units form ribbons rather than sheets (Bailey, 1980; Jones and Galan, 1988). The structural difference between sepiolite and palygorskite is the width of the T–O–T (2:1) ribbons. In sepiolite the ribbons are eight octahedral cation positions wide while in palygorskite the ribbons are five octahedral cation positions wide. The structure of the 2:1 ribbons in palygorskite and sepiolite gives rise to a series of channels running parallel to the length of the 2:1 ribbons. In nature these channels contain 'zeolitic' water, most of which is removed by heating to around 150 °C (Post and Heaney, 2008; Post *et al.*, 2007). In general, sepiolite has mainly Mg in octahedral sites while palygorskite has a significant amount of Al in octahedral sites (Wilson, 2013). Palygorskite and sepiolite have a thin, fibrous habit that imparts strong gelling properties in aqueous suspensions even at relatively low solids content. The strong gelling properties of palygorskite and sepiolite can cause serious issues for solid–liquid separation in mineral processing facilities.

4.2.2 Selected Properties

4.2.2.1 Effect of Interlayer Species

The relatively sparse distribution of the layer-charge centres in smectites and vermiculites means that it is electrostatically advantageous for the interlayer cations to be hydrated. The hydration of the interlayer cations acts to diffuse the charge from both the cations and the layer-charge centres. The degree of hydration of the interlayer cations depends on a number of factors including layer charge, water activity and the species of interlayer cation. For example, when a smectite or vermiculite is in a low relative humidity (RH) environment, the degree of hydration of the interlayer cations is low; in high RH environments the degree of hydration of the interlayer cations is high. The variation in the hydration state of the interlayer cations is manifest as changes in the layer spacing. This, in turn, is seen as a macroscopic swelling or shrinking of the clay.

Over the full range of RH (0–100 per cent), changes in layer spacing of smectites and vermiculites occur in a stepwise fashion (MacEwan and Wilson, 1984). At low RH (below 5 per cent) or when heated to sufficiently high temperatures the interlayer water is almost completely removed and the T–O–T layers of the smectites and vermiculites form a mica-like structure with a layer repeat of 0.95–1.0 nm. As the RH increases, the layer repeat distance jumps to a value between 1.2 and 1.3 nm. Further increase in RH causes the layer repeat distance to jump to a value between 1.48 and 1.6 nm. These increases in layer repeat distance are caused by an increasing degree of water coordination around the interlayer cations and are classified as crystalline swelling. The specific RH values where the jumps in layer repeat distances occur are dependent on interlayer cation and layer charge (MacEwan and Wilson, 1984; Mooney *et al.*, 1952). The 1.2–1.3 nm spacing corresponds to a single layer of interlayer water (one-layer hydrate), while the 1.48–1.6 nm spacing corresponds to two layers of water (two-layer hydrate) in the interlayer. For some interlayer cations (e.g. Li^+ and Na^+) three- and four-layer hydrate configurations can occur in montmorillonite (Norrish, 1954). Although the layer repeat increases in a stepwise fashion, the amount of water in the interlayer varies with RH in continuous fashion. The steps in the layer repeat distance correspond to changes in the coordination of water around the interlayer cations. At RH values between the steps, water populates the gaps between the interlayer cations. For smectites this is manifest by small increases in the layer repeat distance after each step as RH increases.

The RH at which a step in the layer repeat distance takes place is dependent on layer charge, with smectites expanding at lower RH values than vermiculites. The high layer charge of the vermiculites means that several water-layer 'sub-structures' can be observed (de la Calle and Suquet, 1988). Within the smectite group, montmorillonites expand at lower RH values than beidellites and saponites. Additionally, the stepwise increases in the basal spacing of smectites occur at higher RH values when the interlayer cation is monovalent than when the interlayer cation is divalent (MacEwan and Wilson, 1984).

In moist conditions (i.e. in the presence of free water) and depending on the interlayer cations, smectites and vermiculites can expand to very large layer repeats. Norrish (1954) showed that the layer repeat distance for Na-saturated montmorillonite is inversely proportional to the square root of the Na^+ concentration in solution and that at very low Na^+ concentrations the layer repeat can exceed 13 nm. Similar behaviour can be observed for some K^+- and Li^+-saturated smectites, but this behaviour is not observed for smectites saturated with di or trivalent cations. The swelling of smectites under conditions where there is free water is classified as osmotic swelling. In a constrained environment, osmotic swelling of consolidated montmorillonites can generate swelling pressures of up to 10 atm (Mielenz and King, 1955). Consequently, the swelling properties of montmorillonites combined with their high surface area and high absorbance makes these clays ideally suited to being used as containment barriers. In a non-constrained environment (e.g. an agitated suspension) the clay will disperse. The long-range bonding interactions that occur in aqueous suspensions of smectites plus the sheet structure of the smectites imparts these clays with strong gelling and poor flocculation properties (see Chapter 2). Both properties are detrimental when attempting to remove smectites from suspension, as is the case with mine tailings.

Some other layer silicates (e.g. halloysite) also have interlayer water as an integral part of their structure. Unlike the smectites and vermiculites, these layer silicates do not show the reversible hydration–dehydration behaviour. Once the interlayer water is removed, the clay structure remains in a collapsed state even in the presence of excess water. However, these collapsed structures can often be expanded by treatment with highly polar liquids such as hydrazine and formamide. Churchman *et al.* (1984) use a formamide treatment to distinguish dehydrated halloysite from kaolinite and to quantify the relative amounts of these minerals in mixtures. Similar methods for distinguishing dehydrated halloysite from kaolinite have been developed using hydrazine and potassium acetate (Theng *et al.*, 1984).

Water is not the only liquid that will move into the interlayer region of smectites and vermiculites. These clays can be intercalated by many other polar or charged molecules. Ethylene glycol and glycerol are often used for the characterization of smectites and vermiculites (MacEwan and Wilson, 1984). Ethylene glycol intercalated with smectites or vermiculites will expand all but the highest layer charged vermiculites to a layer repeat of around 1.7 nm, irrespective of the interlayer cation. Glycerol intercalation of Mg-saturated smectites and vermiculites can be used to distinguish smectites from vermiculites. Intercalation of smectites with n-alkyl ammonium species can be used to determine the layer charge of 2:1 layer silicates (for example, see Mermut and Lagaly, 2001).

4.2.2.2 Cation Exchange Capacity and Surface Area

All materials when placed in aqueous solution will have some affinity for cations. The number of cations that attach to a material while in solution is a measure of the negative component of the material's surface charge. When there is a sufficient concentration gradient, the cations that are attached to a material's surface can be exchanged. For example, Na^+ ions attached to a surface can be exchanged by Ca^{2+} ions when there is a sufficiently high concentration of Ca^{2+} ions in solution relative to the concentration of Na^+ ions. Cation exchange capacity (CEC) is a measure of the total amount of cations that can be exchanged

Table 4.2. *Typical CEC and surface area (N_2 BET) values for some clays (Dixon and Weed, 1989; Grim, 1968; McBride, 1994; Sumner, 2000; Wilson, 2013).*

Clay	CEC ($cmol_c^+ kg^{-1}$)	Surface area $m^2 g^{-1}$
Talc	0–2	1–10
Kaolinite	3–15	5–20
Sepiolite *	5–15	200–360
Palygorskite *	5–20	100–250
Chlorite	10–40	0–5
Allophane	10–40	140–180
Halloysite	5–50	50–70
Illite	20–40	50–200
Montmorillonite *	80–150	30–130
Vermiculite *	100–150	20–50

* Surface area can be significantly higher when measured with smaller molecules (e.g. water and NH_3).

from the surface of a material. Typical CEC values for clays are given in Table 4.2. Smectites and vermiculites have a high CEC because, for these minerals, the interlayer cations can be displaced by other cations from solution and therefore are exchangeable. Consequently, for the vermiculites and smectites, CEC is a measure of the exchange capacity on both internal and external surfaces. For all clays, CEC is usually quoted in units of centimole of positive charge ($cmol_c^+$) per kilogram of sample. Older texts may use the equivalent but non-SI unit of milliequivalents (meq) per 100 grams. When measuring CEC, it is important to know the initial hydration state of the clay material because, as CEC is a measure of exchangeable cations per unit weight, the degree of hydration of the clay will alter the measured CEC value. This is particularly the case for smectites, for which the amount of interlayer water is dependent on the interlayer cation and RH. Generally, CEC values are referenced against a standard state such as the weight of the clay after drying at 105 °C. Drying at 105 °C will remove the majority of the interlayer water as well as any adsorbed water on the surfaces of clay particles.

Generally CEC is determined using the methods developed for soils (for example, Rayment and Lyons, 2011). These methods involve the removal of soluble salts from the clay and then completely exchanging the clay with a suitable cation (often NH_4^+). The sample is then washed and again completely exchanged with a second cation (often Na^+). The CEC is determined by measuring the amount of the first cation displaced by the second cation (e.g. by measuring the concentration of the first cation in solution after the clay has been exchanged by the second cation). Although this method of double cation exchange is widely used for the measurement of CEC, it is recognized that the method has significant problems (Sumner and Miller, 1996). This is particularly the case for smectites and vermiculites where the cation exchange processes used in the method are often incomplete (Verburg and Baveye, 1994), leading to an underestimation of the CEC.

When the measurement of soluble salts is not required, an alternate method of measuring CEC is to Ba-saturate the clay and use XRF analysis to measure the amount of Ba intercalated into the clay (Battaglia *et al.*, 2006). When using Ba saturation, it is necessary to remove carbonates and sulphates from the clay material, otherwise Ba carbonate and sulphate will be precipitated and the CEC is overestimated.

Another method used to estimate CEC is the absorption of methylene blue. This is a colourimetric method often used in industrial applications involving bentonites. For example, methylene blue absorption is used as a field test for drilling muds (American Petroleum Institute, 2009) and for estimating the CEC of soils (Wang *et al.*, 1996). In practice, the methylene blue absorption method can overestimate CEC by as much as 20 per cent. This overestimation of CEC is attributed to the absorption of methylene blue dimers onto the clay's internal and external surfaces (Bergmann and O'Konski, 1963). Thus the term methylene blue capacity (MBC) is used to distinguish the exchange capacity determined by the methylene blue method from the true CEC.

Values of CEC for smectites and vermiculites are very high (Table 4.2) because the presence of labile water in the interlayer allows cations in the interlayer to be exchanged. Thus, the CEC of smectites and vermiculites is an indirect measure of the layer charge of these clays.

In general, all clays have a high specific surface area because of their inherent small particle size and platy morphology (Table 4.2). The most commonly used technique for measuring surface area is based on the analysis of absorption isotherms of a gas or vapour using the techniques developed by Brunauer *et al.* (1938). This technique is known as the BET method and N_2 is most often used as the absorbing gas. For clays, surface areas measured by the BET method can be highly dependent on whether the absorbing gas or vapour can access the interlayer region of the clay (see, for example, Michot and Villiéras, 2006). This is particularly the case for the expanding clays (smectites and vermiculites) as well as sepiolite and palygorskite. In general, N_2 molecules cannot access the interlayer regions of expanding clays or the structural tunnels of sepiolite and palygorskite; consequently N_2 BET measurements give a relatively low value of surface area for these clays (Table 4.2). Gases consisting of small molecules such as water vapour and ammonia can access the interlayer regions of expanding clays and the structural tunnels of sepiolite and palygorskite and thereby give a much greater value for the measured surface areas of these clays. When a gas that can access the interlayer regions or structural tunnels is used, measured surface areas for montmorillonite and vermiculite can be as high as 800 $m^2\ g^{-1}$ and 700 $m^2\ g^{-1}$, respectively, while for sepiolite and palygorskite the measured surface areas can be as high as 750 $m^2\ g^{-1}$ and 700 $m^2\ g^{-1}$, respectively (Cases *et al.*, 1991; Wilson, 2013).

Cation exchange of clays to make them homo-ionic is useful when quantifying the amount of various clays in a mixed sample. Most natural smectites and vermiculites have a mix of interlayer cations that leads to variable degrees of hydration and, therefore, variable layer repeat distances. Exchanging the interlayer cations so that swelling clays are homo-ionic makes the interlayer repeat distances of these clays more regular. This is particularly the case for quantitative X-ray diffraction (XRD) where exchange with Ca^{2+} or Mg^{2+} is commonly used.

4.2.3 Associated Minerals

Clays are rarely found in a pure form and there is often an intimate association between clays and other fine-grained minerals. For example, kaolinite and the iron oxyhydroxide goethite are often found in association. Although these associated minerals are not the prime interest for this chapter, it is important to be aware of their existence and possible presence when identifying and characterizing clays. Table 4.3 lists some of the minerals more commonly found in association with clays.

4.3 Extraction of Clays from a Mineral Assemblage

Small amounts of clay in a mineral assemblage can have a marked effect on bulk properties such as rheology, sedimentation and hydraulic conductivity (e.g. smectite in oil shales

Table 4.3. *Minerals commonly associated with clays.*

	Mineral name	Formula	Reference
Silica minerals			**29** (1994)
	Quartz	SiO_2	
	Cristobalite	SiO_2	
	Tridymite	SiO_2	
	Opal (-C,-CT,-A)*	$SiO_2.nH_2O$	
Feldspars			**2** (1983)
	K-feldspars (alkali feldspars)	$KAlSi_3O_8 - NaAlSi_3O_8$	
	Plagioclase feldspars	$NaAlSi_3O_8 - CaAl_2Si_2O_8$	
Olivines, pyroxene, amphiboles			**5** (1982), **7** (1980), **9** (1981–1982), **67** (2007)
Iron oxides and oxyhydroxides			Cornell and Schwertmann (2003)
	Hematite	α-Fe_2O_3	
	Goethite	α-FeOOH	
	Lepidocrocite	γ-FeOOH	
	Magnetite	Fe_3O_4	
	Maghemite	γ-Fe_2O_3	
	Ferrihydrite*	~ $Fe_5O_3(OH)_9$	
	Schwertmannite*	~ $Fe_{16}O_{16}(OH)_{12}(SO_4)_2$	

Table 4.3 (*cont.*)

	Mineral name	Formula	Reference
Aluminium oxides and oxyhydroxides			**25** (1991), Hudson *et al.* (2000)
	Corundum	Al_2O_3	
	Böhmite	γ-AlOOH	
	Diaspore	α-AlOOH	
	Gibbsite	γ-$Al(OH)_3$	
	Nordstrandite	β-$Al(OH)_3$	
	Bayerite	α-$Al(OH)_3$	
Titanium and zirconium oxides			**3** (1976), **25** (1991), **53** (2003) Milnes and Fitzpatrick (1989)
	Rutile	TiO_2	
	Anatase	TiO_2	
	Ilmenite	$FeTiO_3$	
	Baddeleyite	ZrO_2	
	Zircon	$ZrSiO_4$	
	Zirconolite	$CaZrTi_2O_7$	
Manganese minerals			Chukhrov *et al.* (1980), Skinner and Fitzpatrick (1992), McKenzie (1989)
	Pyrolusite	MnO_2	
	Birnessite	$Na_4Mn_{14}O_{27}.9H_2O$	
	Lithiophorite	$(Al,Li)MnO_2(OH)_2$	
	Hollandite	$Ba(Mn^{4+},Mn^{2+})_8O_{16}$	
	Todorokite	$(Mn^{2+},Ca,Mg)Mn^{4+}{}_3O_7.H_2O$	
Carbonates			**11** (1983)
	Calcite	$CaCO_3$	
	Aragonite	$CaCO_3$	
	Dolomite	$CaMg(CO_3)_2$	
	Ankerite	$(Ca,Mg,Fe)CO_3$	
	Siderite	$Fe^{2+}CO_3$	
	Rhodochrosite	$MnCO_3$	
	Trona	$Na_3(CO_3)(HCO_3).2H_2O$	
Halides			
	Halite	$NaCl$	
	Sylvite	KCl	
Sulphates			**40** (2000)
	Gypsum	$CaSO_4.2H_2O$	
	Bassanite	$CaSO_4.\frac{1}{2}H_2O$	

Table 4.3 (*cont.*)

	Mineral name	Formula	Reference
	Anhydrite	$CaSO_4$	
	Celestine	$SrSO_4$	
	Jarosite	$KFe_3(SO_4)_2(OH)_6$	
	Natrojarosite	$NaFe_3(SO_4)_2(OH)_6$	
	Alunite	$KAl_3(SO_4)_2(OH)_6$	
	Natroalunite	$NaAl_3(SO_4)_2(OH)_6$	
	Melanterite	$Fe^{2+}SO_4.7H_2O$	
	Siderotil	$Fe^{2+}SO_4.5H_2O$	
	Rozenite	$Fe^{2+}SO_4.4H_2O$	
	Szomolnokite	$Fe^{2+}SO_4.H_2O$	
Sulphides			**1** (1974), **61** (2006)
	Pyrite	FeS_2	
	Pyrrhotite	$Fe_{1-x}S$ (x = 0–0.17)	
	Greigite	$Fe^{2+}Fe^{3+}{}_2S_4$	
	Mackinawite	$(Fe,Ni)_9S_8$	
Phosphates			**48** (2002)
	Plumbogummite	$Pb\,Al_3(PO_4)_2(OH)_6.H_2O$	
	Florencite	$(Ce,La,Nd)Al_3(PO_4)_2(OH)_6$	
	Apatites	$Ca_5(PO_4)_3(F,OH,Cl)$	
Zeolites			**45** (2001)
	Anaclime	$NaAlSi_2O_6.H_2O$	
	Heulandite/ clinoptilolite	$(Na,Ca)_{2-3}Al_3(Al,Si)_2Si_{13}O_{36}.12H_2O$	
	Phillipsite	$(K,Na,Ca)_{1-2}(Si,Al)_8O_{16}.6H_2O$	
Hydroxy-carbonates			Duan and Evans (2006)
	Hydrotalcite	$Mg_6Al_2(CO_3)(OH)_6.4H_2O$	
	Pyroaurite	$Mg_6Fe^{3+}{}_2(CO_3)(OH)_6.4H_2O$	
	Sjogrenite	$Mg_6Fe^{3+}{}_2(CO_3)(OH)_6.4H_2O$	
Poorly crystalline alumino silicates			Wada and Wada (1977), Wada and Yoshinaga (1969), Wada (1989)
	Allophane*	$Al_2O_3(SiO_2)_{1.3-2}(H_2O)_{2.5-3}$	
	Imogolite*	$(Al_2O_3.1.2H_2O)SiO_2.2.3\text{-}2.8H_2O$	

Where a number and date is given as a reference, the reference is to the appropriate volume of *Reviews in Mineralogy and Geochemistry* (Mineralogical Society of America and The Geochemical Society: www.minsocam.org/MSA/RIM). Minerals marked with * are poorly crystalline.

or mine tailings). These aspects are dealt with in detail in Chapters 2, 8 and 9. Invariably the clays in a mineral assemblage will consist of more than one type of clay and may also consist of interstratified clay phases. In order to better identify and quantify the clays in a mineral assemblage it is often necessary to separate (concentrate) the clay from the non-clay materials.

The properties of clays that are exploited to extract clays from a mineral assemblage are the inherently small size of clay particles and, most importantly, the dispersibility of clays in aqueous solutions. For example, the work of Norrish (1954) shows that Na-saturated montmorillonite can be dispersed down to individual structural layers in high-purity water. Clays can be extracted by dispersing the mineral assemblage in an aqueous solution and then collecting the smallest size fraction. Conventionally, the clay fraction is considered to be the less than 2 μm size fraction. This stems from the observation that the vast majority of quartz and feldspar in a soil sample is excluded when the <2 μm size fraction is extracted. However, the extraction of smaller size fractions (<1 μm, <0.5 μm, <0.2 μm and <0.08 μm) may be necessary for specific studies.

The extraction of clays by size separation can be categorized into three stages: (1) dispersion, (2) separation and (3) collection and treatment.

4.3.1 Dispersion

The appropriate methodology for dispersing a mineral assemblage to ensure the liberation of clays is dependent on the starting material. Large particles that do not contain clays or that contain only a minimal amount of clay can be excluded from a mineral assemblage by sieving. For 'rocks', some crushing or grinding is a necessary pre-treatment (see also Chapter 7). The amount of crushing/grinding required is a compromise between over-crushing/grinding and under-crushing/grinding. Over-crushing is where the larger-sized, non-clay minerals are reduced to clay-sized particles; under-crushing results in clays not becoming sufficiently liberated from the rock. In general, a rock can be prepared for dispersion by use of a bench-top jaw crusher with an opening reduced in regular steps down to 0.2 mm. Alternatively, a puck/ring mill can be used provided care is taken not to over-crush the rock by milling for too long a period.

Mineral assemblages that are dried powders or powdered rocks are dispersed in aqueous solution either by mechanical means (i.e. shaking) or ultrasonic treatment. The dispersion of clays may be aided by the use of chemical reagents. If appropriate, treatments such as mild-acid treatment can be used to break up cementing agents. Sodium saturation is commonly used to disperse expanding clays such as smectites and vermiculites. For low layer charge and clays with no layer charge such as kaolinite, poly-ionic dispersants can be used. The most commonly used poly-ionic dispersant is the original formulation of Calgon®,[2] which is a solution of sodium hexametaphosphate pH buffered by the addition of sodium

[2] Calgon® originally consisted of amorphous sodium polyphosphate, but because of environmental problems associated with phosphates, it is now formulated with zeolite and polycarboxylate.

carbonate. The concentrations of sodium hexametaphosphate and sodium carbonate used are not critical, provided the pH of the solution is in the range 8–10 so as to stop the hexametaphosphate hydrolysing to orthophosphate. Concentrations of 1 per cent w/w sodium hexametaphosphate and 1 per cent w/w sodium carbonate are acceptable. When chemical dispersing agents are used it can be necessary to remove excess dispersing agent (e.g. Na^+ cations) by repeated washing in high-purity water. Washing is best achieved by use of high-speed centrifugation or dialysis.

4.3.2 Separation

The most common methods of extracting the small size fraction from a dispersion are gravity sedimentation or centrifugation. Filtration may be used when there is insufficient material for sedimentation/centrifugation. Gravity sedimentation and centrifugation take advantage of dependence of the drag force experienced by a particle moving through a viscous fluid, and therefore the settling velocity of the particle, on particle diameter. The drag force (F_d) experienced by a small particle moving through a viscous fluid is given by Stokes' Law:

$$F_d = 3\pi\mu D v \tag{4.1}$$

where D is the equivalent spherical diameter of the particle, μ is the viscosity of the fluid and v is the velocity of the particle. For gravity sedimentation the particle will reach terminal velocity when the drag force (F_d) plus the buoyancy force of the particle is exactly matched to the gravitational force. Thus, the settling velocity (v_s) of a particle with equivalent spherical diameter D is given by the equation:

$$m_p\, g = 3\,\pi\,\mu\, D\, v_s + \frac{r_f\, m_g\, g}{r_p} \tag{4.2}$$

where m_p is the mass of the particle, g is the gravitational constant, ρ_f is the density of the fluid, and ρ_p is the density of the particle. The buoyancy force is derived from Archimedes' Principle that states that the buoyancy force on an object is equal to the weight of the fluid displaced by the object. In Eq. (4.2), m_p can be replaced by the particle volume multiplied by the particle density ($\pi D^3 \rho_p / 6$) so that

$$v_s = \frac{(\rho_p - \rho_f)\, g\, D^2}{18\,\mu} \tag{4.3}$$

Consequently, if a particle is moving with velocity v_s, the time (t) taken for a particle to fall through a distance H is

$$t = \frac{18\,\mu\, H}{(\rho_p - \rho_f)\, g\, D^2} \tag{4.4}$$

as $v_s = H / t$. Thus, after time t, as given by Eq. (4.4), the top of the suspension to depth H (as illustrated in Fig. 4.2a) will only contain particles with an equivalent spherical diameter less than or equal to D. At depths greater than H, the suspension will contain particles with equivalent spherical diameter greater than D, as well as some particles with equivalent spherical diameters less than D.

Taking g as 9.807 m s^{-2} and using the convenient (but non-SI) units of centipoise (= 1 mPa s) for viscosity, g cm^{-3} for density and micron for the particle diameter, Eq. (4.4) can be written as:

$$t = \frac{305.9\,\mu\,H}{\left(\rho_p - \rho_f\right) D^2} \tag{4.5}$$

for H in centimetres and t in minutes. For clays the value of ρ_p is often taken as 2.65 g cm^{-3}. The terms μ and ρ_f are dependent on temperature and can be determined from tabulated values (Haynes *et al.*, 2012;Tanner and Jackson, 1948).

For centrifugation, the buoyancy force can be neglected while the gravitational force is replaced by the centrifugal force (F_c):

$$F_c = \omega^2 R = 4\pi^2 N^2 R \tag{4.6}$$

where ω is the angular velocity, R is the radius of rotation and N is the rotational speed (e.g. revolutions per minute, RPM). However, the situation for centrifugation is more complicated than for gravity sedimentation because a particle experiences a larger centrifugal force as it moves further way from the axis of rotation. To allow for this variation in force an integrated form of Stokes' Law must be used (Jackson, 1956; Svedberg and Nichols, 1923). Consequently, for centrifugation, the time (t) taken for a particle with an equivalent spherical diameter of D to move from R_1 to R_2, where R_1 and R_2 are as illustrated in Fig. 4.2b, is:

$$t = \frac{9\,\mu\,ln\left(\frac{R_2}{R_1}\right)}{2\,\pi^2\,N^2\left(\rho_p - \rho_f\right) D^2} \tag{4.7}$$

where ln() is the natural logarithm. Again, using the convenient units of centipoise for viscosity, g cm^{-3} for density, micron for the particle diameter and RPM for N, Eq. (4.7) can be written as:

$$t = \frac{2.736\,x\,10^7\,\mu\,ln\left(\frac{R_2}{R_1}\right)}{N^2\left(\rho_p - \rho_f\right) D^2} \tag{4.8}$$

for t in minutes. The units used for R_1 and R_2 are arbitrary because these values are ratioed in Eqs (4.7) and (4.8).

For fixed-angle rotors R_1 and R_2 should be taken as the maximum distance of travel as illustrated in Fig. 4.2c, but are often interpreted as being the relevant distances around the midpoint of the suspension (Jackson, 1956).

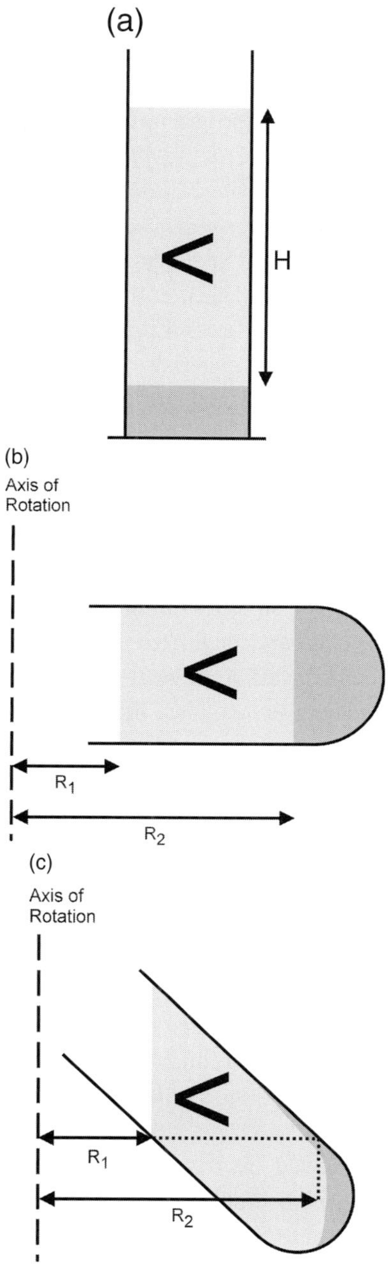

Figure 4.2. Particle separation for (a) gravity separation, (b) swing-out bucket centrifugation and (c) fixed-angle centrifugation. The lightly shaded areas labelled with a < indicate the volume of suspension containing only particles smaller than sizes given by Eqs (4.5) and (4.7). The darker shaded region will contain larger sized particles plus some of the smaller sized particles.

In the derivation of Eq. (4.8) no allowance has been made for spin-up and spin-down times. Poppe and Eliason (2009) show that for constant acceleration and deceleration an additional two-thirds of the sum of the spin-up and spin-down times should be added to the time given in Eq. (4.8). A computer program (CENTRISET) developed by Poppe and Eliason (2009) to calculate settling times for gravity and centrifuge particle size separation can be downloaded from https://woodshole.er.usgs.gov/software/sediment-software.html.

Complete extraction of clays requires repeated dispersions and sedimentations/centrifugations. Ideally, the dispersion and sedimentation/centrifugation process should be repeated until the suspension becomes clear (i.e. free of particles). In practice, the dispersion and sedimentation/centrifugation process is repeated to a point of diminishing returns. The dried weights of the fine and coarse materials recovered during separation can be used as a guide to the relative amounts of clay and non-clay materials in a mineral assemblage.

4.3.3 Collection and Treatment

After sedimentation/centrifugation, the clays (plus other small particle sized phases) are separated from the bulk of the original mineral assemblage but may remain dispersed in a large volume of aqueous solution. In order to minimize the effort required to collect the clays from the dispersion, the clays are coagulated using a suitable reagent and then recovered by sedimentation and centrifugation. In most cases a reasonably high ionic strength (0.1 M) NaCl solution is sufficient to coagulate the clays. Under the influence of gravity sedimentation and/or centrifugation, the coagulated clays should settle to the bottom of the aqueous solution from which the clear supernatant can be discarded. Clay behaviour in terms of coagulation and flocculation is dealt with in detail in Chapter 8.

This point in the clay extraction is suitable to apply chemical treatments to remove any remaining non-clay materials. Many carbonates can be removed by reaction with acetic acid. Double washing using an aqueous solution of acetic acid of $\leq$0.1 M is usually sufficient to remove carbonates such as calcite. Other carbonates such as siderite and, in some cases, dolomite may require further and perhaps more aggressive treatment. The use of acetic acid over stronger acids such as HCl is preferred because, as a weak acid, the pH of an acetic acid solution (especially when buffered with sodium acetate) is more easily controlled than solutions of strong acids (see, for example, Carrado *et al.*, 2006).

Organic matter can be removed by treatment with hydrogen peroxide or sodium hypochlorite or disodium peroxodisulphate (Mikutta *et al.*, 2005). Hydrogen peroxide can also be used to remove sulphides (Chao and Sanzolone, 1977) and manganese oxides (Jackson, 1956). Iron oxides/oxyhydroxides and, to some extent, aluminium oxides are removed with sodium dithionite–sodium citrate solution pH buffered with sodium bicarbonate (Mehra and Jackson, 1958; Rueda *et al.*, 1992). Selectable extraction techniques (e.g. Chao, 1984) can also be applied to remove specific mineral phases. The treatment of the collected material is assisted by vigorous shaking or stirring or, preferably, by ultrasonics.

The final chemical treatment of the recovered clay material is to ensure the clay is in a form suitable for quantitative analysis. For example, cation exchange of smectites to make them mono-ionic (often using a divalent cation such as Ca^{2+}) for XRD analysis. Similarly, exchange with Ba^{2+} is advantageous for CEC determination by XRF (Battaglia *et al.*, 2006).

The final stage in obtaining the separated clay material is washing to remove any excess salts or chemicals that have been used in the treatment of the clays. Washing is usually achieved using ultrasonics and centrifugation initially with high-purity water followed by a final wash using ethanol. As ethanol is a less polar solvent than water, many clays (especially the smectites) tend to flocculate more readily in ethanol than water. Obviously it is necessary to ensure that any chemicals that are being removed are soluble in water and, if used, ethanol. For example, if a clay is being Ba saturated, washes should be carried out using water or methanol as Ba^{2+} is not soluble in ethanol. In cases where the level of excess chemicals needs to be very low (e.g. for XRF analysis) or where the clay is poorly flocculating (e.g. Na-saturated smectite), dialysis may be necessary.

4.4 Techniques for Identification and Quantification

As highlighted in previous sections of this chapter and elsewhere in this book, clays are structurally, chemically and physically diverse. Any technique used to identify and characterize clays must recognize this diversity. Even so, there are many different techniques used to analyse clay minerals. Each technique utilizes the clay's unique properties in order to characterize them for specific purposes. X-ray diffraction analysis has traditionally been the most widely used technique for the identification and quantification of clay minerals and is consequently the main focus of this chapter.

4.4.1 X-ray Diffraction

4.4.1.1 Sample Preparation for X-ray Diffraction

The small size of clays (generally <2 μm) and the variations in the crystal structure of clays make powder XRD an ideal technique for the identification and characterization of clays. Powder XRD analysis relies on the sample that is being examined consisting of a large number of small crystals at random orientations (Cullity, 1978; Dinnebier and Billinge, 2008). The small size of the crystals minimizes micro-absorption effects and facilitates a greater randomness in orientations. Smith (1992) suggests that in order to achieve sufficient randomness for accurate powder XRD analysis, crystal sizes should be of the order of 1 μm or less when the irradiated area of the sample is 1 cm^2. However, by increasing the irradiated area of the sample (2 cm^2 is not uncommon in current powder XRD equipment) and spinning the sample so as to reduce the dimensional degrees of freedom, crystal sizes no larger than 5–10 μm are sufficient for accurate XRD analysis.

In order to achieve crystal sizes below 10 μm, some sample treatment is required. This treatment may be the extraction of the clay-sized fraction from the sample, but for bulk materials

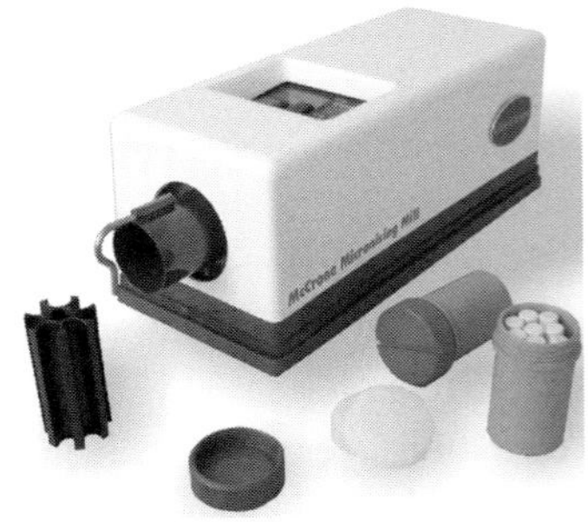

Figure 4.3. McCrone micronizing mill from McCrone Microscopes and Accessories.

initial treatment will usually consist of crushing or grinding down to a particle size of 200 μm or less. For samples that are solely for powder XRD analysis, over-crushing/grinding is not as severe a problem as it is for clay extraction. However, it is important not to crush/grind the sample so severely that crystalline material is converted to poorly crystalline or amorphous material. If only qualitative powder XRD analysis is required (i.e. specification of phase content as ranges rather than exact values) then vigorous hand-grinding of the crushed/ground material is sufficient. For quantitative powder XRD analysis further grinding by mechanical means is necessary. The most commonly used mechanical means of fine grinding is the use of a micronizing mill (e.g. McCrone Micronizing Mill, McCrone Microscopes and Accessories, Illinois, USA; Fig. 4.3). With a micronizing mill, sample material is mixed with a suitable liquid (usually ethanol) to form a slurry; typically a little over 1 g of sample in 10 ml of liquid. The sample material in the slurry is ground by cylindrical grinding elements (agate or corundum) in a vibrating housing. The McCrone Micronizing Mill will typically reduce particle sizes down to less than 10 μm when the grinding time is around ten minutes (O'Connor and Chang, 1986). The milled sample material is recovered by evaporation of the slurry liquid at a suitably low temperature (e.g. 40–60 °C).

For the commonly used powder XRD geometry (Bragg Brentano geometry; see, for example, Dinnebier and Billinge, 2008) the irradiated surface of the sample must be flat and, ideally, the plane of the irradiated surface of the sample should contain the line formed by axis of rotation of the X-ray source and the receiving slit of the X-ray detector (i.e. the theta axis). Variations in height of the irradiated surface of the sample above or below the ideal plane cause a systematic error in the recorded diffraction angle of the XRD pattern (Klug and Alexander, 1974). When a sample is mounted in a holder, care must be taken while creating the required flat surface not to preferentially orient the crystals in the sample. Preferred orientation effects can be particularly problematic for clays because of the layer structure of clays and resultant high aspect ratio. Consequently, if clay-containing samples are pressed into sample holders there will be a tendency for the surfaces of the clays with the larger dimensions to align parallel with the surface of the mount. Preferred orientation problems can also be experienced with anisotropic-shaped particles such as needle- or lath-shaped particles (e.g. halloysite and chrysotile).

A wide range of sample mounts is available for powder XRD. In a standard powder XRD mount, the sample is spread into a cavity in the mount and the surface of the sample to be irradiated is flattened by gentle pressure. Three classes of sample mount are recognized based on how the pressure is applied to flatten the irradiated surface: (1) front-pressed, (2) side-pressed and (3) back-pressed. With front-pressed sample mounts the sample surface is gently pressed using a suitable flat surface such as a microscope slide. This method of sample mounting is suitable for equi-dimensioned particles, but can lead to strong preferred orientation effects as the pressure is applied directly to the surface of the sample to be irradiated. For side-pressed sample mounting, a flat surface is placed over the cavity that is to hold the sample and the sample is loaded into the cavity from the side of the mount via an orifice. Sufficient pressure is applied to the side of the sample to hold it in place. As no direct pressure is applied to the surface of the sample that will be irradiated, preferred orientation effects are reduced relative to those experienced using front-pressed mounts. For back-pressed mounting the sample holder usually consists of two parts; a ring and a backplate. The ring is placed face-down on a flat surface and the sample loaded from the back of the ring. Gentle pressure just sufficient to hold the sample in place is applied to the back of the sample. In order to reduce preferred orientation effects to a minimum, a suitably rough surface (e.g. a piece of coarse filter paper such as Whatman No. 1) can be temporarily placed between the flat surface and the ring. After filling, the ring is clamped to the backplate. Back-pressing is considered to be the best of the three pressing methods for mounting samples that are prone to exhibiting preferred orientation.

Other types of sample mounts, in particular low-background Si wafer mounts, are often used when there is insufficient sample to fill the cavity of a standard powder diffraction sample mount.

The best method of eliminating preferred orientation during sample mounting for powder XRD is spray-drying (Hillier, 1999). In this method droplets of a slurry of sample and a liquid (usually ethanol or water) are sprayed into the top of a warmed chamber and collected dry at the bottom of the chamber. The dried material consists of small spheroidal clusters of particles in which the individual particles are at random orientations. The collected spray-dried material is loaded into the holder without application of any pressure. The spray-drying method of mounting works extremely well for samples that are very susceptible to preferred orientation effects, such as kaolinite and gibbsite.

Often when characterizing clays it is necessary to apply some form of additional treatment so as to distinguish between the various forms of clay. These treatments generally involve cation exchange and/or intercalation of the clays with a suitable compound. The advantages of mono-ionic saturation of clays with a divalent cation such as Ca^{2+} or Mg^{2+} for quantitative powder XRD analysis have already been noted. Intercalation of halloysite with formamide ($HCONH_2$) can be used to quantify the amount of dehydrated halloysite present in a kaolin assemblage (Churchman *et al.*, 1984; Theng *et al.*, 1984). Mg^{2+} saturation combined with glycerol [$C_3H_5(OH)_3$] intercalation can be used to distinguish vermiculites from smectites (MacEwan and Wilson, 1984). Additional treatment of clays is particularly important when characterizing interstratified clays. These treatments involve a combination of cation exchanges, intercalation and heat treatments applied in a very specific order (for example, see Poppe *et al.*, 2001).

As cation exchange and intercalation treatments necessarily cause diagnostic changes in the interlayer spacings of basal planes, it can be advantageous to accentuate diffraction from the basal planes by creating highly oriented clay films where the basal planes of the clays are parallel to the surface of the film (and therefore the irradiated surface of the film): i.e. to deliberately cause preferred orientation. This method is restricted to the clay fraction of a mineral sample, as larger particles will limit the orientability of clays within the film. A simple method of creating a clay film is to evaporate a relatively dilute slurry of clay onto a glass slide or low-background Si wafer. A more efficacious method is to draw the slurry down onto a porous ceramic tile or a polymer filter (generally with openings of 0.22 μm or less) using vacuum suction. The advantage of ceramic tiles and polymer filters is that successive cation exchange, intercalations and, in the case of ceramic tiles, heat treatments can be performed on a single film. It is important when using a polymer filter to choose a polymer that will not cause a detrimental increase in the scattered X-ray background or show strong diffraction peaks. It has been found that PDVF (polyvinylidene fluoride) filters (e.g. Millipore Durapore membrane filters) do not have strong diffraction peaks and give lower XRD background than cellulose or cellulose nitrate filters. Alternatively, the oriented clay can be removed from the filter and placed onto a suitable low-background support using the filter peel method (Drever, 1973).

4.4.1.2 X-ray Diffraction Measurements

The principal requirements for XRD measurements are that the instrument is well aligned and that sample height is level with the diffractometer theta axis. Any misalignment or variation of sample height will affect the positions of the diffraction peaks. This is especially critical for the low-angle region where the dominant clay diffraction peaks occur.

Resolution of the instrument is not of principal concern because clay diffraction peaks are usually broad owing to disorder and small crystallite size. The maximum diffracted intensity from clays is also significantly less than for crystalline non-clays. Therefore, instrument settings, such as selection of divergence and receiving slits can usually be optimized for intensity rather than resolution.

4.4.2 Qualitative Analysis by X-ray Diffraction

XRD detects long-range order in crystal structures and provides information about relative abundance, crystallite size, strain and degree of crystallinity. Modern XRD systems are computer controlled with software specifically designed by the hardware manufacturer. In order to identify the minerals in a sample, the output from the instrument (usually tabulated 2-theta angle and intensity) undergoes automatic or manual peak searching. The angular position of each peak is converted to *d*-space using the Bragg law for the appropriate X-ray tube wavelength. Search/match routines are then applied to the data to compare the measured *d*-spacings with tabulated *d*-spacings of a set of inorganic and/or organic compounds contained within a digital database. Several powder diffraction databases are available with the most commonly used being from the International Centre for Diffraction Data (ICDD) (www.icdd.com), which is a not-for-profit organization that since the 1950s has collated over 300,000 XRD patterns into a searchable database. A second database that is finding increased use is the Predicted Crystallography Open Database (PCOD), which is based on the Crystallography Open Database (COD) (Gražulis *et al.*, 2009; www.crystallography.net).

Automated search/match routines using databases as large as those of the ICDD and PCOD inevitably match phases that cannot possibly be present in natural mineral specimens. Searches restricted to just the minerals data will be significantly faster but still produce a large list of misidentified phases. Users should, therefore, carefully select only those phases that are likely to be present through knowledge of the sample source and/or select only those phases that can be confirmed using other techniques (i.e. using chemical composition). Care should also be taken when identifying associated minerals such as those listed in Table 4.3. Setting up a database that only contains phases that are realistic for the suite of samples being studied will increase both the accuracy and the speed of the search/match process.

Clay minerals, because of their poor ordering, are very difficult to match using automated techniques, so the user must rely on manual methods of analysis. Identification of the broader group of clays can be done using the basal *d*-spacings (i.e. 001 peaks), either from the random powder or oriented patterns. This can be aided by the high-angle region (e.g. the 060 diffraction peak at approximately 1.5 Å) that can be used to distinguish between di-octahedral and tri-octahedral clays.

With the exception of paracrystalline clays (i.e. allophane and imogolite), all clays and clay minerals can be classified into five subgroups based on their basal *d*-spacings (and therefore their characteristic XRD pattern), as detailed in Table 4.4. Preliminary identification can, therefore, be made using either random pressed powder samples or oriented, air-dried samples. Samples containing complex mixtures of clays may require more extensive treatments such as exchange with various cations, ethylene glycol or glycerol solvation and heat treatments (see, for example, Poppe *et al.*, 2001).

4.4.2.1 7 Å Subgroup

The two main clay minerals in the 7 Å subgroup are kaolins and serpentines. While these are the only clay minerals that have their largest *d*-spacing in this subgroup, chlorite, with a largest *d*-spacing of 14 Å, also has a 002 peak at 7 Å, which overlaps the 001 reflections of the kaolin and serpentine minerals. Modern XRD instruments with high resolutions can distinguish between kaolin (7.15–7.17 Å), serpentine (7.3–7.4 Å) and chlorite (7.01 Å). Kaolins are also readily distinguished from serpentines by their *d*060 spacing. Di-octahedral kaolin has a *d*060 spacing around 1.49 Å with the *d*060 spacing for tri-octahedral serpentine being greater than 1.52 Å.

All minerals in this subgroup with the exception of hydrated halloysite are unaffected by ethylene glycol or glycerol treatments and hence these treatments are of little use

Table 4.4. *Classification of clay minerals subgroups based on the largest basal d-space of air-dried samples (adapted from Moore and Reynolds, 1997).*

Largest *d*-space	Clay mineral subgroup	Clay species
~7 Å	Serpentine	Chrysotile, antigorite, lizardite, amesite
	Kaolin	Kaolinite, dickite, nacrite, dehydrated halloysite
~9 Å	Talc	Talc
	Pyrophyllite	Pyrophyllite
10–11 Å	Kaolin	Hydrated halloysite
	Palygorskite	Palygorskite
	Di-octahedral mica/illite	Muscovite, paragonite, margarite
	Tri-octahedral mica/illite	Phlogopite, biotite, lepidolite, clintonite, anandite
	Di-octahedral smectite	Montmorillonite, beidellite, nontronite
12–15 Å	Sepiolite	Sepiolite
	Tri-octahedral smectite	Saponite, hectorite, sauconite
	Vermiculite	Vermiculite
	Tri-octahedral chlorite	Clinochlore, chamosite, nimite
	Di, di-octahedral chlorite	Donbassite
	Di, tri-octahedral chlorite	Cookeite, sudoite
	Random interlayer	Illite–smectite
>20 Å	Regular interlayer	Corrensite, rectorite, tosudite
	Random interlayer	Chlorite–smectite, kaolin–smectite

for distinguishing minerals in this subgroup. Heat treatments can be used to distinguish between kaolin and serpentine, particularly if samples contain mixtures of di-octahedral and tri-octahedral clays. Kaolin minerals heated between 500 and 700 °C for one hour lose their octahedral hydroxides and transform into an X-ray amorphous phase, metakaolin (Fig. 4.4). Further heating to 1000 °C transforms metakaolin into mullite and cristobalite.

Kaolinite can be distinguished from dickite and nacrite by the presence of several additional minor diffraction peaks and the dehydroxylation temperatures of dickite and nacrite that are slightly higher than the dehydroxylation temperature of kaolinite. Depending on disorder, kaolinite dehydroxylates between 500 and 600 °C, whereas dickite and nacrite dehydroxylate at around 700 °C. Well-crystalline kaolinite can be readily distinguished from dehydrated halloysite because of additional sharp diffraction peaks for kaolinite. Poorly crystalline kaolinite, however, can often only be distinguished from dehydrated halloysite by treatment with formamide (Churchman *et al.*, 1984; Theng *et al.*, 1984). Formamide treatment is usually carried out on oriented samples (i.e. samples drawn down onto polymer filters or ceramic plates). If dehydrated halloysite is present in a sample, the formamide treatment will, in a short period of time (<1 hour), cause the basal spacing of halloysite to increase from 7.3 Å to 10.4 Å, while the basal spacing for kaolinite will remain unchanged for a period of greater than four hours (Churchman *et al.*, 1984).

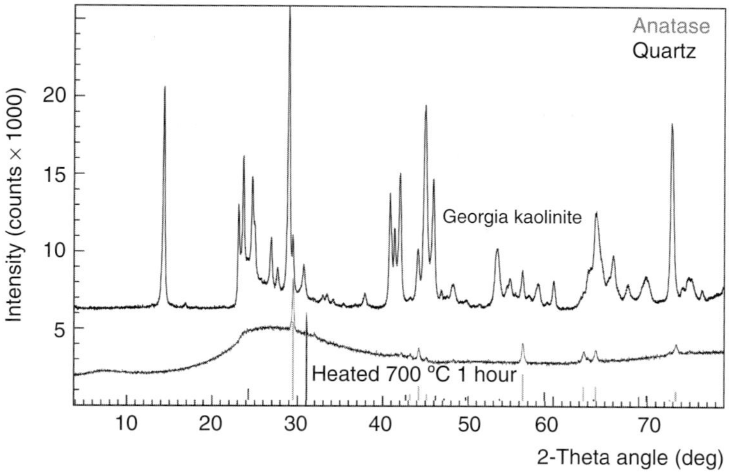

Figure 4.4. XRD pattern (Co Kα radiation) showing effect of heating well-crystalline kaolinite at 700 °C for one hour. The vertical lines at the base of the patterns show the expected positions and relative intensities of diffraction lines from anatase and quartz.

Serpentine minerals are unaffected when heated to 700 °C, but transform at 1100 °C into mixtures of olivine, spinel, hematite and cristobalite, depending on their composition.

4.4.2.2 9 Å Subgroup

Talc and pyrophyllite occupy this subgroup and their identification presents no particular difficulty. Talc at 9.3 Å and pyrophyllite at 9.1–9.2 Å are relatively straightforward to distinguish from XRD patterns with modern instruments. Pyrophyllite has many more sharp XRD peaks than talc, particularly in the *d*-spacing range 4.0 and 4.6 Å. Talc exhibits broad diffraction bands typical of disordered structures in this region-spacing range. In addition, the 060 *d*-spacing is approximately 1.49 and 1.53 Å for pyrophyllite and talc, respectively, and therefore the 060 *d*-spacing may be used to distinguish between these two clays. Some zeolite minerals, laumontite, heulandite–clinoptilolite, also have characteristic low-angle peaks in this region so caution should be taken when the presence of zeolites is suspected.

4.4.2.3 10–11 Å Subgroup

Minerals within this subgroup include micas and illites, hydrated halloysite, and palygorskite. All minerals within this group are virtually unaffected by glycerol or glycol treatments and so these treatments are of no practical use for distinguishing minerals in this subgroup. Hydrated halloysite is readily distinguished from other members of the subgroup by heating to 110 °C. The broad 10 Å peak collapses to approximately 7.2 Å following the loss of water from the structure. The dehydration process is irreversible and can commence at room temperatures under conditions of low RH. Palygorskite is characterized by a broad peak at approximately 10.4 Å, which is unaffected by heating to 110 °C. Heating palygorskite above 800 °C forms spinel.

Di-octahedral and tri-octahedral micas can be distinguished by their 060 *d*-spacings that are in the ranges 1.50–1.52 Å and 1.53–1.54 Å respectively. Their structures are unaffected by heating to 400 °C, but convert to spinel, leucite and iron and aluminium oxides after heating above 1200 °C. Illites are distinguished from micas by heating to 400 °C where the illite 001 peak becomes sharper and more intense as the illite loses water and reverts to mica.

4.4.2.4 12–15 Å Subgroup

Basal *d*-spacings between 12 and 15 Å are characteristic of the clay minerals sepiolite, smectite (including some interstratified smectite), vermiculite and chlorite.

Sepiolite is readily identified from other members in this subgroup because sepiolite has a relatively smaller *d*-spacing of around 12.6 Å that is essentially unaffected by glycerol or glycol treatments. The 12.6 Å peak of sepiolite becomes weak and broad after heating to 400 °C for one hour and transforms to a non-crystalline phase above 700 °C.

Chlorite can be classified into two subgroups based on chemistry. Mg-chlorite is characterized by the first four basal peaks displaying approximately equal intensities, whereas Fe-chlorite generally exhibits intense even-numbered basal peaks (i.e. 002, 004, etc.) and relatively weak odd-numbered basal peak intensities (i.e. 001, 003, etc.). Glycerol and glycol treatments have no effect on chlorite. Heating to 375 °C for two hours has no effect on chlorite; however, the 14 Å peak of vermiculite collapses to 9 Å and 12 Å for di-octahedral and tri-octahedral vermiculite, respectively.

In the Mg-saturated form, most vermiculites are essentially unaffected by ethylene glycol treatments and all are unaffected by glycerol treatments. In the Ca-saturated form the basal peak positions of most vermiculites are expanded from 14 Å to 16.1–16.2 Å (ethylene glycol) or 17.6 Å (glycerol).

Minerals of the smectite subgroup can be the most problematic to accurately quantify from a single preparation procedure. The basal (001) peaks from smectite are often broad and exhibit low- or high-angle asymmetry so other treatments may also be required to characterize smectite content. Peak broadening can be due to changes in RH during data collection (Table 4.5). This is particularly an issue when the clay is in the sodium-saturated form, but it can also be due to uneven cation distribution in the clay interlayer caused by incomplete cation exchange (see Fig. 4.5). In the dry, Na-saturated state most smectites contain single water layers and display basal *d*-spacings of approximately 12.5 Å. They generally remain at 12.5 Å up to 50 per cent RH until they expand to approximately 15 Å (two water layers) above 50 per cent RH. Often this expansion can lead to mixed one and two water layer systems that display ill-defined intermediate *d*-spacings, giving rise to broad diffraction peaks. These intermediate *d*-spacings can be misinterpreted as an unrelated interstratified species. To overcome his confusion, the diffraction pattern should be collected under constant RH or the smectite can be exchanged with a divalent cation, usually Ca or Mg, which results in relatively well-defined *d*-spacings between 14.7 and 15.2 Å.

All smectite subgroup minerals expand with glycerol or ethylene glycol treatment. Ethylene glycol expands the basal *d*-spacings of smectites to approximately 16.9 Å and

Table 4.5. *Effect of interlayer cations on basal spacings (Å) for Wyoming montmorillonite at 32, 52 and 79 per cent RH and immersed in water (Brindley and Brown, 1980: table 3.1).*

Interlayer cation	32 per cent RH	52 per cent RH	79 per cent RH	In water
Na^+	12.5	12.5	14.8 (br)	Ind
NH_4^+	11.7	11.7 (br)	11.9	12.4 (as)
K^+	11.9	11.9	12.1	Ind
Li^+	12.09 (br)	14.1 (br)	15.1 (br)	27.0
Cs^+	12.5	12.4	12.4	12.5
Ca^{2+}	15.2	15.1	15.5	19.0
Mg^{2+}	15.1	15.1	15.2	19.5
Sr^{2+}	15.1	15.3	15.6	19.0
Ba^{2+}	13.2 (br)	14.5 (br)	15.6	18.8

(br) – broad, probably indication of interstratification; (as) – asymmetrical peak shape towards high *d*-spacings; ind – tendency to indefinite swelling.

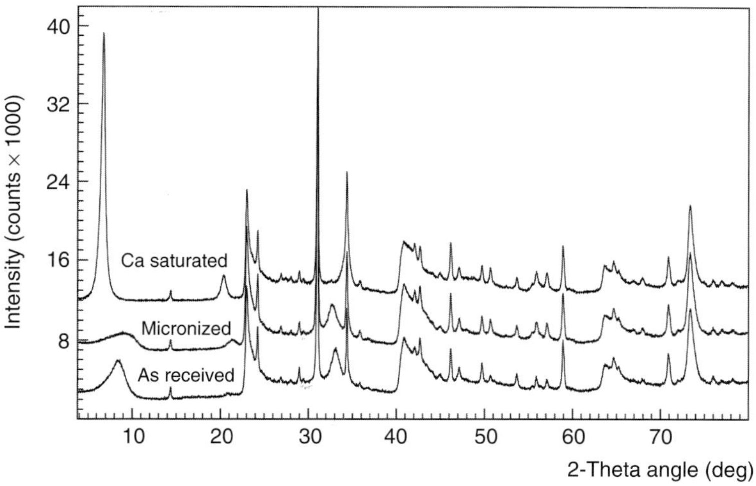

Figure 4.5. XRD pattern of a sodium-saturated bentonite sample showing variation in the 001 peak position and intensity due to different preparation procedures (Co Kα radiation).

glycerol to approximately 17.8 Å. The 060 region of a diffraction pattern can be used to readily distinguish di-octahedral from tri-octahedral smectites because the *d*-spacings that contribute to this region of the diffraction pattern are 1.49 to 1.52 Å for di-octahedral smectite and >1.53 Å for tri-octahedral smectites. The 060 peak of montmorillonite and beidellite are usually closer to 1.49 Å, whereas nontronite is generally around 1.52 Å. Further classification of individual clay species can only be accomplished by extracting

Table 4.6. *Regular 1:1 interlayer clay minerals.*

Layer species	Name
Muscovite–montmorillonite	Rectorite
Mica–beidellite	Allevardite
Di-octahedral chlorite–di-octahedral smectite	Tosudite
Tri-octahedral chlorite–di-octahedral smectite	Corrensite
Mica–vermiculite	Hydrobiotite
Talc–saponite	Aliettite

high-purity material by dispersion and high-speed centrifugation followed by detailed chemical analysis.

Randomly interlayered clays are also found in this subgroup and are characterized by broad, irregular 001 peaks. The most common interlayered clays are illite–smectites and their characteristic basal peak shifts from 14.8 Å for the dominantly smectite-rich endmember to 10 Å for the dominantly illite-rich endmember. Treatments with Mg salts and glycerol or ethylene glycol solvation can enable determination of the relative proportions of the interlayered species. Moore and Reynolds (1997) describe in detail the procedures for determining the ratio of the interlayered species using calculated XRD patterns.

4.4.2.5 >20 Å Subgroup

Regular interlayered clays are those that show regular ordering of the individual clay layers giving rise to basal *d*-spacings that are sums of the discrete components. For example, regular interlayered 50:50 illite–smectite in the air-dried Ca-saturated form will show *d*001 of 10 + 14.8 = 24.8 Å. Common regular interlayered clays are shown in Table 4.6. The clays listed in Table 4.6 are regular 1:1 (50:50) interstratifications and their basal spacing is the sum of the basal spacings of the component clays. The large basal spacings of these regular interlayered clays means that the low-order X-ray diffraction peaks (e.g. 001) of these clays may occur at angles outside the angular range routinely used for the collection of XRD patterns. When these low-order diffraction peaks are not observed, the higher-order diffraction peaks (e.g. 002) can be misinterpreted as arising from the presence of clays from other subgroups.

4.4.3 Quantitative Phase Analysis by X-ray Diffraction

Identification of the mineral phases in a sample is invariably followed by the question 'how much?'. The use of whole-pattern and Rietveld-based techniques in recent years and their application to clay mineralogy has to some degree superseded more traditional methods of quantitative analysis. The traditional quantitative phase analysis techniques are extensively covered in other texts (Cullity, 1978; Dinnebier and Billinge, 2008; Jenkins and Snyder, 1996; Klug and Alexander, 1974; Zevin *et al.*, 1995).

Quantitative phase analysis using XRD can be achieved by using either single-peak or whole-pattern techniques. Single-peak methods rely on the measurement of the intensity of a peak or group of peaks for each phase being quantified and the relative peak intensities are assumed to be representative of the amount of each phase present. This is often not the case because of peak overlaps, preferred orientation and microabsorption (Brindley, 1945; Madsen and Scarlett, 2008). Clays are particularly problematic because of both strong preferred orientation arising from their platy morphology and extensive element substitutions in their structures.

Whole-pattern methods utilize the summation of individual component XRD patterns that are either measured from pure mineral specimens or calculated from published crystal structures. The mathematical formula that relates the intensity of a peak *hkl* reflection for phase α ($I_{(hkl)\alpha}$) in a multi-phase mixture measured on a flat plate sample is given by:

$$I_{(hkl)\alpha} = \left[\frac{I_0\lambda^3}{32\pi r}\frac{e^4}{m_e^2c^4}\right]\cdot\left[\frac{M_{hkl}}{2V_\alpha^2}\left|F_{(hkl)\alpha}\right|^2\left(\frac{1+cos^2 2\theta\, cos^2 2\theta_m}{sin^2\theta\, cos\,\theta}\right)\right]\cdot\left[\frac{W_\alpha}{\rho_\alpha\mu_m^*}\right] \quad (4.9)$$

where I_0 is the incident beam intensity, e is the charge on an electron, m_e is the mass of an electron, r is the distance from the scattering electron to the detector, λ is the wavelength of the diffracting radiation and c is the speed of light. M_{hkl} and F_{hkl} are the multiplicity and structure factor of the *hkl* reflection, V_α is the unit cell volume, θ and θ_m are the diffraction angles for the *hkl* and monochromator (if fitted) respectively. W_α and ρ_α are the weight fraction and density of phase α respectively and μ_m^* is the mass absorption coefficient of the whole sample.

When considering a single peak or a group of peaks, the formula above can be reduced to:

$$I_{i\alpha} = C_{i\alpha}\frac{W_\alpha}{\rho_\alpha\mu_m^*} \quad (4.10)$$

where $C_{i\alpha}$ is a constant for a single peak or group of peaks i of phase α. $C_{i\alpha}$ can be determined by preparing standards of known concentration W_α, measurement of peak intensity I for the standards and estimation of the mass absorption coefficient μ_m^* of the standards. Weight fractions of unknown samples can then be calculated by rearranging Eq. (4.10) and measurements of $I_{i\alpha}$ and estimation of μ_m^* for the unknown samples. The mass absorption coefficient, μ_m^*, can either be measured, usually by XRF, or calculated from tables of theoretical mass absorption coefficients (Ibers and Hamilton, 1974).

The mass absorption coefficient can be eliminated by the addition of an internal standard (s) of known weight fraction W_s so the weight fraction of the unknown W_α is calculated from:

$$W_\alpha = K_{\alpha s}^{ij}\cdot W_s\cdot\frac{I_{i\alpha}}{I_{js}} \quad (4.11)$$

where I_{js} is the intensity of the peak or groups of peaks j of standard s. $K_{\alpha s}^{ij}$ can be determined by making mixtures of the standard and unknown phases.

The reference intensity ratio (RIR) method described by Hubbard *et al.* (1976) and Hubbard and Snyder (1988) is made by rearranging Eq. (4.11). The RIR is the ratio of the intensity of the strongest peak of phase α to the strongest peak of standard s. Corundum is the most widely used standard for quantitative analysis, therefore RIR is equivalent to I / I_c where I_c is the reference intensity for corundum. The weight fraction of unknown W_α is given by the equation:

$$W_\alpha = \frac{I_{i\alpha}}{I_{js}} \cdot \frac{W_s}{RIR_{\alpha s}} \tag{4.12}$$

Values of RIR can be obtained by measuring known mixtures of corundum with the unknown phase or calculated from crystal structure information. The ICDD also provides RIR values for many common phases. However, the values are dependent on the experimental procedure, so accuracy may be compromised.

Chung (1974a, 1974b) introduces a method whereby the sum of the weight fractions of the crystalline components are constrained to 1.0. This method has become known as the 'matrix flushing' method. The weight fraction of component α can be calculated from:

$$W_\alpha = \frac{I_{i\alpha}}{RIR_{\alpha s}} \cdot \left(\sum_{k=1}^{n} \frac{I_{ik}}{RIR_{ks}} \right)^{-1} \tag{4.13}$$

where the sum is over all crystalline phases in the sample. As the total composition is normalized to 100 per cent, the technique only calculates relative phase abundance. In order to determine absolute abundances a known concentration of an internal standard is added and absolute abundances calculated from:

$$W_{\alpha(abs)} = W_\alpha \cdot \frac{W_{std(weighed)}}{W_{std(measured)}} \tag{4.14}$$

and the amounts of non-crystalline and unidentified phases are calculated from:

$$W_\alpha = 1.0 - \sum_{k=1}^{n} W_{k(abs)} \tag{4.15}$$

4.4.4 Whole-Pattern Fitting Methods

The techniques used for single peak methods have also been successfully applied to whole-pattern analysis systems – for example, FULLPAT (Chipera and Bish, 2002) and ROCKJOCK (Eberl, 2003). Many techniques use the RIR technique and utilize Microsoft Excel spreadsheets with the SOLVER add-in to objectively minimize the difference between the measured unknown pattern and a composite pattern made up of patterns from individual pure reference materials.

Whole-pattern analysis techniques have the potential to produce more accurate and precise quantitative analysis results than conventional single-peak methods. This is because of a number of factors such as: all data points in the XRD patterns contribute to the quantitative analysis; peak overlaps are handled effectively; preferred orientation effects are reduced; and differences between measured and calculated backgrounds may indicate the presence of non-crystalline material.

4.4.5 Rietveld Methods

The Rietveld method (Rietveld, 1967, 1969) was originally developed for the refinement of crystal structures from powder patterns using neutron diffraction experiments. The technique was applied to structure refinement using X-ray powder diffraction patterns by Young (1980), then quantitative phase analysis of neutron diffraction data by Hill and Howard (1987) and X-ray diffraction data by O'Connor and Raven (1988). The Rietveld method has several advantages over other whole-pattern techniques because it does not require the measurement of standards and has the ability to refine structural and non-structural parameters. These parameters include: refinement of the atom positions and occupancy of structure sites and unit cell parameters to allow for chemical substitutions, preferred orientation, and changes in crystallite size and strain by refining peak widths and shapes.

The Rietveld method is a non-linear least squares technique whereby specimen and instrument parameters are refined in order to minimize the following sum:

$$R = \sum_i w_i \left(y_i - y_{ci}\right)^2 \tag{4.16}$$

where the sum is over each step (i) in the XRD pattern, w_i is a weighting factor, y_i is the measured step intensity and y_{ci} is the calculated step intensity. O'Connor and Raven (1988) show that by rearranging Eqs (4.9) and (4.10) and substituting the Rietveld scale factor S_α for I_α the concentration of phase α can be determined by:

$$W_\alpha = \frac{S_\alpha \rho_\alpha V_\alpha^2 \mu_m^*}{K} \tag{4.17}$$

where K is an instrument scaling factor that is independent of the sample composition. The value K can be determined from a single XRD pattern of a highly crystalline material (e.g. corundum) and applied to subsequent unknown measurements, provided the instrument conditions are identical to those of the standard.

Hill and Howard (1987) apply the matrix flushing method of Chung (1974a, 1974b) and the relationship:

$$\rho_\alpha = \frac{10^{24}}{N_A} \cdot \frac{Z_\alpha M_\alpha}{V_\alpha} = 1.6605 \cdot \frac{Z_\alpha M_\alpha}{V_\alpha} \tag{4.18}$$

where N_A is the Avogadro constant, Z_α is the number of formula units in the unit cell for phase α, M_α is the molecular mass of the formula unit for phase α, V_α is the unit cell volume (in Å^3) for phase α and ρ_α is expressed in g cm^{-3} to derive the commonly used Rietveld quantitative analysis formula:

$$W_\alpha = \frac{S_\alpha (ZMV)_\alpha}{\sum_{k=1}^{n} S_k (ZMV)_k} \tag{4.19}$$

The sum in Eq. 4.19 represents the sum over all phases in the sample and Z_α, M_α and V_α have been grouped as $(ZMV)_\alpha$. Analogous to the Chung method using single peaks, the Rietveld phase concentrations are normalized to 100 per cent. Therefore, a known amount of internal standard is added to the unknown sample and Eqs (4.14) and (4.15) can then

Table 4.7. *Quantitative analysis of IUCr-CPD QPA round robin sample 1b (Madsen et al., 2001).*

Phase	Weighed (wt%)	Determined (wt%)
Corundum	94.31	94.38
Fluorite	4.33	4.36
Zincite	1.36	1.26

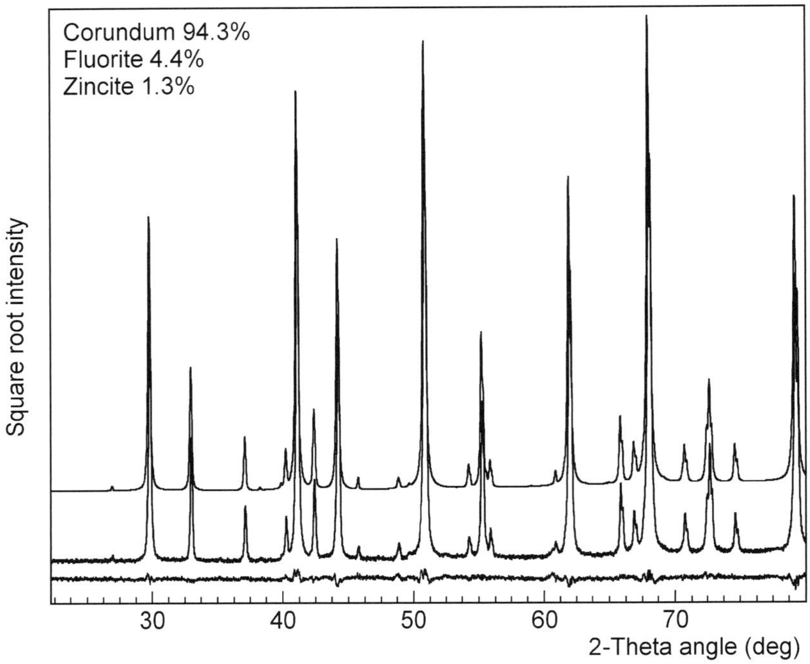

Figure 4.6. XRD patterns (Co Kα radiation) of calculated (uppermost pattern), observed (middle pattern) and difference (lowermost trace) showing excellent fitting of well-crystalline components of sample 1b from the quantitative phase analysis round robin (Madsen *et al.*, 2001) using TOPAS™ (Bruker AXS, 2009). The measured and calculated patterns are shown using a square root intensity scale. The maximum intensity of the measured and calculate patterns is approximately 75 000 counts. The difference pattern is also shown using a square root intensity scale but with the sign of the intensity difference applied to the square root of the absolute value of the difference in intensity. The maximum difference is approximately 50 counts.

be used to determine absolute concentrations. Figure 4.6 shows an example of the excellent fit between the observed and calculated XRD patterns for a well-crystalline sample 1b from the International Union of Crystallography Commission on Powder Diffraction Round Robin on Quantitative Phase Analysis (IUCr-CPD QPA) (Madsen *et al.*, 2001). Quantitative analysis results are also excellent compared with the originally weighed amounts of each phase (Table 4.7).

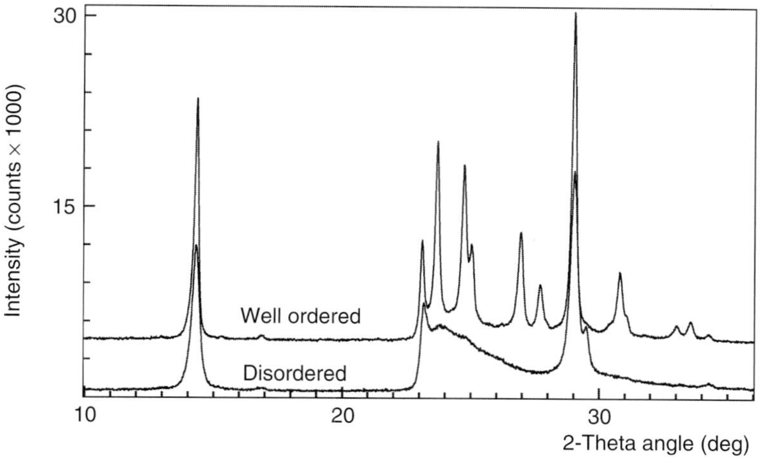

Figure 4.7. XRD patterns (Co Kα radiation) showing comparison of well-ordered and highly disordered specimens.

The main drawback with early implementations of the Rietveld method is the requirements of accurate crystal structure information and XRD patterns of samples containing well-crystalline materials. However, many materials exhibit disordered structures and none more so than clay minerals. Kaolinite, for example, exhibits a wide range of disorder from relatively low disorder to highly disordered (Fig. 4.7). Disorder in kaolinite is mainly because of non-regular displacements of the Al vacancy in the octahedral sheet, while stacking faults, caused by b / 3 displacements, contribute less to disorder (Plançon and Tchoubar, 1977).

Turbostratic disorder is a special case where consecutive layers are stacked with random rotation and/or translation within crystallites (Biscoe and Warren, 1942; Warren, 1941). Minerals of the smectite group experience this type of disorder and calculated XRD patterns using the Rietveld method from crystal structure data show poor correlation with measured patterns (Fig. 4.8).

The problem of disordered structures was first overcome by Taylor and Matulis (1994) using 'observed *hkl* files' that are calculated from crystal structures. The procedure involves generating a series of *hkl* peaks with individual structure factors (i.e. intensities) using the unit cell and atom positions of a known or related crystal structure. Often the space group and relatively small unit cells of most disordered clay minerals are insufficient to generate enough peaks to model the broad diffraction band between 22° and 28° 2-theta. Taylor and Matulis (1994) overcome this limitation by doubling the *a* and *b* unit cells and introducing two refinable half-width terms to allow for sharp and broad peaks. Structure factors are manually edited to fit an XRD pattern measured from a pure clay. This process disrupts the relationship between the Rietveld scale factor and the calibration factor ZMV used in the quantitative analysis Eq. (4.19). Refinements of the unit cell dimensions performed in this way provide the unit cell volume V, but ZM needs to be determined by preparing several mixtures of the pure clay with an internal standard. While the ZM no longer represents the true unit cell mass it does relate the observed structure factors (intensities) to the phase concentration.

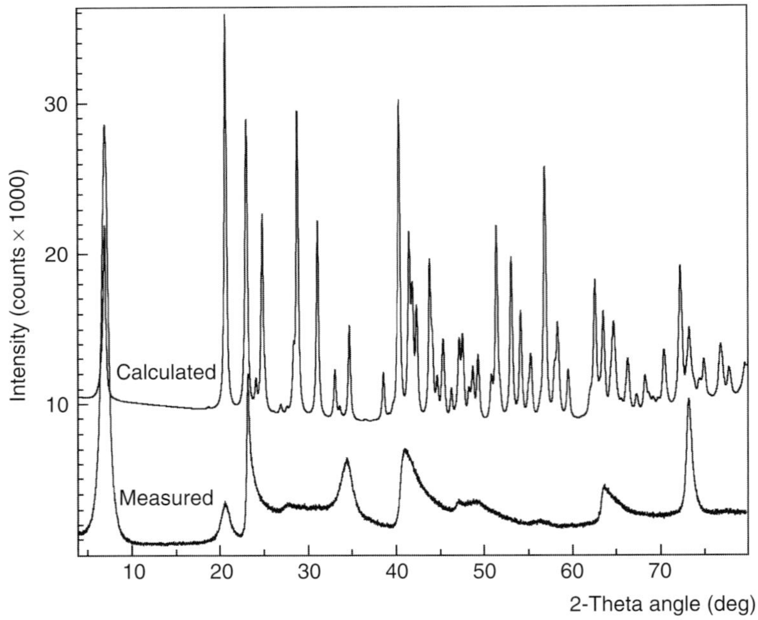

Figure 4.8. XRD patterns (Co Kα radiation) of montmorillonite showing turbostratic disorder in a measured sample compared with the pattern calculated from a crystal structure (Tsipurski and Drits, 1984) using TOPAS V4.2 (Bruker AXS, 2009).

For the crystalline phases, calculated *hkl* files can be generated from published crystal structures and applied unedited. Databases of crystal structures that are commonly used to generate *hkl* files are the Inorganic Crystal Structure Database (ICSD; www2.fiz-karlsruhe.de/icsd_home.html) and the COD database (Gražulis *et al.*, 2009; www.crystallography.net). Quantitative analyses of unknown samples are then applied using standard and 'observed' *hkl* files. Analyses using this method are limited because parameters such as atom positions, elements and occupancies are no longer refineable. However, minor changes in unit cell dimensions because of element substitutions can be adequately modelled for quantitative phase analysis.

The 'observed *hkl* files' technique is employed in the SIROQUANT™ software package (Taylor, 1991) from Sietronics Pty Ltd (www.siroquant.com) and can produce accurate quantitative analyses of complex mineral assemblages containing substantial amounts of disordered clay types. An example of the observed and calculated fit from SIROQUANT for sample RC5-2 of the winning entry of the 2010 Reynolds Cup competition is shown in Fig. 4.9 (Raven and Self, 2011). Quantitative analysis results are shown in Table 4.8. The results listed in Table 4.8 give a good indication of the accuracy that can be achieved using X-ray diffraction for the identification and quantification of clays. Precision and accuracy are discussed further in Section 4.4.6.

Scarlett and Madsen (2006) apply the 'observed' *hkl* file technique of Taylor and Matulis (1994) using TOPAS (Bruker AXS, 2009) to mineral phases with either partial structures (i.e. only space group and unit cell known) or no known structure at all. The partial structure

Table 4.8. *Quantitative XRD analysis results of sample RC5-2 from the 2010 Reynolds Cup competition using SIROQUANT.*

Mineral	Weighed (wt%)	Determined (wt%)	Difference (wt%)
Quartz	2.2	1.7	0.5
Huntite	5.0	4.4	0.6
Anatase	2.3	1.6	0.7
Alunite	10.7	9.5	1.2
Barite	5.8	4.1	1.7
Fluorite	1.9	2.0	0.1
Microcline	8.2	7.7	0.5
Albite	0.5	–	0.5
Gibbsite	9.6	12.0	2.4
Total non-clay	**46.5**	**43.0**	**8.5**
Di-muscovite $2M_1$	5.1	5.6	0.5
Kaolinite	14.3	14.2	0.1
Halloysite	19.9	23.4	3.5
Di-illite–smectite	14.3	13.8	0.5
Total clay	**53.5**	**57.0**	**4.6**
Total	**100**	**100.0**	**13.1**

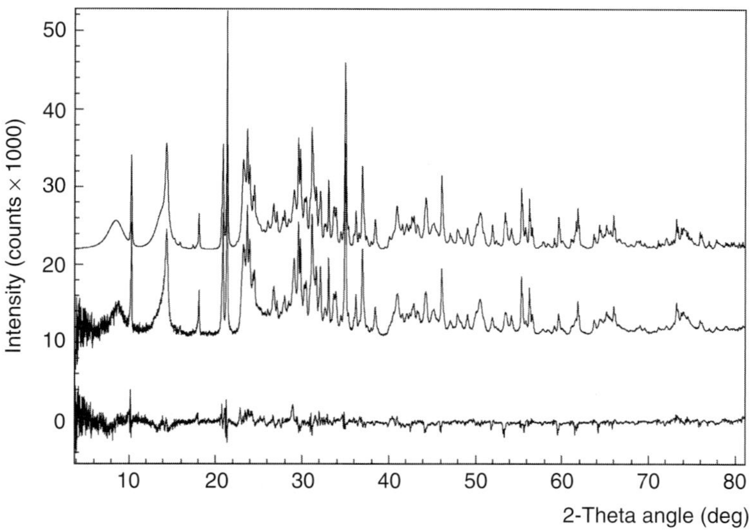

Figure 4.9. XRD patterns (Co Kα radiation) of observed (middle), calculated (upper) and difference showing good fitting of clay and non-clay components of sample RC5-2 from the 2010 Reynolds Cup quantitative analysis contest (Raven and Self, 2011) using SIROQUANT (Taylor, 1991).

type is essentially the same as that used in SIROQUANT, where lists of *hkl* structure factors are generated from the unit cell and space group. The ZM factors are determined by preparing several mixtures of the pure clay and an internal standard. In the case where no structure is available, a series of individual peaks are grouped and scaled together. Because a unit cell is no longer available, the ZMV factor requires determination using mixtures of the unknown and internal standard.

Bergmann and Kleeberg (1998) develop several models to directly handle structural disorder in several different clay mineral types. The models are based on ideal structures with *hkl* dependent peak shifting and line broadening dependent on the type of disorder. The special case of turbostratic disorder was investigated by several groups (Drits and Tchoubar, 1990; Treacy *et al.*, 1991; Yang and Frindt, 1996) and specifically for smectites by Viani *et al.* (2002). The first structurally based application to smectite was performed by Ufer *et al.* (2004). The Rietveld programme BGMN (Bergmann *et al.*, 1998) utilizes a fundamental parameter approach to model peak profiles (Cheary and Coelho, 1992) and allows modelling turbostratic disorder using a programming language-style data input. The model approximates the intensity distribution from the turbostratic stacking order by using a super cell, which is a multiple of the standard unit cell in the *c* direction. The use of this approach for the turbostratic disordered layer structure effectively models the highly asymmetrical broad peaks characteristic of smectite group minerals (Fig. 4.10).

Wang *et al.* (2011, 2012) applied the Ufer *et al.* (2004) turbostratic disorder model to the study of nontronite using the program TOPAS (Bruker AXS, 2009). Figure 4.11 shows excellent agreement between the calculated XRD pattern using a supercell model consisting of nine layers along the *c*-axis and the observed XRD pattern for nontronite.

4.4.6 Precision and Accuracy

Precision is relatively simple to determine by replicating a number of analyses and reporting the standard deviation of the mean values. Accuracy, however, is much more difficult to determine, particularly when measuring unknown samples containing a multitude of phases. The International Union of Crystallography Commission on Powder Diffraction organized the IUCr-CPD QPA in the late 1990s to examine the levels of accuracy, precision and lower limits of detection of relatively 'simple' multi-phase mixtures of known composition (Madsen *et al.*, 2001; Scarlett *et al.*, 2002). The round robin primarily focused on the analysis of powder diffraction data for quantitative phase analysis. In addition, the data could be validated by other techniques (i.e. FTIR, XRF, etc.). A total of 12 synthetic mixtures (samples labelled 1a–1h, 2, 3, 4 and synthetic bauxite) and one unknown sample (granodiorite) were distributed to 128 participants. The principal organizers also made available X-ray diffraction data for all samples in order to compare results independent of sample preparation and data collection issues or for participants without diffraction instruments. Results for samples 1a–1h from user-collected data were returned from 71

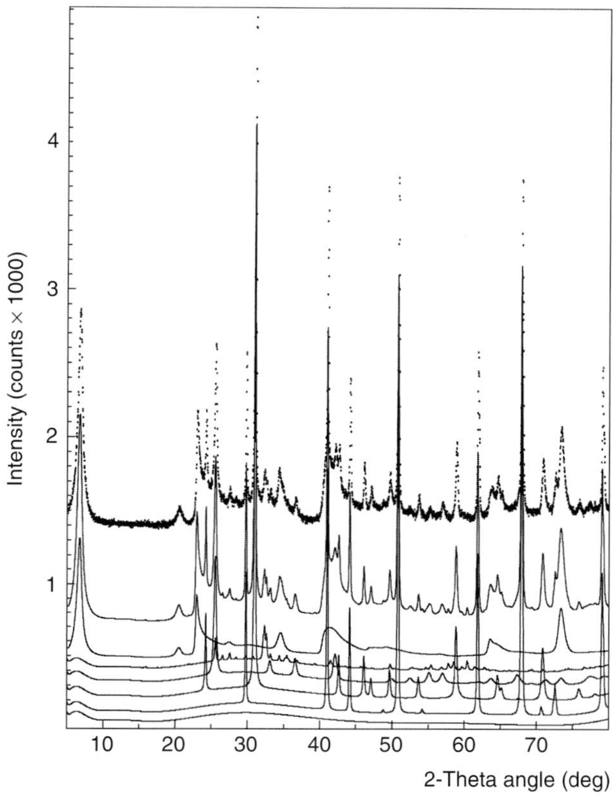

Figure 4.10. XRD patterns (Co Kα radiation) of observed (dots), calculated and individual phase profiles for a bentonite showing excellent fitting of clay and non-clay components using BGMN (Bergmann *et al.*, 1998). The patterns from bottom to top are background, corundum (internal standard), quartz, cristobalite, plagioclase feldspar, montmorillonite and pattern composed of the sum of all phases and background.

participants, with 48 participants returning results from the CPD measured data (Madsen *et al.*, 2001). Results for samples 2, 3, 4, synthetic bauxite and granodiorite were returned from 49 participants, with 27 analysing the CPD data (Scarlett *et al.*, 2002).

The majority of participants (80 per cent) used Rietveld techniques, with few participants using single-peak and whole-pattern RIR techniques. Results obtained for RIR methods suffered mostly from incorrect values used in the calculations. Confusion over how tables of RIR values were determined (peak height, peak area, whole pattern) also contributed to inaccuracies. Accuracy problems with users of Rietveld techniques were mostly because of the use of incorrectly assigned space groups, incorrect use of site occupancy factors and atom thermal parameters, and also a poor understanding of the physical meaning of the parameters being refined. Factors influencing the accuracy of all participants' results included inappropriate X-ray data collection methods, preferred orientation and

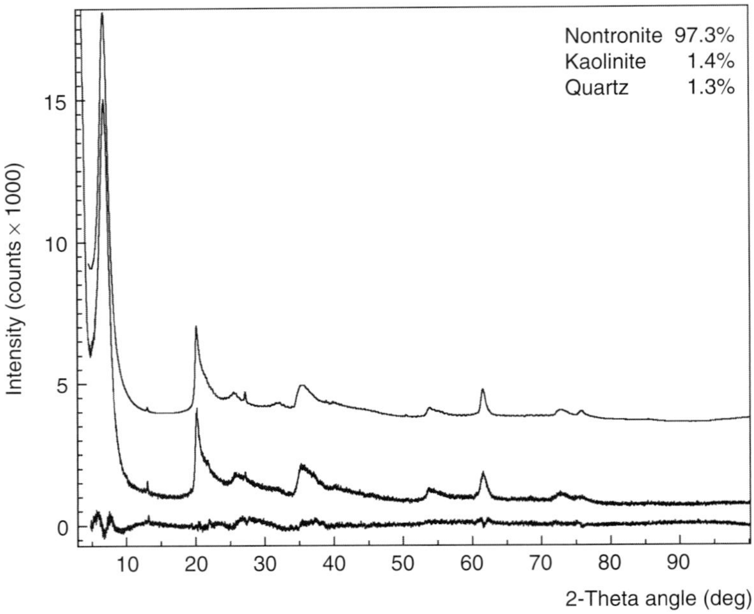

Figure 4.11. XRD patterns (Co Kα radiation) of calculated (uppermost pattern), observed (middle pattern) and difference (lowermost trace) for a nontronite-rich material showing excellent fitting of nontronite using the turbostratic disorder model in TOPAS (Wang *et al.*, 2011, 2012).

microabsorption. The use of 'standard' data collection strategies and better sample preparation practices can address data collection and preferred orientation issues, but microabsorption remains one of the greatest problems for quantitative phase analysis.

The importance of accurate quantitative analysis of disordered clay minerals, particularly for the oil exploration industry, prompted the establishment of the Reynolds Cup contest in 2000 (www.clays.org/Reynolds.html). The biennial contest is named in honour of Professor Bob Reynolds for his pioneering work in quantitative clay mineralogy and his exceptional contributions to clay science. The Reynolds Cup contest differs from other quantitative analysis round robin events (Madsen *et al.*, 2001; Ottner *et al.*, 2000; Scarlett *et al.*, 2002) in that samples are synthetic mixtures of pure mineral standards that represent realistic sedimentary rock compositions (Kleeberg, 2005; McCarty, 2005; Omotoso *et al.*, 2006). The international contest is open to anyone with an interest in quantitative mineralogical analysis, and participants are invited to use any technique available to them to perform their analysis. Three sample mixtures are prepared, containing on average between 12 and 20 clay and non-clay mineral mixtures. The accuracy of a particular quantification is judged by calculating a 'bias' for each phase in an assemblage. This is simply the absolute value of the difference between the measured and calculated mineral percentages. Determining the true amount of a phase in a sample would therefore give a bias of 0 per cent. A summary of the round robin results to date are given in Table 4.9. Average biases per mineral phase for the winners of the first four contests between 2002 and 2008

Table 4.9. *Summary of the seven Reynolds Cup round robin contests 2002 to 2014.*

Year	Participants	Returned results	Return rate (per cent)	No. of mineral phases	First place total bias (per cent)	First place average bias per mineral (per cent)	Average total bias for all participants (per cent)
2002	40	15	37.5	36	32.4	0.9	90.5
2004	60	35	58.3	34	14.0	0.4	102.4
2006	64	37	57.8	42	33.8	0.8	138.1
2008	53	42	79.2	35	35.2	1.0	162.3
2010	76	63	82.9	42	87.8	2.1*	269.2
2012	74	62	83.8	40	72.0	1.8*	213.4
2014	81	67	82.7	61	61.3	1.0*	195.2

* The method for assessing clay biases were changed for the 2010 to 2014 round robins.

remained fairly constant at a little under 1 per cent. Since 2010 the contest was assessed in greater detail, with the clay minerals in particular being judged to a higher level of detail. For example, di-octahedral 2:1 clay minerals (muscovite, illite, illite–smectite and smectite) and tri-octahedral 2:1 clays minerals (saponite, biotite and vermiculite) were grouped together prior to 2010, but were assessed separately from 2010 onward. As expected, average biases per mineral phase for the winners of the 2010 and 2012 contests almost doubled to 2 per cent; however, the most recent contest in 2014 saw this return to 1 per cent bias per mineral phase.

Techniques used for quantification by entrants in the Reynolds Cup included single-peak and whole-pattern XRD methods, as well as grain size separation, chemical analysis, SEM/TEM EDX, TG-DTA-DSC, and Mössbauer spectroscopy. For those using XRD, Rietveld methods remained the technique of choice for most participants (57 per cent) because of the perceived ease of use and non-reliance on mineral standards. Of the seven contests to date, two have been won by contestants using whole-pattern non-Rietveld techniques, with the remaining five winners using Rietveld techniques. Generally the higher-placed participants correctly identified all or most of the mineral phases present. Conversely the worst respondents failed to identify or misidentified phases. Several contestants reported a long list of minor exotic phases, most likely reported by automated search/match programs but mineralogically not feasible. Clay minerals were the greatest source of error reported.

The conclusion that can be drawn from the extensive dataset of the XRD results from the Reynolds Cup is that, if applied correctly, XRD can determine the concentrations of clays in a mineral assemblage to an absolute accuracy of 1 wt% for clays present at low concentrations and around 3 wt% for clays present at high concentrations. The detection limit ranges from 0.5 to 2 wt%, depending on the clay type.

4.5 Other Techniques for Identification and Quantification

While XRD is the most advantageous technique for the identification and quantification of clays and clay minerals, there are, of course, a myriad of other techniques that can be used to characterize clays in mineral assemblages. The more common of these techniques are summarized in Section 1.5 and more comprehensively reviewed in texts such as Wilson (1994), Bergaya *et al.* (2006) and Bergaya and Lagaly (2013). Many of these techniques are specific to a particular characteristic. For example, Mössbauer spectroscopy is mostly limited to the study of Fe-containing phases. In this section of this chapter only the more widely used, readily available laboratory-based techniques will be discussed.

4.5.1 Visible and Infrared Spectroscopy

Infrared spectroscopy is perhaps the only technique that can rival XRD as a method for the identification and quantification of clays in mineral assemblages. Visible and infrared spectroscopies as applied to clays are extensively discussed in Chapter 6. These spectroscopies, often jointly termed as vibrational spectroscopy, measure the absorption of electromagnetic radiation in the wavelength (λ) range from 0.35 to 25 μm. This wavelength range extends from the visible (0.35–0.75 μm) through the near-infrared (NIR) (0.75–2.5 μm) to the mid-infrared (MIR) (2.5–25 μm).[3] Identification of individual minerals is achieved by observing absorption bands at wavelengths characteristic of the mineral (e.g. Hunt, 1977). The theory of vibrational spectroscopy as applied to minerals can be found in Farmer (1974). Discussions of vibrational spectroscopy as applied specifically to clays and clay minerals can also be found in Farmer (1974) as well as Russell and Fraser (1994), Kloprogge (2005), Pansu and Gautheyrou (2006) and Yitagesu *et al.* (2011).

Quantification of the amounts of individual mineral phases in a mineral assemblage using visible and infrared spectroscopy can be achieved by representing an absorbance spectrum (absorbance = $\log(I_0 / I_t)$ where I_0 is the incident intensity and I_t is the intensity transmitted by the sample), or at least critical regions of a spectrum, as the weighted sum of the absorbance spectra of pure phases (Bertaux *et al.*, 1998; Pansu and Gautheyrou, 2006; van der Marel and Beutelspacher, 1976). A similar process can be applied to reflectance spectra (e.g. Viscarra Rossel, 2011). This method of quantification does have limitations (van der Marel and Beutelspacher, 1976) and more sophisticated and robust quantification methods based on principle components analysis (PCA), partial least squares regression (PLS) and discriminant analysis (DA) have been developed (Janik and Keeling, 1993; Ritz *et al.*, 2010; Viscarra Rossel *et al.*, 2006). These methods can be extended to determine a wide range of characteristics of mineral assemblages such as CEC, exchangeable cations and moisture content (Bertrand *et al.*, 2002; Janik *et al.*, 1998; Viscarra Rossel *et al.*, 2006).

[3] Wavenumber is often used as an alternate to wavelength when discussing visible and infrared spectra. Wavenumber ($\overline{V}$) (cm^{-1}) = 10 000 / λ for wavelength (λ) in μm.

The ability of visible and infrared reflectance spectroscopies to rapidly identify and quantify mineral abundances has not only been applied in the laboratory, but has found wide application at the mineral deposit level and even the global level (Kruse, 2010). Visible and infrared spectroscopies are applied at the global level via multi-spectral or hyperspectral satellite images from platforms such as Landsat and ASTER (see Chapter 6, Section 6.5.3). Multispectral images are images taken using several non-contiguous wavelength bands. For example, ASTER records images from 14 non-contiguous wavelength bands covering the wavelength range 0.52 to 11.65 μm at spatial resolutions ranging from 15 m (short wavelengths) to 90 m (long wavelengths) (http://asterweb.jpl.nasa.gov/characteristics.asp). Hyperspectral images are images taken using many contiguous wavelength bands. For example, the Hyperion system images 220 10-nm bands covering the wavelength range from 0.4 to 2.5 μm at a 30 m spatial resolution (Pearlman *et al.*, 2001).

Airborne visible and infrared systems such as HyMap (Cocks *et al.*, 1998) and AVIRIS (http://aviris.jpl.nasa.gov/aviris/index.html) allow mapping over wide regions (e.g. Chabrillat *et al.*, 2002) and are particularly useful for mapping at the deposit level (see Chapter 6 Section 6.5.3). Airborne systems allow greater payloads and more flexibility than satellite systems. For example, HyMap provides hyperspectral images in the wavelength range 0.44 μm to 2.5 μm, with spatial resolution between 3 m and 10 m (Kruse *et al.*, 2000; www.hyvista.com/?page_id=440). At the local level, portable systems such as the Analytical Spectral Devices Inc. instruments (www.asdi.com) can be used *in situ* to rapidly characterize mineral assemblages.

Satellite and airborne visible and infrared systems are necessarily limited to mapping at the Earth's surface. One area where visible and infrared spectroscopies have found great application is the rapid assessment of drill cores using systems such as the HyLogger (www.csiro.au/en/Research/MRF/Areas/Exploration-through-cover/Rapid-resource-characterisation/Hylogging). The HyLogger system uses diagnostic absorption bands in several sub-ranges of the visible and NIR wavelength range. More specifically, the visible to near-infrared (VNIR) (0.35–1.4 μm), the short-wavelength infrared (SWIR) (1.4–3 μm) and the thermal infrared (TIR) (8–15 μm) wavelength ranges. From the diagnostic absorption bands it is possible to quantify the mineralogical composition of drill cores for a suite of minerals, including clays (Huntington *et al.*, 2006;Mauger *et al.*, 2007; Tappert *et al.*, 2011; see also Chapter 6 Section 6.5.4).

Before leaving the subject of infrared, it is worthwhile mentioning Raman spectroscopy (Larkin, 2011). Although Raman spectroscopy can be used for the identification of clays (e.g. Frost, 1995; Frost and Kloprogge, 2000; Joussein *et al.*, 2005), this form of spectroscopy has found most application in characterizing the adsorption and intercalation of clays by various inorganic and organic species (e.g. Frost and Martens, 2005; Johnston, 1990). Consequently, both Raman and infrared spectroscopies are widely used in studies of the interaction of the surfaces of clays with various compounds, such as in studies of the flocculation of clays by organic flocculants.

4.5.2 Electron Microscopy

The electron microscopies, that is, scanning electron microscopy (SEM) and transmission electron microscopy (TEM) can be used to identify and quantify clay mineral content using specialized techniques and equipment such as mineral liberation analysis (MLA) (Gu, 2003) and QEM*SEM (Hagni, 2008; Miller *et al.*, 1983). These techniques combine SEM imaging (in particular, back-scattered electron (BSE) imaging) and elemental analysis (usually using energy-dispersive X-ray (EDX) systems) to identify minerals. The intensity in BSE images of a material is positively correlated with the mean atomic number of the material (Goldstein *et al.*, 1992). For example, an iron sulphide will be much brighter in a BSE image than a phyllosilicate under the same imaging conditions because the mean atomic number of the iron sulphide is greater than the mean atomic number of a phyllosilicate. By combining this contrast difference with elemental analysis, it is possible to clearly identify the mineral phases in a mineral assemblage. Quantification of the content of each phase is achieved by analysing the areas occupied by each of the identified minerals in SEM images. Once the minerals have been identified, standard image analysis techniques can be used to determine various properties such as the morphological properties of the minerals (e.g. shape) and the associations between the various minerals in the mineral assemblage.

Without specialized equipment such as MLA and QEM*SEM, the electron microscopies are not well suited for the characterization of the bulk properties of mineral assemblages. However, these techniques have found wide application in areas such as the associations between various constituents of mineral assemblages (e.g. in weathering studies), the physical structure of mineral assemblages (e.g. particle morphology and flocc structure) and in clay–chemical interactions (e.g. the intercalation of clays). The electron microscopies are also very useful at identifying trace minerals that are present at levels below the detection limits of other techniques.

SEM imaging is particularly useful for characterizing the 3D structure of clay aggregates. Figure 4.12 shows secondary electron SEM images of kaolinite and halloysite. These images clearly show the detail that can be achieved using SEM imaging. In Fig. 4.12 the kaolinite stacks and halloysite tubes are clearly recognizable from their characteristic morphologies. Many other clays can be recognized in SEM images from their characteristic shapes (see, for example, Sudo *et al.*, 1981 and also Figs 2.2 and 2.3 in Chapter 2) and, with the aid of ancillary techniques such as EDX spectroscopy, can be unambiguously identified.

One of the disadvantages of conventional SEM imaging is that the sample is subjected to high vacuum and, therefore, the clays will be in a dehydrated state. This limitation can be overcome by using cryo-SEM, in which a hydrated sample is rapidly frozen and ice is removed by sublimation to reveal the 'as-hydrated' structure (Echlin, 1992). This technique has been very successful in the study of clay flocculation under the influence of various flocculants (McFarlane *et al.*, 2006;Morris and Żbik, 2009;Żbik *et al.*, 2008), enabling images to be obtained such as that shown in Fig. 4.13. This image shows the

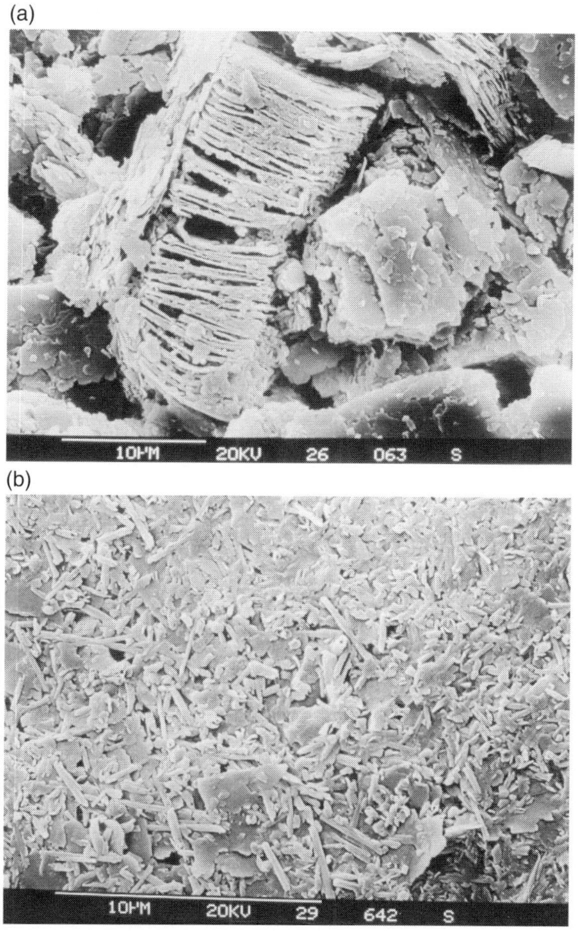

Figure 4.12. SEM images of kaolin from Mt Hope, South Australia. (a) Kaolinite stack among kaolinite fragments. (b) Halloysite sitting on kaolinite platelets. Images courtesy of J. Keeling, DIMTRE, South Australia.

smectite has flocculated in an open 'honeycomb' structure. By using image analysis techniques such as STIMAN (Sergeev *et al.*, 1984a, 1984b) it is possible to quantify the pore parameters.

Another method of imaging hydrated structures is to use environmental SEM (ESEM) (Danilatos, 1991, 1994). The environment around the specimen in an ESEM can be varied from high to low vacuum. Under low-vacuum conditions the environment immediately surrounding the specimen can be maintained at the saturation vapour pressure of water. That is, the specimen can be maintained and imaged in a fully hydrated state. Consequently, ESEM is well suited for the study of the hydration–dehydration properties of clays (Baker *et al.*, 1993; Huggett and Uwins, 1994; Montes-H *et al.*, 2005).

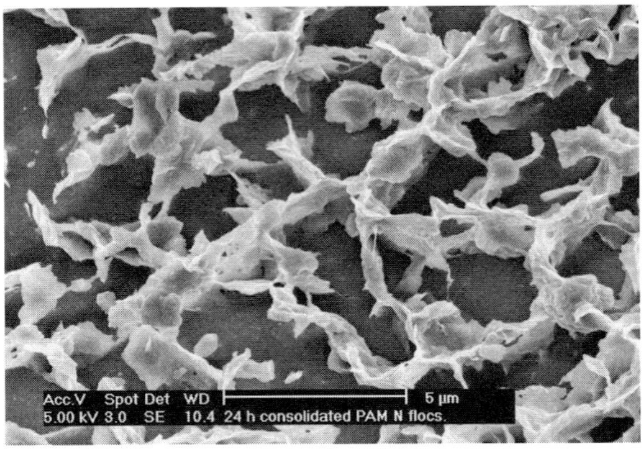

Figure 4.13. Cryo-SEM image of Na-exchanged smectite flocculated under the influence of non-ionic polyacrylamide homopolymer. Image courtesy of A. McFarlane, CSIRO, CESRE (McFarlane *et al.*, 2006).

In a similar manner to SEM, TEM uses characteristic morphology and elemental analysis to unambiguously identify clays (e.g. Sudo *et al.*, 1981; see also Fig. 2.3 in Chapter 2). When used in combination with other techniques, TEM can give very detailed analyses of clays. For example, electron energy loss spectroscopy (EELS) (Buseck and Self, 1992; Egerton, 2011) can be used to determine a number of properties, including elemental concentrations (Egerton, 2011; Laribi *et al.*, 2007), valence (Otten and Buseck, 1987; van Aken *et al.*, 1998) and atomic coordination (Huggett *et al.*, 2006).

One of the most common uses of TEM for the study of clays is high-resolution imaging of layer structure (Gutherie and Veblen, 1990; MacKinnon and Mumpton, 1990). For high-resolution TEM (HRTEM), samples are usually embedded in resin and either ultramicrotomed or ion-beam thinned to a thickness suitable for HRTEM (Elsass *et al.*, 2008). In this way, clays can be viewed edge-on (i.e. parallel to the clay layers). As a consequence it is possible to view in detail the stacking sequence of clays that range from well-crystalline (Gutherie and Veblen, 1990; Sudo *et al.*, 1981) through mixed-layer clays and clay intergrowths (Baronnet, 1992; Klimentidis and MacKinnon, 1986; Środoń, 1999) to highly interstratified clays (Laird and Nater, 1993; Nieto and Arroyo, 2010; Środoń *et al.*, 1990; Vali and Köster, 1986). Similarly, HRTEM can be used to study intercalated clays (Clinard *et al.*, 2003; Ghabru *et al.*, 1989; Vali and Hesse, 1990, 1992; Yu *et al.*, 2006).

4.5.3 Other and Emerging Techniques

There are, of course, many additional techniques used in the analysis of clays but which are not necessarily suited to quantification of the individual components of a mineral assemblage. Many of these techniques are detailed in Chapter 1 and elsewhere in this book. Other techniques that are worth noting are X-ray fluorescence (XRF) (Bain *et al.*, 1994; Norrish and Chappell, 1977; Norrish and Hutton, 1969; Willis *et al.*, 2014) and electron microprobe analysis (Guillemette, 2008; Velde, 1984). These techniques are readily available and provide very detailed information on specific properties of clays.

Other, relatively new, techniques that show great potential in the characterization of clays are scanned-probed microscopy (i.e. AFM and related techniques) (Nagy and Blum, 1994), X-ray absorption spectroscopy (Gates, 2006) and small-angle scattering (Tchoubar and Cohaut, 2006). These techniques have advanced considerably in recent years because of developments such as high-intensity X-ray sources (synchrotron radiation) and high-intensity light sources (i.e. compact high-intensity lasers).

Perhaps the most exciting emerging technique that is relevant to the study of clays and, in particular, the study of pore structure, not only in clay flocs but also in any mineral assemblage, is X-ray, micro, computer-aided tomography (μCT) (Żbik and Frost, 2009; Żbik *et al.*, 2010, 2012). This technique uses standard X-ray tomographic methods to create a full three-dimensional rendering of an object (e.g. Salvo *et al.*, 2003). X-ray μCT equipment is available in various configurations, including bench-top units and as attachments to scanning electron microscopes. The highest resolution μCT images are obtained using the high-intensity X-ray beams obtained from a synchrotron source, and it is from synchrotron-based instruments that much of the future development in the μCT of clays will occur (e.g. beamlines ID19 at the ESRF, 2BM at the APS, BLXU20/49 at SPRING-8).

4.6 Summary

Clays are distinguished by their small particle size, often defined as being less than 2 μm in dimension. Mineralogically, clays are defined as having a crystal structure based on planes of oxygen atoms and OH groups. The O/OH planes that make up the clay structures are stacked in such a way as to form sheets of octahedrally coordinated metals (predominantly Al^{3+}, Mg^{2+} and Fe^{2+}) and sheets of tetrahedrally coordinated metals (Si^{4+}, often substituted by Al^{3+} and Fe^{3+}). In general terms it is the sheet structure of clays (sheet silicates) that is used by methods that identify and quantify clays. In the 1:1 clays the sheets stack in a tetrahedral–octahedral (T–O) sequence and in the 2:1 clays the sheets stack in a tetrahedral–octahedral–tetrahedral (T–O–T) sequence, thereby, forming the base layers of the crystal structures of clays.

Depending on the valence state of the metals in the octahedral sheets, the octahedral sites can either be completely filled (for metals with +2 valence) or only two in every three sites filled (for metals with +3 valence). This gives rise to the descriptors tri-octahedral and di-octahedral for the clay structures. Depending on the metals in the octahedral and tetrahedral sheets, the base layers of the 2:1 clays can have a permanent (negative) charge. This charge ranges from zero (e.g. talc and pyrophyllite), through low charge (e.g. smectites), to intermediate charge (e.g. vermiculites) and high charge (e.g. illite and micas). The base layers of the 1:1 clays (e.g. kaolinite) have zero or very low charge. The charge of the base layers determines how the base layers stack to form the complete structure. For base layers of zero charge, the bonding between base layers is either van der Waals (e.g. talc) or hydrogen bonding (e.g. kaolinite). For base layers carrying a permanent charge the charge must be balanced by positively charged species (e.g. cations) situated between each base layer. For the smectites and vermiculites, under ambient conditions these positively charged interlayer species are usually accompanied by water molecules. The water molecules disperse the charge over the interlayer surfaces of the base layers. The amount of water in the interlayer is dependent on factors such as relative humidity and the energy of hydration of the interlayer cations. The interlayer water can move in and out of the interlayer in a reversible fashion leading to unique shrink/swell characteristics for the smectites and vermiculites. The low charge of the base layers and the presence of water in the interlayer leads to several characteristic properties for smectites and vermiculites. These characteristic properties include high cation exchange capacities and high specific surface areas.

Apart from the smectites and vermiculites interlayer, water can be found in other clays such as halloysite and the illites. However, in most cases the dehydration of these clays is not reversible. Reversible hydration and dehydration is found in sepiolite and palygorskite, in which O/OH sheets are arranged to form tunnel structures.

In addition to 1:1 and 2:1 layer structures a common group of clays is the chlorites that have a 2:1:1 layer structure. In the chlorite structure there is a positively charged T–O–T layer and a negatively charged layer consisting of two planes of OH groups containing octahedrally coordinated metals (cf. brucite). Chlorites are often described as consisting of an alternation of 'talc-like' and 'brucite-like' layers. Although this nomenclature adequately describes the layer structure of the chlorites it is somewhat of a misnomer because of the charge carried by the layers in chlorite.

The symmetry of base layers (approximately hexagonal) means that the stacking from one base layer to the next is not constrained to a particular orientation. Consequently, clay structures can take a variety of polytypes. In broad terms, the stacking within polytypes can be described in terms of layer rotations, layer shifts and changes in the site occupancies for the octahedral cations. In addition to the regular polytypes, clay structures can contain random variations in stacking sequences. The extreme case, and one that is not uncommon in smectites, is turbostratic stacking, in which the base layers stack in totally random orientations. The stacking sequence of base layers can be further complicated by the interstratification of two or more clay base layers. Methods that are used to identify clays must be able to recognize the characteristic crystal structures of clays and be able to distinguish between the variations in clay structures brought about by factors such as atomic substitutions, interlayer cations, hydration state, regularity of stacking and interstratification.

In order to completely identify clays it is often necessary to separate the clays from other minerals in a mineral assemblage. Clay separation methods invariably take advantage of the small size of clay particles by using gravity sedimentation or centrifugation to isolate clays. The separation of clays may be carried out concurrently with other treatments such as acid washing to remove carbonates, dithionite treatment to remove iron oxyhydroxides and cation exchange to make the clays homo-ionic.

The most widely used technique for the identification and quantification of clays is powder X-ray diffraction (XRD). The first stage in the identification and quantification of clays is to recognize the characteristic basal spacings of the clays. These spacings can be divided into five groups ranging from 7 Å for the 1:1 clays through to >20 Å for the interstratified clays. Identification may be aided by automatic search/match programs using databases of powder XRD patterns, but when using such programs it is necessary to recognize misidentified phases and the presence of phases that are often associated with clays, such as iron oxides and carbonates. Search/match programs will often not identify clays because of the variability in the XRD pattern of a given clay caused by factors such as variability in the degree of hydration of interlayer cations. Having correlated basal spacing with a particular clay or group of clays, X-ray reflections corresponding to non-basal spacings can be used to further identify specific clays. The 060 X-ray reflection is of particular use in distinguishing between di-octahedral and tri-octahedral clays.

In some cases, identification of specific clays can only be made by applying a set of determinative treatments to a sample. These treatments include cation exchange, glycerol/ethylene glycol/formamide solvation, heat treatments and preparation of oriented sample mounts.

Three standard methods for quantification of clay content by XRD are commonly used. These three methods are:

(1) comparison of the intensity of a single peak or group of peaks with the intensity of peaks from a standard material;
(2) whole-pattern fitting where the measured XRD pattern from an unknown sample is described as the weighted sum of patterns from individual pure reference materials; and
(3) Rietveld analysis, which uses a least squares fit of XRD patterns calculated from known crystal structures or standard XRD patterns of pure phases to quantify the amount of each phase in a mineral assemblage.

Of these methods, the Rietveld method is generally accepted as being the most comprehensive. The Rietveld method can also be used to refine structural parameters such as unit cell dimensions, atomic coordinates and site occupancies. Non-crystalline material content is determined by difference, with the diffracted intensities referenced against either an internal or external standard. Round robin tests show that, if applied correctly, XRD can determine the concentrations of clays in a mineral assemblage to an absolute accuracy of 1 wt% for clays present at low concentrations and around 3 wt% for clays present at high concentrations. The detection limit ranges from 0.5 wt% to 2 wt%, depending on the clay type. The biggest source of error in quantification by XRD arises from preferred orientation, microabsorption, stacking disorder and interstratified phases.

Apart from XRD, there are many other methods for the identification of clays. These methods are discussed in Chapter 1. Vibrational spectroscopy (visible and infrared) has found wide use in the study of clays and can be used to quantify clay content. Electron microscopy is particularly useful for detailed morphological studies and for characterizing the associations that occur between different types of clays, as well as associations between clays and other minerals. Other emerging techniques for clay studies are scanned-probe microscopy, small-angle scattering and micro, computer-aided tomography.

Computer-aided tomography using synchrotron X-ray sources shows considerable promise for three-dimensional study of flocs and the distribution of clays in mineral assemblages.

References

Alcover, J. F., Gatineau, L. & Mering, J. 1973. Exchangeable cation distribution in nickel-vermiculites and magnesium-vermiculites. *Clays and Clay Minerals*, 21 (2), 131–136.

American Petroleum Institute. 2009. *ANSI/API RP 13B-1. Recommended Practice for Field Testing Water-Based Drilling Fluids*, 4th edition (Identical to ISO 10414-1:2008). Washington, DC: American Petroleum Institute.

Bailey, S. W. 1988a. Polytypism of 1:1 layer silicates. In: Bailey, S. W. (ed.) *Hydrous Phyllosilicates (Exclusive of Micas)*. Washington, DC: Mineralogical Society of America.

Bailey, S. W. 1988b. Chlorites: Structure and crystal chemistry. In: Bailey, S. W. (ed.) *Hydrous Phyllosilicates (Exclusive of Micas)*. Washington, DC: Mineralogical Society of America.

Bailey, S. W. 1980. Structures of layer silicates. In: Brindley, G. W. & Brown, G. (eds) *Crystal Structures of Clay Minerals and their X-ray Identification*, new edition. London: Mineralogical Society.

Bailey, S. W., Hurley, P. M., Fairbairn, H. W. & Pinson, W. H. 1962. K–Ar dating of sedimentary illite polytypes. *Geological Society of America Bulletin*, 73 (9), 1167–1170.

Bailey, S. W., Brindley, G. W., Fanning, D. S., Kodama, H. & Martin, R. T. 1984. Report of the Clay Mineral Society Nomenclature Committee for 1982 and 1983. *Clays and Clay Minerals*, 32 (3), 239–240.

Bain, D. C., McHardy, W. J. & Lachowski, E. E. 1994. X-ray fluorescence spectroscopy and microanalysis. In: Wilson, M. J. (ed.) *Clay Mineralogy: Spectroscopic and Chemical Determinative Methods*. Amsterdam: Springer.

Baker, J. C., Uwins, P. J. R. & MacKinnon, I. D. R. 1993. ESEM study of illite/smectite freshwater sensitivity in sandstone reservoirs. *Journal of Petroleum Science and Engineering*, 9 (2), 83–94.

Baronnet, A. 1992. Polytypism and stacking disorder. In: Buseck, P. R. (ed.) *Minerals and Reactions at the Atomic Scale: Transmission Electron Microscopy*. Washington, DC: Mineralogical Society of America.

Barshard, I. 1948. Vermiculite and its relation to biotite as revealed by base exchange reactions, X-ray analyses, differential thermal curves and water content. *American Mineralogist*, 33 (11–12), 655–678.

Battaglia, S., Leoni, L. & Sartori, F. 2006. A method for determining the CEC and chemical composition of clays via XRF. *Clay Minerals*, 41 (3), 717–725.

Bergaya, F. & Lagaly, G. (eds) 2013. *Handbook of Clay Science*. Amsterdam: Elsevier B.V.

Bergaya, F., Theng, B. K. G. & Lagaly, G. (eds) 2006. *Handbook of Clay Science*. Amsterdam: Elsevier.

Bergmann, J. & Kleeberg, R. 1998. Rietveld analysis of disordered layer silicates. *Materials Science Forum*, 278 (2), 300–305.

Bergmann, K. & O'Konski, C. T. 1963. A spectroscopic study of methylene blue monomer, dimer and complexes with montmorillonite. *Journal of Physical Chemistry*, 67, 2169–2177.

Bergmann, J., Friedel, P. & Kleeberg, R. 1998. BGMN a new fundamental parameter based Rietveld program for laboratory X-ray sources, its use in quantitative analysis and structure investigations. *Commission of Powder Diffraction: International Union of Crystallography, CPD Newsletter*, 20, 5–8.

Bertaux, J., Frohlich, F. & Ildefonse, P. 1998. Multicomponent analysis of FTIR spectra: Quantification of amorphous and crystallized mineral phases in synthetic and natural sediments. *Journal of Sedimentary Research*, 68 (3), 440–447.

Bertrand, I., Janik, L. J., Holloway, R. E., Armstrong, R. D. & McLaughlin, M. J. 2002. The rapid assessment of concentrations and solid phase associations of macro- and micronutrients in alkaline soils by mid-infrared diffuse reflectance spectroscopy. *Soil Research*, 40 (8), 1339–1356.

Biscoe, J. & Warren, B. E. 1942. An X-ray study of carbon black. *Journal of Applied Physics*, 13 (6), 364–371.

Bradley, W. F. 1950. The alternating layer sequence of rectorite. *American Mineralogist*, 35 (3–4), 278.

Brindley, G. W. 1945. The effect of grain or particle size on x-ray reflections from mixed powders and alloys, considered in relation to the quantitative determination of crystalline substances by x-ray methods. *Philosophical Magazine Series 7*, 36 (256), 347–369.

Brindley, G. W. 1956. Allevardite, a swelling double-layer mica mineral. *American Mineralogist*, 41 (1–2), 91–103.

Brindley, G. W. & Brown, G. (eds) 1980. *Crystal Structures of Clay Minerals and Their X-ray Identification.* London: Mineralogical Society.

Brown, B. E. & Bailey, S. W. 1962. Chlorite polytypism:.1. Regular and semi-random 1-layer structures. *American Mineralogist*, 47 (7–8), 819–850.

Bruker AXS. 2009. *Topas v4.2: General Profile and Structure Analysis Software for Powder Diffraction Data*. Karlsruhe: Bruker AXS.

Brunauer, S., Emmett, P. H. & Teller, E. A. 1938. Adsorption of gases in multimolecular layers. *Journal of the American Chemical Society*, 60, 309–319.

Buseck, P. R. & Self, P. G. 1992. Electron energy-loss spectroscopy (EELS) and electron channelling (ALCHEMI). In: Buseck, P. R. (ed.) *Minerals and Reactions at the Atomic Scale: Transmission Electron Microscopy.* Washington, DC: Mineralogical Society of America.

Capitani, G. C. & Mellini, M. 2007. High-resolution transmission electron microscopy (HRTEM) investigation of antigorite polysomes (m = 15 to 18). *American Mineralogist*, 92 (1), 64–71.

Carrado, K. A., Decarreau, A., Petit, S., Bergaya, F. & Lagaly, G. 2006. Synthetic clay minerals and the purification of natural clays. In: Bergaya, F., Theng, B. K. G. & Lagaly, G. (eds) *Handbook of Clay Science.* Amsterdam: Elsevier.

Cases, J. M., Grillet, Y., François, M., *et al.* 1991. Evolution of the porous structure and surface area of palygorskite under vacuum thermal treatment. *Clays and Clay Minerals*, 39, 191–201.

Chabrillat, S., Goetz, A. F. H., Krosley, L. & Olsen, H. W. 2002. Use of hyperspectral images in the identification and mapping of expansive clay soils and the role of spatial resolution. *Remote Sensing of Environment*, 82 (2), 431–445.

Chao, T. T. 1984. Use of partial dissolution techniques in geochemical-exploration. *Journal of Geochemical Exploration*, 20 (2), 101–135.

Chao, T. T. & Sanzolone, R. F. 1977. Chemical dissolution of sulphide minerals. *Journal of Research of the US Geological Survey*, 5, 409–412.

Cheary, R. W. & Coelho, A. 1992. A fundamental parameters approach to x-ray line-profile fitting. *Journal of Applied Crystallography*, 25 (2), 109–121.

Chipera, S. J. & Bish, D. L. 2002. FULLPAT: A full-pattern quantitative analysis program for X-ray powder diffraction using measured and calculated patterns. *Journal of Applied Crystallography*, 35 (6), 744–749.
Chukhrov, F. V., Gorshkov, A. I., Rudnitskaya, E. S., Beresovskaya, V. V. & Sivtsov, A. V. 1980. Manganese minerals in clays: A review. *Clays and Clay Minerals*, 28 (5), 346–354.
Chung, F. H. 1974a. Quantitative interpretation of X-ray-diffraction patterns of mixtures: 1. Matrix-flushing method for quantitative multicomponent analysis. *Journal of Applied Crystallography*, 7 (6), 519–525.
Chung, F. H. 1974b. Quantitative interpretation of X-ray-diffraction patterns of mixtures: 2. Adiabatic principle of X-ray-diffraction analysis of mixtures. *Journal of Applied Crystallography*, 7 (6), 526–531.
Churchman, G. J., Whitton, J. S., Claridge, G. G. C. & Theng, B. K. G. 1984. Intercalation method using formamide for differentiating halloysite from kaolinite. *Clays and Clay Minerals*, 32 (4), 241–248.
Churchman, G. J., Pasbakhsh, P., Lowe, D. J. & Theng, B. K. G. 2016. Unique but diverse: Some observations on the formation, structure and morphology of halloysite. *Clay Minerals*, 51 (3), 395–416.
Clinard, C., Mandalia, T., Tchoubar, D. & Bergaya, F. 2003. HRTEM image filtration: Nanostructural analysis of a pillared clay. *Clays and Clay Minerals*, 51 (4), 421.
Cocks, T., Jenssen, R., Stewart, A., Wilson, I. & Shields, T. 1998. The HyMap airborne hyperspectral sensor: The system, calibration and performance. In: Schaepman, M., Schläpfer, D. R. & Itten, K. (eds) *1st EARSEL Workshop on Imaging Spectroscopy, 6–8 October 1998*. Zurich: SUI. EARSel, 1–7.
Cornell, R. M. & Schwertmann, U. 2003. *The Iron Oxides: Structure, Properties, Reactions, Occurrences, and Uses.* Weinheim: Wiley-VCH.
Cullity, B. D. 1978. *Elements of X-ray Diffraction.* Reading, MA: Addison-Wesley.
Danilatos, G. D. 1991. Review and outline of environmental SEM at present. *Journal of Microscopy*, 162 (3), 391–402.
Danilatos, G. D. 1994. Environmental scanning electron microscopy and microanalysis. *Mikrochimica Acta*, 114–115, 143–155.
de la Calle, C. & Suquet, H. 1988. Vermiculite. In: Bailey, S. W. (ed.) *Hydrous Phyllosilicates (Exclusive of Micas).* Madison, WI: Mineralogical Society.
Dinnebier, R. E. & Billinge, S. J. L. (eds) 2008. *Powder Diffraction: Theory and Practice.* Cambridge: Royal Society of Chemistry.
Dixon, J. B. & Weed, S. B. (eds) 1989. *Minerals in Soil Environments.* Madison, WI: Soil Science Society of America.
Dódony, I. & Buseck, P. R. 2004. Serpentines close-up and intimate: An HRTEM view. *International Geology Review*, 46 (6), 507–527.
Dódony, I., Pósfai, M. & Buseck, P. R. 2002. Revised structure models for antigorite: An HRTEM study. *American Mineralogist*, 87 (10), 1443–1457.
Drever, J. I. 1973. The preparation of oriented clay mineral specimens for X-ray diffraction analysis by a filter-membrane peel technique. *American Mineralogist*, 58 (5–6), 553–554.
Drits, V. A. & Tchoubar, C. 1990. *X-ray Diffraction by Disordered Lamellar Structures: Theory and Applications to Microdivided Silicates and Carbons.* Berlin: Springer-Verlag.
Duan, X. & Evans, D. G. (eds) 2006. *Layered Double Hydroxides.* Berlin: Springer.
Eberl, D. D. 2003. User's guide to ROCKJOCK – a program for determining quantitative

mineralogy from powder X-ray diffraction data. US Geological Survey Open-File Report [Online]. Available: ftp://brrcrftp.cr.usgs.gov/pub/ddeberl.

Echlin, P. 1992. *Low-Temperature Microscopy and Analysis.* New York: Springer.

Egerton, R. F. 2011. *Electron Energy-loss Spectroscopy in the Electron Microscope.* New York: Springer.

Elsass, F., Chenu, C. & Tessier, D. 2008. Transmission electron microscopy for soil samples: Preparation methods and use. In*:* Ulery, A. L. & Richard Drees, L. (eds) *Methods of Soil Analysis, Part 5 – Mineralogical Methods, 5.5.* Madison, WI: Soil Science Society of America.

Farmer, V. C. (ed.) 1974. *The Infrared Spectra of Minerals.* London: Mineralogical Society.

Frost, R. L. 1995. Fourier transform Raman spectroscopy of kaolinite, dickite and halloysite. *Clays and Clay Minerals*, 43 (2), 191–195.

Frost, R. L. & Kloprogge, J. T. 2000. Vibrational spectroscopy of ferruginous smectite and nontronite. *Spectrochimica Acta Part A: Molecular and Biomolecular Spectroscopy*, 56 (11), 2177–2189.

Frost, R. L. & Martens, W. N. 2005. Raman spectroscopy of kaolinite and selected intercalates. In*:* Kloprogge, J. T. (ed.) *The Application of Vibrational Spectroscopy to Clay Minerals and Layered Double Hydroxides.* Aurora, CO: The Clay Mineral Society.

Gates, W. P. 2006. X-ray absorption spectroscopy. In*:* Bergaya, F., Theng, B. K. G. & Lagaly, G. (eds) *Handbook of Clay Science.* Amsterdam: Elsevier.

Gates, W. P., Slade, P. G., Manceau, A. & Lanson, B. 2002. Site occupancies by iron in nontronites. *Clays and Clay Minerals*, 50 (2), 223–239.

Ghabru, S. K., Mermut, A. R. & St. Arnaud, R. J. 1989. Layer-charge and cation-exchange capacity characteristics of vermiculite (weathered biotite) isolated from a gray luvisol in northeastern Saskatchewan. *Clays and Clay Minerals*, 37 (2), 164–172.

Giese Jr., R. F. & Costanzo, P. M. 1987. Behavior of water on the surface of kaolin minerals. In*:* Davis, J. A. & Hayes, K. F. (eds) *Geochemical Processes at Mineral Surfaces.* Washington, DC: American Chemical Society.

Goldstein, J. I., Newbury, D. E., Echlin, P., *et al.* 1992. *Scanning Electron Microscopy and X-ray Microanalysis: A Text for Biologists, Material Scientists, and Geologists.* New York: Plenum Press.

Gražulis, S., Chateigner, D., Downs, R. T., *et al.* 2009. Crystallography Open Database: An open-access collection of crystal structures. *Journal of Applied Crystallography*, 42 (4), 726–729.

Grim, R. E. 1968. *Clay Mineralogy.* New York: McGraw-Hill.

Grim, R. E., Bray, R. H. & Bradley, W. F. 1937. The mica in argillaceous sediments. *American Mineralogist*, 22 (7), 813–829.

Gu, Y. 2003. Automated scanning electron microscope based mineral liberation analysis. *Journal of Minerals and Materials Characterization and Engineering*, 2 (1), 33–41.

Guggenheim, S., Adams, J. M., Bain, D. C., *et al.* 2006. Summary of recommendations of nomenclature committees relevant to clay mineralogy: Report of the Association Internationale pour l'Etude des Argiles (AIPEA) Nomenclature Committee for 2006. *Clay Minerals*, 41 (4), 863–877.

Guggenheim, S., Brown, R., Daniels, E., *et al.* 2011. The Clay Minerals Society Glossary for Clay Science Project. The Clay Minerals Society [Online]. Available: www.clays.org/GLOSSARY/GlossIntro.html.

Guillemette, R. N. 2008. Electron microprobe techniques. In*:* Ulery, A. L. & Richard Drees, L. (eds) *Methods of Soil Analysis Part 5 – Mineralogical Methods, 5.5.* Madison, WI: Soil Science Society of America.

Gutherie Jr, G. D. & Veblen, D. R. 1990. High-resolution transmission electron microscopy applied to clay minerals. In: Coyne, L. M., Mckeever, S. W. S. & Blake, D. F. (eds) *Spectroscopic Characterisation of Minerals and Their Surfaces*. Washington, DC: American Chemical Society.

Güven, N. 1988. Smectites. In: Bailey, S. W. (ed.) *Hydrous Phyllosilicates (Exclusive of Micas)*. Washington, DC: Mineralogical Society of America.

Hagni, A. M. 2008. Phase identification, phase quantification, and phase association determinations utilizing automated mineralogy technology. *JOM*, 60 (4), 33–37.

Haynes, W. M., Lide, D. R. & Bruno, T. J. (eds) 2012. *CRC Handbook of Chemistry and Physics*. Boca Raton, FL: CRC Press.

Hill, R. J. & Howard, C. J. 1987. Quantitative phase-analysis from neutron powder diffraction data using the Rietveld method. *Journal of Applied Crystallography*, 20 (6), 467–474.

Hillier, S. 1999. Use of an air brush to spray dry samples for X-ray powder diffraction. *Clay Minerals*, 34 (1), 127–135.

Hubbard, C. R. & Snyder, R. L. 1988. RIR: Measurement and use in quantitative XRD. *Powder Diffraction*, 3 (2), 74–77.

Hubbard, C. R., Evans, E. H. & Smith, D. K. 1976. The reference intensity ratio, I/Ic, for computer simulated powder patterns. *Journal of Applied Crystallography*, 9 (2), 169–174.

Hudson, L. K., Misra, C., Perrotta, A. J., Wefers, K. & Williams, F. S. 2000. Aluminum oxide. In: *Ullmann's Encyclopedia of Industrial Chemistry*. New York: Wiley & Sons.

Huggett, J. M. & Uwins, P. J. R. 1994. Observations of water–clay reactions in water-sensitive sandstone and mudrocks using an environmental scanning electron microscope. *Journal of Petroleum Science and Engineering*, 10 (3), 211–222.

Huggett, J. M., McCarty, D. K., Calvert, C. C., Gale, A. S. & Kirk, C. 2006. Serpentine–nontronite–vermiculite mixed-layer clay from the Weches formation, Claiborne group, middle Eocene, northeast Texas. *Clays and Clay Minerals*, 54 (1), 101.

Hunt, G. R. 1977. Spectral signatures of particulate minerals in the visible and near infrared. *Geophysics*, 42 (3), 501–513.

Huntington, J. F., Mauger, A. J., Skirrow, R. G., *et al.* 2006. Automated mineralogical core logging at the Emmie Bluff iron oxide–copper–gold prospect. *MESA Journal*, 41, 38–44.

Ibers, J. A. & Hamilton, W. C. 1974. *International Tables for X-ray Crystallography*. Birmingham: Kynoch Press.

Jackson, M. L. 1956. *Soil Chemical Analysis: Advanced Course*. Madison, WI: Department of Soils, University of Wisconsin.

Janik, L. J. & Keeling, J. L. 1993. FT-IR partial least-squares analysis of tubular halloysite in kaolin samples from the Mount Hope kaolin deposit. *Clay Minerals*, 28 (3), 365–378.

Janik, L. J., Merry, R. H. & Skjemstad, J. O. 1998. Can mid infrared diffuse reflectance analysis replace soil extractions? *Australian Journal of Experimental Agriculture*, 38 (7), 681–696.

Jenkins, R. & Snyder, R. L. 1996. *Introduction to X-ray Powder Diffractometry*. New York: Wiley & Sons.

Johnston, C. T. 1990. Raman and FT-IR spectra of kaolinite–hydrazine intercalate. In: Coyne, L. M., Mckeever, S. W. S. & Blake, D. F. (eds) *Spectroscopic Characterization of Minerals and Their Surfaces*. Washington, DC: American Chemical Society.

Jones, B. F. & Galan, E. 1988. Sepiolite and palygorskite. In: Bailey, S. W. (ed.) *Hydrous Phyllosilicates (Exclusive of Micas)*. Madison, WI: Mineralogical Society.

Joussein, E., Petit, S., Churchman, J., *et al.* 2005. Halloysite clay minerals: A review. *Clay Minerals*, 40 (4), 383–426.

Kleeberg, R. 2005. Results of the second Reynolds Cup contest in quantitative mineral analysis. *IUCr CPD Newsletter*, 30, 22–26.

Klimentidis, R. E. & MacKinnon, I. D. R. 1986. High-resolution imaging of ordered mixed-layer clays. *Clays and Clay Minerals*, 34 (2), 155–164.

Kloprogge, J. T. (ed.) 2005. *The Application of Vibrational Spectroscopy to Clay Minerals and Layered Double Hydroxides.* Aurora, CO: The Clay Minerals Society.

Klug, H. P. & Alexander, L. E. 1974. *X-ray Diffraction Procedures for Polycrystalline and Amorphous Materials.* New York: Wiley.

Kruse, F. A. 2010. Mineral mapping using spectroscopy: From field measurements to airborne and satellite-based imaging spectrometry. *Art, Science and Applications of Reflectance Spectroscopy Symposium, 23–25 February 2010.* Boulder, CO: ASD Inc., 1–15.

Kruse, F. A., Boardman, J. W., Lefkoff, A. B., *et al.* 2000. HyMap: An Australian hyperspectral sensor solving global problems – results from USA HyMap data acquisitions. *10th Australasian Remote Sensing and Photogrammetry Conference, 21–25 August 2000.* Adelaide, SA: Causal Productions, 1–16.

Laird, D. A. & Nater, E. A. 1993. Nature of the illitic phase associated with randomly interstratified smectite/illite in soils. *Clays and Clay Minerals*, 41 (3), 280–287.

Laribi, S., Jouffrey, B. & Fleureau, J. M. 2007. Experimental electron energy loss spectroscopy of clays. *European Physical Journal: Applied Physics*, 39 (3), 257–265.

Larkin, P. 2011. *Infrared and Raman Spectroscopy: Principles and Spectral Interpretation.* Boston, MA: Elsevier.

MacEwan, D. M. C. & Wilson, M. J. 1984. Interlayer and intercalation of complexes of clay minerals. In: Brindley, G. W. & Brown, G. (eds) *Crystal Structures of Clay Minerals and Their X-ray Identification.* London: Mineralogical Society.

MacKinnon, I. D. R. & Mumpton, F. A. (eds) 1990. *Electron-Optical Methods in Clay Science.* Boulder, CO: The Clay Minerals Society.

Madsen, I. C. & Scarlett, N. V. Y. 2008. Quantitative phase analysis. In: Dinnebier, R. E. & Billinge, S. J. L. (eds) *Powder Diffraction: Theory and Practice.* London: The Royal Society of Chemistry.

Madsen, I. C., Scarlett, N. V. Y., Cranswick, L. M. D. & Lwin, T. 2001. Outcomes of the International Union of Crystallography Commission on Powder Diffraction round robin on quantitative phase analysis: Samples 1a to 1h. *Journal of Applied Crystallography*, 34 (4), 409–426.

Mauger, A. J., Keeling, J. L. & Huntington, J. F. 2007. Alteration mapping of the Tarcoola Goldfield (South Australia) using a suite of hyperspectral methods. *Applied Earth Science*, 116 (1), 2–12.

McBride, M. B. 1994. *Environmental Chemistry of Soils.* New York: Oxford University Press.

McCarty, D. K. 2005. Quantitative mineral analysis of clay-bearing mixtures: The 'Reynolds Cup' contest. *IUCr CPD Newsletter*, 27, 12–16.

McFarlane, A., Bremmell, K. & Addai-Mensah, J. 2006. Improved dewatering behavior of clay minerals dispersions via interfacial chemistry and particle interactions optimization. *Journal of Colloid and Interface Science*, 293 (1), 116–127.

McKenzie, R. M. 1989. Manganese oxides and hydroxides. In: Dixon, J. B. & Weed, S. B. (eds) *Minerals in Soil Environments*, 2nd edition. Madison, WI: Soil Science Society of America.

Mehra, O. P. & Jackson, M. L. 1958. Iron oxide removal from soils and clays by a dithionite–citrate system buffered with sodium bicarbonate. In: Swineford, A. (ed.)

Clays and Clay Minerals: Proceedings of the Seventh National Conference on Clays and Clay Minerals. Washington, DC: Pergamon Press, 317–327.

Méring, J. 1975. Smectites. In: Gieseking, J. E. (ed.) *Soil Components.* Berlin: Springer-Verlag.

Mermut, A. R. & Lagaly, G. 2001. Baseline studies of the Clay Minerals Society source clays: Layer-charge determination and characteristics of those minerals containing 2:1 layers. *Clays and Clay Minerals*, 49 (5), 393–397.

Michot, L. J. & Villiéras, F. 2006. Surface area and porosity. In: Bergaya, F., Theng, B. K. G. & Lagaly, G. (eds) *Handbook of Clay Science.* Amsterdam: Elsevier.

Mielenz, R. C. & King, M. E. 1955. Physical-chemical properties and engineering performance of clays. In: Pask, J. A. & Turner, M. D. (eds) *Clays and Clay Technology: Proceedings of the First National Conference on Clays and Clay Technology*. San Francisco, CA. Division of Mines & Geology, 196–254.

Mikutta, R., Kleber, M., Kaiser, K. & Jahn, R. 2005. Review: Organic matter removal from soils using hydrogen peroxide, sodium hypochlorite, and disodium peroxodisulfate. *Soil Science Society of America Journal*, 69 (1), 120–135.

Miller, P. R., Reid, A. F. & Zuiderwyk, M. A. 1983. QEM* SEM image analysis in the determination of modal assays, mineral associations, and mineral liberation. In: Maltby, P. D. R. (ed.) *Proceedings of the XIV International Mineral Processing Congress, 17–23 October, 1982*. Toronto, ON: Canadian Institute of Mining and Metallurgy, 3.

Milnes, A. R. & Fitzpatrick, R. W. 1989. Titanium and zirconium minerals. In: Dixon, J. B. & Weed, S. B. (eds) *Minerals in Soil Environments*, 2nd edition. Madison, WI: Soil Science Society of America.

Montes-H, G., Geraud, Y., Duplay, J. & Reuschlé, T. 2005. ESEM observations of compacted bentonite submitted to hydration/dehydration conditions. *Colloids and Surfaces A: Physicochemical and Engineering Aspects*, 262 (1), 14–22.

Mooney, R. W., Keenan, A. G. & Wood, L. A. 1952. Adsorption of water vapor by montmorillonite: II. Effect of exchangeable ions and lattice swelling as measured by X-ray diffraction. *Journal of the American Chemical Society*, 74, 1371–1374.

Moore, D. M. & Reynolds, R. C. 1997. *X-ray Diffraction and the Identification and Analysis of Clay Minerals.* Oxford: Oxford University Press.

Morris, G. E. & Żbik, M. S. 2009. Smectite suspension structural behaviour. *International Journal of Mineral Processing*, 93 (1), 20–25.

Nagy, K. L. & Blum, A. E. (eds) 1994. *Scanning Probe Microscopy of Clay Minerals.* Boulder, CO: The Clay Minerals Society.

Nieto, F. & Arroyo, X. 2010. Identification and characterisation with microscopic methods. In: Fiore, S., Cuadros, J. & Huertas, F. J. (eds) *AIPEA Educational Series, Pub. No. 1.* Bari, IT: Digilabs.

Norrish, K. 1954. The swelling of montmorillonite. *Discussions of the Faraday Society*, 18, 120–134.

Norrish, K. & Hutton, J. T. 1969. An accurate X-ray spectrographic method for the analysis of a wide range of geological samples. *Geochimica et Cosmochimica Acta*, 33 (4), 431–453.

Norrish, K. & Chappell, B. W. 1977. X-ray fluorescence spectrometry. In: Zussman, J. (ed.) *Physical Methods in Determinative Mineralogy.* London: Academic Press.

O'Connor, B. H. & Chang, W. J. 1986. The amorphous character and particle-size distributions of powders produced with the micronizing mill for quantitative X-ray-powder diffractometry. *X-Ray Spectrometry*, 15 (4), 267–270.

O'Connor, B. H. & Raven, M. D. 1988. Application of the Rietveld refinement procedure in assaying powdered mixtures. *Powder Diffraction*, 3 (1), 2–6.

Omotoso, O., McCarty, D. K., Hillier, S. & Kleeberg, R. 2006. Some successful approaches to quantitative mineral analysis as revealed by the 3rd Reynolds Cup contest. *Clays and Clay Minerals*, 54(6), 748–760.

Otten, M. T. & Buseck, P. R. 1987. The oxidation state of Ti in hornblende and biotite determined by electron energy-loss spectroscopy, with inferences regarding the Ti substitution. *Physics and Chemistry of Minerals*, 14 (1), 45–51.

Ottner, F., Gier, S., Kuderna, M. & Schwaighofer, B. 2000. Results of an inter-laboratory comparison of methods for quantitative clay analysis. *Applied Clay Science*, 17 (5–6), 223–243.

Pansu, M. & Gautheyrou, J. 2006. *Handbook of Soil Analysis: Mineralogical, Organic and Inorganic Methods*. Berlin: Springer-Verlag.

Pearlman, J., Carman, S., Segal, C., *et al.* 2001. Overview of the Hyperion Imaging Spectrometer for the NASA EO-1 mission. In: *Geoscience and Remote Sensing Symposium, 2001. IGARSS '01. IEEE 2001 International, 2001*, 3036–3038.

Plançon, A. & Tchoubar, C. 1977. Determination of structural defects in phyllosilicates by X-ray diffraction: Part II. Nature and proportions of defects in natural kaolinites. *Clays and Clay Minerals*, 25 (6), 436–450.

Poppe, L. J. & Eliason, A. E. 2009. A BASIC program to calculate gravitational and centrifugal parameters. In *Geological Society of America, Northeastern Section – 44th Annual Meeting*. Portland, ME: Geological Society of America, 26.

Poppe, L. J., Paskevich, V. F., Hathaway, J. C. & Blackwood, D. S. 2001. A laboratory manual for X-ray powder diffraction. US Geological Survey Open-File Report [Online]. Available: http://pubs.usgs.gov/of/2001/of01-041/index.htm.

Post, J. E. & Heaney, P. J. 2008. Synchrotron powder X-ray diffraction study of the structure and dehydration behavior of palygorskite. *American Mineralogist*, 93 (4), 667–675.

Post, J. E., Bish, D. L. & Heaney, P. J. 2007. Synchrotron powder X-ray diffraction study of the structure and dehydration behavior of sepiolite. *American Mineralogist*, 92 (1), 91–97.

Radoslovich, E. W. 1960. The structure of muscovite, $KAl_2(Si_3Al)O_{10}(OH)_2$. *Acta Crystallographica*, 13 (11), 919–932.

Raven, M. D. & Self, P. G. 2011. Secrets to winning the 5th Reynolds Cup competition. In: *48th Annual Meeting of The Clay Minerals Society*. Lake Tahoe, NV: The Clay Minerals Society, 85.

Rayment, G. E. & Lyons, D. J. 2011. *Soil Chemical Methods : Australasia*. Collingwood, VIC: CSIRO Publishing.

Reynolds, R. C. 1980. Interstratified clays. In: Brindley, G. W. & Brown, G. (eds) *Crystal Structures of Clay Minerals and Their X-ray Identification*, new edition. London: Mineralogical Society.

Rieder, M. G. C., D'yakonov, Y. S., Frank-Kamenetskii, V. A., *et al.* 1998. Nomenclature of micas. *The Canadian Mineralogist*, 36 (2), 905–912.

Rietveld, H. M. 1967. Line profiles of neutron powder-diffraction peaks for structure refinement. *Acta Crystallographica*, 22, 151.

Rietveld, H. M. 1969. A profile refinement method for nuclear and magnetic structures. *Journal of Applied Crystallography*, 2 (2), 65–71.

Ritz, M., Vaculikova, L. & Plevova, E. 2010. Identification of clay minerals by infrared spectroscopy and discriminant analysis. *Applied Spectroscopy*, 64 (12), 1379–1387.

Rueda, E. H., Ballesteros, M. C., Grassi, R. L. & Blesa, M. A. 1992. Dithionite as a dissolving reagent for goethite in the presence of EDTA and citrate: Application to soil analysis. *Clays and Clay Minerals*, 40 (5), 575–585.

Russell, J. D. & Fraser, A. R. 1994. Infrared methods. In: Wilson, M. J. (ed.) *Clay Mineralogy: Spectroscopic and Chemical Determinative Methods.* Amsterdam: Springer.
Salvo, L., Cloetens, P., Maire, E., *et al.* 2003. X-ray micro-tomography an attractive characterisation technique in materials science. *Nuclear Instruments and Methods in Physics Research Section B: Beam Interactions with Materials and Atoms*, 200, 273–286.
Sawhney, B. L. 1989. Interstratification of layer silicates. In*:* Dixon, J. B., Weed, S. B. & Dinauer, R. C. (eds) *Minerals in Soil Environments*, 2nd edition. Madison, WI: Soil Science Society of America.
Scarlett, N. V. Y. & Madsen, I. C. 2006. Quantification of phases with partial or no known crystal structures. *Powder Diffraction*, 21 (4), 278–284.
Scarlett, N. V. Y., Madsen, I. C., Cranswick, L. M. D., *et al.* 2002. Outcomes of the International Union of Crystallography Commission on powder diffraction round robin on quantitative phase analysis: Samples 2, 3, 4, synthetic bauxite, natural granodiorite and pharmaceuticals. *Journal of Applied Crystallography*, 35 (4), 383–400.
Sergeev, Y. M., Spivak, G. V., Sasov, A. Y., *et al.* 1984a. Quantitative morphological analysis in a SEM-microcomputer system – I: Quantitative shape analysis of single objects. *Journal of Microscopy*, 135 (1), 1–12.
Sergeev, Y. M., Spivak, G. V., Sasov, A. Y., *et al.* 1984b. Quantitative morphological analysis in a SEM-microcomputer system – II: Morphological analysis of complex SEM images. *Journal of Microscopy*, 135 (1), 13–24.
Shirozu, H. & Bailey, S. W. 1966. Crystal structure of a 2-layer Mg-vermiculite. *American Mineralogist*, 51 (7), 1124.
Singh, B. 1996. Why does halloysite roll? A new model. *Clays and Clay Minerals*, 44 (2), 191–196.
Skinner, H. C. W. & Fitzpatrick, R. W. 1992. *Biomineralization: Processes of Iron and Manganese – Modern and Ancient Environments.* Cremlingen: Catena Verlag.
Skipper, N. T., Soper, A. K. & McConnell, J. D. C. 1991. The structure of interlayer water in vermiculite. *Journal of Chemical Physics*, 94 (8), 5751–5760.
Smith, D. K. 2001/1992. Particle statistics and whole-pattern methods in quantitative X-ray powder diffraction analysis. *Powder Diffraction*, 16 (4), 186–191. (Reprinted from *Advances in X-ray Analysis*, 35A)
Smith, J. V. & Yoder, H. S. 1955. Experimental and theoretical studies of the mica polymorphs. *American Mineralogist*, 40 (3–4), 343–344.
Środoń, J. 1999. Nature of mixed-layer clays and mechanisms of their formation and alteration. *Annual Review of Earth and Planetary Sciences*, 27, 19.
Środoń, J., Andreoli, C., Elsass, F. & Robert, M. 1990. Direct high-resolution transmission electron microscopic measurement of expandability of mixed-layer illite/smectite in bentonite rock. *Clays and Clay Minerals*, 38 (4), 373–379.
Sudo, T., Shimoda, S., Yotsumoto, H. & Aita, S. (eds) 1981. *Electron Micrographs of Clay Minerals*. Amsterdam: Elsevier.
Sumner, M. E. (ed.) 2000. *Handbook of Soil Science*. Washington, DC: CRC Press.
Sumner, M. E. & Miller, W. P. 1996. Cation exchange capacity and exchange coefficients. In: Sparks, D. L., Page, A. L., Helmke, P. A., *et al.* (eds) *Methods of Soil Analysis Part 3 – Chemical Methods*. Madison, WI: Soil Science Society of America.
Svedberg, T. & Nichols, J. B. 1923. Determination of size and distribution of size of particle by centrifugal methods. *Journal of the American Chemical Society*, 45, 2910–2917.
Tanner, C. B. & Jackson, M. L. 1948. Nomographs of sedimentation times for soil particles under gravity or centrifugal acceleration. *Soil Science Society of America Journal*, 12 (C), 60–65.
Tappert, M., Rivard, B., Giles, D., Tappert, R. & Mauger, A. J. 2011. Automated drill core logging using visible and near-infrared reflectance spectroscopy: A case study from Olympic Dam IOGG deposit, South Australia. *Economic Geology*, 106 (2), 289–296.

Taylor, J. C. 1991. Computer programs for standardless quantitative analysis of minerals using the full powder diffraction profile. *Powder Diffraction*, 6 (1), 2–9.

Taylor, J. C. & Matulis, C. E. 1994. A new method for Rietveld clay analysis: Part I. Use of a universal measured standard profile for Rietveld quantification of montmorillonites. *Powder Diffraction*, 9 (2), 119–123.

Tchoubar, D. & Cohaut, N. 2006. Small-angle scattering techniques. In: Bergaya, F., Theng, B. K. G. & Lagaly, G. (eds) *Handbook of Clay Science*. Amsterdam: Elsevier.

Theng, B. K. G., Churchman, G. J., Whitton, J. S. & Claridge, G. G. C. 1984. Comparison of intercalation methods for differentiating halloysite from kaolinite. *Clays and Clay Minerals*, 32 (4), 249–258.

Treacy, M. M. J., Newsam, J. M. & Deem, M. W. 1991. A general recursion method for calculating diffracted intensities from crystals containing planar faults. *Proceedings of the Royal Society of London Series A: Mathematical Physical and Engineering Sciences*, 433 (1889), 499–520.

Tsipursky, S. I. & Drits, V. A. 1984. The distribution of octahedral cations in the 2:1 layers of dioctahedral smectites studied by oblique-texture electron diffraction. *Clay Minerals*, 19(2), 177.

Ufer, K., Roth, G., Kleeberg, R., *et al.* 2004. Description of X-ray powder pattern of turbostratically disordered layer structures with a Rietveld compatible approach. *Zeitschrift Fur Kristallographie*, 219 (9), 519–527.

Vali, H. & Hesse, R. 1990. Alkylammonium ion treatment of clay minerals in ultrathin section: A new method for HRTEM examination of expandable layers. *American Mineralogist*, 75 (11–12), 1443–1446.

Vali, H. & Hesse, R. 1992. Arrangement of n-alkylammonium ions in phlogopite and vermiculite: An XRD and TEM study. *Clays and Clay Minerals*, 40 (2), 240–245.

Vali, H. & Köster, H. M. 1986. Expanding behaviour, structural disorder, regular and random irregular interstratification of 2:1 layer-silicates studied by high-resolution images of transmission electron microscopy. *Clay Minerals*, 21 (5), 827–859.

van Aken, P. A., Liebscher, B. & Styrsa, V. J. 1998. Quantitative determination of iron oxidation states in minerals using Fe L2,3-edge electron energy-loss near-edge structure spectroscopy. *Physics and Chemistry of Minerals*, 25 (5), 323–327.

van der Marel, H. W. & Beutelspacher, H. 1976. *Atlas of Infrared Spectroscopy of Clay Minerals and Their Admixtures*. Amsterdam: Elsevier.

Velde, B. 1984. Electron microprobe analysis of clay minerals. *Clay Minerals*, 19 (2), 243–247.

Verburg, K. & Baveye, P. 1994. Hysteresis in the binary exchange of cations on 2:1 clay minerals: A critical review. *Clays and Clay Minerals*, 42 (2), 207–220.

Viani, A., Gaultieri, A. F. & Artioli, G. 2002. The nature of disorder in montmorillonite by simulation of X-ray powder patterns. *American Mineralogist*, 87 (7), 966–975.

Viscarra Rossel, R. A. 2011. Fine-resolution multiscale mapping of clay minerals in Australian soils measured with near infrared spectra. *Journal of Geophysical Research: Earth Surface*, 116 (F4), F04023.

Viscarra Rossel, R. A., Walvoort, D. J. J., McBratney, A. B., Janik, L. J. & Skjemstad, J. O. 2006. Visible, near infrared, mid infrared or combined diffuse reflectance spectroscopy for simultaneous assessment of various soil properties. *Geoderma*, 131 (1), 59–75.

Wada, K. 1989. Allophane and imogolite. In: Dixon, J. B., Weed, S. B. & Dinauer, R. C. (eds) *Minerals in Soil Environments*, 2nd edition. Madison, WI: Soil Science Society of America.

Wada, S. I. & Wada, K. 1977. Density and structure of allophane. *Clay Minerals*, 12 (4), 289–298.

Wada, K. & Yoshinaga, N. 1969. Structure of imogolite. *American Mineralogist*, 54 (1–2), 50.

Wang, M. K., Wang, S. L. & Wang, W. M. 1996. Rapid estimation of cation-exchange capacities of soils and clays with methylene blue exchange. *Soil Science Society of America Journal*, 60 (1), 138–141.

Wang, X. D., Li, J., Hart, R. D., van Riessen, A. & McDonald, R. 2011. Quantitative X-ray diffraction phase analysis of poorly ordered nontronite clay in nickel laterites. *Journal of Applied Crystallography*, 44 (5), 902–910.

Wang, X. D., Hart, R. D., Li, J., McDonald, R. G. & van Riessen, A. 2012. Quantitative analysis of turbostratically disordered nontronite with a supercell model calibrated by the PONKCS method. *Journal of Applied Crystallography*, 45 (6), 1295–1302.

Warren, B. E. 1941. X-ray diffraction in random layer lattices. *Physical Review*, 59 (9), 693–698.

Wicks, F. J. & O'Hanley, D. S. 1988. Serpentine minerals: Structure and petrology. In: Bailey, S. W. (ed.) *Hydrous Phyllosilicates (Exclusive of Micas)*. Madison, WI: Mineralogical Society.

Willis, J., Feather, C. & Turner, K. 2014. *Guidelines for XRF Analysis*. Cape Town: James Willis Consultants.

Wilson, M. J. 1994. *Clay Mineralogy: Spectroscopic and Chemical Determinative Methods*. Dordrecht, NL: Springer Netherlands.

Wilson, M. J. 2013. *Sheet Silicates: Clay Minerals*. London: The Geological Society.

Yang, D. & Frindt, R. F. 1996. Powder X-ray diffraction of turbostratically stacked layer systems. *Journal of Materials Research*, 11 (7), 1733–1738.

Yitagesu, F. A., van der Meer, F., van der Werff, H. & Hecker, C. 2011. Spectral characteristics of clay minerals in the 2.5–14μm wavelength region. *Applied Clay Science*, 53 (4), 581–591.

Young, R. A. 1980. Structural analysis from X-ray powder diffraction with the Rietveld method. In: Block, S. & Hubbard, C. R. (eds) *Accuracy in Powder Diffraction*. Gaithersburg, MD: National Bureau of Standards, 143–162.

Yu, X., Zhao, L., Gao, X., Zhang, X. & Wu, N. 2006. The intercalation of cetyltrimethylammonium cations into muscovite by a two-step process: II. The intercalation of cetyltrimethylammonium cations into Li-muscovite. *Journal of Solid State Chemistry*, 179 (5), 1525–1535.

Żbik, M. S. & Frost, R. L. 2009. Micro-structure differences in kaolinite suspensions. *Journal of Colloid and Interface Science*, 339 (1), 110–116.

Żbik, M. S., Smart, R. C. & Morris, G. E. 2008. Kaolinite flocculation structure. *Journal of Colloid and Interface Science*, 328 (1), 73–80.

Żbik, M. S., Song, Y.-F. & Frost, R. L. 2010. Kaolinite flocculation induced by smectite addition: A transmission X-ray microscopic study. *Journal of Colloid and Interface Science*, 349 (1), 86–92.

Żbik, M. S., Song, Y.-F., Frost, R. L. & Wang, C.-C. 2012. Transmission X-ray microscopy: A new tool in clay mineral floccules characterization. *Minerals*, 2 (4), 283–299.

Zevin, L. S., Kimmel, G. & Mureinik, I. 1995. *Quantitative X-ray Diffractometry*. New York: Springer.

5

Clays by Deposit Type

Martin Wells, Erick Ramanaidou, Carsten Laukamp, Kai Yang and Mark Pownceby

5.1 Introduction

Clay minerals occur across a diverse range of geological settings and are the ubiquitous products of alteration and weathering where these processes take place at low temperatures and pressures (refer to Chapter 1). Consequently, from this perspective the following discussion provides a description of the geological settings and mineralogical associations under which the kaolinite and smectite clay groups occur for a range of commodities (Fig. 5.1). The commodities described herein are not intended nor should be considered as comprehensive, nor should those discussed be viewed as the only occurrence or association of clays with economic concentrations of metals or other commodities of economic importance. Rather, this chapter presents a review of the predominant or economically more significant commodities with which these clay minerals may be associated. It must be noted that kaolinite and smectite are important industrial minerals in their own right. Examples of the varied range of industrial applications in which smectite and kaolinite are used are provided in the reviews by Odom (1984), Jepson (1984), Bundy (1993) and the references contained therein, with more recent uses of kaolinite given by Glover *et al.* (2012). For example, the most important use of kaolinite is as a filler and coating in the manufacture of paper (Bundy, 1993). However, examination of the formation of industrially significant kaolinite and smectite deposits is beyond the scope of the present review and is not further discussed.

To place some context around the following discussion, the value of major export commodities to the Western Australian (WA) economy for the 2013–2014 financial year is presented in Fig. 5.2. Within a global context the production of selected WA and Australian commodities is compared to the total percentage of world production (Fig. 5.3) and shows that Australia is a significant producer of a number of important commodities.

Hence, the following discussion has been structured to broadly reflect the general order and economic significance of the mineral and energy commodities listed in Figs. 5.2 and 5.3, and also includes those commodities for which clays play a significant role, good or bad. For each of the commodities examined, a brief definition and description of the geological setting is provided to familiarize the reader, and to provide some context of the association in which kaolinite and/or smectite occur.

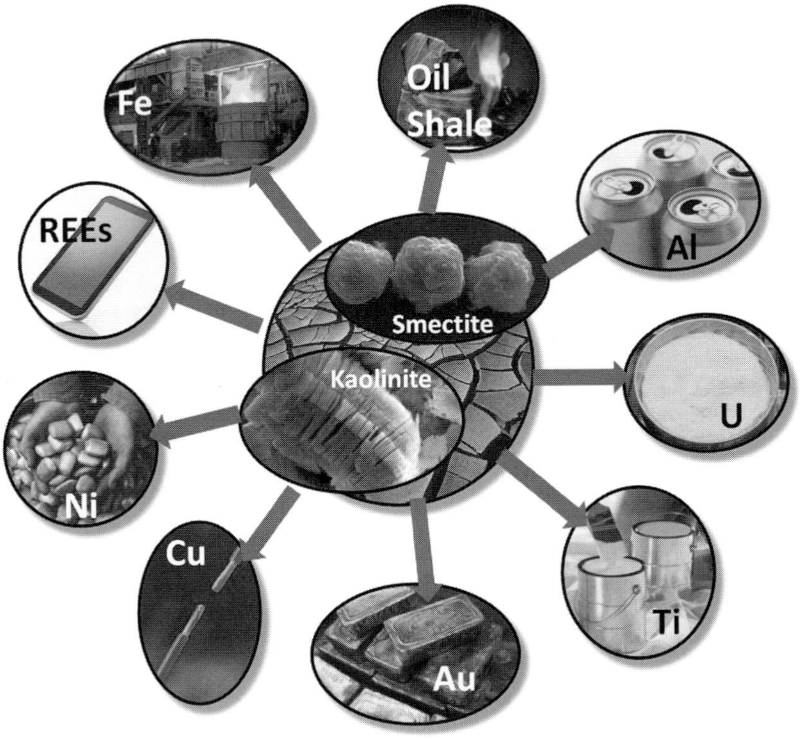

Figure 5.1. Kaolinite and smectite may be associated with a range of economically important commodities, typically as deleterious phases that can adversely affect the ore grade, processing and recovery of the commodity of interest. A black and white version of this figure will appear in some formats. For the colour version, please refer to the plate section.

5.2 Iron Ore

5.2.1 Introduction

The bulk of iron ore mined today is extracted from Precambrian iron formation that represents 90 per cent of the 220 iron ore mines currently in operation. The remaining 10 per cent include ooidal ironstones, such as the very large channel iron deposits (CID) of Western Australia and in Kazakhstan, and also in Savage River in Australia, El Romeral in Chile, La Perla in Mexico, Marcona in Peru, Faleme in Senegal, Kiruna and Malberget in Sweden, and magmatic and skarn deposits (Ramanaidou and Wells, 2014).

In 2012, a direct shipping ore (DSO) or direct to smelter ore is required to have >55 per cent Fe, <5 per cent SiO_2, <5 per cent Al_2O_3 and <0.075 per cent P, and very low levels of environmentally damaging elements such as chlorine and sulphur. Iron ores with chemical composition and metallurgical characteristics outside of these boundaries can still be mined profitably but will attract financial penalties. In iron ore, both kaolinite and smectite are common contaminants and they have damaging effects on grade and processing.

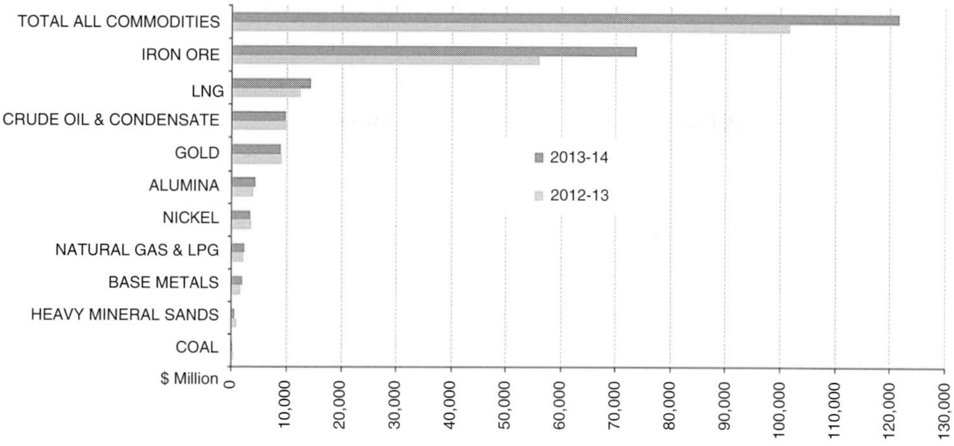

Figure 5.2. Value ($ million) of major commodities for Western Australia for the 2012–2013 and 2013–2014 financial years.
Source: Department of Mines and Petroleum (2015).

Kaolinite and Al- and Fe-smectite have been identified in both Precambrian banded iron-formation-hosted (BIF-hosted) ores and Phanerozoic ooidal ironstones (POI) ores, the only two types currently mined in large volumes (>120 Mt annually).

5.2.2 Clays in Precambrian Banded Iron-Formation-Hosted Iron Ore

In BIF-hosted iron ores, the weathering of three potential sources can generate clay minerals: (1) shaly sequences intercalated within the BIF; (2) dolerite dyke and sill intrusions in the BIF; and (3) aluminous rocks stratigraphically close to the BIF.

The stratigraphic succession of the Hamersley Province in Western Australia shows a classic example of the internal subdivision of units, including intrusive series, sills and dykes, carbonate units and BIF (Fig. 5.4). Shaly sequences are prominent as intercalations throughout the major BIF (except the Joffre Member) and are the dominant lithology in the McRae and Whaleback Shale Members (Fig. 5.4). The dominant mineralogy of the shale includes stilpnomelane, chlorite and minnesotaite with minor chert and carbonates. Weathering of the shale generates secondary minerals dominated by goethite, and kaolinite with minor minerals, particularly gibbsite and nontronite.

The enriched BIF-hosted iron ores fall into two main types: (1) the bedded iron deposits (BID) that provide the bulk of the high-grade iron ore; and (2) the minor detrital iron deposits (DID) that are the result of the breakdown of BID. Bedded iron deposits exhibit both additive and subtractive aspects and they have been conventionally classified into three subtypes, namely: (1) residual iron ore; (2) martite–goethite ores; and (3) hematite (martite–microplaty) ores. Secondary weathering processes affect all pre-existing iron ore deposits around the world; the effects include textural obliteration, dissolution

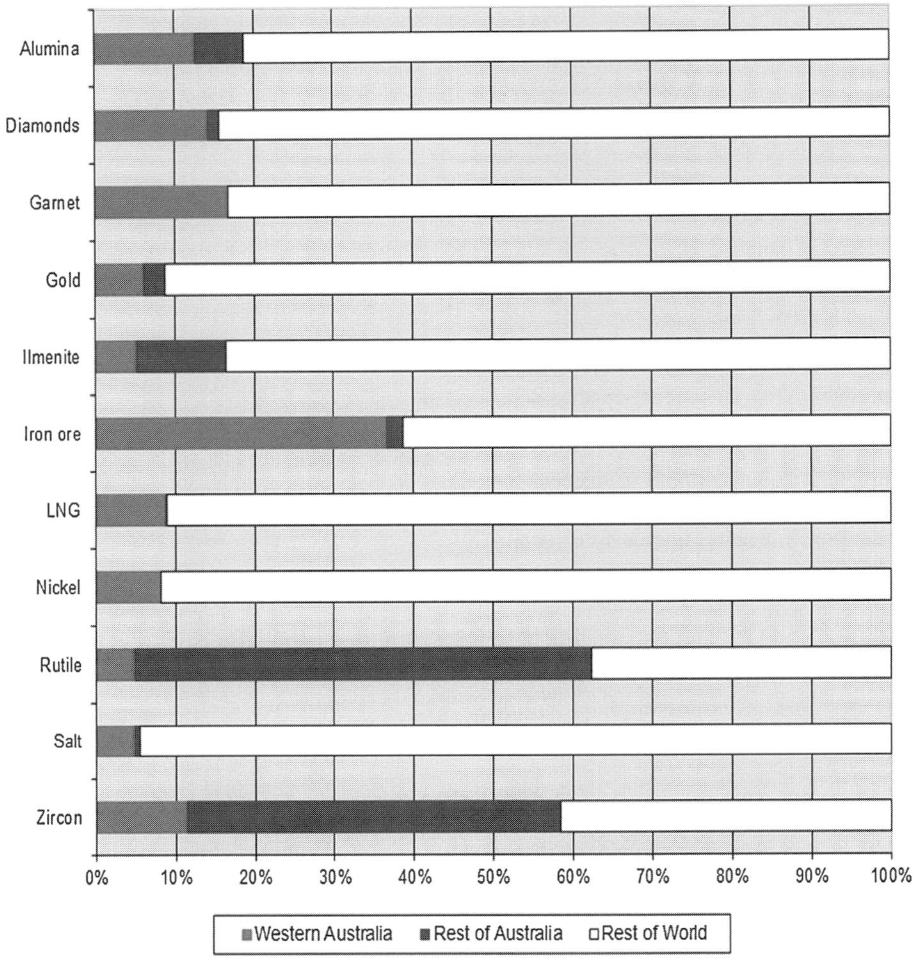

Figure 5.3. Production of selected Western Australian and Australian commodities compared to world production (ending 2014).
Source: Department of Mines and Petroleum (2015).

and precipitation processes. The result is a classic, three-tiered profile consisting from bottom to top of: (1) the iron ore proper; (2) a ‘hydrated zone’ with a thickness varying from a few metres up to 75 m; and (3) a 1–2 m hematite-rich carapace that has the textural feature of the primary iron ore (Morris, 1985). The hydrated zone is dominated by goethite as either vitreous or ochreous types often associated with kaolinite and gibbsite. Identified clay minerals include primarily kaolinite with some halloysite and Ca,Fe-montmorillonite.

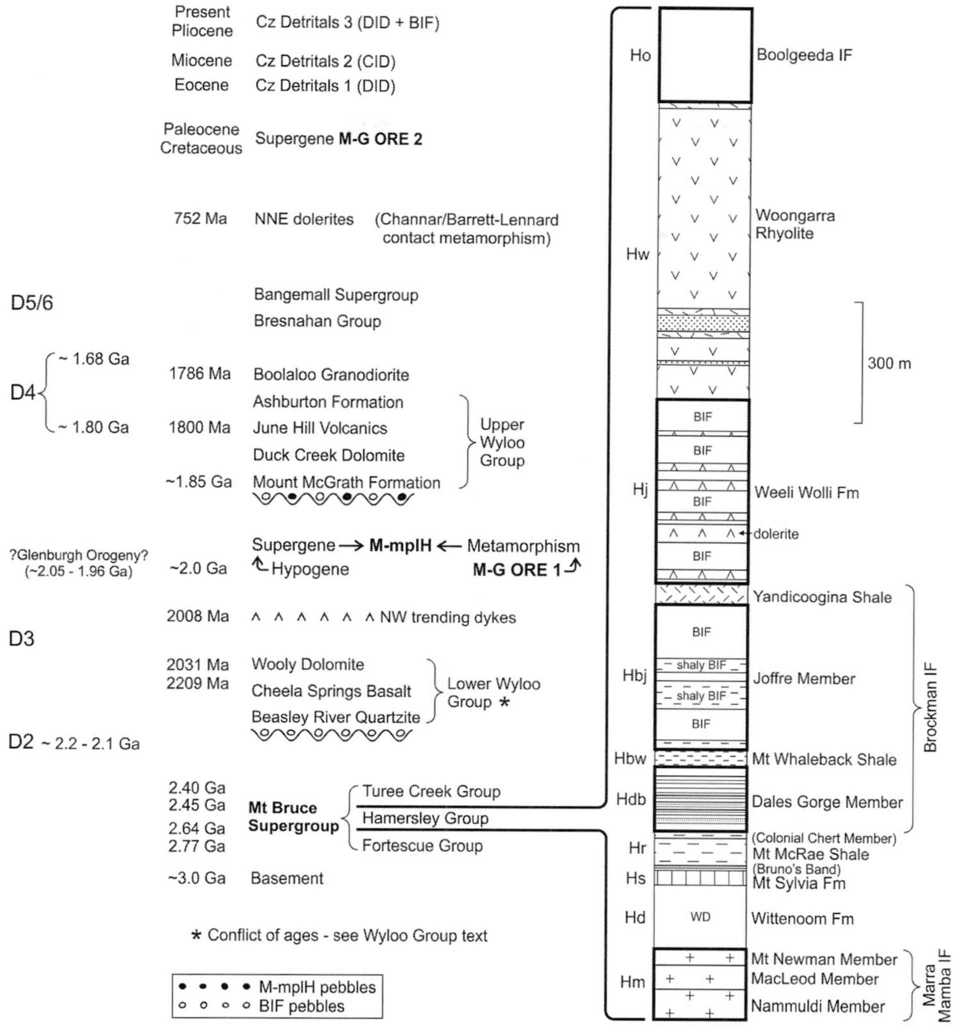

Figure 5.4. Stratigraphy for the major iron ore-related events of the Hamersley Province, WA, with an expanded stratigraphy for the Hamersley Group shown at right. Banded iron formation (BIF) in the Hamersley Group host the major BIF-derived iron ore deposits (redrawn from Morris and Kneeshaw, 2011), with permission of Taylor & Francis Ltd, www.tandfonline.com). Shale bands are shown as dark, horizontal lines – for example in the Dales Gorge Member of the Brockman iron formation.

5.2.3 Clays in Phanerozoic Ironstone Iron Ore

Petránek and Van Houten (1997) defined POI as 'distinctive non-cherty, sandy clayey, siliclastic or siliclastic-carbonate sedimentary rock with more than 5 per cent of ferruginous ooids and more than 15 per cent of iron'. Ramanaidou *et al.* (2003) described the only POI currently mined (also known as CIDs) as iron-rich fluvial deposits of Oligocene to Miocene age occupying generally meandering paleochannels. The deposits,

Figure 5.5. Mine bench face of mainly vertical clay pipes (white) exposed in the Yandicoogina CID. Photograph by Erick Ramanaidou. A black and white version of this figure will appear in some formats. For the colour version, please refer to the plate section.

anomalously high in iron, range from <1 m to 100 m thick, with channel widths typically <1 km, but ranging up to several kilometres. The general term CID includes the Western Australian Robe, Marillana and Poondano Formations and the Turgai and Aral'sk deposits in Kazakhstan. A recent review of sedimentary iron ore deposits by Kholodov *et al.* (2012) offers some additional information on the occurrence of clays in the Kazakhstan deposits.

In the POI ores and, in particular, the CID of the Hamersley Province in Western Australia (Kneeshaw, 2003; Ramanaidou *et al.*, 2003), clays occur as metre-thick units at the base or top of the deposits. For example, at Yandi vertical clay 'pipes' starting from the surface and up to 10 m deep can crosscut the CID (Fig. 5.5). The Marillana Formation of the Western Australian CID is a classic example of the clay distribution in CID. It consists of three members and includes, from top to bottom: (1) Iowa Eastern Member, with the upper Iowa Eastern CID, and the lower Iowa Eastern Clay; (2) Barimunya Member, with the Upper CID, and the Lower CID; and (3) Munjina Member with the upper Basal Clay and the lower Basal Conglomerate (Fig. 5.6). The clay layers in the CID probably originate from weathered shaly BIF and deeply weathered dolerite sills occurring in the area. According to

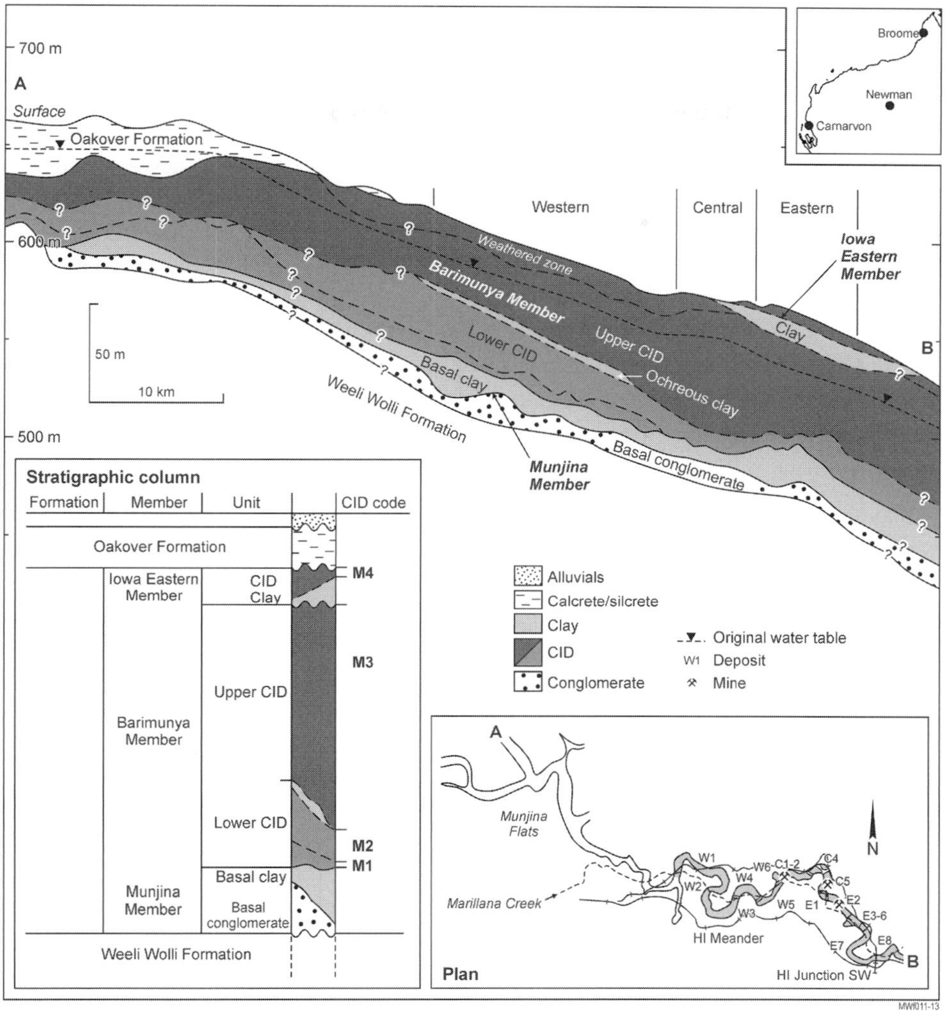

Figure 5.6. Schematic cross-section of the Marillana Formation CID and constituent members in the Marillana Creek area (Ramanaidou *et al.*, 2003). Clay layers occur with the lower Iowa Eastern Member, at the base of the Barimunya Member and are found in the Basal Clay unit of the Munjina Member.

Burnham and Hawke (2002), clay forms through two main processes; *in situ* alteration and cavity in-fill.

Clay minerals, identified by X-ray diffraction and electron microprobe analyses, typically occur in voids, but kaolinite, the dominant species of this group, can also occur as fine layers alternating with hematite or goethite in the pelletoids. Halloysite, gibbsite, boehmite, illite, Ca,Fe-montmorillonite show a similar habit to kaolinite. Invariably the clays remain part of the mined material.

5.3 Gold

Gold deposits can be classified into three major types according to the main processes that concentrate gold: (1) igneous, (2) hydrothermal and (3) supergene (sedimentary/surficial processes, e.g. Robb (2005). However, igneous and hydrothermal processes are inherently linked in a number of gold deposits and are, therefore, grouped as magmatic-hydrothermal deposits, *sensu stricto* (Robb, 2005). Kaolinite and smectite can occur in both magmatic-hydrothermal and supergene systems, either as syngenetic or epigenetic alteration mineral phases, forming a halo around the ore-body. Clay minerals can also host nanoparticulate gold (Hough *et al.,* 2008), and kaolinite and smectite can adsorb aqueous gold species (Cohen and Waite, 2004) in, for example, placer gold deposits (see also Chapters 1 and 3).

In magmatic-hydrothermal systems clay minerals are diagnostic for (advanced) argillic alteration, and can form during the interaction between magmatic, hydrothermal and meteoric fluids and suitable country rocks. Mainly due to their near-surface development (i.e. low PT) and likely interaction with meteoric fluids, epithermal and porphyry systems are classic examples of gold deposits where kaolinite and smectite are formed, but they can even be found in skarn systems due to retrograde alteration (Thompson and Thompson, 1996). Mars and Rowan (2006), among others, showed that hydrothermal footprints related to argillic and phyllic alteration associated with epithermal and porphyry deposits may be remotely detected using satellite systems such as ASTER (see Chapter 6). Furthermore, argillic alteration is described from sediment-hosted, disseminated gold deposits (Wilson and Parry, 1995).

In supergene gold deposits, clays can occur as major mineral phases in the sediment, either formed prior to, during or after the precipitation of gold. The characteristics of the source (i.e. primary mineralization) and environmental conditions (Williams-Jones *et al.*, 2009), control the supergene concentration of gold and influences the formation and alteration/weathering of clay minerals.

5.3.1 Magmatic-Hydrothermal

Porphyry deposits are genetically and spatially associated with porphyry intrusions of calc-alkaline to alkaline composition. In most cases, gold represents a substantial by-product in porphyry Cu deposits. Gold-rich porphyry Cu deposits, such as Grasberg (Indonesia) and Ridgeway (Australia), have average gold contents of >0.4 g t^{-1} (Sillitoe, 2000). Interaction of high-temperature fluids with the igneous rocks leads to a diagnostic potassic alteration, comprising quartz, K-feldspar, biotite, anhydrite and magnetite. In late-stage magmatic emplacement, the cooling and changing physicochemical conditions lead to further alteration dominated by quartz and sericite (Tosdal *et al.*, 2009). In the upper (epithermal) levels of the system, advanced argillic alteration assemblages are formed due to interaction of localized ascending magmatic-hydrothermal fluids and resulting H_2S-rich groundwaters. This acid leaching leads to the formation of quartz, kaolinite/dickite or pyrophyllite, aluminium hydroxides and alunite, producing a ‘lithocap’ at the top of the mineral system, that can be barren or can contain gold as in the Mankayan Intrusion-Centered Cu–Au district, Luzon (Chang *et al.*, 2011).

Epithermal deposits can be subdivided into high- and low-sulphidation epithermal deposits, according to the oxidation state of sulphur in the mineralizing fluid, the chemistry and pH, which all find their expression in typical alteration mineral assemblages associated with the ore deposit (Robb, 2005; Simmons *et al.*, 2005; Tosdal *et al.*, 2009). Advanced argillic alteration, comprising proximal to distal quartz, alunite, kaolinite, illite, montmorillonite and chlorite, results from the interaction of the acidic fluids with rocks, which is characteristic of high-sulphidation deposits, such as the Nansatsu-type deposits on Kyushu, Japan (Hedenquist *et al.*, 1994). The more acidic conditions in this type of magmatic-hydrothermal system enable the transport of large quantities of precious metals (Williams-Jones and Heinrich, 2005). Low-sulphidation epithermal deposits form in neutral pH environments and show in most cases a characteristic adularia–sericite alteration, with quartz/chalcedony, calcite, adularia, sericite and chlorite from proximal to distal (Berger and Henley, 1989;Heald *et al.*, 1987). However, a number of low-sulphidation epithermal deposits commonly lack adularia, such as in the Philippines (Mitchell and Leach, 1991; Sillitoe, 1989). Sonntag *et al.* (2012) described the presence of interlayered illite–smectite clays and higher amounts of kaolinite in alteration proximal to gold in a low-potassic (adularia-free), low-sulphidation epithermal system in Eastern Mindanao (Philippines). They interpreted the interlayered illite–smectite clays to be syngenetic to formation of the deposit, indicating lower temperatures and possibly even less acidic conditions when compared to other low-sulphidation epithermal systems that contain illite. Carrillo-Rosúa *et al.* (2009) described native gold associated with interstratified illite–smectite clays in gold–copper deposits in south-eastern Spain. Kaolinite associated with epithermal systems in Eastern Mindanao may be related to supergene alteration (Sonntag *et al.*, 2012), but is also described from other low-sulphidation deposits as the result of interaction of localized ascending magmatic-hydrothermal fluids and resulting H_2S-rich groundwaters (Tosdal *et al.*, 2009).

5.3.2 Supergene

Supergene concentration of gold depends on the characteristics of the primary mineralization and the environmental factors (Williams-Jones *et al.*, 2009). Climatic conditions, triggering the weathering rate as well as the run-off and groundwater chemistry, steer the mobilization of gold from its source (i.e. primary mineralization) and the re-precipitation close to the surface in favourable country rocks. The same climatic conditions, together with the physicochemical characteristics of the country rocks, guide the formation of clay minerals associated with secondary gold concentration. In addition, clay minerals can form prior to and after the emplacement of the supergene concentration of gold, leading in some cases to a complex assemblage of different clay types.

Alluvial (or 'placer') gold deposits are naturally rich in clay minerals, where gold can be concentrated above clay layers and cemented gravels (Boyle, 1987). Seeley and Senden

(1994) suggested a genetic relationship of alluvial gold associated with clay layers within quartz–zircon sands or gravels in Kalimantan, Indonesia. Nucleation of gold may have been driven by chemical gradients within the clay layers or by adsorption of negatively charged colloids onto positively charged clay platelets (Seeley and Senden, 1994).

5.4 Oil Shale and Oil Sands

5.4.1 Oil Shale

The term 'oil shale' describes a diverse range of fine-grained sedimentary rocks containing a complex assemblage of organic matter in predominantly solid form, as kerogen. Oil shale deposits are known on all continents (Fig. 5.7) and occur throughout the Phanerozoic, with the largest deposits of Cambrian, early-mid Ordovician, late Devonian, late Jurassic and Paleogene ages (Ots, 2007). Examples of significant oil shale deposits include the Tertiary Condor deposit (Queensland); the Permian Iratí Formation, Brazil; the Ordovician Collingwood Shale, Ontario; Republic of Congo; deposits in Italy and the United States, particularly the Tertiary Green River Formation in Colorado, Wyoming and Utah, which at an estimated resource of ~1.5 trillion barrels represents the largest oil shale resource in the world (Dyni, 2006).

Known previously by a variety of names, such as cannel coal, boghead coal, alum shale, stellarite, bituminite, gas coal, torbanite and kukersite (Dyni, 2006; Hutton, 1987), to a name a few, oil shale represents a significant and important alternative energy source to conventional flow oil resources. For example, oil shale provides the main energy needs of Estonia (Ots, 2007), and in Turkey oil shale is the second largest energy source after lignite (Altun *et al.*, 2006).

The composition and classification of oil shales has been extensively investigated and recent reviews by Altun *et al.* (2006), Dyni (2006), Knaus *et al.* (2010) and Boak (2011), and the references contained therein, provide a detailed examination of the geological settings and resource characteristics of many of the world's important oil shale deposits. This is beyond the scope of the present discussion and is not repeated here. However, one characteristic feature of oil shales is the low solubility of the organic matter in ordinary organic solvents (Dyni, 2006; Ots, 2007). Bitumen, soluble in organic solvents, is only a minor component of some oil shales, e.g. Ots (2007). Recovery of shale oil and combustible gas from the organic matter, which typically has a greater H_2 content and lower O_2 content than that of lignite and bituminous coal, is achieved by distillation during the thermal decomposition (i.e. pyrolysis) of oil shale, a process commonly known as 'retorting' (Dyni, 2006; Ots, 2007).

Although known as an unconventional oil source from the mid-1800s, it is only recently that a classification scheme relating maceral type to the oil shale depositional environment (e.g. Hutton, 1987) has been developed and is now widely adopted (Dyni, 2006), although other classification schemes are known (e.g. Altun *et al.*, 2006). Viewed as one of three groups of organic-rich sedimentary rocks including humic coal, carbonaceous shale and

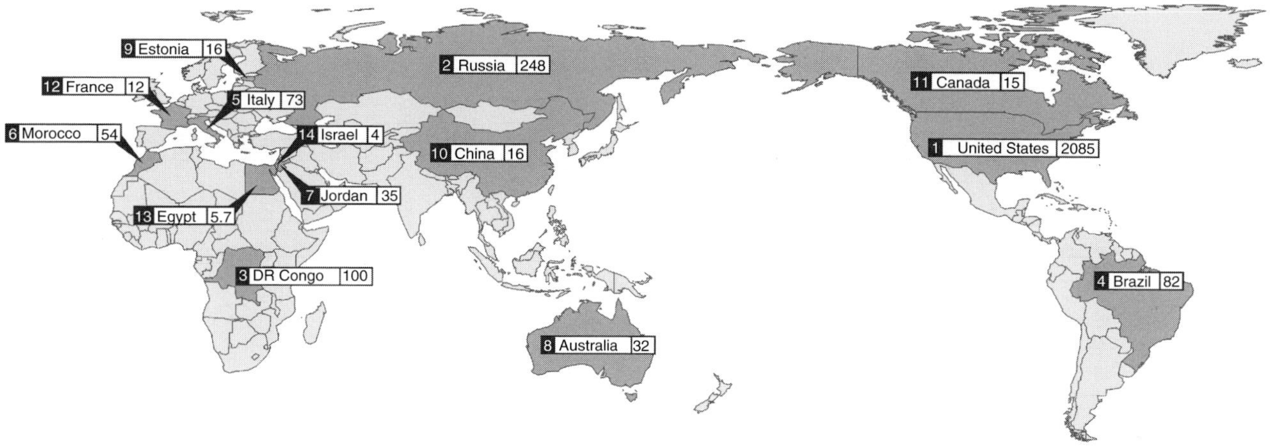

Figure 5.7. Global distribution of the top 14 oil shale resources (resource estimate in billions of barrels, Bbls – white box; global ranking – black box. Reprinted (adapted) with permission fromKnaus *et al.* (2010). Copyright (2015) American Chemical Society.

bitumen-impregnated rock, oil shales may be classed depending upon their environment of deposition as either terrestrial, lucastrine or of marine origin (Hutton, 1987). These three classes can be further divided into six types depending on the type and source of organic matter (Dyni, 2006; Hutton, 1987). Organic matter in oil shales comprises principally the degraded remains of marine and lucastrine algal matter, but also includes spores, pollen, bacterial remains and other terrestrial, marine and lucastrine plant debris (Dyni, 2006).

The inorganic component, which includes organically derived carbonates and phosphates (e.g. fossil shells and bone fragments), typically comprises 60–90 per cent of the oil shale, significantly greater than in coal, which is defined as containing <40 per cent mineral matter (Dyni, 2006). Oil shale mineral components typically include carbonates (calcite, dolomite and siderite), quartz, feldspar and clay minerals. Sulphides, such as pyrite and marcasite, constitute a minor combustible but ubiquitous component of the inorganic mineral fraction, and are indicative of the anoxic conditions under which oil shale forms (Dyni, 2006).

During oil shale pyrolysis, dehydroxylation and decarbonation (i.e. endothermic) reactions of the mineral constituents can influence the heat or energy requirements during retorting, which affects the thermal degradation of kerogen and, hence, the release and yield of shale oil (e.g. Patterson, 1994; Patterson *et al.*, 1990). These thermal reactions also have implications in the evaluation of the 'heating value' and oil shale grade, which is typically based on the yield of oil per tonne of rock (Dyni, 2006). This led recent work to develop DRIFT (diffuse reflectance infrared Fourier transform spectroscopy), X-ray diffraction (XRD) and thermally based calibration curves to more reliably predict the kerogen content, and hence the yield, of shale oil (e.g. Bhargava *et al.*, 2005).

In examining Tertiary oil shales of eastern Queensland, Patterson *et al.* (1990) and Patterson and Henstridge (1990) identified smectite, kaolinite, Fe-carbonates (e.g. siderite) and pyrite as the inorganic phases most important to oil shale processing. Kaolinite (up to 7 wt%) and smectite (18–42 wt%) are typically the main inorganic constituents in the Rundle and Stuart oil shale deposits (Patterson and Henstridge, 1990). Expandable clays (i.e. smectite) were recently identified as a significant constituent (up to 25 wt%) in the Mahogany Zone of the Green River Formation oil shale (Wigand *et al.*, 2007). Smectite, examined by Patterson *et al.* (1990) in eight eastern Queensland oil shale deposits, was identified as a member of the montmorillonite/nontronite series with ≈10 per cent smectite/illite interlayering. Compositionally, the smectites examined contained relatively high Fe contents and showed only small differences in the Fe + Mg substitution for Al in the octahedral positions, whereas kaolinite showed no significant differences in composition compared to other kaolinites (Patterson, 1994; Patterson *et al.*, 1990).

Bulk oil shale composition is a reflection of the geological setting of the deposit, and oil shales can be enriched with potentially environmentally hazardous elements including As, Hg and Se, which are of particular concern as these elements are partially mobilized to the oil and retort gases during pyrolysis (e.g. Patterson, 1994; Qing *et al.*, 2007). Mobile trace elements most likely derive from porphyrins or other organo-metallic complexes in oil shales, but As and Se can also originate in sulphides, such as pyrite (Patterson, 1994; Patterson and Henstridge, 1990).

Geochemical cluster analysis of the Kerosene Creek member in the Rundle and Stuart oil shales (eastern Queensland) demonstrated an association of Ti, V and Cr (plus Si, Mg, Na and K) with aluminosilicate phases (Patterson and Henstridge, 1990). Enrichment of Cr, V and Ni in the Lopanta oil shale, Nigeria, is associated with the high kaolinite and illite–smectite mixed-layer clays identified in the deposit (Sonibare *et al.*, 2011). The mixed-layer montmorillonite–illite clay in the Cretaceous Julia Creek oil shale, Queensland, is highly enriched in V and contains ~60 per cent of the total oil shale V with contents of 5–10 wt% V_2O_3 to give an idealized formula of $K(Al,V,Mg)_4Si_7AlO_{20}(OH)_4$ (Patterson, 1994). The Julia Creek oil shale also contains elevated contents of elements of environmental concern, including As, S, Hg, Mo, Cd, Tl and U (Patterson, 1994). Recent work by Fu *et al.* (2012) demonstrated elevated contents of Ga, Cs, Li, Rb, Sc, Th, Zr and REEs associated with high Al_2O_3 contents due to abundant kaolinite (and illite) in oil shales from northern Tibet.

The by-products of oil shale pyrolysis, notably oil shale ash, have potential economic applications in the generation of a range of products including Na_2CO_3, $(NH_4)_2SO_4$ and K_2SO_4, as well as the recovery of V and Mo from vanadiferous clays (e.g. Patterson, 1994; Resheidat *et al.*, 2007). Spent oil shale may also serve as the raw material for cement production (e.g. Dyni, 2003, 2006; Heinstand, 2006). Retention of potentially hazardous heavy metals within the ash by amorphous and crystalline Ca–Al-silicates, that form through reactions between lime and the dehydroxylation products of clays (e.g. Yoffe *et al.*, 2002) may improve the suitability for the intended final end-use of oil shale by-products.

5.4.2 Oil Sands

Unlike oil shale, oil sands contain hydrocarbons, including bitumen, that are soluble in certain organic solvents (e.g. Dyni, 2006), and are typically hosted in unconsolidated sands, though reservoirs may also include porous sandstone or carbonate (Hein, 2006; Higley and Hein, 2011). Known and used since antiquity, for example as sealants, oil sands or tar sands as they are also known, occur throughout the Phanerozoic with some 598 deposits found in 23 countries (Attanasi and Meyer, 2010; Hein, 2006). The most notable oil sand deposits are in the Western Canada sedimentary basin, northern Alberta (Fig. 5.8). Here, three oil sands areas, the Athabasca, Peace River and Cold Lake oil sand deposits, contain a resource of 1804 billion barrels (BB) or 286.6 billion cubic metres (Higley and Hein, 2011). This represents ~65–70 per cent of the world's known oils sand deposits (Attanasi and Meyer, 2010; Hein, 2006) and is the source of almost all commercial bitumen production in North America (Higley and Hein, 2011). Other significant oil sand deposits occur in the Orinoco Oil Belt in the eastern Venezuelan Basin, Venezuela, and in Kazakhstan (Attanasi and Meyer, 2010). Australia has no reported oil sand deposits.

Previously defined as oil that was too viscous to flow into a well bore under natural reservoir conditions, an oil sand or bitumen deposit is characterized as having a viscosity of >10 000 centipoise (cP) and an API (American Petroleum Institute) gravity of 10°, whereas

Figure 5.8. Location of major, global oil sands deposits in Athabasca, Canada, and the Orinoco Oil belt, Venezuela (modified from Attanasi and Meyer, 2010; used by permission of the World Energy Council, London, www.worldenergy.org).

heavy oil has a viscosity of 10 to <10,000 cP, but typically a viscosity of 100 cP and an API gravity of 10–25° API (Hein, 2006; Higley and Hein, 2011). Viscosity, not gravity, is the main issue for the recovery of oil sands (Hein, 2006). For comparison, water has a room temperature (i.e. 25 °C) viscosity of 0.890 cP (CRC, 2013).

Natural bitumen, the lowest grade of crude oil, forms through the action of bacterial degradation, coupled with mild leaching by brackish or meteoric water and oxidation of conventional oils that have accumulated in subsurface reservoirs (e.g. Attanasi and Meyer, 2010; Hein, 2006). Depending on the influence of stratigraphic-structural controls to bitumen formation, bitumen (and heavy oil) deposits can be grouped into large basin-margin/unconformity related or medium–small fault/unconformity related types (Hein, 2006). During the degradation of conventional oil through heavy oil to form bitumen there is a progressive loss of light hydrocarbons (e.g. paraffins) and an enrichment in heavy hydrocarbons (e.g. asphaltenes) that results in an increase in viscosity and density (Hein, 2006). Associated with this is an increase in heavy metals such as V and Ni, and non-metals including N, S and O in bitumen (Hein, 2006; Higley and Hein, 2011).

Due to their significance and sheer size as an unconventional source of oil, both the Athabasca and Orinoco Oil Belt oil sand deposits have been intensively investigated in reviews of alternative energy sources, though most work has focused on the technological developments to process these oil sands (e.g. Attanasi and Meyer, 2010; Dusseault, 2001; Hein, 2006, 2009; Hein and Cotterill, 2006; Higley and Hein, 2011; Martinez, 1987, and the references contained therein). Consequently, the following discussion relies heavily on examples from both of these oil sand fields, particularly the Athabasca oil sand field.

In North America, most oil sands occur as subsurface deposits (<200 m depth) and are mined using conventional surface mining (open pit or quarrying) techniques (e.g. George, 1998) or *in situ* methods (i.e. underground) such as steam assisted gravity drainage (SAGD) and cyclic steam stimulation (CSS) (e.g. Higley and Hein, 2011; Topf, 2010). Though described as oil sands, mining of most deposits intersects clay lenses/pods interspersed throughout the deposit. For example, in the Athabasca field almost all bitumen and heavy oil deposits are hosted within the Manville Group, a series of sand–silt–shale arenites, with the main deposit hosted by the McMurry Formation, a quartzose, uniform, fine- to medium-grained sand (Berkowitz and Speight, 1975; Dusseault, 2001; Scott and Ozum, 2010). Compositionally, the McMurray Formation mineralogy comprises 90–98 per cent quartz, with 1–5 per cent feldspar, 0–3 per cent muscovite and 0–4 per cent clay minerals; with an overall bitumen content range in fine to coarse-grained sands of 8–16 per cent (Dusseault and Morgenstern, 1978). The clay mineralogy is predominantly kaolinite and illite, which account for 80–85 per cent of the clay fraction (Scott *et al.*, 1985), with more recent XRD analysis confirming the presence of interstratified kaolinite–smectite (K/S) and illite–smectite (I/S) with minor amounts of chlorite (Omotoso *et al.*, 2001). The presence of K/S and I/S supports the very high surface areas of 124–149 $m^2\ g^{-1}$ for the <2 μm fraction of oil sands tailings examined by Omotoso *et al.* (2001), which could not be accounted for by the presence of kaolinite and illite alone.

Bitumen occurs within the interstitial pore space between the sand grains, which are sheathed by a thin film of water within which the majority of the fines occur (Fig. 5.9). Processing of oil sands is conventionally undertaken by first mixing the crushed ore with hot water (50–80 °C) at alkaline pH (7–8.5) and then separating the bitumen by flotation (George, 1998; Omotoso *et al.*, 2001; Scott and Ozum, 2010). Processing of oil sands

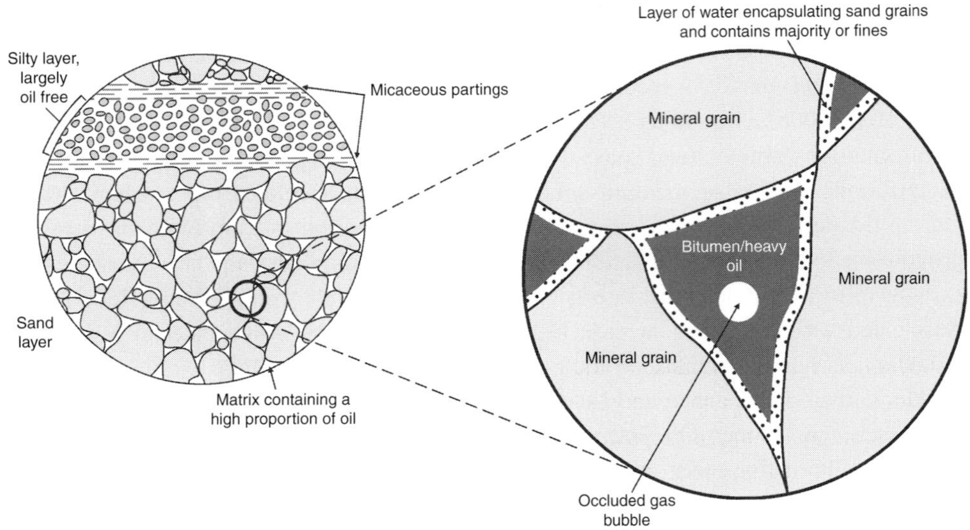

Figure 5.9. *In-situ* structure of quartzose oil sand and diagrammatic example of the occurrence of bitumen (modified from Dusseault and Morgenstern, 1978; with permission by © Canadian Science Publishing or its licensors). Bitumen occurs within the pore space between sand grains which are coated by a thin film of water. It is this thin film of water that contains the majority of fines (clays).

releases the entrained clays within the water film that encapsulates the sand grains and also from the intersected seams and lenses of weak, clay-rich shales. For the McMurray Formation, the tailings stream averages 17 per cent solids of silt and clay (Scott and Ozum, 2010).

Tailings are pumped to settling ponds where the initial separation of sand occurs quickly, whereas the flocculation of fines, defined as solids <44 μm in diameter (Liu *et al.*, 2004), may take 2–3 years to consolidate to form mature fine tailings (MFT) at the ~30–33 per cent solids level (Scott and Ozum, 2010; Scott *et al.*, 1985). Further consolidation to attain the required shear strength for subsequent utilization of the fines solids takes prohibitively longer (decades to centuries). With the sheer size of the MFT inventory in Alberta alone at more than 800 million cubic metres (Scott and Ozum, 2010), understanding the fines (i.e. clay) rheology in relation to reducing water usage and economic utilization of the fines solids for reclamation purposes is a major issue within the industry. Tailing separation is considered in more depth in Chapter 8.

The presence of clay (notably montmorillonite) within oil sands has long been known to significantly affect the extraction and recovery of bitumen during processing (e.g. Johnston, 1955). Bitumen recovery generally decreases with increasing fines content and divalent metal concentration (e.g. Ca) in the process water (Liu *et al.*, 2004). The presence of low concentrations of Ca to about 30–40 ppm (about 1 mM Ca) compresses the diffuse double layer, which reduces the hydrophobic nature of montmorillonite, enabling clay particles to strongly adhere to the bitumen surface (Czarnecka and Gillott, 1980; Liu *et al.*, 2004).

Adhesion of montmorillonite to the bitumen surface creates a steric layer that prevents attachment of air bubbles to the droplets of bitumen during flotation and also reduces the quality of the bitumen froth for subsequent treatment through the introduction of deleterious material (Liu *et al.*, 2004). These effects are not reported for kaolinite, where only weak adhesive forces were measured between kaolinite and bitumen, either in the presence or absence of Ca (Liu *et al.*, 2004).

5.5 Porphyry Copper and Related Epithermal Deposits

Porphyry copper deposits are the most significant type of copper resource (Cooke *et al.*, 2005; Sinclair, 2007) because of the large tonnages, e.g. usually >100 Mt at 0.3–2.0 wt% Cu (John *et al.*, 2010) and up to 1248 Mt at 0.63 per cent Cu for El Teniente (Camus, 2002). Porphyry deposits, and the genetically/spatially related epithermal deposits, especially in the high-sulphidation category (Hedenquist, 1986; Sillitoe, 2010), are formed in convergent plate margins (Arribas, 1995; Hedenquist *et al.*, 1998; Tosdal *et al.*, 2009).

These deposits are characterized by extensive hydrothermal alteration from relatively high (>300 °C) to low (<200 °C) temperatures (e.g. Tosdal *et al.*, 2009) resulting in various alteration mineral assemblages, as shown in Fig. 5.10a. Hydrothermal alteration at <300 °C produces abundant clay minerals in advanced and intermediate argillic, phyllic and, to a lesser extent, propylitic zones. Clay minerals are principally formed by hydrolysis of feldspar and plagioclase and, to a lesser extent, by alteration of Fe–Mg minerals, such as amphibole, biotite and chlorite, in association with quartz, carbonate or sulphides. As in many other types of ore deposits, the occurrence of clay minerals as hydrothermal alteration products in porphyry and epithermal deposits is primarily controlled by:

(1) composition of host rocks and fluids (e.g. Cooke *et al.*, 1998; Sinclair, 2007); and
(2) environmental conditions, mainly temperature and fluid pH (e.g. Tosdal *et al.*, 2009).

This is because: (1) porphyries commonly have intermediate to felsic compositions (e.g. Berger *et al.*, 2008; Sinclair, 2007); (2) in a magmatic to hydrothermal process, hydrothermal alteration involves high through intermediate to low temperatures; and (3) magmatic volatiles in fluids form acids under hydrothermal conditions (Hedenquist *et al.*, 1998), clay minerals in porphyry and related epithermal deposits are dominantly kaolinite, smectite and illite groups, each representing a particular environmental condition of the hydrothermal system (Fig. 5.10b).

Clay assemblages in porphyry and related epithermal deposits are commonly complex, both from overprinting and replacement of clay minerals at different stages due to evolving environmental conditions and through episodes of hydrothermal activity (e.g. Braxton *et al.*, 2009; Hedenquist *et al.*, 1998; Tosdal *et al.*, 2009). As can be inferred from Fig. 5.11, in the hydrothermal process a later stage of alteration may spatially overprint an earlier assemblage, resulting in the co-existence of clay assemblages formed under different conditions. For example, the dickite–pyrophyllite–alunite (advanced argillic) assemblage overprints

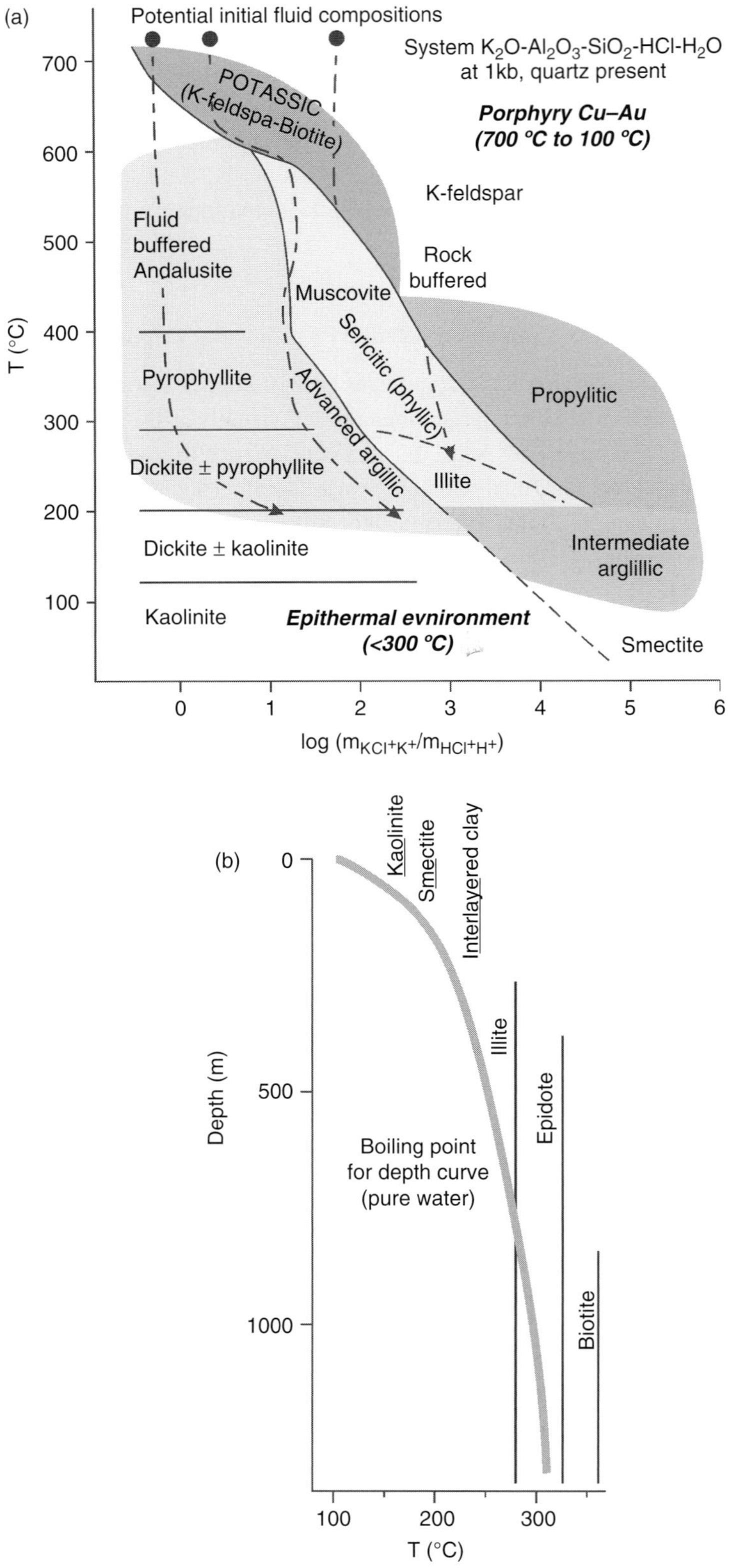

Figure 5.10. (a) Controls on hydrothermal alteration assemblages of porphyry and epithermal deposits by environmental conditions. (b) The dominant clay mineral groups that form under hydrothermal conditions (after Tosdal *et al.*, 2009).

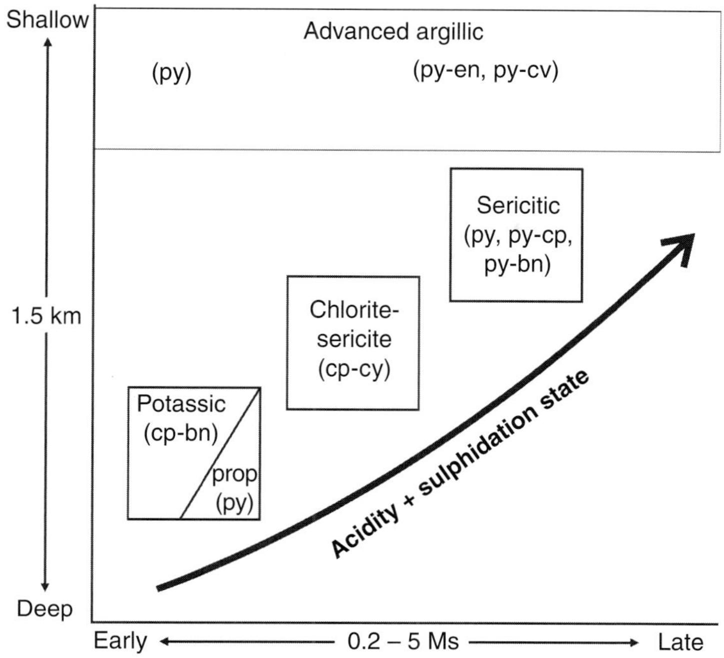

Figure 5.11. Evolving hydrothermal alteration in a porphyry mineralization system (after Sillitoe, 2010).

the deeper and central potassic (feldspar–biotite–magnetite–anhydrite) alteration as commonly observed in many porphyry deposits (e.g. Escondida; Garza *et al.*, 2001).

Zoning of clay minerals in and around a porphyry deposit documents the history of hydrothermal activities and mineralization processes involved, and is therefore used for assisting mineral exploration (refer to Section 5.5.4 and Chapter 6). Conversely (advanced) argillic assemblage can overprint copper mineralization in many copper porphyry deposits, and can result in clay enrichment in the ore and, therefore, impose difficulties in ore processing. Kaolinite, illite and smectite are the most common clay species present in porphyry copper deposits.

5.5.1 *Kaolinite*

The kaolin group forms under acidic conditions most commonly from precursor minerals such as feldspar, muscovite and plagioclase (e.g. Cruz, 2007), and comprises dickite, kaolinite and halloysite, which represents a series with decreasing crystallinity and implies a decrease in formation temperature (e.g. Ehrenberg *et al.*, 1993; Zotov *et al.*, 1998). Often, a change with depth in kaolinite species from dickite to kaolinite (e.g. Fig. 5.10a) reflects the evolution of hydrothermal fluids from ascending magmatic components to descending groundwater. Dickite usually results from hypergene acidic alteration, whereas halloysite

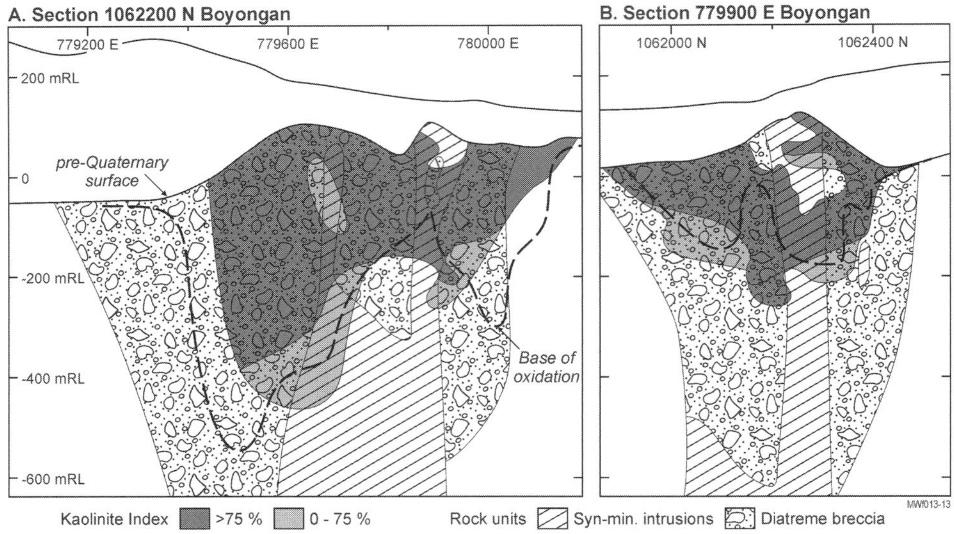

Figure 5.12. Supergene kaolinite on cross-sections of the Bayugo porphyry copper–gold deposits (after Braxton *et al.*, 2009).

is typical of supergene alteration. Kaolinite can be hypergene or supergene, or both, in a deposit. Although supergene kaolinite (halloysite or kaolinite, or both) is generally formed near the surface, it may, under some circumstances, occur at considerable depth. For example, kaolinite is reported at depths of 600 m below the surface in the Bayugo porphyry copper–gold deposits (Braxton *et al.*, 2009), due to deep penetration of cool groundwater probably at the waning stage of a hydrothermal mineralization process (see Fig. 5.12). In the supergene, argillic alteration, overprinted part of a system, a decrease in kaolinite content is directly related to the intensity of oxidation (e.g. Braxton *et al.*, 2009). More recently, part of the late stage, hydrolytic alteration at the Red Chris porphyry Cu–Au deposit, northern British Columbia, was characterized by alteration assemblages dominated by kaolinite down to the 1300 m elevation level (Rees *et al.*, 2015).

5.5.2 Smectite

Di-octahedral members of the smectite series, such as montmorillonite and beidellite, are the common smectite minerals in porphyry deposits (e.g. Franchini *et al.*, 2007). Smectite is formed mainly from alteration of feldspar or plagioclase (e.g. Fournier, 1965) at close to neutral pH conditions, or from Fe and Mg minerals like amphibole and biotite (Beane and Bodnar, 1995) in slightly mafic rocks.

Smectite is an important component of (intermediate) argillic and phyllic zones (Parry *et al.*, 2002), but diminishes in advanced argillic alteration due to low pH. With increasing temperature, smectite is transformed to illite when K is available (Pollastro, 1993; Velde and Vasseur, 1992) (Fig. 5.10b).

5.5.3 Illite

Illite, a non-expanding micaceous phyllosilicate, is another common clay species in porphyry deposits. As a result of the difficulty in differentiating this mineral from other clays such as muscovite or smectite, sericite is used as a field term for white mica minerals in the so-called phyllic and (intermediate, advanced) argillic alteration assemblages (Hedenquist *et al.*, 1998).

Illite is a di-octahedral, mica-like mineral with a crystal structure and chemical composition similar to muscovite. However, illite differs from muscovite in that it has vacant interlayer cation sites, i.e. interlayer charge <1 per formula, and more Si, Mg, Fe and water and slightly less tetrahedral Al and interlayer K than muscovite (Bailey, 1980). Illite crystallinity is sensitive to temperature, so that with increasing temperature illite tends to have decreasing interlayer charge vacancy and contains less water, i.e. in transition towards muscovite.

In porphyry deposits, illite is often associated with smectite, as part of the intermediate argillic alteration and phyllic assemblages (e.g. Cooke *et al.*, 1998). The relationship between illite and smectite can be complex, and intermediate members, i.e. interlayered illite–smectite, may form at low temperatures (Fig. 5.10b).

5.5.4 Exploration Significance

Associations of clay and other minerals such as phyllosilicates, anhydrous silicates (e.g. quartz and feldspar), oxides, carbonates and sulphides form particular alteration assemblages, i.e. potassic, propylitic, (intermediate) argillic, and advanced argillic (Fig. 5.10a), each with particular environmental implications (e.g. John *et al.*, 2010). Clay minerals of hydrothermal origin associated with porphyry copper deposits are commonly zoned at the deposit or prospect scale (e.g. the Campana Mahuida porphyry Cu deposit; Franchini *et al.*, 2007), indicating changes in chemical and physical conditions, or pulses of hydrothermal activity. Therefore, the spatial zoning of clay minerals bears significant information on the architecture, fluid flow and temporal evolution of a hydrothermal system, and is significant for understanding the hydrothermal mineralization system and, thus, guiding exploration (e.g. Cooke *et al.*, 2005; Sinclair, 2007).

Copper mineralization is either temporally related to, and spatially associated with, potassic alteration zones partially overprinted by phyllic alteration (e.g. Sillitoe, 1973, 2010), or occurs across almost the entire hydrothermal evolution from early potassic, through phyllic to later advanced argillic stages (e.g. Escondida; Garza *et al.*, 2001). In rare circumstances, copper mineralization does not occur in the potassic zone but is hosted in intermediate argillic alteration (e.g. the Cerro Colorado porphyry copper deposit; Bouzari and Clark, 2006).

The detail of clay alteration zoning in a porphyry mineralization system varies from one deposit to another, but a general broad pattern emerges (e.g.Seedorff *et al.*, 2008; Sillitoe, 1973, 2010) that shows a high-temperature, potassic alteration at either the centre or upper part of a porphyry-driven hydrothermal system, laterally surrounded and vertically overlaid by various lower-temperature assemblages, i.e. phyllic, propylitic and argillic. The zoning pattern generalized by individual workers or for individual porphyry deposits may vary, but the controlling mechanisms are similar to those depicted in Fig. 5.10a. Understanding the underlying mechanisms in a hydrothermal system is the key to assisting exploration. Exploration issues are covered in more detail in Chapter 6.

5.6 Bauxite

5.6.1 Introduction

The term bauxite was coined by Berthier in 1821 and named after the deposit of Les Baux de Provence in France. Bauxite, the main source of aluminium ore, is a mixture of oxides and hydroxides of aluminium, such as gibbsite and, to some extent, boehmite, diaspore, and semi-amorphous phases together with silicon, iron and titanium (Tardy, 1997).

Australia, with 81 Mt of bauxite mined in 2014 (USGS, 2015) at an estimated export value of about $4.6 billion (alumina value) (Department of Mines and Petroleum, 2015), is the top producer of bauxite, with more than one-third of the world's production, followed by China, Brazil, Guinea and India. Five bauxite mines currently operate in Australia: Boddington, Huntly and Willowdale in Western Australia, Gove in the Northern Territory and Weipa in Queensland. The latter northern deposits are high grade, with around 50 per cent accessible alumina, whereas the southern Western Australian deposits contain about 30 per cent available alumina and are also high in silica, which is present as quartz. Most of the Brazilian bauxite mines are located in the Goiás state (Porto Trombetas, Niquelândia and Barro Alto), including the largest bauxite mine in Brazil. The Paragominas bauxite mine is located in the east of the state of Pará, northern Brazil. The major bauxite deposits of India are located in Orissa, Andhra Pradesh, Bihar, Madhya Pradesh, Maharashtra, Karnataka and Tamil Nadu. In contrast, coastal bauxite deposits are found along the coastal tracts of Gujarat, Karnataka and Kerala. In Guinea, operations are located in the west, close to the border with Guinea-Bissau. Three main ore regions are mined, Sangarédi, Bidikoum and Silidara, with further resources at the N'danga, Boundou Waade and Paravi deposits. In bauxites, kaolinite, the only clay identified, is regarded as a contaminant as it has a deleterious effect on both grade and processing via the Bayer method (see Section 3.2.1).

5.6.2 Classification and Global Distribution of Bauxite

Bardossy (1982) and Bardossy and Aleva (1990) divided bauxites into two groups based on the parent rock and the genesis of the deposits: (1) karst bauxite deposits and (2) lateritic bauxite deposits. Other classifications also exist and are based on chemical and mineralogical parameters (Valeton, 1972) or petrological features (Boulangé, 1984). The minor (10 per cent) karst bauxite deposits are the result of the dissolution of carbonates and the residual concentration of Al derived from the weathering of associated Al silicates from interbedded volcanics. Recently, both laterite and karst-type bauxite deposits were reported in the Bolkardagi bauxite province in the Tauride Belt, southern Turkey (Hanilçi, 2013). Here, the karstic bauxites occur in carbonate units within the stratigraphic sequence of the Tauride Belt as clastic accumulations of material derived from the laterite bauxites in the sequence (Hanilçi, 2013). Other examples of more recently described karstic bauxites include the uneconomic, Mid-Late Cretaceous deposits in the southern Apennines, Abruzzi Mountains, Apulia (Boni *et al.*, 2013; Mongelli *et al.*, 2015) and in Campania, southern Italy (Boni *et al.*, 2013). Economic bauxite deposits (100 Mt of both laterite and karst bauxite) were described in the Wuchuan–Zheng'an–Daozhen area, Guizhou Province in central China (Gu *et al.*, 2013), with karstic deposits also reported in the Wulong–Nanchuan area, Guizhou Province (Li *et al.*, 2013). In Greece, Laskou and Economou-Eliopoulos (2013) provided evidence of the role biogeochemical processes played in the beneficiation of low-grade, Upper Jurassic–Middle Cretaceous karstic bauxites in the Parnassos-Ghiona mountains, central Greece. Abedini and Calagari (2013) described the geochemical characteristics of older (Permian and Triassic–Jurassic) bauxite horizons in Iran as part of the Irano-Himalayan karst bauxite belt. Most recently, Cretaceous karst bauxites of the Az Zabirah area, central northern Saudi Arabia, were characterized byAl-Mutairi *et al.* (2015).

However, laterite bauxite deposits represent the predominant (90 per cent) bauxite style and are developed through lateritic weathering under subtropical to tropical conditions over a range of aluminous-rich rocks. It is the general consensus that most of the important bauxite deposits were discovered following a period of extensive exploration activity from 1960 to 1990 (Freyssinet *et al.*, 2005). All bauxites formed in the last 65 My occur in tropical and subtropical areas. Bauxites from higher latitudes are older (Paleozoic to Mesozoic) but were formed under similar conditions (Fig. 5.13). Five major bauxite provinces are defined: south American; west African; Indian; south-east Asian; and north and south-western Australian. Localized deposits occur in south-east Brazil, south-east Africa, south-east Australia, southern United States, and in Russia and Kazakhstan. For example, irregular occurrences of small, potentially economic, pisolitic bauxite in the Southern Highlands of New South Wales, Australia were very recently described by Taylor and Eggleton (2015).

The characteristics of the bauxite weathering profile mimic those of ferruginous lateritic profiles (Nahon, 1991; Tardy, 1997) and include, from bottom to top: (1) a parent rock; (2) a 2–100 m thick saprolite horizon with preserved primary texture composed of kaolinite, Fe oxyhydroxides and residual primary minerals; (3) a bauxite horizon (ore) where kaolinite is dissolved and replaced by gibbsite with high Al and low Fe contents; and (4) a bauxitic or ferruginous duricrust (Fe or Al oxyhydroxides with kaolinite) to soft clay-rich horizons. The nature of the top horizons is related to the paleoclimatic history of the profile.

The vast majority of studies of the economic geology of lateritic bauxite were carried out more than 20 years ago (Valeton, 1972; Bardossy and Aleva, 1990). In 1997, Tardy developed a three-tiered classification of bauxites based on a combination of genesis and parent rock; namely orthobauxite, metabauxite and cryptobauxite (Freyssinet *et al.*, 2005).

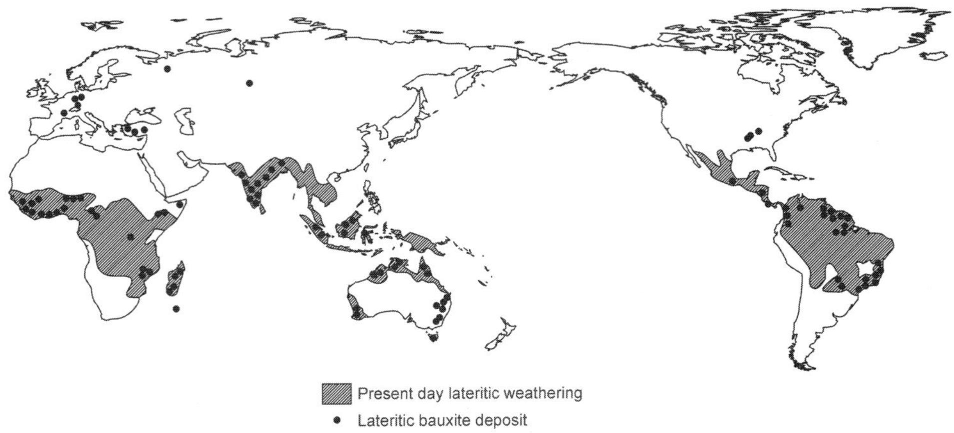

Figure 5.13. Location of the main lateritic bauxite deposits and present-day lateritic weathering (after Bardossy and Aleva, 1990 and Tardy and Roquin, 1998).

Orthobauxites are characterized by a single weathering stage and the profile consists of a saprolite horizon overlain by a light-coloured, indurated bauxite horizon capped by a pink-red, bauxitic duricrust (Bardossy and Aleva, 1990; Boulangé, 1984;Varajao *et al.*, 1989). From the base of the saprolite to the top of the profile, aluminium appears to 'migrate' from halloysite to kaolinite and then as gibbsite as the silica is leached (Fig. 5.14). Iron occurs as goethite in the saprolite but is present as hematite in the bauxitic horizon. The bauxitic duricrust consists predominantly of gibbsite and hematite, with rare kaolinite and residual quartz. Tardy (1997) correlated these mineralogical sequences with hydration conditions: hydrated minerals (halloysite and goethite) form at the base of the profile, whereas the less hydrated minerals (kaolinite and hematite) develop towards the top of the saprolite. Rare orthobauxite profiles, developed under very humid and well-drained sites, are entirely gibbsitic (Bardossy and Aleva, 1990; Boulangé, 1984). The gibbsite/kaolinite ratio in metabauxites increases upward. The surface horizons are freely drained and strongly leached by silica-poor water, so that kaolinite is transformed into gibbsite and dissolved silica: $Al_2Si_2O_5(OH)_4 + 3H_2O + O_2 = 2Al(OH)_3 + 2SiO_2$.

Metabauxites show many similarities with orthobauxites, although they contain more aluminium, as gibbsite and boehmite, and less iron. Metabauxites are not as common as orthobauxites. Tardy (1997) suggested that metabauxites formed by the transformation of orthobauxites by dehydration through a change to more arid climate. At the top of the profile, goethite and gibbsite are dehydrated to hematite and boehmite, respectively (Tardy and Nahon, 1985). Boehmite is preferentially developed in the superficial pisolitic horizon while massive hematite is found at the base of the duricrust.

Cryptobauxites (from the Greek *kruptos*, i.e. hidden) are concealed by a thick, clay-rich cover and are predominantly restricted to the Amazonia region (Manaus, Juriti and Trombetas in Brazil). A typical cryptobauxite profile consists of three main horizons (Fig. 5.15): (1) a kaolinite-dominated saprolite with rare Fe oxyhydroxides and

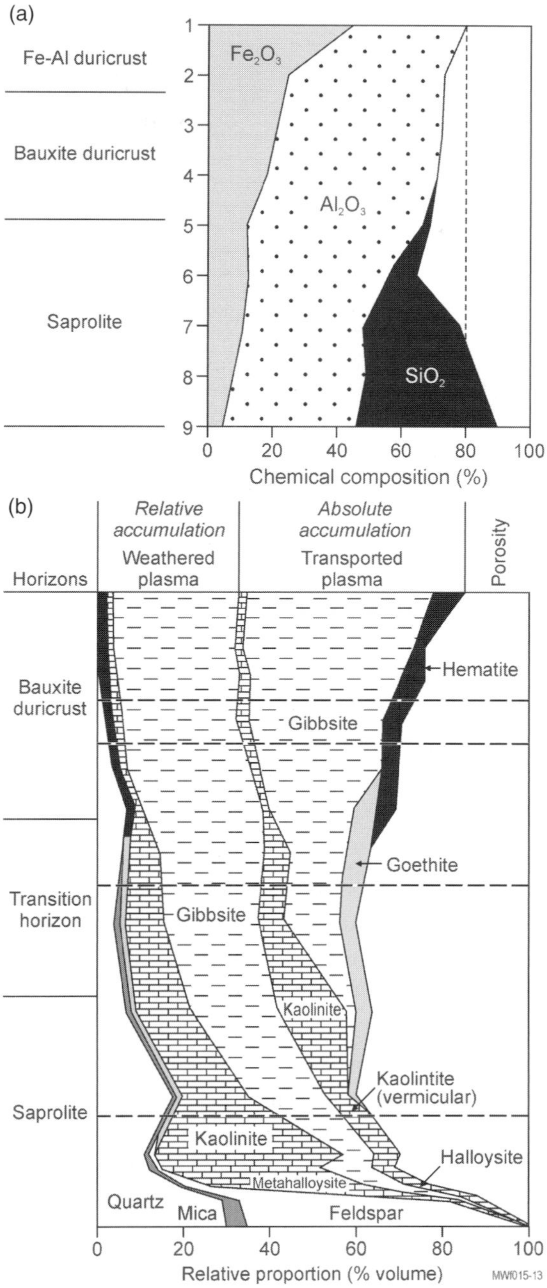

Figure 5.14. Orthobauxite profiles. (a) Chemical composition of a profile of the Bakhuis deposit in Suriname (after Aleva, 1979). (b) Evolution of the mineralogical composition in the Mont Tato deposit, Côte d'Ivoire (modified after Boulangé, 1984; as published in Freysinnet *et al.*, 2005).

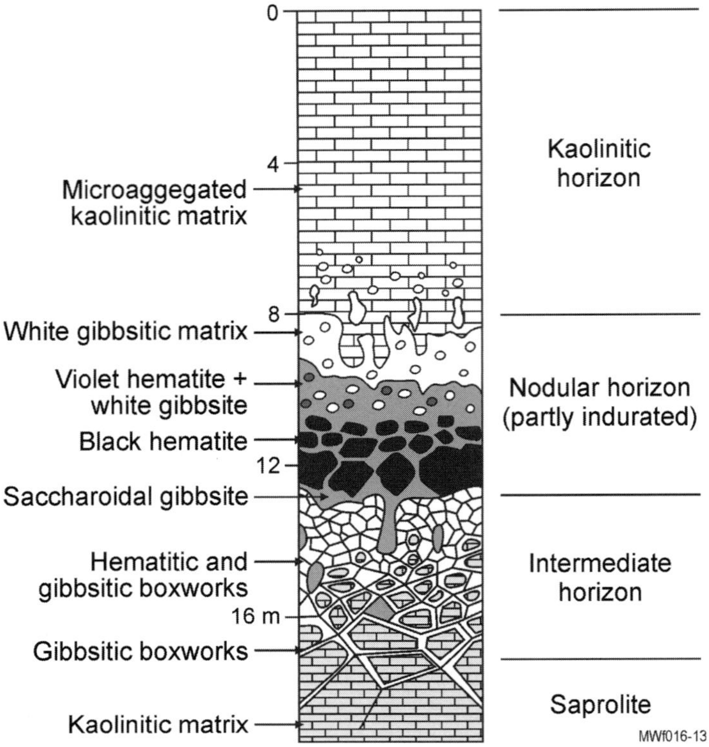

Figure 5.15. Schematic of a weathered profile representative of a cryptobauxite in the Juriti region of Amazonia (modified after Lucas, 1989; as published by Freyssinet *et al.*, 2005).

quartz that grades upward through a kaolinitic matrix crosscut by a thin gibbsitic boxwork and capped by indurated facies with Fe-rich yellow-red mottles in a gibbsitic matrix; (2) a 1–8 m thick gibbsite-rich, ferruginous nodular horizon merging upward to a duricrust consisting of hematitic pisoids embedded in a soft matrix of goethite, gibbsite, and minor kaolinite; and (3) a soft upper yellow to red homogeneous horizon (latosol) mainly composed of kaolinite, with rare gibbsite and goethite. The formation of the upper horizon is the result of a complex genesis generated by climatic evolution (Lucas, 1989). Beginning with a classic lateritic profile (Fe accumulation in a ferruginous duricrust) developed under a seasonally humid tropical climate, formation of a clay-rich surface horizon takes place via the transformation of the duricrust under a more humid climate.

5.6.3 Controls on the Formation of Bauxite Deposits

The bedrock lithology influences the grade and quality of the bauxite; 97 per cent of the global reserves of lateritic bauxite are derived from quartz-free or quartz-poor rocks (Bardossy and Aleva, 1990). Unexpectedly, the original aluminium content (10–35 per cent) of the

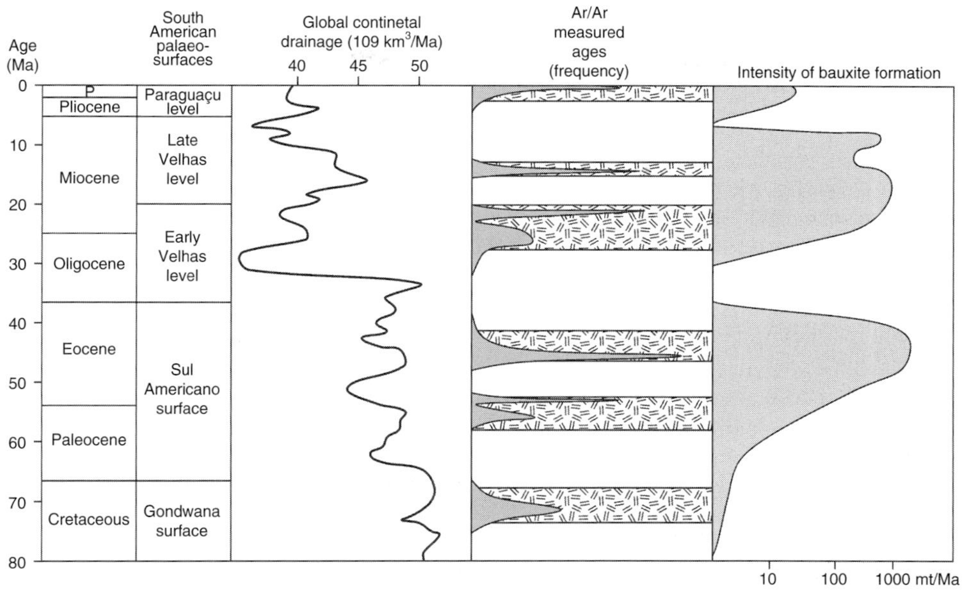

Figure 5.16. Comparison through time of the global continental drainage (Tardy and Roquin, 1998), the frequency of $^{39}Ar–^{40}Ar$ dates from Mn duricrusts in Brazil (Vasconcelos *et al.*, 1994) and the global bauxite reserves (Bardossy and Aleva, 1990; as published by Freyssinet *et al.*, 2005).

parent rock is a minor factor in the bauxite formation process (Bardossy and Aleva, 1990; Tardy, 1997) and the Al enrichment is mostly controlled by the primary Al/Si ratio, the rate of weathering and the Fe content. A large proportion of bauxite deposits are derived from sediments such as arkosic sandstones and shales, and granite, basalt and dolerite.

Most bauxite deposits occur on ancient lateritic surfaces in the higher parts of the present-day landscape as large plateaus, commonly remnants of ancient surfaces. Bauxite grade formation can be controlled by the local geomorphological setting in freely drained sites, such as upper slopes and crests (e.g. Darling Range in Australia and Madhya Pradesh in India).

The mineralogical paragenesis of the upper part of weathering profiles is controlled by long-term atmospheric humidity. The modern optimum conditions for bauxite formation are created by a humid tropical climate with an average temperature >22 °C and an annual rainfall >1200 mm, distributed over 9–11 months (Bardossy and Aleva, 1990). The location of numerous known bauxite deposits is no longer suitable for contemporary bauxite formation. Hence, these are simply vestigial (see metabauxites and cryptobauxites).

All mined bauxite deposits are Cretaceous or younger (Valeton, 1983). The formation of all of the present bauxite deposits started during the Late Cretaceous, with peaks in the Paleocene–Eocene and the Middle Miocene (Fig. 5.16).

A summary of the bauxite formation events for India, Australia and Africa encompassing fluctuations of global climate and continental drift (Tardy and Roquin, 1998) showed that the Eocene was the most favourable period (Fig. 5.16). Bauxites developed from the tropics to relatively high latitudes around the pre-Tethys domain and in India, Western Australia and southern Brazil. A second major event of formation occurred during the Miocene in subtropical zones of India, Australia and Amazonia.

5.7 Laterite and Sulphide-Hosted Ni Deposits

5.7.1 Nickel Sulphides

Nickel sulphide deposits are broadly classified according to the dominant Ni–S mineralization style, with five Ni–S mineralization types (Type 1–5) recognized (Barnes, 2006). In the Yilgarn Craton (Australia) Ni–S mineralization is predominantly of Type 1 (massive and net-textured or matrix ores, where sulphides form a continuous matrix) and Type 2 (low-grade disseminated sulphide within the ultramafic host), with the sequence: Type 4 (sediment-hosted stratiform sulphides), Type 3 (stratiform disseminated sulphides in fractionated ultramafic units) and Type 5 (sheared or folded massive sulphide ore) mineralization of lesser importance (Barnes, 2006). The long period of subaerial exposure, likely since the mid-Proterozoic but certainly since before glaciation in the Permian (Butt *et al.*, 2006), to which Ni–S deposits in the Yilgarn have been exposed, has resulted in the development of a thick, widespread and intensely weathered regolith with dating indicating some weathering in the Paleozoic, but most taking place from the late Cretaceous (Anand and Paine, 2002, as cited by Butt *et al.*, 2006).

This protracted period of subaerial exposure and alteration of Ni–S deposits under variable climatic conditions since the Cretaceous has produced deep weathering profiles, with the zone of oxidation extending to 100 m or more below the water table (Butt *et al.*, 2006). Supergene alteration of Type 1 and Type 2 Ni–S occurs via an electrochemical process where the upper part of the sulphide body acts as the cathode and the lower part, at or below the water table, acts as the anode. Progressive electrochemical oxidation results in Ni being released from the primary sulphide mineralogy (e.g. pyrrhotite, pentlandite) to precipitate as a range of secondary sulphides (e.g. violarite and millerite), sulphates (e.g. morenosite, carrboydite) and carbonates (e.g. gaspeite, takovite) at or close to the water table (Butt *et al.*, 2006). Though the previous discussion deals with komatiite-related Ni mineralization, many magmatic Ni–Cu ± PGE sulphide deposits undergo similar supergene mineral reactions to those described for komatiitic Ni deposits, and produce supergene mineral assemblages comparable to WA deposits (Schulz *et al.*, 2010).

Textural differences between Type 1 and Type 2 deposits that relate to the primary sulphide mineral distribution and silicate mineral to sulphide ratio influence the final distribution of Ni during oxidation. The low pH conditions generated during sulphide oxidation coupled with the dearth of Al in the primary Ni–S minerals and associated ultramafic host rocks are not conducive to the formation and stability of Al–Mg–Ni silicates. Thus, kaolinite and smectite are not associated with the primary Ni–S mineralization or present in the supergene zone. Only 7 Å and 10 Å type, hydrous Ni–Mg-silicates phases, such as nepouite, pecoraite and pimelite ('garnierite'), were listed as the main Ni-bearing silicates identified in weathered Ni–S deposits of the Yilgarn (e.g. see table 2 in Butt *et al.*, 2006). In Type 2 sulphides, Ni released from supergene sulphides, secondary Ni-carbonates and serpentine is taken up by goethite, or by chalcedonic quartz (with Si released from wall-rock silicates) (Butt *et al.*, 2006).

In comparing the mean composition of weathering profiles developed on Yilgarn ultramafic rocks that typically host Ni–S deposits, Butt *et al.* (2006) reported elevated contents for Al_2O_3 of 6.2 and 18–21 wt% in the ferruginous lag and lateritic residuum developed over peridotite and pyroxenite, when compared to fresh peridotite (1.8 wt% Al_2O_3) and pyroxenite (3.3 wt% Al_2O_3). This increase partly represents relative enrichment through the loss of mobile elements such as Mg (Butt *et al.*, 2006). However, elevated Al_2O_3 contents can also result from the input of exotic material introduced during development of the weathering profile. For example, at the giant, disseminated (Type 2) Mt Keith MKD5, Ni–S deposit (WA), elevated Al_2O_3 contents, as kaolinite (plus gibbsite), within the upper surficial hard-pan clays (alluvial–colluvial sediments and lateritic gravels) and extending down to the mixed ferruginous zone (with Al_2O_3 contents to 25 wt%), reflect exotic input from the erosion of arkoses and granitic rocks to the west of the deposit (Brand and Butt, 1998; Butt and Nickel, 1981). Elevated contents of Ti, V, Zr and Cr, incompatible with ultramafic Ni–S geochemistry within the surficial zones, were similarly interpreted as indicative of introduced material (Butt and Nickel, 1981). The occurrence of smectite, as montmorillonite, at Mt Keith (Brand and Butt, 1998; Grguric, 2003) is much less common than kaolinite being restricted within the quartz–dolomite zone immediately below the ferruginous zone and likely formed through later dissolution of dolomite (Butt and Nickel, 1981).

Similar associations may also be inferred in weathering profiles developed on Ni–(Cu)–S deposits elsewhere. For example, regional gneissic basement rocks to the north-east and south-west of the Fortaleza De Minas greenstone that hosts the O'Toole Ni deposit (Minas Gerais, Brazil) are the likely source of the argillitic component in argillitic, limonitic soils that cover soil profiles in the vicinity of the deposit (Taufen and Brenner, 1987). In the Thompson nickel belt (TNB) in Manitoba, Canada, Quaternary glaciation has resulted in extensive till deposits sourced from basement Archaean granites and gneisses to the east of the TNB (McClenaghan *et al.*, 2011). At the Aguablanca Ni–Cu (PGE) deposit in Extremadura (south-west Spain), alteration of chlorite in the high Al-bearing (to 8.22 wt% Al_2O_3) host gabbronorite unit has resulted in extensive smectite formation throughout the weathering profile (Suárez *et al.*, 2011). Smectite, present mainly as the di-octahedral varieties beidellite and minor nontronite, with tri-octahedral saponite, showed only minor scavenging capabilities for Ni (and Cu) incorporating up to 3.9 wt% NiO and 2.9 wt% CuO, respectively, within octahedral sites (Suárez *et al.*, 2011).

This is consistent with the Ni tenor exhibited for some lateritic nickel deposits, where smectite represents one of the principal host phases in some deposits of this type. As will also be discussed in the following section, the contribution of Al-bearing material and the influence of variably Al-depleted, ultramafic host rock lithologies from which lateritic nickel deposits are formed strongly influences the resulting Ni-host mineralogy.

5.7.2 Lateritic Nickel Deposits

Nickel laterites are residual products derived from the chemical and physical weathering of olivine-rich cumulate rocks, and their metamorphic derivatives, which have primary nickel contents of 0.2–0.4 per cent. Indeed, most Ni laterites of commercial significance appear to have formed from host rocks with a minimum Ni content of 0.25 per cent (Burger, 1996).

The general processes and controls of lateritic nickel formation, and the economics of their processing, have been comprehensively reviewed (e.g. Alcock, 1988; Brand *et al.*, 1998; Golightly, 1979, 1981; Kühnel *et al.*, 1978; Schellmann, 1983) and more recently by Elias (2001, 2002, 2006), Gleeson *et al.* (2003), Dalvi *et al.* (2004), Freyssinet *et al.* (2005), Marsh and Anderson (2011), Thorne *et al.* (2012b) and Butt and Cluzel (2013), and the references contained therein. Viewed simply, the main geochemical processes operating during lateritization that result in the formation of an Ni laterite profile are the dissolution and leaching of Mg (± Si and Ca), residual concentration of Fe, Al, Cr and Ti, with supergene enrichment of Ni, Co and Mn (Golightly, 1981; Kühnel *et al.*, 1978). In general terms, the characteristics of an Ni laterite, including grade, tonnage and mineralogy, are controlled by the dynamic interaction of climatic and geological factors, such as geomorphological history, drainage, structure and lithology, which enables nickel and cobalt to concentrate in the regolith.

About 85 per cent of Ni laterites are located in accretionary terrains, developed in the weathered mantle of obducted Miocene and Pliocene ophiolite (harzburgite–dunite olivine cumulates) complexes and occur generally in areas of marked topographic relief, such as in New Caledonia, the Philippines, Indonesia, Colombia, Dominican Republic, Brazil, western USA and Australia (Queensland) (de Oliveira *et al.*, 1992; Kühnel *et al.*, 1978; Mercado, 1979; Ogura, 1986).

The remaining resources are developed from komatiites and layered ultramafic (peridotite–dunite) complexes in stable cratonic platforms in Archaean and Proterozoic greenstone belts, such as the Yilgarn Craton with deposits including Murrin Murrin, Cawse and Bulong (Yilgarn) and Ravensthorpe (Albany-Fraser Province, WA), and others in Brazil, west Africa, Greece and the Ukraine (Bernadelli *et al.*, 1982; Eliopoulos and Economou-Eliopoulos, 2000; Nahon *et al.*, 1982; Trescases *et al.*, 1979).

Nickel laterite profiles are zoned and consist of a general sequence from bedrock to surface of: unweathered serpentinized ultramafic; saprolite zone, clay (i.e. smectite-rich) zone; ferruginous zone and soil cover. Based on the dominant nickel-host minerals present, three main types of deposit (Fig. 5.17) are recognized (Brand *et al.*, 1998):

(1) Hydrous Ni–Mg-silicates or ‘saprolitic’ deposits. Dominated by hydrated Mg–Ni-silicates (e.g. Ni-bearing serpentine-like and talc-like phases, such as ‘garnierite’) generally occurring deep in the saprolite. These are typified by high Ni grades (locally 2.5–5.0 per cent Ni), with a global mean in the range ≈1.8–2.5 per cent Ni, in deposits developed over ophiolitic rocks in areas of high relief in seasonally humid, tropical environments (i.e. ‘wet’ laterite). Examples include the Thio-Nakety districts, New Caledonia; Cerro Matoso, Colombia; Marlborough, QLD; Rio Tuba, Philippines;

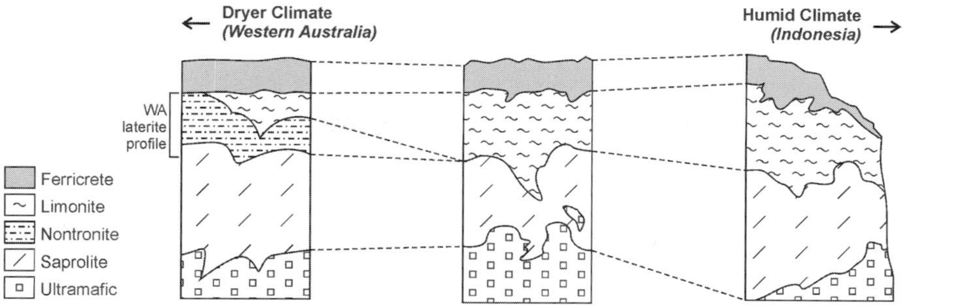

	%Ni	%Co	%Mg	%Fe	%Ni	%Co	%Mg	%Fe	%Ni	%Co	%Mg	%Fe
Ferricrete	0.2 - 0.5	0.2	0.6	35+	0.2 - 0.5	0.02	0.6	35+	0.2 - 0.5	0.02	0.6	35+
Limonite	0.6 - 1.4	0.1 - 0.2	1 - 2	45	1.2 - 1.7	0.1 - 0.2	1 - 2	45	1.2 - 1.7	0.1 - 0.2	1 - 4	45
Nontronite	1.2	0.08	3.5	18								
Saprolite	0.4	0.02	12	9	1.5 - 3	0.05 - 0.1	10 - 20	10 - 25	1.5 - 3	0.05 - 0.1	10 - 30	10 - 20

MWf004-13

Figure 5.17. Schematic of the general profile characteristics and composition of the three main Ni laterite types: smectite-hosted (left), oxide- or limonite-hosted (centre), hydrous Ni–Mg-silicate or 'garnierite'-hosted (right) (modified from Dalvi *et al.*, 2004).

Liberdade, Brazil; and Bitincke, Albania (Esson, 1983; Foster and Eggleton, 2002; Gleeson *et al.*, 2004; Golightly, 1979, 1981; Ogura *et al.*, 1987; Ouangrawa *et al.*, 1996; Thorne *et al.*, 2012a).

(2) Smectite or 'clay' silicate deposits. Dominated by smectitic clays (e.g. nontronite), with mineralization commonly occurring in the upper saprolite and/or pedolith. These are generally low-grade deposits with a global deposit mean of ≈1.21 per cent Ni, now occurring in semi-arid regions of low relief (i.e. 'dry' laterite), such as the Murrin Murrin and Bulong deposits (Burger, 1996; Elias *et al.*, 1981; Gaudin *et al.*, 2005).

(3) Oxide or 'limonitic' deposits. Dominated by the Fe oxy-hydroxide mineral, goethite, with mineralization commonly occurring in the upper saprolite and/or pedolith. Like the 'smectitic' deposits, these are also low-grade deposits with a global deposit mean of ≈1.06 per cent Ni, also occurring in now semi-arid regions of low relief (i.e. 'dry' laterite). Examples include Goro, New Caledonia; Moa Bay, Cuba; Cawse and Ravensthorpe, Western Australia; Orissa, India; and Caldag in western Turkey (Brand *et al.*, 1996; de Oliveira *et al.*, 2001; Rutherford *et al.*, 2001; Sahu and Venkateswaran, 1989; Som and Joshi, 2002; Thorne *et al.*, 2009).

In all lateritic nickel types Ni is also hosted by Mn-oxides but accounts only for a minor proportion of the total nickel content; Mn-oxides are the principal host of cobalt (Butt and Nickel, 1981;Elias *et al.*, 1981; Llorca, 1993).

Globally, clay silicate or 'smectitic' Ni laterites comprise about 5–10 per cent of the total lateritic resource, with 'saprolitic' (hydrous Ni–Mg-silicate) and 'limonitic' (oxide) deposits approximately equally accounting for the remainder (Freyssinet *et al.*, 2005). In Australia, the garnieritic silicate deposits have a somewhat limited expression, with the smectitic and oxide type deposits being the main laterite types developed.

Though most nickel laterite deposits are generally a composite of the three lateritic types described, since smectite comprises the dominant Ni-host in the 'smectitic' lateritic Ni deposits, the following discussion focuses mainly on, but is not necessarily restricted to, the occurrence and association of smectite in lateritic nickel deposits of this type. Deposits where this style of mineralization is found include Brolga, Queensland (Golightly, 1981), São João do Piauí, Brazil (de Oliveira *et al.*, 1992) and Indonesia (Kühnel *et al.*, 1978). However, the following discussion makes particular reference to two of the more well-known deposits of this type, Murrin Murrin and Bulong.

Protore lithology and composition is an important control on the type of Ni laterite that forms. Although both the smectite-silicate and oxide type laterite deposits develop from komatiitic cumulate rocks, smectite type deposits such as Murrin Murrin and Bulong have formed predominantly from peridotite, whereas oxide type deposits, such as Cawse and Ravensthorpe, are developed on dunitic rocks. The relatively high Al contents of clino-pyroxene-bearing peridotites (up to 3–4 per cent Al_2O_3) compared to the almost barren (<1 per cent Al_2O_3) dunitic komatiites (e.g. Freyssinet *et al.*, 2005), partly accounts for the smectite-dominated mineralogy of the Murrin Murrin and Bulong deposits over the oxide-dominated mineralogy of the Cawse and Ravensthorpe deposits. At Murrin Murrin, variation in cumulate type is also shown to influence regolith mineralogy, with the smectite zone developed from orthocumulates having a higher chlorite:nontronite ratio due to the higher Al bedrock content compared to adcumulates where the smectite zone develops a lower chlorite:nontronite ratio, i.e. a higher nontronite content (Elias, 2006; Wells and Butt, 2000).

Aluminium is a necessary precursor for the precipitation and formation of clay silicates such as some smectite group minerals. However, development of an extensive smectite-hosted clay zone also requires limited cation leaching due to episodic or impeded drainage associated with a wet–dry climate with seasonal leaching under mildly alkaline (pH 9) conditions (Elias *et al.*, 1981; Golightly, 1979, 1981;). The smectite zone in 'smectitic' lateritic Ni deposits can reach a considerable thickness, with the smectite zone at, for example, Bulong and Murrin Murrin reaching ~10–15 m thick (Elias *et al.*, 1981), within which the highest mineralization grades, typically 1–2 wt% Ni, occur associated with smectite (e.g. Elias *et al.*, 1981; Kühnel *et al.*, 1978).

Until recently, the Ni-host smectite phase was generally and collectively described as 'nontronite' in a broad sense. For example, early work at Murrin Murrin described the predominant Ni-host smectite as nontronite (e.g. Camuti and Riel, 1996). Recent detailed XRD, chemical and spectroscopic (e.g. FTIR and XAS) analyses have demonstrated that the smectite at Murrin Murrin is compositionally complex and occurs as intermediate to four endmember, di-octahedral types (Fig. 5.18), Al–Fe beidellite and Al–Fe montmorillonite, with a minor tri-octahedral component (Mg–Ni saponite) (Gaudin *et al.*, 2004a). Although being dominantly di-octahedral in character (i.e. montmorillonite-like), layer charge was shown to originate from both the tetrahedral and octahedral sheets.

Iron was shown to be the dominant octahedral cation, although the large variations in composition of the smectites examined was due mainly to substitution of Al/Fe/Mg

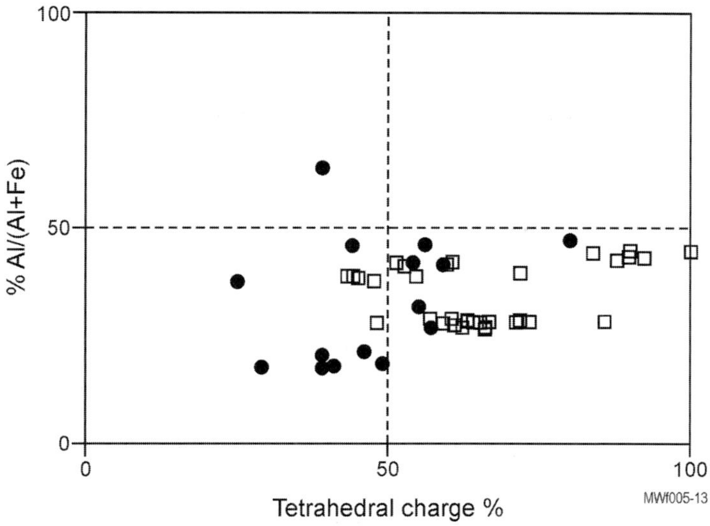

Figure 5.18. Charge and Al–Fe contents of Murrin Murrin smectite. The Al–Fe contents are total abundances averaged from TEM-EDX examination of samples, with the percentage tetrahedral charge estimated using FTIR analysis. The figure was reconstructed using the compositional data of (•) (Gaudin *et al.*, 2004b: table 1) for Murrin Murrin, and from data presented for smectites (□) identified in a weathering profile developed from serpentinite (Caillaud *et al.*, 2004).

within octahedral sheets, and was considered to represent a continuous solid–solution series (Gaudin *et al.*, 2004a, 2004b). For Murrin Murrin Ni-bearing smectites, spectroscopic refinement demonstrated that Ni did not occur at the interlayer position (i.e. was not exchangeable) nor was it present as a distinct clay phase, such as Ni-kerolite, but that Ni was contained within the octahedral sheet occurring as small, tri-octahedral clusters (Gaudin *et al.*, 2004b).

The large range in the Fe–Al–Mg composition of the Murrin Murrin smectites examined by Gaudin *et al.* (2004a, 2004b) fills or completes the so-called 'chemical-gap', previously thought to exist between di-octahedral Al-bearing smectites and Fe-rich smectites (e.g. Güven, 1988) as shown in Fig. 5.19. For smectite compositions with combined Mg + Ni contents of ~0.25 per unit-cell, a complete solid–solution series of Fe–Al substitution was shown to occur (Gaudin *et al.*, 2004a).

5.7.3 Kaolinite in Lateritic Nickel Deposits

As previously described in the discussion regarding Ni–S deposits, the low levels of Al in the primary, high-Mg protolith from which Ni laterites develop is not conducive to the formation of high, Al-bearing phases such as kaolinite. Yet, contrary to this, kaolinite is often reported as a significant phase in a number of 'smectitic' and 'limonitic' lateritic Ni deposits. Kaolinite is a dominant component of the ferruginous zone at Bulong, particularly

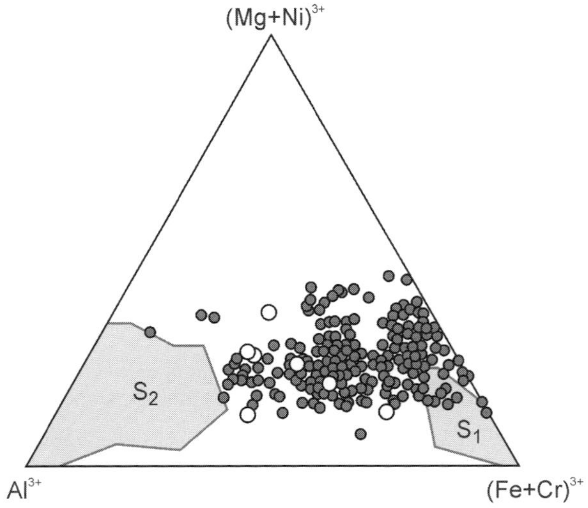

Figure 5.19. Composition distribution of cations in the octahedral sheet of Murrin Murrin smectite (●) examined by TEM-EDX analysis, plotted as an Al, (Fe + Cr), (Mg + Ni) ternary diagram (Figure reprinted from Gaudin *et al.* (2004a), reproduced with the kind permission of the Mineralogical Society of Great Britain and Ireland). Compositional fields denoted as S_1 and S_2 represent the extent of chemical variability of Fe-rich (S_1) and aluminous di-octahedral (S_2) smectites, as described by Güven (1988); the so-called 'chemical-gap' between these smectites. Previously, smectite compositions similar to those reported for the Murrin Murrin smectite were not commonly reported in the literature (○) (Brigatti, 1983). The data of Caillaud *et al.* (2004) as presented in Fig. 5.17 is not shown here.

between ~6 m and 14 m depth (Elias *et al.*, 1981), and is also a significant component of the ferruginous zone at Murrin Murrin (Wells and Butt, 2000). This is particularly the case for profiles of the northern deposits where felsic material derived from granite/adamellite outcrops to the north-west and contributes to the sediment flux of the Lake Carey drainage system to the north-east (Wells and Butt, 2000).

In oxide-style Yilgarn deposits, such as Goongarrie South (Heron Resources, Kalgoorlie Nickel Project), the top of the ferruginous zone is marked by an increase in kaolinite (and hematite), which reflects the greater influence of pedogenic processes and allochthonous input of an exotic Al source (Elias, 2006). Similarly, at Ravensthorpe a generalized compositional profile of the laterite sequence highlights the decrease in Al content from about 5 wt% at the near-surface to <1 per cent Al (i.e. the 'alumina discontinuity') at about 20 m depth, and reflects the decreasing contribution of introduced surface cover and lateritic residuum (Elias, 2006).

A similar situation is reported for other worldwide, oxide-style lateritic nickel deposits. In the oxide-style Ni laterites of the Sukinda ultramafic belt (Orissa, India), significant amounts of kaolinite (40–60 wt%) exist in the near surface (12–18 m), clayey ferruginous zone, whereby under presumably greater leaching conditions at the surface (<12 m) kaolinite is altered to gibbsite (Das *et al.*, 1999). Elevated alumina contents (to ~20 wt%)

were explained by either a chemical or mechanical enrichment from laterally transported Al-bearing material sourced from gabbroic dykes within the parent rock (Das *et al.*, 1999). In the Lomie region (south-east Cameroon) abundant kaolinite (and gibbsite) detected in the upper limonite, ferricrete breccia and extending into the upper part of the lower breccia zones of the oxide-style lateritic deposits (Dzemua and Gleeson, 2012) were attributed to felsic material sourced from high-grade gneisses and granulites in close proximity to the lateritic Ni deposits (Yongue-Fouateu *et al.*, 2006).

5.8 Palaeochannel Calcrete Uranium Deposits

Calcretes are chemically precipitated carbonate rocks that form within the phreatic and vadose zones of drainage systems (Arakel, 1988). Calcrete-hosted uranium deposits, classified as surficial U mineralization (IAEA, 2009; Skirrow *et al.*, 2009), are an important U resource formed mainly in palaeochannels and playa lake sediments in Precambrian granitic terranes, e.g. the Yilgarn Craton (Western Australia) (Butt *et al.*, 1977) and the Late Proterozoic Damaran Belt of Namibia (Bowell *et al.*, 2009).

The geological characteristics of palaeochannel calcrete U deposits are summarized as follows (Arakel, 1988; Butt and Gray, 2007; Butt *et al.*, 1977; Hobday and Galloway, 1999; Misra *et al.*, 2011; Noble *et al.*, 2011):

(1) Developed in deeply weathered Precambrian terranes under arid climate conditions and of subdued topography, with uraniferous granites as the primary U source.
(2) Formed in Tertiary to Quaternary times and hosted in valley-fill, channel sediments.
(3) Interplay of physiographic, hydrologic, hydrogeochemical, and depositional factors as controlling mechanisms for U enrichment.
(4) Mineralization occurs as semi-continuous, thin, elongate and tabular sheets, usually 1 to 15 m thick, 0.5 to 4 km wide and up to 150 km long, with U grade varying from tens of ppm to over 1000 ppm. The mineralized horizon is located in the upper part of a palaeochannel sedimentary sequence, i.e. at the base of the calcrete unit (Fig. 5.20).
(5) U is present mainly as carnotite, ideally $K_2(UO_2)_2(VO_4)_2{\cdot}1\text{–}3H_2O$, which precipitates from saline and alkaline groundwater at or near the water table, in association with carbonates (calcite and dolomite), quartz and Mg-rich, clay minerals.

Uranium dissolution as uranyl carbonate complexes or precipitation as carnotite from groundwater is the result of the interplay between several factors, including Eh, pH, groundwater composition (e.g. K, CO_2, V concentrations), physiographic (topography, weathering, evaporation, etc.) and hydrologic conditions of local drainage systems (Arakel, 1988; Bowell *et al.*, 2009; Noble *et al.*, 2011).

Carnotite occurs in the calcrete as: (1) subhedral crystals in pore spaces of the host, or coated by other minerals like quartz and clay, formed at or below the water table; and (2) as microcrystalline powder disseminated in the matrix of the host and coating voids or fractures, probably formed above the water table (Arakel, 1988).

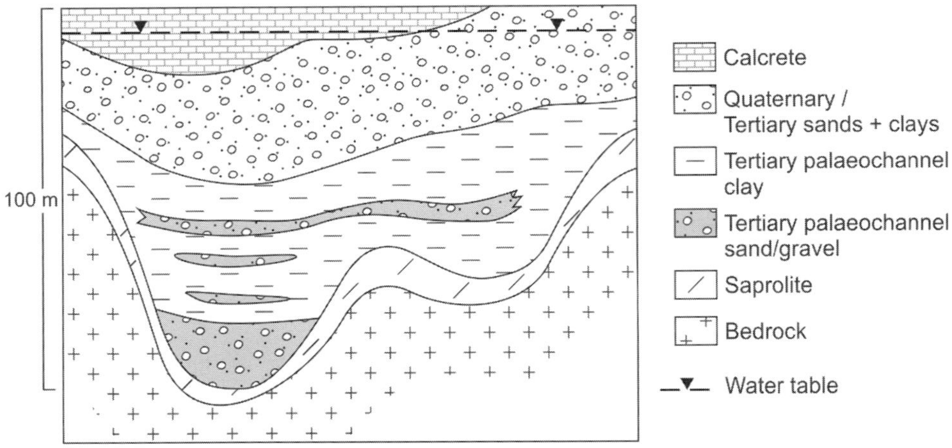

Figure 5.20. Generalized cross-section of palaeochannel that can host U (after Noble *et al.*, 2011). U mineralization (carnotite) is located at the base of the calcrete unit.

5.8.1 Clay Mineralogy and Zoning

Clay minerals are common in valley-fill, sand–silt sediments. As generalized in Fig. 5.20, clay is a significant component in almost all units of the palaeochannel sequence, including the upper calcrete, where carnotite mineralization occurs. They are also zoned in a palaeochannel sequence. For example, in the Yeelirrie deposit, the largest calcrete U deposit so far known, Mg-smectite and kaolinite, the two dominant clay minerals, are respectively enriched in the upper and lower part of the channel sequence, whereas palygorskite is present almost exclusively in the lower part of the sequence (Fig. 5.21).

The vertical mineral zoning of the palaeochannel sediments at Yeelirrie show a top, thin layer (<1 m) of kaolinite–quartz–palygorskite (loam), an upper 3–4 m thick layer of smectite-carbonate (i.e. the calcrete), and a lower kaolinite–quartz–palygorskite unit (Fig. 5.21). Also shown in Fig. 5.21, U mineralization occurs in a transition zone (a few metres thick) between the upper carbonate-smectite (calcrete) and the lower kaolinite–quartz–palygorskite sediments.

The vertical change in lithology through a palaeochannel sequence, as summarized in Fig. 5.20 and in the mineralogy (Fig. 5.21), is due to groundwater precipitation and metasomatic replacement processes. Chemical and mineralogical changes of the original valley-fill sediments at the water table are caused by laterally flowing saline and alkaline groundwater, which introduces Mg, Ca and CO_3^{2-}, and removes some of the Si, Al, K and Fe from the original sand and silt-rich sediments (e.g. Wang *et al.*, 1994). As carbonates precipitate and replace the silicates of the original sand–silt sediment at the water table, the calcrete body grows upwards. Carnotite precipitation happens later in this calcrete formation process (Airey and Roman, 1981; Butt and Gray, 2007). Therefore, the (clay) mineral zoning of palaeochannel sediments reflects significant changes in the chemical composition of the sediments.

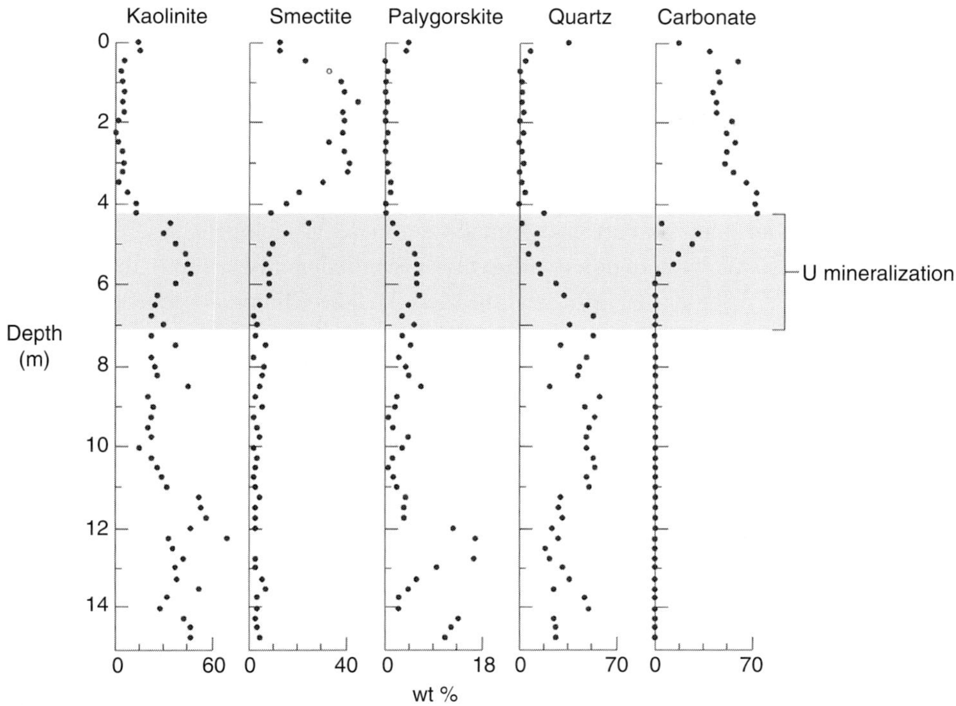

Figure 5.21. Mineral zoning at Yeelirrie (modified from Lower *et al.*, 2011), and with data from Yang *et al.* (2010).

5.8.2 Kaolinite

Kaolinite is a common clay constituent in the entire channel sequence, but is relatively more abundant in the lower, sand–silt sediments than in the upper calcrete (e.g. Fig. 5.21). This is because clay in the calcrete is dominated by (Mg-) smectite due to the abovementioned intense Mg metasomatic replacement process.

At Yeelirrie, kaolinite in both the valley-fill sediments unaltered by groundwater and the calcrete is of low crystallinity as revealed by infrared spectroscopy (unpublished CSIRO data), possibly indicative of its transported nature.

5.8.3 Smectite

Smectite is a major mineral component in association with carbonates in the upper calcrete unit of the palaeochannel sequence (e.g. Wang *et al.*, 1994). In the Yeelirrie calcrete U deposit, tri-octahedral, Mg-rich smectite (e.g. saponite and hectorite) is the dominant smectite species identified, whereas montmorillonite occurs only as a minor constituent (Yang *et al.*, 2010). The predominance of tri-octahedral Mg-smectite agrees with the Mg-rich chemical composition of the calcrete and groundwater, similar to that formed typically in evaporative salt lake sediments under arid climate conditions (e.g.Hover *et al.*, 1999).

5.8.4 Palygorskite

Palygorskite is present as a common clay mineral in calcrete (Verrecchia and Le Coustumer, 1996), formed from saline and alkaline groundwater under arid climate conditions (Saied *et al.*, 2010), e.g. in the western desert of Egypt (El-Sherbini and Issa, 1989), the Provence Basin of France (Colson *et al.*, 1998) and the Xeric–Aridic Climotoposequence of south central Iran (Moazallahi and Farpoor, 2012). At Yeelirrie, this Mg clay occurs in almost the entire palaeochannel sequence, except the upper calcrete unit (Fig. 5.20).

Like Mg-smectite, the Mg-rich composition of palygorskite is related to the surficial environmental conditions of arid and semi-arid climates. Various mechanisms are proposed for the origin of palygorskite in calcrete, e.g. precipitation from saline groundwater, transformation of smectite and other Mg-rich silicates, and a detrital origin as summarized by Verrecchia and Le Coustumer (1996).

5.8.5 Metallurgical Implications

Due to the abundance of carbonates, an alkaline leaching technique is normally considered for processing calcrete U ore (Bowell *et al.*, 2011; IAEA, 2009). Since clay minerals as major constituents of calcrete U ore directly influence fluid rheology (Bowell *et al.*, 2011), the correct identification and quantification of clay minerals is critical to the technical success of processing calcrete U ore and, ultimately, the economics of a mining project (IAEA, 2009).

5.9 Mineral Sands

Mineral sands form a class of ore deposit that contain suites of minerals with high specific gravity, known as 'heavy minerals'. They include economically important minerals rich in titanium, zirconium and rare earths, but also include industrial minerals such as diamond, sapphire, garnet, staurolite, kyanite and occasionally precious metals or gemstones.

The principal heavy minerals of commercial interest are zircon, rutile, ilmenite, and leucoxene:

(1) Zircon ($ZrSiO_4$), specific gravity (SG) 4.85, a colourless to off-white mineral that is the world's major source of ZrO_2 used in ceramic glazes, specialized ceramics and foundry sand moulds.
(2) Rutile (TiO_2), SG 4.85, is a red to black, naturally occurring titanium dioxide with a theoretical TiO_2 content of 100 per cent, but impurities such as Fe_2O_3, MnO, MgO and Cr_2O_3 reduce this to 93–95 per cent. About 90 per cent of the world's titanium mineral production is used in the manufacture of white titanium dioxide pigment.

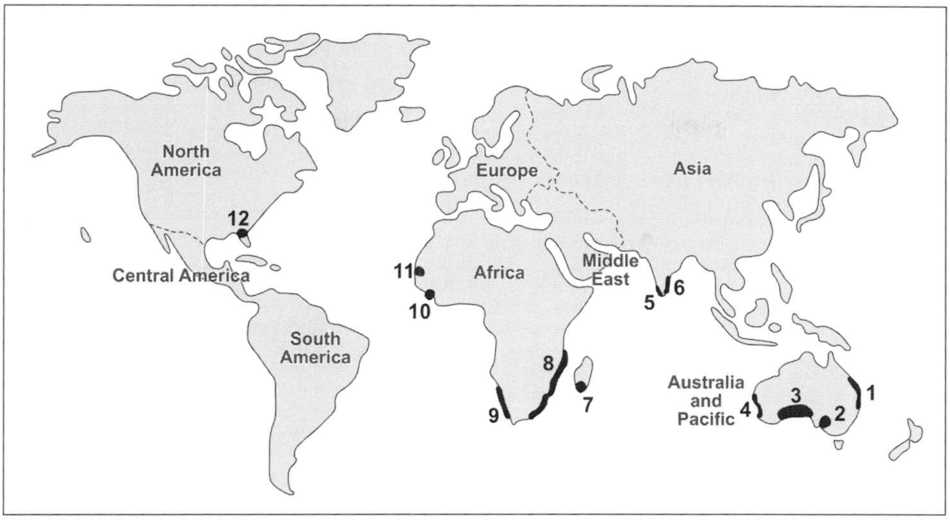

Figure 5.22. World distribution of major heavy mineral sand deposits. 1 = East Australia District; 2 = Murray Basin; 3 = Eucla Basin; 4 = Southwest Australia (Eneabba, Capel Districts); 5 = Kerala; 6 = Tamil Nadu; 7 = Madagascar; 8 = Southeast Africa (South Africa, Mozambique, Kenya Districts); 9 = Southwest Africa; 10 = Sierra Leone; 11 = Senegal; 12 = Jacksonville District, Florida.

(3) Ilmenite ($FeTiO_3$), SG 4.70–4.80, is black and opaque when unaltered, but has typically undergone some weathering and iron removal, so TiO_2 contents are between 45 and 65 per cent. Ilmenite is typically upgraded to a synthetic rutile (SR) via removal of iron and other impurities.

(4) Leucoxene is the name given to highly altered ilmenite. Grains are typically brown or grey with a waxy lustre and TiO_2 content of >68 per cent. The SG of leucoxene is variable but close to ~4.0.

The Earth hosts at least 12 significant sites of titanium- and/or zircon-heavy mineral sands enrichment, together with at least as many smaller areas. The location of these deposits is shown in Fig. 5.22.

The source rocks that provide the heavy mineral sands determine the composition of the mineral suite. For example, zircon, monazite, rutile, and some ilmenite are usually sourced from granite terrains, while ilmenite, garnet, sapphire and diamond are typically from ultramafic and mafic rocks (e.g. kimberlite or basalt). Garnet is also sourced commonly from metamorphic rocks, such as amphibolite schists.

Mineral sand deposits are formed following the erosion of suitable host rocks, the transport of material usually as river sediment to a low-energy depositional site (e.g. river mouth or coastal environment), the subsequent concentration of the heavy mineral fractions, and finally the protection of the deposit from later erosion. Deposits can be broadly classified into five main types: strandline placers, coastal dune, offshore marine, fluvial and channel. They may be further subdivided depending on the degree of consolidation, which is usually dependent on age. For example, recent deposits are largely unconsolidated, whereas older ('fossil') deposits are usually well consolidated and often covered by significant amounts of barren overburden. The five main groups of mineral sands deposits are summarized in Table 5.1.

Table 5.1. *Classification of heavy mineral sand deposits.*

Type of deposit	Characteristics of deposits	Selected deposits
Strandline placers		
Present-day shoreline deposits	Typically thin, coarse-grained, low grade deposits that develop on beach face depending on favourable supply, wave regime, longshore drift, coast configuration and sea-floor morphology (Force, 1991; Roy, 1999). Common after intense storm activity.	Common on east coast of Australia and south-west WA (Roy, 1999).
Tertiary fossil shoreline deposits	Generally high-grade deposits with finite, well-defined edges. Strands may be several kilometres long and hundreds of metres wide and are often covered by overburden. Can occur in large numbers of stacked sequences marking the gradual retreat of the sea. The heavy mineral assemblage is typically heterogeneous and coarse-grained (90–150 μm). Deposits may occur large distances inland from the shoreline (e.g. Murray Basin, south-eastern Australia).	Deposits are mainly found on trailing margins of continents, at latitudes lower than 35° (Force, 1991). Main sources of production in south-west WA (e.g. Eneabba north of Perth and Yoganup south of Perth (Baxter, 1990), the Murray Basin in south-eastern Australia (e.g. Ginkgo, NSW and Douglas, Vic (Roy and Whitehouse, 2003), and the Eucla Basin in SA and WA (Hou and Warland, 2005).
Coastal dune sand deposits		
Aeolian	Coastal dune sand deposits are usually broad, low-grade, recent deposits that occur at or near surface (minimal overburden). The deposits typically have a more homogeneous distribution of heavy minerals but a greater quantity of fine material such as clay (slimes). May be further subdivided into foredune, transgressive dune or stationary dune systems.	Examples of this type of heavy mineral sand deposits occur in West Africa, South Africa, Kenya, India, New Jersey and Florida in the US, Mozambique, Malawi and Queensland.

Table 5.1 (*cont.*)

Type of deposit	Characteristics of deposits	Selected deposits
Offshore marine deposits		
Present-day offshore deposits	Low-grade, fine-grained deposits.	Offshore heavy mineral sand deposits are known to occur off the southern Queensland and northern NSW coast.
Fossil offshore deposits	Typical features include: broad, lobate plan geometry, lateral continuity of heavy mineral lenses, low grade, uniformity of heavy mineral assemblage, fine grain size (typically 40–50 μm) of heavy mineral and trash minerals, high slimes content. Often buried under several metres of overburden (Williams, 1990).	Mainly in the Murray Basin in south-eastern Australia ('WIM' type deposits; Williams, 1990).
Channel deposits		
	New type of heavy mineral deposit recently encountered in Western Australia. High grade (4.4 per cent), of Tertiary or Quaternary in age and are derived from source areas that include granitic rocks and/or high-grade metamorphic rocks (Towner *et al.*, 1996).	Cooljarloo region in Western Australia.
Fluvial deposits		
	Occur in plateau, coastal plain and rift valley settings, usually close to the source. Grain sizes tend to be coarser that strandline placer and coastal dune deposits and also grains tend to have high angularity (less altered compositions; Force, 1991).	Irsha, Ukraine and Gbangbama in Sierra Leone (Stanaway, 1992).

The grade of a typical mineral sand ore deposit is usually low and the valuable components of the total heavy mineral (THM) concentrate rarely exceed 30 per cent. The lowest heavy mineral cut-off grade for most deposits is generally around 1–3 per cent. Within a heavy mineral concentrate, the components are typically

- zircon, from 1 per cent of THM to upwards of 50 per cent of THM;
- ilmenite, generally of 10 per cent to 60 per cent of THM;
- rutile, from 5 per cent to 25 per cent of THM;

- leucoxene, from 1 per cent to 10 per cent of THM;
- trash minerals, typically quartz, magnetite, garnet, chromite and kyanite, which usually account for the remaining bulk of the THM content; and
- slimes, typically the minerals previously described as well as clay minerals too fine to be economically extracted.

5.9.1 Clays in Mineral Sands

The presence of very fine material, typically <50–60 μm, collectively known as slimes, can significantly impact both the mining and separation of heavy mineral deposits and reduce heavy mineral recoveries (Balderson *et al.*, 2001). Due to the dynamic nature of their formation, most recent mineral sand deposits contain only a low level of fine clay minerals. In older, inland deposits, however, slimes can be a serious problem; a notable example is the now discontinued BHP-Billiton Beenup operation in Australia. In situations with more than 10–15 per cent fines, specific slimes handling circuits are required, and crucially more tailings dam capacity. Typical slimes contents in prominent deposits include: Kwale (Kenya) ~25 per cent; Hillendale (South Africa) ~23 per cent; Jacinth-Ambrosia (South Australia) <10 per cent; Donald (Murray Basin) 14–15 per cent; and Grand Cote (Senegal) <1 per cent.

The slimes in mineral sand deposits are generally made of fine-grained silicate, aluminosilicate and Fe-rich oxides/hydroxides, but they can contain a significant proportion of clay minerals. Clays in heavy mineral sands deposits are typically derived from different sources: (1) erosion of soils and weathered rocks, and (2) crystallization by reaction between solutions and silicates. The former type whose granulometric and mineralogical characteristics depend on transport and deposit processes typically do not form a high component of mineral sand deposits as the processes of erosion and transport (density sorting) removes most of the lighter clay fraction (Ware *et al.*, 2001). In some ancient deposits covered by clay-rich overburden, clays are often introduced through contamination during processing. In this case their composition is variable and depends on the surface environment, but are largely kaolinite, illite or smectite in composition (e.g. Corridor sands, Mozambique – Mercer and Pein, 2001; Taylor *et al.*, 2003; Old Hickory, USA – Newton and Romeo, 2006; Van Wormhoudt, 1993). For the second clay source, clays replace minerals or seal pores in rocks that are already formed. Mineral sand deposits are characterized by a high porosity favouring the chemical exchanges between pore solutions and fresh or oceanic water. The common presence of organic debris imposes particular Eh–pH–pCO_2 conditions that trigger dissolution and precipitation reactions of solid phases. In most deposits, the clays are primarily kaolinite-rich, and result from the breakdown of feldspar due to chemical weathering (Masters, 1990). However, a comprehensive study of clays in mineralized dunes of Northern Kwazulu-Natal (South Africa) by Ware *et al.* (2001) indicated that the distribution of clay assemblages also included illite and smectite (along with kaolinite). These clays occurred throughout the dunal deposits although the kaolinite content increased with depth, suggesting a more mature weathering profile. In comparison, smectite was more common in the near-surface environment.

5.9.2 Metallurgical Implications of Clays in Mineral Sands

The clay fraction is problematic for mineral processing as the slimes can interfere with mining and mineral separation activities. For example, strandline deposits, which usually have a very low moisture but higher slimes content, can be mined using dry mining methods, whereas wet mining using a dredge is typically employed on larger and low-slime dunal deposits. In the latter mining method, after being pumped as a slurry to the wet mill for processing, much of the clay is removed via cyclones. In the remaining sands, the quartz and heavy minerals are then separated from one another in a bank of spirals. Land reclamation efforts are also complicated by the need to remix the sand and clay tailings from the different stages of processing in the proper proportions. The presence of clays is, thus, typically detrimental to the cost of mineral sands extraction. As traditional mineral sands deposits are becoming exhausted, and companies are looking to deposits that may have higher slimes contents, mining and processing solutions that address higher slimes content will become increasingly important.

5.10 Rare Earth Elements

Rare earth elements (REEs) comprise 17 transition metals, including scandium (Sc), yttrium (Y) and the lanthanide series (lanthanum to lutetium). The seven lanthanides of low atomic number from La (57) to europium, Eu (63), are grouped as light REEs (LREEs) and the eight lanthanides with higher atomic numbers are named heavy REEs (HREEs). The major types of ore deposits containing economic amounts of REEs are: (1) monazite-bearing placer deposits, (2) igneous (i.e. carbonatites and peralkaline silicate rocks) REE deposits and (3) ‘ionic clays’ in intensely chemical weathered granites (Chakhmouradian and Wall, 2012), where clay minerals are major constituents. In addition, pegmatites, replacement veins and breccias as well as metasomatically altered rocks can be enriched in REEs, but are in many cases sub-economic, mainly because of low bulk REE content (e.g. Shmakin, 2008). One of the largest REE deposits in the world, Bayan Obo in Mongolia (Chao *et al.*, 1997; Drew *et al.*, 1990; Kynicky *et al.*, 2012), does not fit any of the above classifications, but rather shows characteristics of a number of the abovementioned deposit styles.

5.10.1 Placer Deposits

Monazite-bearing placer deposits were initially of high economic interest because of their association with other industrial minerals, such as zircon and ilmenite, and low processing costs, particularly the low amount of crushing required (Chakhmouradian and Wall, 2012). However, placer deposits containing monazite and xenotime, for example, are also enriched in U and Th, which impose a major radioactivity hazard during mining, remediation and even acceptable limits in the final product. Clay minerals naturally occur in distinct layers in placer deposits, defining the less endowed parts of these deposits, the heavy mineral components preferentially depositing upstream. They are seen as contaminants as their close association prevents selective mining. The abundance and composition of the respective clay minerals is determined by the source rocks and physicochemical conditions under which the placer deposit formed.

5.10.2 Carbonatites and Peralkaline Igneous Rocks

REEs are especially enriched in carbonatites and peralkaline igneous rocks. This enrichment can generally only be explained by remobilization of REEs in a late stage of emplacement (Williams-Jones *et al.*, 2012) or through weathering processes (Chakhmouradian and Wall, 2012). The latter processes of concentrating REEs during weathering are highly efficient, leading to the formation of significant REE deposits, such as Mt Weld (Australia) (Lottermoser, 1990), Araxa (Brazil) (Traversa *et al.*, 2001) and Tomtor (Russia) (Kravchenko and Pokrovsky, 1995). At Mt Weld, REEs are accumulated in abundant secondary phosphates, which formed in a deeply weathered Proterozoic carbonatite. Kaolinite and montmorillonite form only minor constituents in the iron oxide-rich, carbonatite laterite (Lottermoser, 1990).

5.10.3 Ionic Clays

Ion-adsorption ('ionic') clay REE deposits form due to lateritic weathering of granitoids (e.g. southern China (Kynicky *et al.*, 2012), or 'China clay' deposits in south-west England). Although of generally low grade, open-cast mining and easy processing make them a viable resource. REEs adsorbed on clay minerals can be easily extracted during leaching with monovalent salt solutions under ambient conditions (Moldoveanu and Papangelakis, 2012). Ionic clay deposits are also typically enriched in HREEs, which are of high economic interest.

Intense chemical weathering of granites can lead to adsorption of REEs on to the surface of Al-clays, such as kaolinite and halloysite (Chakhmouradian and Wall, 2012). Cation exchange on clay mineral basal surfaces and chemisorption of anions at crystal edges are the main drivers (Kynicky *et al.*, 2012). Even though weathering is the main driver for the concentration of the REEs, an initial involvement of low-temperature hydrothermal fluids in argillic alteration of granites cannot be excluded (Alderton *et al.*, 1980). Kaolinite, together with quartz ± muscovite, is the major constituent of argillic alteration mineral assemblages and is derived mainly from the breakdown of feldspar with the REEs being stripped from weathered fluorocarbonates (e.g. bastnäsite), phosphates, oxides or epidote-group minerals (Kynicky *et al.*, 2012). Montmorillonite, dickite and illite have also been described in the rare earth context, e.g. REE-bearing granitoids in south-west England (Alderton *et al.*, 1980). A detailed regolith profile of an ionic clay REE deposit in southern

China is summarized by Kynicky *et al.* (2012). This has upwards increasing kaolinitization of K-feldspars in Mesozoic granitoids that are overlain by a deeply weathered zone, where REEs are adsorbed on kaolinite, halloysite and smectite crystal surfaces.

5.11 Summary

This chapter has explored a diverse range of commodities in which clay, particularly kaolinite and smectite, occur, mainly within the context of those commodities of economic importance to Australia. This Australia-centric view, however, should not be seen as a distraction but rather highlights the almost ubiquitous occurrence and importance of the role and the impact clays have in a diverse range of geological settings. For almost all of the commodities herein discussed, the presence and role of clays is generally viewed in a negative sense; as diluents that introduce chemical contaminants (e.g. increased Al, Si and Mg contents) that decrease ore grade, as inhibitants to selective mining and introduce problems with ore processing and back-end tailings disposal. The only exceptions to this are REE and, in particular, the lateritic nickel deposits, specifically Type B or smectite-hosted deposits where smectite is the dominant Ni-host and main Ni-bearing mineral of economic concern. In the case of REE deposits, clays may contain economic concentrations that can be easily leached and extracted under ambient conditions. However, clays can be used as important vector minerals for hydrothermal ore deposits (e.g. porphyry Cu or epithermal Au), and can, for example, provide valuable information regarding the alteration history of porphyry Cu deposits.

Accurate identification of clay type (e.g. interlayered illite/smectite, I/S) is important to evaluating the thermal maturity of organic matter in sedimentary basins, and knowledge of the clay type also impacts on the extraction and recovery of bitumen from oil sands deposits. Hence, reliable and, importantly, rapid, large-scale identification of clay mineralogy would yield valuable information that could feed directly into decisions regarding the economic viability and resource management of the mineral commodities herein discussed. As a step-change to the traditional, small-scale and often time-consuming methods used for clay mineral quantification (e.g. see Chapter 4), automated scanning systems based on reflectance spectroscopy, such as CSIRO's HyLogger™ system, offer the opportunity for routine quantification of CID ore and gangue (i.e. clay) mineralogy on a large scale (e.g. Haest *et al.*, 2012). The challenge is to ensure that reliable and robust clay mineral identification can be translated from the nascent application to the relatively simple mineralogy of the Pilbara channel iron deposits to large-scale application to other, compositionally complex mineral systems.

Acknowledgements

The authors would like to thank Travis Naughton and Angelo Vartesi (formerly CSIRO Visual Resources Unit) for the high-quality drafting of the figures, and Lionel Fonteneau (formerly CSIRO) for his assistance with the bibliography. The authors would also like to acknowledge CSIRO, the various industry sponsors for supporting this work and the referees whose time and effort greatly improved earlier drafts of this chapter.

References

Abedini, A. & Calagari, A. A. 2013. Geochemical characteristics of bauxites: The Permian Shahindezh horizon, NW Iran. *Neues Jahrbuch für Geologie und Paläontologie – Abhandlungen*, 270 (3), 301–324.

Airey, P. & Roman, D. 1981. Uranium series disequilibria in the sedimentary uranium deposit at Yeelirrie, Western Australia. *Journal of the Geological Society of Australia*, 28 (3–4), 357–363.

Alcock, R. A. 1988. The character and occurrence of primary resources available to the nickel industry. In: Tyroler, G. P. & Landolt, C. A. (eds) *Extractive Metallurgy of Copper, Nickel and Cobalt*. Phoenix, AZ: The Minerals, Metals and Materials Society, 67–89.

Alderton, D. H. M., Pearce, J. A. & Potts, P. 1980. Rare earth element mobility during granite alteration: Evidence from southwest England. *Earth and Planetary Science Letters*, 49 (1), 149–165.

Aleva, R. A. 1979. Bauxitic and other duricrusts in Suriname: A review. *Geologie Minjbouw*, 58, 321–336.

Al-Mutairi, A. N., Galmed, M. A. & Aldamegh, K. S. 2015. Petrogenesis of the Az Zabirah south zone bauxite ore deposits, central northern Saudi Arabia. *Arabian Journal of Geosciences*, 8 (4), 2327–2339.

Altun, N. E., Hiçyilmaz, C., Hwang, J.-H., Bağci, A. S. & Kök, M. V. 2006. Oil shales in the world and Turkey: Reserves, current situation and future prospects – a review. *Oil Shale*, 23 (3), 211–217.

Anand, R. R. & Paine, M. 2002. Regolith geology of the Yilgarn Craton, Western Australia: Implications for exploration. *Australian Journal of Earth Sciences*, 49 (1), 3–162.

Arakel, A. V. 1988. Carnotite mineralization in inland drainage areas of Australia. *Ore Geology Reviews*, 3 (1), 289–311.

Arribas, A. J. 1995. Characteristics of high-sulfidation epithermal deposits, and their relation to magmatic fluid. In: Thompson, J. F. (ed.) *Magmas, Fluids, and Ore Deposits*. Québec: Mineralogical Association of Canada.

Attanasi, E. D. & Meyer, R. F. 2010. Natural bitumen and extra-heavy oil. In: *2010 Survey of Energy Resources*. London: World Energy Council.

Bailey, S. 1980. Summary of recommendations of AIPEA Nomenclature Committee. *Clay Minerals*, 15 (1), 85–93.

Balderson, G., Ross, N. & Goldup, N. 2001. Pilot plant objectives and considerations for evaluating the waste disposal of high slimes heavy mineral sand deposits. In: Ausimm, T. (ed.) *International Heavy Minerals Conference*. Fremantle, WA: Australasian Institute of Mining and Metallurgy.

Bardossy, G. 1982. *Karst Bauxites: Bauxite Deposits on Carbonate Rocks*. Amsterdam: Elsevier.

Bardossy, G. & Aleva, G. J. J. 1990. *Lateritic Bauxites*. Amsterdam: Elsevier.

Barnes, S. J. 2006. Komatiite-hosted nickel sulfide deposits: Geology, geochemistry, and genesis. In: Barnes, S. J. (ed.) *Nickel Deposits of the Yilgarn Craton: Geology, Geochemistry, and Geophysics Applied to Exploration*. Littleton, CO: Society of Economic Geologists.

Baxter, J. L. 1990. Heavy mineral sand deposits of Western Australia. In: Hughes, F. E. (ed.) *Geology of the Mineral Deposits of Australia and Papua New Guinea*. Melbourne, VIC: The Australasian Institute of Mining and Metallurgy.

Beane, R. E. & Bodnar, R. J. 1995. Hydrothermal fluids and hydrothermal alteration in porphyry copper deposits. In: Pierce, F. W. & Bohm, J. G. (eds) *Porphyry Copper Deposits of the American Cordillera.* Tucson, AZ: Arizona Geological Society.

Berger, B. R. A., Robert A., Wynn, J. C., Seal, R. R. 2008. *Preliminary Model of Porphyry Copper Deposits.* Version 1.0. Reston, VA: US Geological Survey.

Berger, B. & Henley, R. W. 1989. Advances in the understanding of epithermal gold–silver deposits, with special reference to the western United States. In: Keays, R. R., Ramsay, W. B. H. & Groves, D. I. (eds) *Economic Geology Monographs.* New Haven, CT: The Economic Geology Publishing Company.

Berkowitz, N. & Speight, J. G. 1975. The oil sands of Alberta. *Fuel*, 54, 138–149.

Bernadelli, A., Melfi, A. J., Oliveira, S. M. B. & Trescases, J. J. 1982. The Carajas nickel deposits. In: Melfi, A. J. & Carvalho, A. (eds) *Proceedings of the II International Seminar on Lateritisation Processes, 4–12 July, 1982.* Sao Paulo: Balkema, 107–118.

Bhargava, S., Awaja, F. & Subasinghe, N. D. 2005. Characterisation of some Australian oil shale using thermal, X-ray and IR techniques. *Fuel*, 84 (6), 707–715.

Boak, J. 2011. Oil shale developments. In: American Association of Petroleum Geologists, Unconventional energy resources: 2011 review. *Natural Resources Research*, 20 (4), 279–328.

Boni, M., Rollinson, G., Mondillo, N., Balassone, G. & Santoro, L. 2013. Quantitative mineralogical characterization of Karst bauxite deposits in the Southern Apennines, Italy. *Economic Geology*, 108 (4), 813–833.

Boulangé, B. 1984. Les formations bauxitiques et latéritiques de Côte d'Ivoire: Les faciès, leur transformation, leur distribution et l'évolution du modelé. PhD, L'Université Paris VII

Bouzari, F. & Clark, A. H. 2006. Prograde evolution and geothermal affinities of a major porphyry copper deposit: The Cerro Colorado hypogene protore, I Región, Northern Chile. *Economic Geology*, 101 (1), 95–134.

Bowell, R. J., Barnes, A., Grogan, J. & Dey, M. 2009. Geochemical controls on uranium precipitation in calcrete palaeochannel deposits of Namibia. In: *International Applied Geochemistry Symposium.* Fredericton, NB: The Association of Applied Geochemists, 425.

Bowell, R., Grogan, J., Hutton-Ashkenny, M., *et al.* 2011. Geometallurgy of uranium deposits. *Minerals Engineering*, 24 (12), 1305–1313.

Boyle, R. W. 1987. *Gold: History and Genesis of Deposits.* New York: Springer.

Brand, N. W. & Butt, C. R. M. 1998. Regolith geology of the Mt. Keith MKD5 nickel sulphide deposit, Western Australia. In: Taylor, G. & Pain, C. (eds) *New Approaches to an Old Continent, Proceedings of Regolith '98, 2–9 May.* Kalgoorlie, WA: CRCLEME, Perth, 335–341.

Brand, N. W., Butt, C. R. M. & Hellsten, K. J. 1996. Structural and lithological controls in the formation of the Cawse Nickel laterite deposits, Western Australia: Implications for supergene ore formation and exploration in deeply weathered terrains. In: Grimsey, E. J. & Neuss, I. (eds) *Nickel '96: Mineral to Market, 27–29 November.* Kalgoorlie, WA: Australasian Institute of Mining and Metallurgy, 185–190.

Brand, N. W., Butt, C. R. M. & Elias, M. 1998. Nickel laterites: Classification and features. *AGSO Journal of Australian Geology and Geophysics*, 17 (4), 81–88.

Braxton, D. P., Cooke, D. R., Ignacio, A. M., Rye, R. O. & Waters, P. J. 2009. Ultra-deep oxidation and exotic copper formation at the Late Pliocene Boyongan and Bayugo porphyry copper–gold deposits, Surigao, Philippines: Geology, mineralogy, paleo-altimetry, and their implications for geologic, physiographic, and tectonic controls. *Economic Geology*, 104 (3), 333–349.

Brigatti, M. F. 1983. Relationships between composition and structure in Fe-rich smectites. *Clay Minerals*, 18 (2), 177–186.

Bundy, W. M. 1993. The diverse industrial applications of kaolin. In: Murray, H. H., Bundy, W. M. & Harvey, C. C. (eds) *Kaolin Genesis and Utilization*. Boulder, CO: The Clay Minerals Society.

Burger, P. A. 1996. Origins and characteristics of lateritic nickel deposits. In: Grimsey, E. J. & Neuss, I. (eds) *Nickel '96: Mineral to Market, 27–29 November*. Kalgoorlie, WA: Australasian Institute of Mining and Metallurgy, 179–183.

Burnham, G. F. & Hawke, P. J. 2002. Smart solution to a sticky problem: In-mine clay mapping using high-resolution geophysics. In: Ausimm, T. (ed.) *Iron Ore 2002 Conference, 9–11 September*. Perth, WA: The Australasian Institute of Mining and Metallurgy, 189–194.

Butt, C. R. M. & Cluzel, D. 2013. Nickel laterite ore deposits: Weathered serpentinites. *Elements*, 9, 123–128.

Butt, C. R. M. & Gray, D. J. 2007. Palaeochannel uranium in southern western Australia: Déjà vu all over again. In: Bierlein, F. P. & Knox-Robinson, C. M. (eds) *Kalgoorlie '07; Old Ground, New Knowledge, 25 September*. Kalgoorlie, WA: Geoscience Australia, 115–117.

Butt, C. R. M. & Nickel, E. H. 1981. Mineralogy and geochemistry of the weathering of the disseminated nickel sulfide deposit at Mt Keith, Western Australia. *Economic Geology*, 76 (6), 1736–1751.

Butt, C. R. M., Horwitz, R. C. & Mann, A. W. 1977. Uranium occurrences in calcrete and associated sediments in Western Australia. *Report*. Perth, WA: CSIRO Division of Mineralogy.

Butt, C. R. M., Nickel, E. H. & Brand, N. W. 2006. The weathering of nickel sulfide deposits and implications for geochemical exploration. In: Barnes, S. J. (ed.) *Nickel Deposits of the Yilgarn Craton: Geology, Geochemistry, and Geophysics Applied to Exploration*. Littleton, CO: Society of Economic Geologists.

Caillaud, J., Proust, D., Righi, D. & Martin, F. 2004. Fe-rich clays in a weathering profile developed from serpentinite. *Clays and Clay Minerals*, 52 (6), 779–791.

Camus, F. 2002. The Andean porphyry systems. In: Cooke, D. R. & Pongratz, J. (eds) *Giant Ore Deposits: Characteristics, Genesis and Exploration*. Hobart, TA: University of Tasmania, Centre for Ore Deposit Research (CODES).

Camuti, K. S. & Riel, R. G. 1996. Mineralogy of the Murrin Murrin nickel laterites. In: Grimsey, E. J. & Neuss, I. (eds) *Nickel '96: Mineral to Market, 27–29 November*. Kalgoorlie, WA: The Australasian Institute of Mining and Metallurgy, 209–210.

Carrillo-Rosúa, J., Morales-Ruano, S., Esteban-Arispe, I. & Hach-Alí, P. F. 2009. Significance of phyllosilicate mineralogy and mineral chemistry in an epithermal environment: Insights from the Palai-Islica Au–Cu deposit (Almeria, SE Spain). *Clays and Clay Minerals*, 57 (1), 1–24.

Chakhmouradian, A. R. & Wall, F. 2012. Rare earth elements: Minerals, mines, magnets (and more). *Elements*, 8 (5), 333–340.

Chang, Z., Hedenquist, J. W., White, N. C., *et al.* 2011. Exploration tools for linked porphyry and epithermal deposits: Example from the Mankayan intrusion-centered Cu–Au district, Luzon, Philippines. *Economic Geology*, 106 (8), 1365–1398.

Chao, E. C. T., Back, J. M., Minkin, J. A., *et al.* 1997. *The Sedimentary Carbonate-Hosted Giant Bayan Obo REE-Fe-Nb Ore Deposit of Inner Mongolia, China: A Cornerstone Example for Giant Polymetallic Ore Deposits of Hydrothermal Origin*. Washington, DC: US Geological Survey, United States Government Printing Office.

Cohen, D. & Waite, T. 2004. Interaction of aqueous Au species with goethite, smectite and kaolinite. *Geochemistry: Exploration, Environment, Analysis*, 4 (3), 279–287.

Colson, J., Cojan, I. & Thiry, M. 1998. A hydrogeological model for palygorskite formation in the Danian continental facies of the Provence Basin (France). *Clay Minerals*, 33 (2), 333–347.

Cooke, D. R., Heithersayz, P. S., Wolfel, R. & Calderon, A. L. 1998. Australian and western Pacific porphyry Cu–Au deposits. *AGSO Journal of Australian Geology & Geophysics*, 17 (4), 97–104.

Cooke, D. R., Hollings, P. & Walshe, J. L. 2005. Giant porphyry deposits: Characteristics, distribution, and tectonic controls. *Economic Geology*, 100 (5), 801–818.

CRC. 2013. Viscosity of liquids. In: Haynes, W. M. (ed.) *CRC Handbook of Chemistry and Physics*, 93rd edition (Internet version 2013). Boca Raton, FL: CRC Press.

Cruz, M. D. R. 2007. *Genesis and Evolution of the Kaolin-Group Minerals During the Diagenesis and the Beginning of Metamorphism*. Malaga: Departamento de Química Inorgánica, Cristalografía y Mineralogía. Facultad de Ciencias.

Czarnecka, E. & Gillott, J. E. 1980. Formation and characterization of clay complexes with bitumen from Athabasca oil sand. *Clays and Clay Minerals*, 28 (3), 197–203.

Dalvi, A. D., Bacon, W. G. & Osborne, R. C. 2004. The past and the future of nickel laterites. In: *PDAC 2004 International Convention, Trade Show and Investors Exchange, March 7–10, 2004*. Toronto, ON: Prospectors & Developers Association of Canada, 1–27.

Das, S. K., Sahoo, R. K., Muralidhar, J. & Nayak, B. K. 1999. Mineralogy and geochemistry of profiles through lateritic nickel deposits at Kansa, Sukinda, Orissa. *Journal of the Geological Society of India*, 53, 649–668.

de Oliveira, S. M. B., Trescases, J. J. & Melfi, A. J. 1992. Lateritic nickel deposits of Brazil. *Mineralium Deposita*, 27 (2), 137–146.

de Oliveira, S. M. B., Partiti, C. S. d. M. & Enzweiler, J. 2001. Ochreous laterite: A nickel ore from Punta Gorda, Cuba. *Journal of South American Earth Sciences*, 14 (3), 307–317.

Department of Mines and Petroleum 2015. *Mineral and Petroleum Statistics Digest 2014*. Perth, WA: Western Australian Department of Mines and Petroleum. Accessed at: www.dmp.wa.gov.au/Documents/About-Us-Careers/statsdigest2014.pdf (viewed 5 August 2015).

Drew, L. J., Qingrun, M. & Weijun, S. 1990. The Bayan Obo iron–rare-earth–niobium deposits, Inner Mongolia, China. *Lithos*, 26 (1), 43–65.

Dusseault, M. B. & Morgenstern, N. R. 1978. Shear strength of Athabasca oil sands. *Canadian Geotechnical Journal*, 15 (2), 216–238.

Dusseault, M. D. 2001. Comparing Venezuelan and Canadian heavy oil and tar sands. In: *International Petroleum Conference 2001, 12–14 June*. Calgary, AB: Petroleum Society of Canada, 20.

Dyni, J. R. 2003. Geology and resources of some world oil-shale deposits. *Oil Shale*, 20 (3), 193–252.

Dyni, J. R. 2006. Geology and resources of some world oil-shale deposits. In: *U.S. Geological Survey Scientific Investigations*. Reston, VA: United States Geological Survey.

Dzemua, G. L. & Gleeson, S. A. 2012. Petrography, mineralogy, and geochemistry of the Nkamouna serpentinite: Implications for the formation of the cobalt–manganese laterite deposit, southeast Cameroon. *Economic Geology*, 107 (1), 25–41.

Ehrenberg, S., Aagaard, P., Wilson, M., Fraser, A. & Duthie, D. 1993. Depth-dependent transformation of kaolinite to dickite in sandstones of the Norwegian continental shelf. *Clay Minerals*, 28 (3), 325–352.

Elias, M. 2001. Global laterite resources. *Australian Journal of Mining*, September/October, 64–65.

Elias, M. 2002. Nickel laterite deposits: Geological overview, resources and exploitation. In: Cooke, D. R. & Pongratz, J. (eds) *Giant Ore Deposits: Characteristics, Genesis and Exploitation*. Hobart, TA: CODES.

Elias, M. 2006. Lateritic nickel mineralization of the Yilgarn Craton. In: Barnes, S. J. (ed.) *Nickel Deposits of the Yilgarn Craton: Geology, Geochemistry, and Geophysics Applied to Exploration*. Boulder, CO: Society of Economic Geologists.

Elias, M., Donaldson, M. J. & Giorgetta, N. 1981. Geology, mineralogy, and chemistry of lateritic nickel–cobalt deposits near Kalgoorlie, Western Australia. *Economic Geology*, 76 (6), 1775–1783.

Eliopoulos, D. G. & Economou-Eliopoulos, M. 2000. Geochemical and mineralogical characteristics of Fe–Ni- and bauxitic-laterite deposits of Greece. *Ore Geology Reviews*, 16 (1), 41–58.

El-Sherbini, M. & Issa, G. I. 1989. Composition and origin of some calcrete deposits in south Western Desert of Egypt. *Journal of African Earth Sciences (and the Middle East)*, 9 (3), 461–466.

Esson, J. 1983. Geochemistry of a nickeliferous laterite profile, Liberdade, Brazil. In: Wilson, R. C. L. (ed.) *Residual Deposits: Surface Related Weathering Processes and Materials*. Oxford: Blackwell Scientific Publications.

Force, E. R. 1991. *Geology of Titanium-Mineral Deposits*. Boulder, CO: Geological Society of America.

Foster, L. & Eggleton, R. A. 2002. The Marlborough nickel laterite deposits. In: Roach, I. C. (ed.) *Regolith Landscapes in Eastern Australia*. Perth, WA: Cooperative Research Center for Landscape Evolution and Mineral Exploration.

Fournier, R. O. 1965. Montmorillonite pseudomorphic after plagioclase in a porphyry copper deposit. *American Mineralogist*, 50 (5–6), 771–777.

Franchini, M., Impiccini, A., Meinert, L., Grathoff, G. & Schalamuk, I. B. A. 2007. Clay mineralogy and zonation in the Campana Mahuida porphyry Cu deposit, Neuquén, Argentina: Implications for porphyry Cu exploration. *Economic Geology*, 102 (1), 27–54.

Freyssinet, P., Butt, C. R. M., Morris, R. C. & Piantone, P. 2005. Ore-forming processes related to lateritic weathering. *Economic Geology*, 100th Anniversary (4), 681–722.

Fu, X., Wang, J., Tan, F., *et al.* 2012. Geochemistry of terrestrial oil shale from the Lunpola area, northern Tibet, China. *International Journal of Coal Geology*, 102 (1), 1–11.

Garza, R. A. P., Titley, S. R. & Pimentel, F. 2001. Geology of the Escondida porphyry copper deposit, Antofagasta region, Chile. *Economic Geology*, 96 (2), 307–324.

Gaudin, A., Petit, S., Rose, J., *et al.* 2004a. The accurate crystal chemistry of ferric smectites from the lateritic nickel ore of Murrin Murrin (Western Australia): II. Spectroscopic (IR and EXAFS) approaches. *Clay Minerals*, 39 (4), 453–467.

Gaudin, A., Grauby, O., Noack, Y., Decarreau, A. & Petit, S. 2004b. Accurate crystal chemistry of ferric smectites from the lateritic nickel ore of Murrin Murrin (Western Australia): I. XRD and multi-scale chemical approaches. *Clay Minerals*, 39 (3), 310–315.

Gaudin, A., Deacrreau, A., Noack, Y. & Grauby, O. 2005. Clay mineralogy of the nickel laterite ore developed from serpentinised peridotites at Murrin Murrin, Western Australia. *Australian Journal of Earth Sciences*, 52 (2), 231–241.

George, R. L. 1998. Mining for oil. *Scientific American*, 278 (3), 84–85.

Gleeson, S. A., Butt, C. R. M. & Elias, M. 2003. Nickel laterites: A review. *SEG Newsletter*.

Gleeson, S. A., Herrington, R. J., Durango J., Velásquez, C. A. & Koll, G. 2004. The mineralogy and geochemistry of the Cerro Matoso S. A. Ni laterite deposit, Montelíbano, Colombia. *Economic Geology*, 99, 1197–1213.

Glover, A. S., Rogers, W. Z. & Barton, J. E. 2012. Granitic pegmatites: Storehouses of industrial minerals. *Elements*, 8, 269–273.

Golightly, J. P. 1979. Nickeliferous laterites: A general description. In: Evans, D. J. I., Shoemaker, R. S. & Veltman, H. (eds) *International Laterite Symposium, 19–21 February*. New Orleans, LA: Society of Mining Engineers, 3–23.

Golightly, J. P. 1981. Nickeliferous laterite deposits. *Economic Geology*, 75th Anniversary (3), 710–735.

Grguric, B. A. 2003. Minerals of the MKD5 nickel deposit, Mount Keith, Western Australia. *Australian Journal of Mineralogy*, 9 (2), 55–71.

Gu, J., Huang, Z., Fan, H., *et al.* 2013. Mineralogy, geochemistry, and genesis of lateritic bauxite deposits in the Wuchuan–Zheng'an–Daozhen area, Northern Guizhou Province, China. *Journal of Geochemical Exploration*, 130, 44–59.

Güven, N. 1988. Smectites. In: Bailey, S. W. (ed.) *Reviews in Mineralogy*.Washington, DC: Mineralogical Society of America.

Haest, M., Cudahy, T., Laukamp, C. & Gregory, S. 2012. Quantitative mineralogy from infrared spectroscopic data: II. Three-dimensional mineralogical characterization of the Rocklea Channel Iron Deposit, Western Australia. *Economic Geology*, 107 (2), 229–249.

Hanilçi, N. 2013. Geological and geochemical evolution of the Bolkardaği bauxite deposits, Karaman, Turkey: Transformation from shale to bauxite. *Journal of Geochemical Exploration*, 133, 118–137.

Heald, P., Foley, N. K. & Hayba, D. O. 1987. Comparative anatomy of volcanic-hosted epithermal deposits: Acid–sulfate and adularia–sericite types. *Economic Geology*, 82 (1), 1–26.

Hedenquist, J. W. 1986. Mineralization associated with volcanic-related hydrothermal systems in the Circum-Pacific basin. In: Horn, M. K. (ed.) *Transactions of the Fourth Circum-Pacific Energy and Mineral Resources Conference, 17–22 August*. Singapore: American Association of Petroleum Geologists, 513–524.

Hedenquist, J. W., Matsuhisa, Y., Izawa, E., *et al.* 1994. Geology, geochemistry, and origin of high sulfidation Cu–Au mineralization in the Nansatsu district, Japan. *Economic Geology*, 89 (1), 1–30.

Hedenquist, J. W., Arribas, A. & Reynolds, T. J. 1998. Evolution of an intrusion-centered hydrothermal system: Far Southeast-Lepanto porphyry and epithermal Cu–Au deposits, Philippines. *Economic Geology*, 93 (4), 373–404.

Hein, F. J. 2006. Heavy oil and oil (tar) sands in North America: An overview and summary of contributions. *Natural Resources Research*, 15 (2), 67–84.

Hein, F. J. 2009. Oil sands. In: American Association of Petroleum Geologists, Unconventional energy resources: 2007–2008 review. *Natural Resources Research*, 18 (2), 65–83.

Hein, F. J. & Cotterill, D. K. 2006. The Athabasca oil sands: A regional geological perspective, Fort McMurray area, Alberta, Canada. *Natural Resources Research*, 15 (2), 85–102.

Heinstand, R. N. 1981. Retorted oil shale disposal research. In: Stauffer, H. (ed.) *Oil Shale, Tar Sands, and Related Materials*.Washington, DC: American Chemical Society.

Higley, D. K. & Hein, F. J. 2011. Oil sands. In: American Association of Petroleum Geologists, Unconventional energy resources: 2011 review. *Natural Resources Research*, 20 (4), 279–328.

Hobday, D. & Galloway, W. 1999. Groundwater processes and sedimentary uranium deposits. *Hydrogeology Journal*, 7 (1), 127–138.

Hou, B. & Warland, I. 2005. Heavy mineral sands potential of the Eucla Basin in South Australia: A world-class palaeo-beach placer province. *Primary Industries and Resources, South Australia, MESA Journal*, 37 (1), 4–12.

Hough, R. M., Noble, R. R. P., Hitchen, G. J., *et al.* 2008. Naturally occurring gold nanoparticles and nanoplates. *Geology*, 36 (7), 571–574.

Hover, V. C., Walter, L. M., Peacor, D. R. & Martini, A. M. 1999. Mg-smectite authigenesis in a marine evaporative environment, Salina Ometepec, Baja California. *Clays and Clay Minerals*, 47 (3), 252–268.

Hutton, A. C. 1987. Petrographic classification of oil shales. *International Journal of Coal Geology*, 8 (3), 203–231.

International Atomic Energy Agency 2009. World distribution of uranium deposits (UDEPO) with uranium deposit classification. In: *IAEA-TECDOC-1629, Division of Nuclear Fuel Cycle.*Vienna, AT: International Atomic Energy Agency.

Jepson, W. B. 1984. Kaolins: Their properties and uses. *Philosophical Transactions of the Royal Society of London A: Mathematical, Physical and Engineering Sciences*, 311 (1517), 411–432.

John, D. A., Ayuso, R. A., Barton, M. D., *et al.* 2010. Porphyry copper deposit model. In: *Mineral Deposit Models for Resource Assessment.* Reston, VA: US Geological Survey.

Johnston, N. 1955. Role of clay in oil reservoirs. In: Pask, J. A. & Turner, M. D. (eds) *Clays and Clay Technology, Proceedings of 1st National Conference*. San Francisco, CA: Department of Natural Resource, Division of Mines, 306–313.

Kholodov, V. N., Nedumov, R. I. & Golubovskaya, E. V. 2012. Facies types of sedimentary iron ore deposits and their geochemical features: Communication 1. Facies groups of sedimentary ores, their lithology, and genesis. *Lithology and Mineral Resources*, 47 (6), 447–472.

Knaus, E., Killen, J., Biglarbigi, K. & Crawford, P. 2010. An overview of oil shale resources. In: Ogunsola, O. (ed.) *Oil Shale: A Solution to the Liquid Fuel Dilemma.*Washington, DC: American Chemical Society.

Kneeshaw, M. 2003. Appendix 1: Marillana formation, channel deposits in the Marillana Creek palaeochannel of the central Hamersley Ranges, Western Australia. *Australian Journal of Earth Sciences*, 50 (5), 669–690.

Kravchenko, S. & Pokrovsky, B. 1995. The Tomtor alkaline ultrabasic massif and related REE-Nb deposits, northern Siberia. *Economic Geology*, 90 (3), 676–689.

Kühnel, R. A., Roorda, H. J. & Steensma, J. J. S. 1978. Distribution and partitioning of elements in nickeliferous laterites. *Bulletin du BRGM*, II (3), 191–206.

Kynicky, J., Smith, M. P. & Xu, C. 2012. Diversity of rare earth deposits: The key example of China. *Elements*, 8 (5), 361–367.

Laskou, M. & Economou-Eliopoulos, M. 2013. Bio-mineralization and potential biogeochemical processes in bauxite deposits: Genetic and ore quality significance. *Mineralogy and Petrology*, 107 (4), 471–486.

Li, Z., Din, J., Xu, J., *et al.* 2013. Discovery of the REE minerals in the Wulong–Nanchuan bauxite deposits, Chongqing, China: Insights on conditions of formation and processes. *Journal of Geochemical Exploration*, 133, 88–102.

Liu, J., Xu, Z. & Masliyah, J. 2004. Role of fine clays in bitumen extraction from oil sands. *American Institute of Chemical Engineers Journal*, 50 (8), 1917–1927.

Llorca, S. M. 1993. Metallogeny of supergene cobalt mineralization, New Caledonia. *Australian Journal of Earth Sciences*, 40 (4), 377–385.

Lottermoser, B. G. 1990. Rare-earth element mineralisation within the Mt. Weld carbonatite laterite, Western Australia. *Lithos*, 24 (2), 151–167.

Lower, C., Ehrig, K., Macmillan, E., *et al.* 2011. Quantitative clay mineralogy for a calcrete-hosted uranium deposit: Innovative application of existing technology on an unprecedented scale. In: Ausimm, T. (ed.) *The First AusIMM International Geometallurgy Conference, 5–7 September, 2011*. Brisbane, QLD: The Australasian Institute of Mining and Metallurgy, 173–180.

Lucas, Y. 1989. Systèms pédologiques en Amazonie brésilienne: Equilibres, Déséquilibres et Transformations. PhD, University of Poitiers.

Mars, J. C. & Rowan, L. C. 2006. Regional mapping of phyllic- and argillic-altered rocks in the Zagros magmatic arc, Iran, using Advanced Spaceborne Thermal Emission and Reflection Radiometer (ASTER) data and logical operator algorithms. *Geosphere*, 2 (3), 161–186.

Marsh, E. & Anderson, E. 2011. *Ni-Co Laterites: A Deposit Model*. Reston, VA: US Geological Survey.

Martinez, A. R. 1987. The Orinoco oil belt, Venezuela. *Journal of Petroleum Geology*, 10 (2), 125–134.

Masters, B. K. 1990. Heavy minerals deposits in the Yoganup formation. In: Hughes, F. E. (ed.) *Geology of the Mineral Deposits of Australia and Papua New Guinea*. Melbourne, VIC: The Australasian Institute of Mining and Metallurgy.

McClenaghan, M. B., Layton-Matthews, D. & Matile, G. 2011. Till geochemical signatures of magmatic Ni–Cu deposits, Thompson nickel belt, Manitoba, Canada. *Geochemistry: Exploration, Environment and Analysis*, 11 (2), 145–159.

Mercado, J. M. O. 1979. Geochemistry of the laterites in Nonoc Islands, Surigao Province, Philippines. In: *Proceedings of the International Seminar on Lateritization Processes, 11–14 December*. Trivandrum: Oxford and IBH Publishing Co., 45–57.

Mercer, K. G. & Pein, H. R. 2001. Geological and geotechnical characteristics of the corridor sands deposits in Mozambique. In: Ausimm, T. (ed.) *International Heavy Minerals Conference, 18–19 June, 2001*. Fremantle, WA: The Australasian Institute of Mining and Metallurgy, 9–16.

Misra, A., Pande, D., Kumar, K. R., *et al.* 2011. Calcrete-hosted surficial uranium occurrence in playa-lake environment at Lachhri, Nagaur District, Rajasthan, India. *Current Science*, 101 (1), 84–88.

Mitchell, A. H. G. & Leach, T. M. 1991. *Epithermal Gold in the Philippines: Island Arc Metallogenesis, Geothermal Systems and Geology*. London: Academic Press.

Moazallahi, M. & Farpoor, M. H. 2012. Soil genesis and clay mineralogy along the xeric–aridic climotoposequence, south central Iran. *Journal of Agricultural Science and Technology*, 14 (3), 683–696.

Moldoveanu, G. A. & Papangelakis, V. G. 2012. Recovery of rare earth elements adsorbed on clay minerals: I. Desorption mechanism. *Hydrometallurgy*, 117, 71–78.

Mongelli, G., Buccione, R. & Sinisi, R. 2015. Genesis of autochthonous and allochthonous Apulian karst bauxites (Southern Italy): Climate constraints. *Sedimentary Geology*, 325, 168–176.

Morris, R. C. 1985. Genesis of iron ore in banded iron-formation by supergene and supergene-metamorphic processes: A conceptual model. In: Wolf, K. H. (ed.) *Handbook of Strata-Bound and Stratiform Ore Deposits*. New York: Elsevier.

Morris, R. C. & Kneeshaw, M. 2011. Genesis modelling for the Hamersley BIF-hosted iron ores of Western Australia: A critical review. *Australian Journal of Earth Sciences*, 112 (B56–B67), 417–451.

Nahon, D. 1991. *Introduction to the Petrology of Soils and Chemical Weathering*. New York: Wiley & Sons.

Nahon, D., Paquet, H. & Delvigne, J. 1982. Lateritic weathering of ultramafic rocks and the concentration of nickel in the western Ivory Coast. *Economic Geology*, 77 (5), 1159–1175.

Newton, M. C. & Romeo, A. J. 2006. Geology of the Old Hickory mineral sand deposit, Dinwiddie and Sussex Counties, Virginia. In: Jeffrey, C. (ed.) *Proceedings of the 42nd Forum on the Geology of Industrial Minerals, May 7–13, 2006*. Asheville, NC: North Carolina Geological Survey, 464–480.

Noble, R., Gray, D. & Reid, N. 2011. Regional exploration for channel and playa uranium deposits in Western Australia using groundwater. *Applied Geochemistry*, 26 (12), 1956–1974.

Odom, I. E. 1984. Smectite clay minerals: Properties and uses. *Philosophical Transactions of the Royal Society of London A: Mathematical, Physical and Engineering Sciences*, 311 (1517), 391–409.

Ogura, Y. 1986. Mineralogical studies on the profiles of nickeliferrous latrite deposits in the southwestern Pacific area. In: Geological Society of India (ed.) *Memoirs of the Geological Survey of India: Laterisation Processes*. Calcutta: Geological Society of India.

Ogura, Y., Murata, K. & Iwai, M. 1987. Relation between chemical composition and particle-size distribution of ores in the profile of nickeliferous laterite deposits of the Rio Tube mine, Philippines. *Chemical Geology*, 60 (1–4), 259–271.

Omotoso, O., Mikula, R. J. & Stephens, P. W. 2001. Surface area of interstratified phyllosilicates in Athabasca oil sands from synchrotron XRD. *Advances in X-ray Analysis*, 45, 391–396.

Ots, A. 2007. Estonian oil shale properties and utilization in power plants. *Energetika*, 53 (2), 8–18.

Ouangrawa, M., Trescases, J.-J. & Ambosi, J.-P. 1996. Behaviour of iron oxides during the weathering of nickeliferous ultramafic rock from New Caledonia. *Comptes Rendus de l'Academie de Sciences de Paris*, 323, 243–249.

Parry, W., Jasumback, M. & Wilson, P. N. 2002. Clay mineralogy of phyllic and intermediate argillic alteration at Bingham, Utah. *Economic Geology*, 97 (2), 221–239.

Patterson, J. H. 1994. A review of the effects of minerals in processing of Australian oil shales. *Fuel*, 73 (3), 321–327.

Patterson, J. H. & Henstridge, D. A. 1990. Comparison of the mineralogy and geochemistry of the Kerosene Creek Member, Rundle and Stuart oil shale deposits, Queensland, Australia. *Chemical Geology*, 82, 319–339.

Patterson, J. H., Hurst, H. J., Levy, J. H. & Killingley, J. S. 1990. Mineral reactions in the processing of Australian Tertiary oil shales. *Fuel*, 69 (9), 1119–1123.

Petránek, J. & Van Houten, F. B. 1997. Phanerozoic ooidal ironstones. *Czech Geological Survey Special Papers*, 7, 71.

Pollastro, R. M. 1993. Considerations and applications of the illite/smectite geothermometer in hydrocarbon-bearing rocks of Miocene to Mississippian age. *Clays and Clay Minerals*, 41 (2), 119–133.

Qing, W., Jingru, B., Yan, C., *et al.* 2007. Research on release of trace elements on retorting from Haudian oil shale. In: *27th Oil Shale Symposium, 15–19 October, 2007*. Golden, CO: Colorado Energy Research Institute, 1.

Ramanaidou, E. R. & Wells, M. A. 2014. 13.13 – Sedimentary Hosted Iron Ores. In: Turekian, K. K. (ed.) *Treatise on Geochemistry of Mineral Resources*, 2nd edition. Oxford: Elsevier.

Ramanaidou, E. R., Morris, R. C. & Horwitz, R. C. 2003. Channel iron deposits of the Hamersley Province, Western Australia. *Australian Journal of Earth Sciences*, 50 (5), 669–690.

Rees, C., Riedell, K. B., Proffett, J. M., Macpherson, J. & Robertson, S. 2015. The Red Chris porphyry copper–gold deposit, Northern British Columbia, Canada: Igneous phases, alteration, and controls of mineralization. *Economic Geology*, 110 (4), 857–888.

Resheidat, M., Al-Shyoukh, A. & Al-Nawasrah, M. K. 2007. On the use of oil shale beyond production of oil and gas. In: *27th Oil Shale Symposium, 15–19 October, 2007*. Golden, CO: Colorado Energy Research Institute, 21.

Robb, L. 2005. *Introduction to Ore-Forming Processes*. Oxford: Blackwell Publishing.

Roy, P. S. 1999. Heavy mineral beach placers in southeastern Australia: Their nature and genesis. *Economic Geology*, 94 (4), 567–588.

Roy, P. S. & Whitehouse, J. 2003. Changing Pliocene sea levels and the formation of heavy minerals beach placers in the Murray Basin, southeastern Australia. *Economic Geology*, 98 (5), 975–983.

Rutherford, J., Munday, T., Meyers, J. & Cooper, M. 2001. Relationships between regolith materials, petrophysical properties, hydrogeology and mineralisation at the Cawse Ni laterite deposits, Western Australia: Implications for exploring with airborne EM. In: *15th Geophysical Conference and Exhibition: Geophysical Odyssey*. Brisbane, QLD: Australian Society of Exploration Geophysicists, 13.

Sahu, K. C. & Venkateswaran, D. 1989. Characterization of nickeliferrous laterite of Sukinda in Cuttack district, Orissa. *Journal of the Geological Society of India*, 33 (2), 116–131.

Saied, H., Khademi, H. & Cano, A. F. 2010. Palygorskite formation under the influence of saline and alkaline groundwater in Central Iranian soils. *Soil Science*, 175 (6), 303–312.

Schellmann, W. 1983. Geochemical principles of lateritic nickel ore formation. In: Melfi, A. J. & Carvalho, A. (eds) *Proceedings of the 2nd International Seminar of Laterisation Processes*. Sao Paulo: Instito Agronomico Geofisico, University of Sao Paulo, Brazil.

Schulz, K. J., Chandler, V. W., Nicholson, S. W., *et al.* 2010. *Magmatic Sulfide-Rich Nickel–Copper Deposits Related to Picrite and (or) Tholeiitic Basalt Dike-Sill Complexes: A Preliminary Report*. Reston, VA: US Geological Survey.

Scott, J. D. & Ozum, B. 2010. Oil sands tailings: What needs to be done. Mining.com. [Online] Available: www.mining.com/oil-sands-tailings-what-needs-to-be-done

Scott, J. D., Dusseault, M. B. & Carrier, W. D. I. 1985. Behaviour of the clay/bitumen/water sludge system from oil sands extraction plants. *Applied Clay Science*, 1 (1–2), 207–218.

Seedorff, E., Barton, M. D., Stavast, W. J. A. & Maher, D. J. 2008. Root zones of porphyry systems: Extending the porphyry model to depth. *Economic Geology*, 103 (5), 939–956.

Seeley, J. B. & Senden, T. J. 1994. Alluvial gold in Kalimantan, Indonesia: A colloidal origin? *Journal of Geochemical Exploration*, 50 (1), 457–478.

Shmakin, B. 2008. Diversity of accessory minerals in rare-metal–rare earth pegmatites. *Geology of Ore Deposits*, 50 (7), 518–523.

Sillitoe, R. H. 1973. The tops and bottoms of porphyry copper deposits. *Economic Geology*, 68 (6), 799–815.

Sillitoe, R. H. 1989. Gold deposits in the western Pacific island arcs: The magmatic connection. In: Keays, R. R., Ramsay, W. B. H. & Groves, D. I. (eds) *Economic Geology Monographs*. New Haven, CT: The Economic Geology Publishing Company.

Sillitoe, R. H. 2000. Gold-rich porphyry deposits: Descriptive and genetic models and their role in exploration and discovery. In: Hagemann, S. G. & Brown, P. E. (eds) *Gold in 2000: Reviews in Economic Geology*. Littleton, CO: Society of Economic Geologists, Inc.

Sillitoe, R. H. 2010. Porphyry copper systems. *Economic Geology*, 105 (1), 3–41.

Simmons, S. F., White, N. C. & John, D. A. 2005. Geological characteristics of epithermal precious and base metal deposits. In: Hedenquist, J. W., Thompson, J. F. H., Goldfarb, R. J. & Richards, J. P. (eds) *Economic Geology One Hundredth Anniversary Volume: 1905–2005*. Littleton, CO: Society of Economic Geologists.

Sinclair, W. D. 2007. Porphyry deposits. In: Goodfellow, W. D. (ed.) *Mineral Deposits of Canada: A Synthesis of Major Deposit-Types, District Metallogeny, the Evolution of Geological Provinces, and Exploration Methods*. Vancouver, BC: Geological Association of Canada, Mineral Deposits Division.

Skirrow, R. G., Jaireth, S., Huston, D. L., *et al.* 2009. *Uranium Mineral Systems: Processes, Exploration Criteria and a New Deposit Framework*. Canberra, ACT: Geoscience Australia.

Som, S. K. & Joshi, R. 2002. Chemical weathering of serpentinite and Ni enrichment in Fe oxide at Sukinda area, Jajpur district, Orissa, India. *Economic Geology*, 97 (1), 165–172.

Sonibare, O. O., Jacob, D. E., Ward, C. R. & Foley, S. F. 2011. Mineral and trace element composition of the Lokpanta oil shales in the Lower Benue Trough, Nigeria. *Fuel*, 90 (9), 2843–2849.

Sonntag, I., Laukamp, C. & Hagemann, S. G. 2012. Low potassium hydrothermal alteration in low sulfidation epithermal systems as detected by IRS and XRD: An example from the Co–O mine, Eastern Mindanao, Philippines. *Ore Geology Reviews*, 45 (1), 47–60.

Stanaway, K. J. 1992. Heavy-mineral placers. *Mining Engineering*, 44 (4), 352–358.

Suárez, S., Nieto, F., Velasco, F. & Martín, F. J. 2011. Serpentine and chlorite as effective Ni–Cu sinks during weathering of the Aguablanca sulphide deposit (SW Spain): TEM evidence for metal-retention mechanisms in sheet silicates. *European Journal of Mineralogy*, 23 (2), 179–196.

Tardy, Y. 1997. *Petrology of Laterites and Tropical Soils*. Rotterdam: Balkema.

Tardy, Y. & Nahon, D. 1985. Geochemistry of laterites, stability of Al-goethite, Al-hematite and Fe-kaolinite in bauxites and ferricretes: An approach to the mechanism of concretion formation. *American Journal of Science*, 285 (10), 865–903.

Tardy, Y. & Roquin, C. 1998. Dérive des continets, latérites et paléoclimates tropicaux. *Bulletin de l'Association de Géographes Français* 77 (4), 373–383.

Taufen, P. M. & Brenner, T. L. 1987. Geochemical orientation survey of the Fortaleza de Minas O'Toole Ni deposit southwestern Minas Gerais, Brazil. *Geochimica Brasiliensis*, 1 (1), 1–18.

Taylor, B., Mazzoni, P., Brown, F., Kubank, C. & Rademeyer, L. 2003. Process flow sheet development for the Corridor Sands project. In: *Heavy Minerals Conference: Current Challenges in Heavy Mineral Exploitation 6–8th October 2003*. Cape Town: Southern African Institute of Mining and Metallurgy, 8.

Taylor, G. & Eggleton, R. A. 2015. Bauxites of the NSW Southern Highlands. *Australian Journal of Earth Sciences*, 62 (3), 341–363.

Thompson, J. F. H. & Thompson, A. J. B. (eds) 1996. *Atlas of Alteration: A Field and Petrographic Guide to Hydrothermal Alteration Minerals*. Vancouver, BC: Geological Association of Canada.

Thorne, R., Herrington, R. & Roberts, S. 2009. Composition and origin of the Çaldağ oxide nickel laterite, W. Turkey. *Mineralium Deposita*, 44 (5), 581–595.

Thorne, R., Roberts, S. & Herrington, R. 2012a. The formation and evolution of the Bitincke nickel laterite deposit, Albania. *Mineralium Deposita*, 47 (8), 933–947.

Thorne, R. L., Roberts, S. & Herrington, R. 2012b. Climate change and the formation of nickel laterite deposits. *Geology*, 40, 331–334.

Topf, A. 2010. In situ: Oil sands mining goes underground. Mining.com. [Online]. Available: www.mining.com/insitu-oil-sands-mining-goes-underground.

Tosdal, R. M., Dilles, J. H. & Cooke, D. R. 2009. From source to sinks in auriferous magmatic-hydrothermal porphyry and epithermal deposits. *Elements*, 5 (5), 289–295.

Towner, R. R., Ewers, G. R. & Cassidy, K. F. 1996. Rare earths and mineral sands in Australia. In: *Commercial-in-Confidence Report*. Canberra, ACT: Australian Geological Survey Organisation and Bureau of Resources Sciences.

Traversa, G., Gomes, C. B., Brotzu, P., *et al.* 2001. Petrography and mineral chemistry of carbonatites and mica-rich rocks from the Araxá complex (Alto Paranaíba Province, Brazil). *Anais da Academia Brasileira de Ciências*, 73 (1), 71–98.

Trescases, J.-J., Melfi, A. J. & de Oliverira, S. M. B. 1979. Nickeliferous laterites of Brazil. In: Venkatesh, S. V. & Raman, P. K. (eds) *Proceedings of the International Seminar on Lateritization Processes, 11–14 December*. Trivandrum: Oxford and IBH Publishing Co., 170–184.

US Geological Survey 2015. *Mineral Commodity Summaries 2015*. Reston, VA: US Geological Survey.

Valeton, I. 1972. *Bauxites*. Amsterdam: Elsevier.

Valeton, I. 1983. Paleoenvironment of lateritic bauxites with vertical and lateral differentiation. In: Wilson, R. C. L. (ed.) *Residual Deposits: Surface Related Weathering Processes and Materials*. Oxford: Blackwell Scientific Publications.

Van Wormhoudt, A. 1993. *Soil Mineralogy of an Upper Coastal Plain Landscape in Virginia*. Blacksburg, VA: Virginia Polytechnic Institute and State University.

Varajao, C. A. C., Boulangé, B. & Carvalho, A. 1989. The bauxites of Quadrilaterro Ferrifero, Minas Gerais, Brazil. In: L'académie Yugoslave Des Sciences Et Des Arts (ed.) *Proceedings of the 6th International Congress of the International Committee for the Study of Bauxite, Alumina and Aluminium (ICSOBA), 11–20 May*. Pocos de Caldas: Travaux ICSOBA, 127–136.

Vasconcelos, P. M., Renne, P. R., Brimhall, G. H. & Becker, T. A. 1994. Direct dating of weathering phenomena by $^{40}Ar/^{39}Ar$ and $^{40}K/^{40}Ar$ analysis of supergene K–Mn oxides. *Geochimica et Cosmochimica Acta*, 58 (6), 1635–1665.

Velde, B. & Vasseur, G. 1992. Estimation of the diagenetic smectite to illite transformation in time–temperature space. *American Mineralogist*, 77 (9–10), 967–976.

Verrecchia, E. & Le Coustumer, M. 1996. Occurrence and genesis of palygorskite and associated clay minerals in a Pleistocene calcrete complex, Sde Boqer, Negev Desert, Israel. *Clay Minerals*, 31 (2), 183–202.

Wang, Y., Nahon, D. & Merino, E. 1994. Dynamic model of the genesis of calcretes replacing silicate rocks in semi-arid regions. *Geochimica et Cosmochimica Acta*, 58 (23), 5131–5145.

Ware, C. I., Whitmore, G. P. & Uken, R. 2001. Evolution of the northern KwaZulu-Natal coastal dune Cordon, South Africa: evidence from clays. In: Ausimm, T. (ed.) *International Heavy Minerals Conference, 18–19 June*. Fremantle, WA: The Australasian Institute of Mining and Metallurgy, 3–7.

Wells, M. A. & Butt, C. R. M. 2000. *Geology, Geochemistry and Mineralogy of the Murrin*

Murrin Nickel Laterite Deposit. Perth, WA: CSIRO.

Wigand, M., Chipera, S., Woldegabriel, G., *et al.* 2007. Unique mineralogy of oil shale from the Piceance Basin, Colorado. In: *27th Oil Shale Symposium, 15–19 October*. Golden, CO: Colorado Energy Research Institute, 1.

Williams, V. (ed.) 1990. *WIM 150 Detrital Heavy Mineral Deposit*. Melbourne, VIC: The Australian Institute of Mining and Metallurgy.

Williams-Jones, A. E. & Heinrich, C. A. 2005. 100th anniversary special paper: Vapor transport of metals and the formation of magmatic-hydrothermal ore deposits. *Economic Geology*, 100 (7), 1287–1312.

Williams-Jones, A. E., Bowell, R. J. & Migdisov, A. A. 2009. Gold in solution. *Elements*, 5 (5), 281–287.

Williams-Jones, A. E., Migdisov, A. A. & Samson, I. M. 2012. Hydrothermal mobilisation of the rare earth elements: A tale of 'Ceria' and 'Yttria'. *Elements*, 8 (5), 355–360.

Wilson, P. N. & Parry, W. T. 1995. Characterization and dating of argillic alteration in the Mercur Gold District, Utah. *Economic Geology*, 90 (5), 1197–1216.

Yang, K., Cardy, M., Mason, P. & Hacket, A. 2010. *HyLogging for Characterisation of Clay Minerals and Carbonates in Drill-Hole Samples of the Yeelirrie Uranium Deposit*. Perth, WA: CSIRO.

Yoffe, O., Nathan, Y., Wolfarth, A., Cohen, S. & Shoval, S. 2002. The chemistry and mineralogy of the Negev oil shale ashes. *Fuel*, 81 (9), 1101–1117.

Yongue-Fouateu, R., Ghogomu, R. T., Penaye, J., *et al.* 2006. Nickel and cobalt distribution in the laterites of the Lomié region, south-east Cameroon. *Journal of African Earth Sciences*, 45 (1), 33–47.

Zotov, A., Mukhamet-Galeev, A. & Schott, J. 1998. An experimental study of kaolinite and dickite relative stability at 150–300 degrees C and the thermodynamic properties of dickite. *American Mineralogist*, 83 (5–6), 516–524.

6

Exploration and Discovery

Carsten Laukamp, Erick Ramanaidou, Martin Wells and Kai Yang

6.1 Introduction

Understanding the spatial variations of clay abundance and composition can be used to characterize a mineral system. Kaolinite, smectite and other phyllosilicate minerals are found in a number of hydrothermal systems, such as high- and low-sulphidation epithermal, porphyry Cu ± Au and Carlin-type deposits (van der Meer *et al.*, 2012) (see also Chapter 5) and can represent important vector minerals. Hydrothermal alteration in the Comstock mining district (Nevada, USA) for example, shows a quartz–alunite core, enveloped by an inner kaolinite-rich zone and an outer illite–smectite zone (Yang *et al.*, 1999). In Eastern Mindanao (Philippines), interlayered illite–smectite minerals occur in distinct alteration zones in epithermal gold systems (Sonntag *et al.*, 2012).

Kaolin and smectite are also main constituents of regolith covering concealed mineral deposits. They can be used to characterize the origin of the regolith, e.g. transported vs *in situ* (Cudahy *et al.*, 2005), or can be related back to weathering products of vector minerals and/or host rocks (e.g. kaolinite types in weathered profiles (White *et al.*, 2010)); Mg-clays in mafic rocks hosting orogenic gold deposits (Velde and Meunier, 2008). Clay minerals can also host economic amounts of rare earth elements (REE), such as in ion-adsorption clay REE deposits (Chapter 5, this volume). Furthermore, kaolinite and smectite are common contaminants in, for example, iron ore deposits (Wells and Ramanaidou, 2007). The integration of surface and subsurface measurements in 3D software packages can be used to create a comprehensive mineral system model, showing for example the role/impact of kaolin and smectite group minerals within the geological architecture or as a potential vector or contaminant (Haest *et al.*, 2012a, 2012b).

While X-ray diffraction (XRD, see Chapter 4) is the most widely used method for identifying and quantifying kaolin and smectite, they can also be detected directly or inferred by a number of other technologies, ranging from satellite remote sensing data, for example argillic alteration detected using the Advanced Spaceborne Thermal Emission and Reflection Radiometer (ASTER) (Mars and Rowan, 2010), to field portable spectroscopic techniques (Haest *et al.*, 2013). Indirect methods of detection comprise geochemical and geophysical techniques. This chapter provides an introduction to common technologies (excluding XRD) used for the detection of kaolinite and smectite group minerals in exploration and mining, with a focus on visible and infrared spectroradiometry. For the purpose of detecting clays during exploration, multi- and hyperspectral remote sensing systems such as the ASTER (Yamaguchi *et al.*, 1999) and the airborne HyMap™ system (Cocks *et al.*, 1998) are becoming increasingly important as they represent one of the more efficient methods to date to map directly the abundance, composition and/or crystallinity of clay minerals spatially. The technologies described here represent a selection of those readily available, with a focus on equipment that can be used in the field, therefore complementing Chapter 4. Prior to describing the technologies, the characteristics of kaolinite and smectite group minerals in regolith environments and selected hydrothermal alteration systems are briefly summarized in order to emphasize the importance of geological models as a base for a successful application of the discussed technologies. These applied geological models build on the more detailed descriptions of mineral systems in Chapter 5.

6.2 Clays in Regolith and Hydrothermal Mineral Systems

6.2.1 Argillic Alteration

Clay minerals can form during hydrothermal alteration associated with formation of ore deposits. Classic examples are porphyry Cu–Au and epithermal gold deposits (Chapter 5), where clay minerals form during interaction between magmatic, hydrothermal and meteoric fluids with country rock, manifesting in argillic and phyllic alteration. Kaolinite and smectite, together with quartz and alunite, are major minerals of argillic alteration (Fig. 6.1). Advanced argillic alteration (quartz, kaolinite/dickite or pyrophyllite, aluminium hydroxides, alunite) develops due to extended acid leaching near the surface, commonly forming a lithocap at the top of porphyry Cu–Au systems (Fig. 6.1). Phyllic alteration (quartz, white mica, pyrite) develops at higher temperature and more saline conditions (Tosdal *et al.*, 2009).

6.2.2 Regolith

Clay minerals develop in certain levels of the regolith stratigraphy, depending on the source/bed rocks and the physicochemical conditions of the environment (e.g. groundwater chemistry). Kaolinite and halloysite, for example, are typical weathering products of feldspars (and micas) and are more abundant in regolith on mafic and felsic rocks than on ultramafic rocks (Anand and Paine, 2002). This is in contrast to smectites, which form especially at the base of deeply weathered profiles in dry or poorly drained conditions (Anand and Paine, 2002). Smectites are weathering products of 'primary' minerals (e.g. chlorite), with high iron–aluminium smectites forming from mafic rocks. In upper levels of the regolith, smectite weathers to kaolinite and goethite (Fig. 6.2). Transported sediments are in many cases characterized by poorly crystalline kaolinite, whereas well-crystalline kaolinite is typical of residual materials (Cudahy, 1997). However, Churchman and Gilkes (1989) observed the opposite distribution in weathering profiles on dolerites, with halloysite forming close

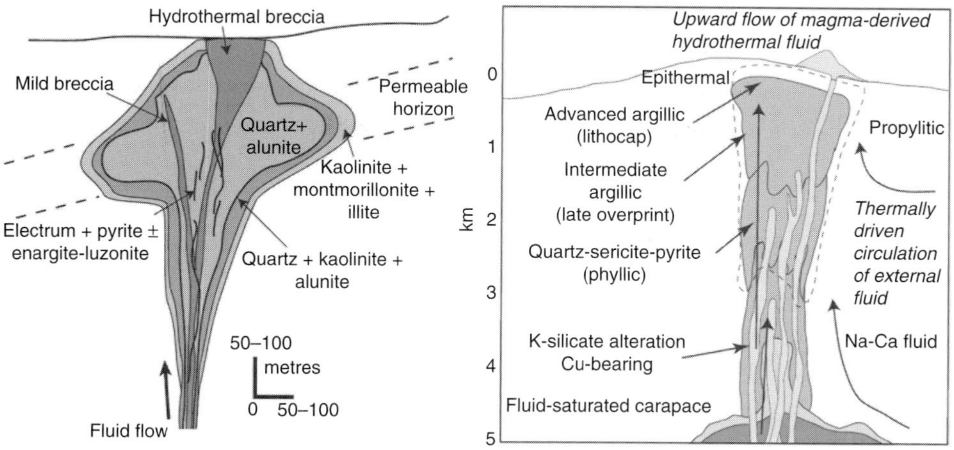

Figure 6.1. Schematic cross-section of a quartz–alunite epithermal (left) and porphyry Cu–Au (right) deposit (Tosdal *et al.*, 2009) showing argillic (i.e. kaolinite + montmorillonite + illite), advanced argillic alteration (i.e. quartz + kaolinite + montmorillonite) and phyllic alteration. A black and white version of this figure will appear in some formats. For the colour version, please refer to the plate section.

to the parent rock and kaolinite in more intensely weathered regolith. The formation of clay minerals is dependent on numerous factors, which are outlined in Chapters 1 and 2, and Velde and Meunier (2008).

6.3 Clays in Geochemical Exploration

6.3.1 Introduction

Clays form or transform in various geological processes, ranging from hydrothermal alteration processes at depth to diagenetic processes in sedimentary basins and weathering processes at the Earth's surface (Velde and Meunier, 2008). Clays are largely formed under low-temperature conditions which limits their applicability for exploration in mineral systems that develop under higher PT conditions. In near-surface hydrothermal systems, such as porphyry Cu or epithermal Cu–Au systems (Chapter 5), clays develop during interaction of hydrothermal fluids with the host rocks, producing characteristic alteration halos that can be used in mineral exploration as vectors towards the mineralized zone (Robb, 2005; Simmons *et al.*, 2005; Tosdal *et al.*, 2009). However, rather than the chemical composition of clays, it is more the crystallinity, hydration state or type of mixed clay layers that are examined in hydrothermal systems in order to describe the intensity of fluid–rock interaction and the presence of characteristic alteration halos around the mineralization (Cooke *et al.*, 2005;Hedenquist *et al.*, 1998;Sonntag *et al.*, 2012). In sedimentary basins, clays can be used to characterize the diagenetic overprint of the geological province, which can be of particular interest in exploration for hydrocarbon reservoirs. In oil shale provinces,

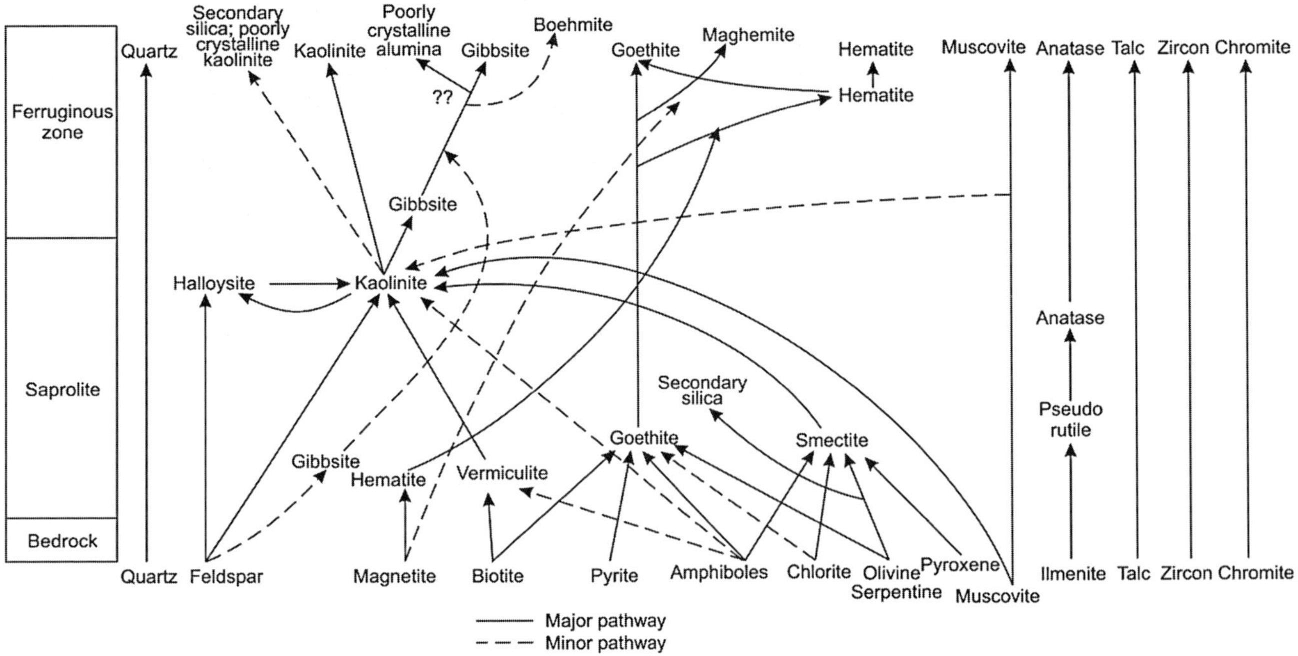

Figure 6.2. Pathways of formation of secondary minerals in weathering minerals (afterAnand and Paine 2002, and references therein).

Table 6.1. *Element mobility during deep weathering under humid conditions (after Butt et al. 2000).*

Host minerals	Leached	Partly retained in secondary minerals
Released at weathering front		
Sulphides	As, Cd, Co, Cu, Mo, Ni, Zn, S	As, Cu, Ni, Pb, Sb, Zn (Fe oxides)
Carbonates	Ca, Mg, Mn, Sr	
Released in lower saprolite		
Aluminosilicates	Ca, Cs, K, Na, Rb	Si, Al (kaolinite), Ba barite
Ferromagnesians (pyroxene, olivine, amphiboles, chlorite, biotite)	Ca, Mg	Fe, Ni, Co, Cr, Ga, Mn, Ti, V (Fe and Mn oxides)
Released in upper saprolite		
Aluminosilicates (muscovite)	Cs, K, Rb	Si, Al (kaolinite)
Ferromagnesians (chlorite, talc, amphibole)	Mg, Li	Fe, Ni, Co, Cr, Ga, Mn, Ni, Ti, V (Fe oxides)
Smectites	Ca, Mg, Na	Si, Al (kaolinite)
Released in mottled and ferruginous zones		
Aluminosilicates (muscovite, kaolinite)	K, Rb, Cs	Si, Al (kaolinite)
Fe oxides	Trace elements	
Retained in stable minerals (all horizons)	B, Cr, Fe, Hf, K, Nb, Rb, REE, Th, Ti, V, W, Zr	

clay-rich stratigraphies can be associated with, for example, high values of Ti, V, Cr and/or Ni (Patterson, 1993; Patterson and Henstridge, 1990; Sonibare *et al.*, 2011).

6.3.2 Clays in Surficial Sedimentary Materials

Clays play an important role in geochemical exploration in surficial sedimentary materials, due to their impact on element distributions in the regolith, but also as hosts of mineralization, such as in smectite-hosted Ni-laterite deposits (Burger, 1996;Elias *et al.*, 1981;Gaudin *et al.*, 2005) (Chapter 5). An overview of the elements retained in clays and other minerals (Table 6.1) is given by Scott (2008) andButt *et al.* (2000). Kaolinite is the weathering product of a number of minerals (Fig. 6.2) and retains especially Si and Al, but also Fe, Ti and Cr (Butt *et al.*, 2000). Smectites are mainly weathering products of mafic primary minerals, reflecting the parent rock type (e.g. iron–magnesium smectites derived from mafic/ultramafic rocks) and can incorporate numerous indicator elements in the octahedral sheet, including Zn, Mn, Co, Ni and Cu (Anand, 2005; Paquet *et al.*, 1987). However, eventually

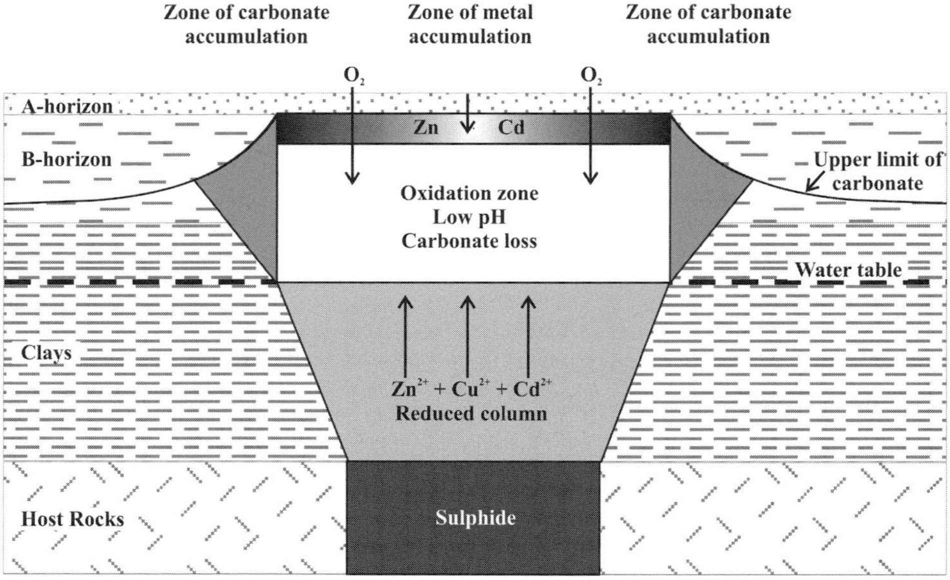

Figure 6.3. Redox model showing development of a reduced column over sulphides in transported regolith (from Hamilton 1998; Hamilton and Christensen 2000; modified after Cameron *et al.* 2004; Cohen *et al.* 2010). Metals that are mobile under reducing conditions (e.g. Zn, Cu, Cd) migrate vertically along redox gradients. Above the watertable, metals are oxidized, creating a zone of low pH and carbonate loss. Carbonate precipitates at the edge of the reduced column. Metals such as Zn and Cd accumulate near the surface.

smectites are destroyed during further weathering in the mid to upper saprolite and the indicator elements may be leached from the regolith (Anand, 2005).

Clays are major constituents of surficial sedimentary materials, which are often well-stratified environments covering potentially mineralized bedrock. The horizontal layering and groundwater levels suggest that any geochemical anomaly related to concealed ore deposits would be dispersed laterally and impossible to detect on the surface. However, a number of models exist, which try to explain geochemical anomalies at the top of thick overburden (Anand and Robertson, 2012; Mann *et al.*, 2005; and references therein). Hamilton (1998), among others, have proposed the electrochemical cell model, where a redox gradient between the reduced lower regolith and the hanging wall, the oxidized zone, leads to vertical migration of mobile metals (Fig. 6.3). The groundwater table represents a sharp redox boundary in this model and clays play an important role, as they may impart an electrical conductivity to the groundwater, thus enhancing the redox gradient (Hamilton and Christensen, 2000). Furthermore, clays are major constituents of the soil or regolith and their capability to 'store' indicator elements is crucial for the interpretation of the surface geochemical anomalies. The electrochemical cell can develop in a few thousand years, which suggests that this model can also be applied to younger transported overburden.

Regolith profiles on top of mineralized bedrock may include palaeoweathering profiles. Anand and Robertson (2012) demonstrated that in those cases, where older sediments covering the bedrock are intensively weathered, indicator elements are more likely to be trapped than in unweathered sedimentary cover. High palaeowatertables as well as distinct changes in the electronegativity within those palaeoweathering profiles enable vertical migration of indicator elements (Anand and Robertson, 2012).

The formation of geochemical footprints due to dispersion of elements in surficial sedimentary materials, such as regolith, lake and stream sediments or glacial till, have been observed in numerous ore deposits. Sampling for trace elements depends on the respective elements of concern and regolith environment. Partial extraction techniques have been developed to target those components of the regolith that host indicator elements (Anand and Robertson, 2012; Butt *et al.*, 2000; Cohen *et al.*, 1998; Ferbey, 2010; Gray *et al.*, 1999; Grunsky, 2010; Mokhtari *et al.*, 2009). For example, in the case of lode gold deposits in Western Australia, clay and iron-rich fine fractions of regolith were used for detection of pathfinder trace elements in gold exploration (Robertson, 2001). This detection method was based on the assumption that the high surface area of the fine fraction was favourable for the retention of hydromorphically dispersed trace elements (Robertson, 2001). However, kaolinite and smectite have distinctly different physicochemical properties. Due to a low cation exchange capacity of kaolinite when compared to smectite (Emerson and Yang, 1997), a kaolinite-rich clay sample should in general contain lesser amounts of indicator elements such as Cu and Zn (Anand and Robertson, 2012). On the other hand, it has to be considered that Cu and Zn can form surface precipitates with AsO_4 or PO_4 in kaolinite-rich environments.

6.3.3 Sampling Methods and Analytical Technologies

The type of geochemical analysis of clays depends on the indicator elements that are targeted in the respective mineral system and/or on the clay-bearing mineral assemblage of interest. For example, the geochemical analysis of soil/regolith requires a standardized sampling and sample preparation strategy. The main points to consider are: (1) sample spacing, (2) selection of appropriate soil/regolith horizon, (3) soil fraction, (4) sample digestion method, and (5) analytical technique (Butt *et al.*, 2000). A case in point is the comprehensive overview of partial extraction methods applied to Cu–Zn soil anomalies as given by Noble and Stanley (2009).

Technologies for detailed chemical analyses of clay minerals, such as the study of substitutions in interlayered sheets of illite–smectite, or the relative Fe^{2+}–Fe^{3+} content in smectites, include: (1) absorption spectrometry in the near, mid and far infrared (see also Section 6.5); (2) high-resolution nuclear magnetic resonance (NMR) spectroscopy; (3) extended X-ray absorption fine structure spectroscopy (XAFS); and (4) Mössbauer spectroscopy (Meunier, 2005).

Scanning electron microscopy (SEM) coupled with energy-dispersive X-ray spectroscopy (EDS) provides information about the configuration, texture and fabric of clay minerals (Bohor and Hughes, 1971), but also their chemical composition (Clayton and Pearce, 2007). Also, the local environment where the clay mineral has formed can be analysed in detail using SEM. This may help to identify, for example, the parent mineral, which in turn can answer questions about the intensity of fluid–rock interactions in hydrothermal systems and identify elements retained in the respective clay mineral during alteration or weathering processes.

Techniques for multi-element analysis of clay-bearing substrates (e.g. soil samples) include inductively coupled plasma optical emission spectroscopy (ICP OES, e.g. Bettinelli *et al.*, 2000) and inductively coupled plasma mass spectrometry (ICP MS; e.g. Kogel and Lewis, 2001). For these methods, sample material has to be treated with an appropriate acid to completely digest, partial leach or undergo selective extraction. Instrumental neutron activation analysis (INAA; e.g. Ferbey, 2010) or X-ray fluorescence spectrometry (XRF) do not require preparation of an aqueous solution and provide a comprehensive multi-element analysis. During the last decade, portable XRF has become readily available to exploration companies (Fisher *et al.*, 2014). Portable XRF acquires large amounts of geochemical data in a short period of time; however, its accuracy is limited when compared to conventional laboratory-based XRF, especially for light elements such as Al or Si (e.g. Fisher *et al.*, 2014). Low gold levels can be determined using graphite furnace atomic absorption spectroscopy (GF AAS). For calculation of the structural formulae of most pure clay minerals, the following elements are essential: Si, Al, Fe^{3+}, Fe^{2+}, Mg, Ti, Mn, P, Ca, Na, K and H_2O (Mermut and Canot, 2001). An overview of the analytical chemistry of clays and clay-bearing materials is provided by Amonette and Zelazny (1994).

6.4 Geophysical Methods for Clay Detection

6.4.1 Introduction

Applied geophysical methods during exploration are generally divided into gravitational, magnetic, seismic, electrical, electromagnetic, radioactivity and logging. Some are downhole probes, or directly measure drill cores or chips extracted from drilling and brought to the surface. In most cases, a combination of geophysical tools and the fusion of geophysical data provide a more complete understanding of the mineralogy and stratigraphy of a deposit. The most suitable geophysical tools for the detection of clays include downhole calliper, spontaneous potential, and nuclear (natural-gamma, gamma–gamma and neutron–gamma) methods, as well as ground-penetrating radar and electromagnetic induction methods. These may be complemented by geophysical remote sensing technologies such as airborne EM and radiometrics. However, as McCann and Forster (1990) discussed, the success of geophysical techniques are reliant on four controlling factors (1) geophysical property contrasts within the ore-body and host rock; (2) resolution and penetration of the methods; (3) calibration of geophysical techniques; and (4) their signal-to-noise ratio.

6.4.2 Calliper Logs

Calliper logs are used to continuously measure the diameter of the drill hole. Two types of calliper logs are used: (1) a mechanical probe made of three or four metal arms kept in contact with the drill hole wall by spring pressure linked to a potentiometer; and (2) acoustic or sonic callipers that calculate the diameter of the drill hole by measuring return time of high-frequency acoustic pulses. The variation of the diameter of the drill hole is connected to hardness; consequently, soft materials like clay minerals are often 'washed out' during drilling, and can be easily detected where the drill hole diameter increases (Schlumberger, 1989).

6.4.3 Spontaneous Potential

Spontaneous potential (SP) log (Bettinelli *et al.*, 2000) measures the electrical potential (voltage) generated by the interaction of waters, conductive drilling fluids and certain ion-selective minerals such as clays (Schlumberger, 1989). One of the main applications of SP logs is the detection of clay layers.

6.4.4 Nuclear Logging

Nuclear techniques are appropriate for drill hole logging of clays because neutrons and gamma-rays can penetrate deeply into the rock, hence providing better sampling statistics in real time (Borsaru *et al.*, 2002).

6.4.5 Natural-Gamma: Radioactivity

All rocks and soils contain a number of radioactive elements that emit gamma radiation. Only selected natural radio-isotopes produce high-energy gamma-rays with sufficient energy to be mapped, including K, U and Th (IAEA, 2003). Clay minerals are high in these radio-isotopes, so high gamma values are a proxy for clay-rich rock formation (Telford *et al.*, 1990). The gamma radiation detector probe uses a scintillometer to detect gamma radiation from the drilled rock units. The penetration depth of borehole gamma-ray spectroscopy ranges from 10 to 30 cm, depending on the density of the rock (IAEA, 2003).

6.4.6 Gamma–Gamma: Density

Density is useful for detecting changes in rock type or the presence of cracks or pores in the rock formation (Schlumberger, 1989). The density probe identifies changes in rock density caused either by (1) minerals containing heavy elements or (2) compaction and cementing between mineral grains. Clay detection falls in the second category. The probe contains an on-board radiation source, either Co-60 or Cs-137, continually sending gamma radiation into the rock. The back-scattered gamma-rays, detected by a sensor located in the probe are inversely proportional to the rock density. Clay minerals such as kaolinite and smectite often occupy the interstices between minerals, increasing the density of the rock, and as such can be detected by density probes.

6.4.7 Neutron–Gamma: Porosity

The porosity of a rock is the amount of space between solid mineral grains, with low-porosity rocks typically containing clays. Neutron logs measure the amount of hydrogen in a rock and determine the porosity of the formation, assuming the majority of the void space is occupied by water. A source of americium-beryllium generates high-energy neutrons, which travel into the rock formation. The maximum energy loss occurs when the neutrons collide with a nucleus of similar mass, for example, a hydrogen nucleus. The neutrons that are slowed down are captured by the nuclei of other atoms, which then emit high-energy gamma-rays (Schlumberger, 1989). Low neutron counts indicate a high water content, such as in hydrated clays (e.g. smectite). A very low neutron count associated with a high gamma log indicates a clay layer.

6.4.8 Ground-Penetrating Radar

Ground-penetrating radar (GPR) is a high-frequency radio wave (10–1000 MHz) imaging tool similar to seismic reflection. When the propagating wave comes across variation in electrical properties, it is either reflected or scattered back to the surface (Davis and Annan, 1989). Changes in the dielectric constant of the GPR signal occur between clay and other materials. Deeper penetration is not always possible as some attenuation of the radio wave energy and its returned reflections occurs in conductive and magnetic environments. In moist, clay-rich materials, for example, the penetration depth may decrease to a few centimetres, whereas dry, sandy soils allow a detection penetration depth of 10–25 m (Munday, 2008). Davis and Annan (1989), among others, successfully applied GPR for studying alluvial systems.

6.4.9 Electromagnetic Induction

Electromagnetic (EM) induction is based on the operation of a transmitter coil conducting a varying electric current, on the ground, or in case of airborne EM, at distance above the ground, from an aeroplane or helicopter, which produces an associated time-varying primary magnetic field. The resulting induced electric eddy currents respond to the varying EM fields within the ground, producing a secondary magnetic field, recorded by a receiver coil in the EM tool. The apparent conductivity near the receiver is determined by the ratio of the magnitude of the secondary magnetic field to that of the primary magnetic field (McNeill, 1980). In channel iron deposits, Burnham and Hawke (2003) showed, using the

Geonics EM31, that the variation of electrical conductivity can easily map the clay content. Depending on their cation exchange capacity (CEC), the abundance and composition of clay minerals can be inferred. Kaolinite has a low CEC (Emerson and Yang, 1997), in contrast to smectites (and vermiculite), which have a high CEC.

Airborne EM technologies cover large areas at comparably low cost, but have a lower spatial resolution (horizontally and vertically) and a lower penetration depth (Munday, 2008). EM technologies are extensively used for mineral exploration (Macnae, 2007) and regolith studies (Thomson *et al.*, 2007) – for example, for mapping palaeochannels (Munday *et al.*, 2001).

6.4.10 Radiometrics

Radiometrics (also 'gamma-ray spectrometry') is used to measure the concentration of three radioactive isotopes of K, Th and U down to a penetration depth of about 30–45 cm (up to 1 m in low-density organic-rich soils). The natural radioactive decay of these elements and their daughter products emits gamma radiation that can be measured with a sodium iodide detector. The concentration of the respective element can be calculated from acquired data, on the basis of distinct spectral peaks that are diagnostic for each decay series. Clay (and iron oxide) rich regolith is typically depleted in K, where kaolinite forms a major component (e.g. kaolinitization of granites in lateritic weathering conditions). In contrast, smectites lead to elevated K values in the regolith. Regolith processes play an important role in the redistribution of the radioactive elements, which is also related to the types of clay minerals formed in different regolith environments (e.g. Cook *et al.*, 1996; Dickson and Scott, 1997). Radiometric data can be acquired with airborne systems (Wilford *et al.*, 1997) as well as downhole systems (i.e. gamma log) (Telford *et al.*, 1990), and both techniques are routinely used in mineral exploration. The Th/K ratio derived from gamma-ray logs, for example, is related to changes in the relative proportions of clay minerals, including mixed-layer illite–smectite, smectite and kaolinite, and can be applied to determine sequence stratigraphy in sedimentary basins, important for hydrocarbon exploration (Hesselbo, 1996).

6.5 Visible and Infrared Spectroscopy

Visible and infrared spectroscopy (termed 'spectroscopy' in the following) is an optical sensing technique that utilizes the 'spectral signature' of a given material in the visible (390–750 nm) and infrared (750–1 000 000 nm) part of the electromagnetic spectrum to determine its presence, abundance, composition and other characteristics. Spectroscopy is very fast (a few seconds) and provides identification and quantification of both kaolinite and smectite. Concerning clay minerals in general (Cariati *et al.*, 1983; Frost and Johansson, 1998; Hunt and Salisbury, 1970; Madejova and Komadel, 2001; Martinez-Alonso *et al.*, 2002; Post and Noble, 1993), the spectral signature is mainly related to (1) visible and infrared functional groups (i.e. molecules causing absorption features, such as Al_2–O–H, or H_2O bound/unbound in the crystal lattice), and (2) crystal defects. Electronic processes, involving the transfer of electrons from lower to higher energy states (e.g. in ferric iron) are negligible in clay minerals when compared to commonly associated mineral phases, such as iron oxides. For detailed introduction to the structure and composition of clays, the reader is referred to Chapters 1, 2 and 4.

Spectroscopy is a non-invasive technology that, depending on the wavelength region, penetrates the upper few microns or only reflects from the surface of the analysed material with minimal interaction. In remote sensing applications it is therefore considered as a surface technology. However, when considering geological models describing the regolith or mineral system of concern (Chapter 5), this surface information can be used to infer the geological situation below the surface.

Presence, abundance and composition of clays can be inferred from multispectral and hyperspectral data in various ways. Among the most widely used approaches is spectral unmixing, where a match filtering technique is applied to separate the spectral components of reflectance and emissivity spectra of each pixel into estimated contributions of minerals or 'endmembers' (Kruse *et al.*, 1993). An alternative procedure to generate mineral abundance maps is the relative band depth method for processing, for example, ASTER data (Crowley *et al.*, 1989; Rowan and Mars, 2003), which is based on the important correlation between the depth of a diagnostic absorption feature and the abundance of the respective functional group or mineral. Band ratios are also useful for minimizing mineral-independent scene effects, such as normalization of illumination differences due to variable atmospheric conditions, topography and sun angle between scenes in larger-scale mosaics that comprise, for example, multi-temporal ASTER scenes (Cudahy *et al.*, 2012; Hewson *et al.*, 2005; Ninomiya *et al.*, 2005).

Clay quantification using reflectance spectroscopy has been demonstrated in the literature (e.g. Flehmig and Kurze, 1973; Wells and Ramanaidou, 2007). The comparison of the reflectance spectra for KGa2 and SAz1 (Fig. 6.4) illustrates that the absorption feature of kaolinite at 2165 nm is valuable in determining the kaolinite content because montmorillonite shows no absorption feature at this wavelength. Hence the depth of the 2165 nm ('D2165') absorption is positively correlated to the kaolinite content in the prepared kaolinite–montmorillonite powdered mixtures (Fig. 6.5). Determination of the montmorillonite abundance is made indirectly by subtracting the percentage Al_2O_3 content due only to kaolinite from the alumina content associated with all clays in the sample (Wells and Ramanaidou, 2007). It should be noted that the physicochemistry of white micas is similar to kaolinite and smectite and, therefore, a similar spectral signature can be expected. Therefore, in many cases of, for example, soil or regolith analysis, all of these three major Al-sheet silicates have to be taken into account when interpreting reflectance spectroscopic data.

The composition of clay minerals can be inferred from the occurrence and wavelength position of diagnostic absorption features. In kaolinite, for example, a weak absorption feature is evident in the 2240 nm wavelength region, when the ferric iron content increases (Cudahy, 1997; Delineau *et al.*, 1994). In aluminium smectites, the Tschermak exchange

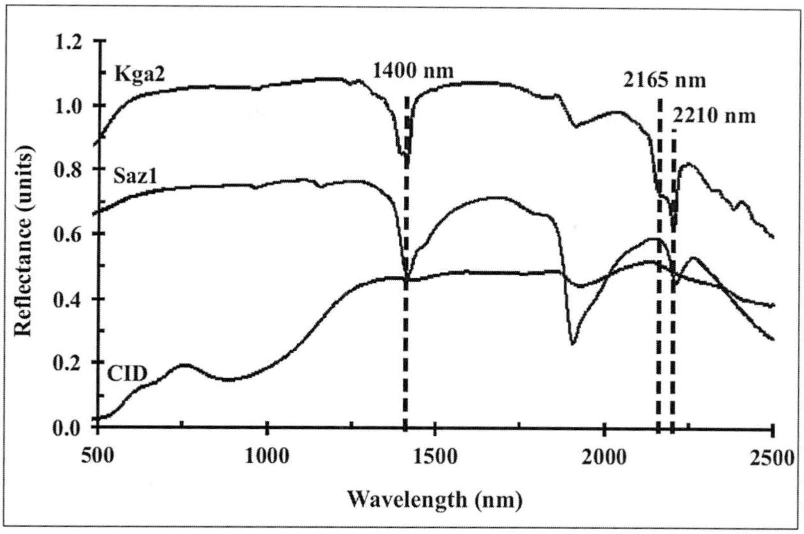

Figure 6.4. Reflectance spectra for standards KGa2 (kaolinite), SAz1 (smectite) and CID (iron oxide from a channel iron deposit) (Wells and Ramanaidou, 2007).

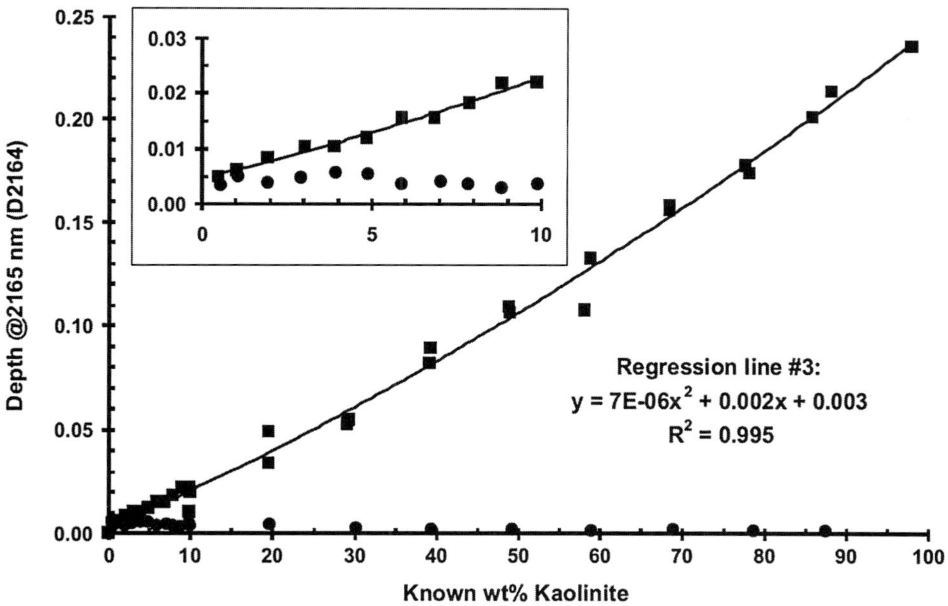

Figure 6.5. Depth of the absorption feature at 2165 nm (D2165) versus the kaolinite abundance in the prepared mixtures. Inset plot shows detail of the regression line for mixtures with <10 wt% kaolinite. Kaolinite (squares); montmorillonite (circles) (Wells and Ramanaidou, 2007).

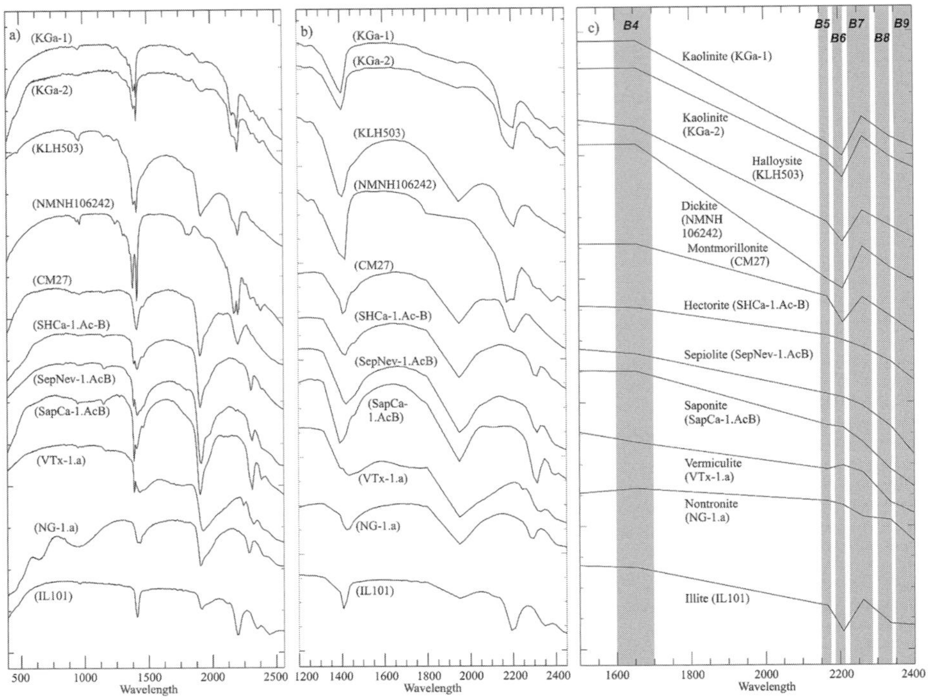

Figure 6.6. Spectral resolution of hyperspectral (a) lab, (b) airborne and (c) multispectral satellite reflectance spectra (wavelength in nm) of clay minerals selected from the USGS spectral library (http://speclib.jpl.nasa.gov.au). ASTER band widths of bands 4–9 in grey.

($Al^{VI}Al^{IV}[Mg,Fe]_{-1}Si_{-1}$) can be estimated by observing the wavelength position of the major absorption feature at around 2200 nm, which is located at shorter wavelengths in beidellite ('Al-rich') and at longer wavelengths in montmorillonite ('Al-poor') (Post and Noble, 1993). The relative amount of ferric and ferrous iron or Mg in the iron–magnesium smectites (e.g. saponite) may be estimated from the occurrence and wavelength position of absorption features in the 2230–2410 nm wavelength region (e.g. Bishop *et al.*, 2008; Cariati *et al.*, 1983; Post and Noble, 1993).

6.5.1 Wavelength Regions

Every material has a characteristic spectral signature, but only certain parts of the electromagnetic spectrum contain absorption bands that are diagnostic for this material. Aluminous clays, for example, show characteristic absorption features in the 1.4, 2.2, 2.7, 10, 14 and 20 μm regions (Fig. 6.6). These can be separated into (1) bending (δ) and stretching (ν) vibrations, their combinations (ν + δ) and overtones (2ν, 3ν, etc.) of the fundamental stretching vibration related to the hydroxyl group; and (2) SiO-related bending (δSiO) and stretching (νSiO) vibrations (Fig. 6.6).

Frequencies of the fundamental vibrations are located in the mid (MIR) and far infrared (FIR) according to mineral spectroscopy literature and in the thermal infrared and towards lower frequencies when using the terminology of the remote sensing literature (Fig. 6.6). In the following, the terminology of the remote sensing literature is applied, as used predominantly in economic geology. Within most literature, the electromagnetic spectrum is subdivided into the ultraviolet (UV, <0.39 μm), visible-near (VNIR, 0.39 to *c.* 1 μm), shortwave (SWIR, 1–2.5 μm), mid (MIR, 2.5–5 μm), thermal (TIR, 5–16 μm) and far (FIR, >16 μm) infrared wavelength regions.

The wavelength regions sampled with remote sensing technologies and field spectroradiometers are largely aligned with atmospheric windows, where the influence of the atmosphere (e.g. H_2O, CO_2) is limited (Fig. 6.7). Absorption features of most functional groups of rock-forming minerals are located in these atmospheric windows, which enables us to detect, for example, the combination feature of the hydroxyl-related bending and stretching in clay minerals located in the 2.2 μm region of the SWIR even with remote sensing technologies. In contrast, the first overtone of the hydroxyl stretching in clay minerals (*c.* 1.4 μm) overlaps significantly with a major water feature and is therefore not extractable from reflectance spectra using a remote sensing technology. A limited number of studies have successfully trialled the detection of clay minerals using TIR (Crowley and Hook, 1996). Crystallinity, particle size, orientation and chemical composition all have influences on the spectral signature of clay minerals. The impact of the respective characteristics on the depth and location of single absorption bands can vary across the electromagnetic spectrum. Smaller particle size of clay minerals, for example, will lead to more pronounced combination features in the SWIR. The crystal orientation, on the other hand, can heavily influence the spectral signature in the TIR.

6.5.2 Spectral Resolution

In order to determine the presence, abundance, composition and crystallinity of aluminium clays accurately, a sufficient spectral resolution is required. Spectral resolution is defined as the minimum spectral separation required for discriminating two spectral features (King *et al.*, 2004). Commonly, spectroscopic technologies are classified as multispectral versus hyperspectral. Multispectral technologies, such as ASTER (Yamaguchi *et al.*, 1999) and Landsat (Short, 1982), feature a few bands that cover selected parts of the electromagnetic spectrum (Figs 6.7 and 6.8). In contrast, hyperspectral technologies (e.g. AVIRIS) record up to hundreds of bands across the spectrum. Typically the width of each band of a multispectral system (e.g. ASTER) is considerably larger when compared to, for example, airborne hyperspectral systems (Fig. 6.7). The result is that many bands of the multispectral ASTER comprise a large number of different absorption features that are due to various minerals, leading to a different level of detail when it comes down to identification of mineral groups or species (Fig. 6.7). In contrast to this, the higher spectral resolution of hyperspectral remote sensing technologies allows the identification of mineral species and their compositional changes.

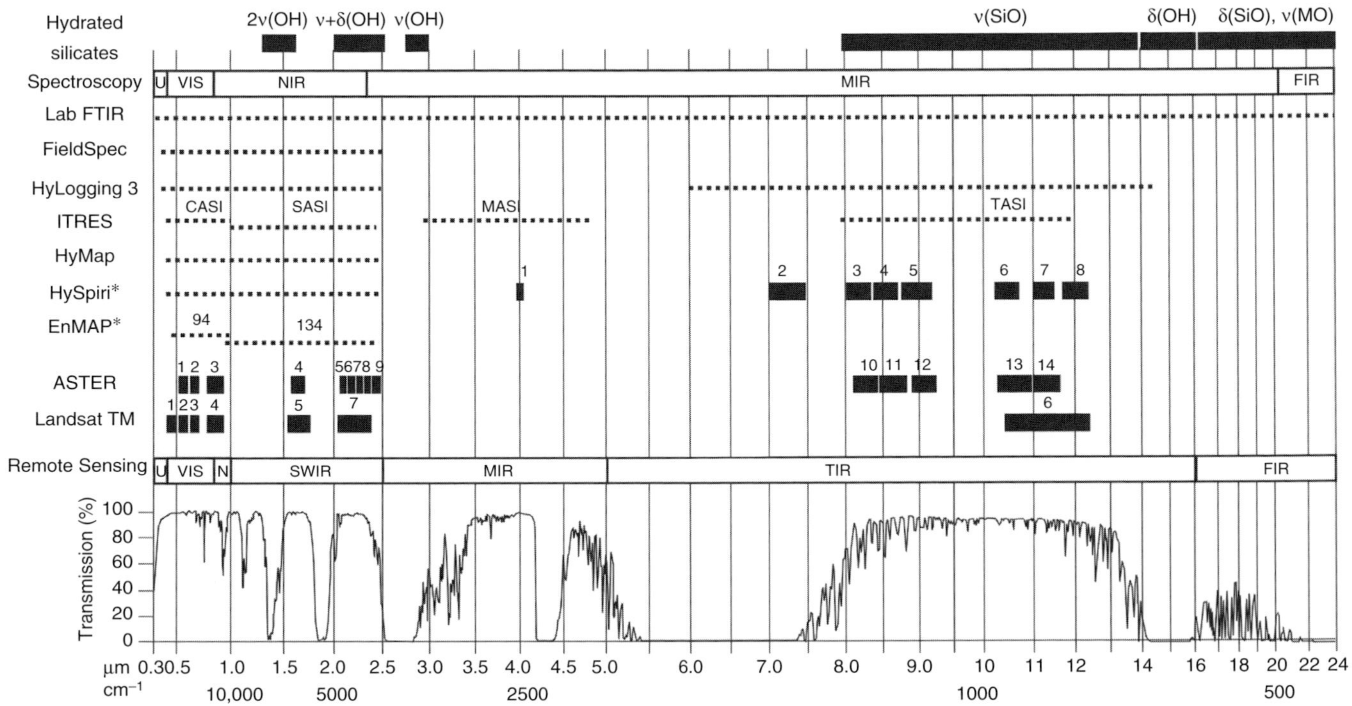

Figure 6.7. Functional groups of clays, wavelength regions according to spectroscopy literature, band coverage of remote and proximal VIS technologies, wavelength regions according to remote sensing literature, and generalized atmospheric transmission spectrum of 1 km horizontal air path at sea level conditions (15 °C air temperature, 46 per cent relative humidity, 1013 MB atmospheric pressure).

*HySpiri and EnMap are the planned band positions of systems that will be available in the near future.

Table 6.2. *Major SWIR spectral parameters of kaolinite (Frost and Johansson, 1998; Hunt and Salisbury, 1970), Al-smectite (e.g. beidellite and montmorillonite; Martinez-Alonso et al., 2002; Post and Noble, 1993) and Fe-smectite (e.g. nontronite; Bishop et al., 2008; Cariati et al., 1983; Post and Noble, 1993) in the 2150–2400 nm wavelength region (ν + δ – combination of stretching and bending vibration related to hydroxyl, OH_i – inner hydroxyl, OH_o – outer hydroxyl).*

Clay	Wavelength centre position	Assignment	Intensity
Kaolinite	2159	$\nu + \delta(Al)_2OH_o$	Medium–weak
	2160	$\nu + \delta(Al)_2OH_o$	Medium–weak
	2194	$\nu + \delta(Al)_2OH_o$	Medium–weak
	2206	$\nu + \delta(Al)_2OH_i$	Strong
	*c.*2320	Possibly $\nu(Al)_2OH$ + lattice	Weak
	*c.*2380	deformation vibration	Weak
Al-smectite	2185–2215	$\nu + \delta(Al)_2OH$	Strong
	2237–2250	$\nu + \delta(Al,Fe^{3+}\vert Al,Mg)_2OH$	Weak, shoulder
Fe-smectite	*c.*2236/2243	$\nu + \delta(Al,Fe^{3+})_2OH$	Medium–strong
	*c.*2288/2292	$\nu + \delta(Al,Fe^{3+})_2OH$	Medium–strong
	*c.*2380–2400	Not assigned	Medium–weak

Kaolinite group minerals, for example, have four characteristic absorption bands in the 2150–2210 nm wavelength region (Table 6.2). However, only some of the hyperspectral technologies have a sufficient spectral resolution to enable a deconvolution of the four diagnostic absorption bands. In most cases (e.g. airborne hyperspectral sensors) the spectral resolution is sufficient to determine a doublet feature with two troughs centred at around 2160 and 2206 nm, respectively. The crystallinity of the kaolinite will control the relative depth of the four features, with a weakening of the shorter wavelength features in poorly crystalline kaolinite (Crowley and Vergo, 1988).

6.5.3 Remote Sensing

Apart from mineral exploration and mining (Clark, 1999; Perry, 2004; Thompson *et al.*, 1999), the applications of multispectral and hyperspectral remote sensing of clay minerals are numerous, ranging from agriculture, forestry and range resources (Crosta *et al.*, 1998; Roberts *et al.*, 1993) to water resource management (Rodriguez *et al.*, 2011), baseline monitoring (Richter *et al.*, 2008) and mapping of geological hazards (Chabrillat *et al.*, 2002). Spectroscopic satellite technologies have even been applied to detect clay minerals on other planets (Loizeau *et al.*, 2007; Poulet *et al.*, 2005). Multispectral remote sensing data and derived mineral maps of the Earth's surface in association with other geoscience data are becoming increasingly available to potential users via online portals such as the AuScope Discovery Portal (Woodcock *et al.*, 2010).

6.5.3.1 Satellite Remote Sensing and ASTER

In geosciences, ASTER has been used extensively for regional geological mapping (Ninomiya *et al.*, 2005) and district-scale mapping of hydrothermal alteration (Rowan *et al.*, 2003; Witt *et al.*, 2013). Ducart *et al.* (2006) mapped advanced argillic alteration and silicic alteration associated with an epithermal system in Patagonia using the SWIR and TIR ASTER bands. In Australia, ASTER data have been assessed in various areas for geological mapping and detection of hydrothermal alteration (Cudahy *et al.*, 2005, 2008; Hewson and Cudahy, 2011).

In contrast to the numerous publications about application of ASTER for mineral exploration, there are only a few publications (Hewson *et al.*, 2005) and reports (Cudahy *et al.*, 2008) that touch on the evaluation of ASTER for regolith characterization, though the theoretical potential for identifying regolith-forming minerals (especially clay minerals) is obvious. Hewson *et al.* (2005) inferred the abundance of AlOH group minerals and quartz to map regolith areas in the Broken Hill-Curnamona province of Australia. The same publication is also one of the first to provide seamless regional geological maps, which allow the comparison of different geological provinces and tracking of large-scale changes of regolith domains. This is in contrast to most other ASTER related publications, where only single or a few scenes were consulted for deposit-scale mapping. The concept of large-scale seamless mineral maps has improved since then, leading to the continental-scale Australian ASTER Geoscience Products (Cudahy *et al.*, 2012).

ASTER collects 14 spectral bands in the VNIR, SWIR and TIR, which are positioned in 'atmospheric windows' to minimize impact of atmospheric effects, such as atmospheric water or aerosols (Fig. 6.7). The spatial resolution is 15 m/pixel in the VNIR, 30 m/pixel in the SWIR and 90 m/pixel in the TIR, with a total swath width of about 60 km for all wavelength regions (Table 6.3). ASTER, being a multispectral system with only 14 bands distributed over the visible and infrared part of the electromagnetic spectrum, can only be used to identify the relative abundance and composition of the mineral groups (Hewson *et al.*, 2005).

Currently available satellite-based systems only feature multispectral resolution; in ASTER, the position of these bands is only designed to separate clay mineral groups from other hydrated silicates and carbonates (Section 5.1; Fig. 6.6). A deconvolution of reflectance and emission spectra into distinct clay mineral species is therefore highly problematic. However, selected band ratios can be used to qualitatively to semi-quantitatively estimate the abundance of Al-clays (Hewson *et al.*, 2005), and also help to map domains of prominent kaolinite and Al-rich smectites (e.g. beidellite) and white micas (e.g. paragonite and muscovite) versus areas characterized by Al-poor smectites (e.g. montmorillonite) and white micas (e.g. phengite). On a continental to regional scale, this in turn can help to characterize, for example, geological provinces, regolith landforms and drainage systems (Laukamp *et al.*, 2015).

Table 6.3. *Specifications of ASTER, modified after Yamaguchi et al. (1999).*

Instrument	VNIR	SWIR	TIR
Bands	b1–b3	b4–b9	b10–b14
Spectral range (μm)	0.52–0.60	1.600–1.700	8.125–8.475
	0.63–0.69	2.145–2.185	8.475–8.825
	0.78–0.86	2.185–2.225	8.925–9.275
		2.235–2.285	10.25–10.95
		2.295–2.365	10.95–11.65
		2.360–2.430	
Spatial resolution (m)	15	30	90
Swath width (km)	60	61	62

Case study: The AlOH group content map in Fig. 6.8a shows the continental-scale distribution of Al-clays (kaolinite group, Al-smectites and white micas). The most prominent areas comprise the Eromanga Basin in eastern Australia and a large area ranging from the northern Yilgarn Craton to the southern margin of the Hamersley Province. These two domains are connected by the Officer Basin in Western Australia and South Australia, and overlap with the regions of lowest rainfall in Australia (Cudahy *et al.*, 2016), providing important information about desertification. Other major Australian basins, such as the Canning, Georgina and Eucla Basins, are characterized by a low Al-clay abundance.

The AlOH group composition map highlights the prominent type of Al-clays across the Australian continent (Fig. 6.8b). The Eucla and Eromanga Basins are both dominated by Al-poor smectites and white micas, whereas the Eromanga Basin contains significantly more Al-clays. A distinct approximate north–south trending boundary separates the Eromanga Basin in the east from the Officer Basin in the west, independent of the generally high Al-clay content in both of these basins (Fig. 6.8a). When focusing in detail on each of these geological provinces, regional-scale differences become clear, such as the alternation of east–west trending basins and inliers in the Northern Territory (NT, Fig. 6.8b). The AlOH group composition map also shows the catchment and river beds of large drainage systems. The Darling River (D), for example, is sourced from the Surat Basin in the east, dominated by Al-poor smectites and/or white micas, which sets it apart from the northern tip of the Lachlan Fold Belt, which shows predominantly kaolinite and/or Al-rich smectites and white micas.

6.5.3.2 Airborne Hyperspectral

Airborne hyperspectral systems have been applied since the 1970s for mapping of hydrothermal alteration. Abrams *et al.* (1977) used a Bendix 24-channel scanner on-board a NASA C-130 aircraft to detect hydrothermal alteration in the Cuprite Mining District (Nevada, USA). Since then, numerous airborne hyperspectral sensors have been deployed for mapping lithologies and alteration patterns, including the Airborne Imaging Spectrometer (AIS) (Kruse, 1988), the Airborne Visible/Infrared Imaging Spectrometer (AVIRIS)

Figure 6.8. Continental-scale ASTER Geoscience Product showing geological provinces in the AlOH group content (a) and composition (b) maps (Cudahy *et al.*, 2012). Pixels containing no colour information are masked out due to absence of mapped material or significant non-mineral related influence (e.g. clouds). C – Canning Basin, D – Darling River; EU – Eucla Basin, ER – Eromanga Basin, G – Gascoyne Province, GE – Georginia Basin, H – Hamersley Basin, O – Officer Basin, NT – Northern Territory Provinces (e.g. Amadeus Basin, Arunta Inlier), P – Pilbara, Y – Yilgarn Craton. A black and white version of this figure will appear in some formats. For the colour version, please refer to the plate section.

(Crosta *et al.*, 1998; Vane *et al.*, 1993), the HyMap™ airborne hyperspectral sensor (Cocks *et al.*, 1998; Cudahy *et al.*, 2008; Laukamp *et al.*, 2011a, 2011b); and the SEBASS system (Vaughan *et al.*, 2003). A recent overview of airborne hyperspectral imaging spectrometers is given in Coulter *et al.* (2007).

The previous section described the use of continental-scale mineral maps derived from multispectral ASTER data for characterizing geological provinces and major drainage systems based on the abundance of major Al-clay mineral groups. Airborne hyperspectral data enable us to map the abundance of hydrated phyllosilicates (including kaolinite and smectites) spatially, in detail down to metre resolution (Laukamp *et al.*, 2011a, 2011b). They can also be used to determine changes in the composition and crystallinity of clay minerals. This has applications, for example, in the characterization of regolith, where the distinction between *in situ* (erosional) and transported material is of major importance to estimate the thickness or source of the overburden (Cudahy *et al.*, 2005). Swelling clays (i.e. smectites) are often a geological hazard (e.g. damage to residential development, slope stability in open pits, difficult to dewater) and can be spatially mapped using airborne hyperspectral imaging systems (Chabrillat *et al.*, 2002).

Case study: Kaolinite group minerals are the weathering products of a wide range of primary and secondary minerals (Anand and Paine, 2002). The composition and crystallinity of kaolinite is related to the source rocks and the sedimentary regime they're forming under (Anand and Paine, 2002). The crystallinity of kaolinite group minerals can provide hints, if the regolith consists of transported or residual material (Cudahy, 1997) and can be inferred from hyperspectral data (Crowley and Vergo, 1988). This model was applied by Cudahy *et al.* (2005) to map transported and residual regolith on a regional scale in the Eastern Goldfields of Western Australia. Airborne hyperspectral (HyMap) data were processed into a mineral map showing the disorder (i.e. crystallinity) of kaolinite group minerals (Fig. 6.9). For validation, reflectance spectra were collected using a field spectrometer and selected samples were analysed using XRD (sample points shown in Fig. 6.9). The three independent datasets were compared with each other and show a significant correlation (Fig. 6.10), suggesting that the crystallinity of kaolinite can be inferred from remote sensing data.

The derived kaolinite disorder map shows good agreement with a regolith landform map published earlier by Craig and Anand (1993). Breakaway zones (i.e. erosional landform) show high-crystallinity kaolinite, whereas the valley floors (i.e. transported landform) contain poorly crystalline kaolinite. This study shows that time-consuming field-mapping could effectively be complemented by using mineral maps derived from hyperspectral remote sensing data. This would help to target fieldwork and to efficiently design drilling programmes based on the information about the inferred composition and thickness of the overburden.

6.5.4 Proximal Sensing Technologies

A wide range of proximal sensing spectroradiometers are available, designed for various applications and covering various parts of the electromagnetic spectrum (Fig. 6.7; ASD's FieldSpec, CSIRO's HyLogging™). Field spectroradiometers, such as the FieldSpec from Advanced Spectral Devices Inc. and the PIMA (Portable Infrared Mineral Analyser, from Integrated Spectronics), have been successfully used in the last two decades to characterize regolith material and hydrothermal alteration, such as argillic alteration (Chang and Yang,

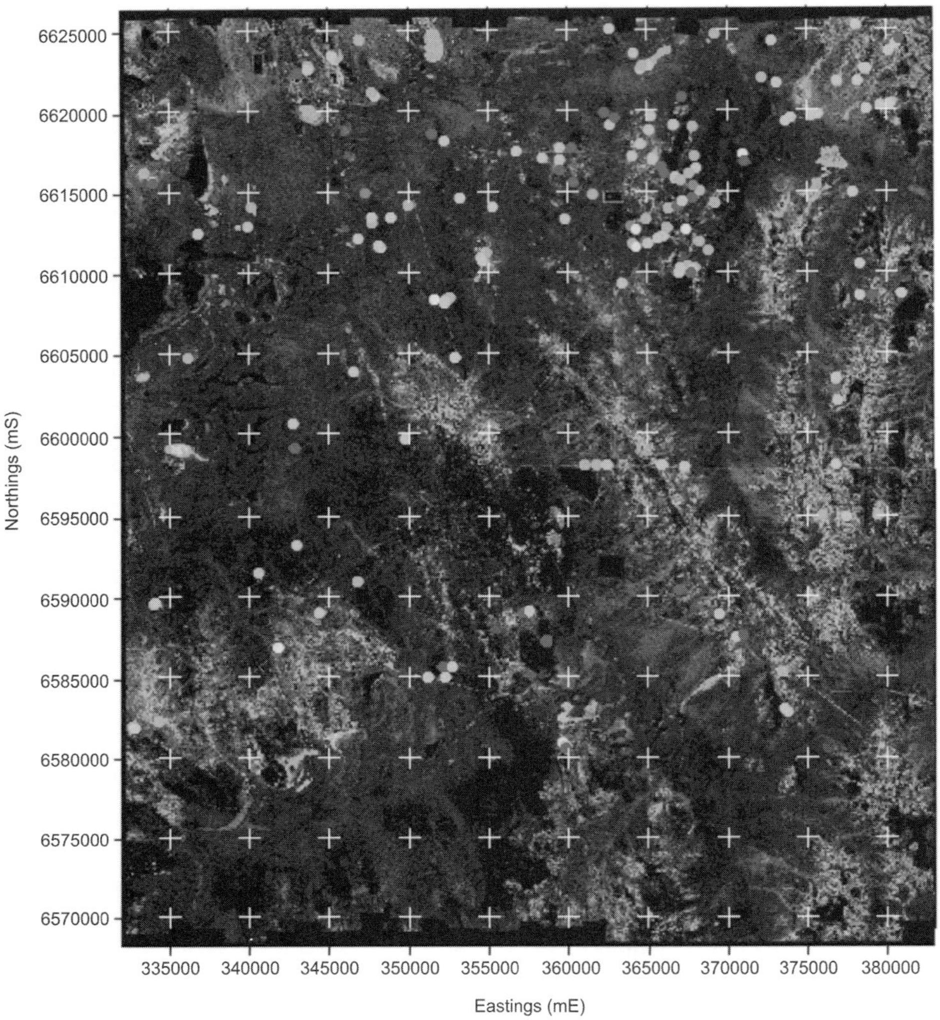

Figure 6.9. Kaolin disorder spectral index (i.e. kaolin crystallinity) derived from airborne hyperspectral (HyMap) data (Cudahy *et al.*, 2005). Coloured points indicate sample locations of field spectrometer and XRD measurements. In both datasets, the kaolin disorder is displayed using a rainbow stretch, with blue colours showing low crystalline order and red colours showing high crystalline order. Black pixels in the map are masked out due to absence of kaolinite or high influence of non-mineralogical material (e.g. vegetation). A black and white version of this figure will appear in some formats. For the colour version, please refer to the plate section.

2012; Sonntag *et al.*, 2012; Yang *et al.*, 2005, 2011). These instruments typically cover the SWIR wavelength region (PIMA) in addition to the VNIR (ASD's FieldSpec) at a high spectral resolution and can be used to determine the abundance and to a certain degree the composition and crystallinity of most clay mineral species (i.e. hydrated phyllosilicates).

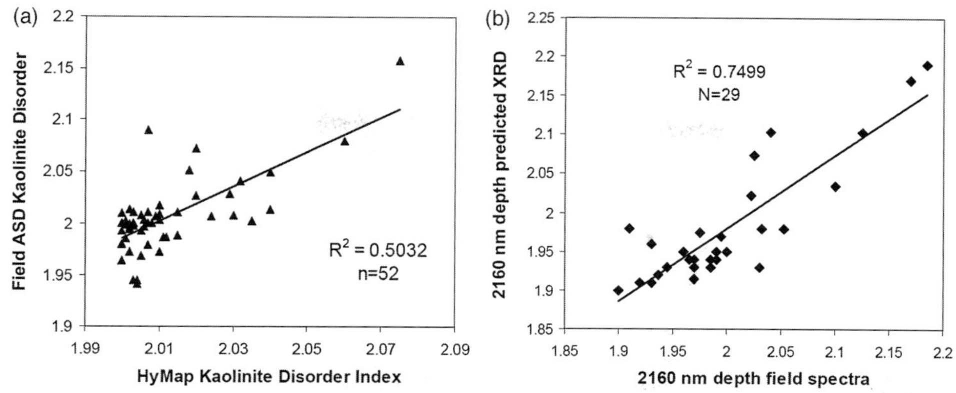

Figure 6.10. Scattergrams showing (a) the depth of the 2160 nm derived from HyMap versus field spectrometer data; and (b) the kaolinite disorder based on the depth of the 2160 nm absorption feature from the field spectrometer data versus an XRD index based on the partial least squares (PLS) regression analysis of both of these data (Cudahy *et al.*, 2005).

Zhang *et al.* (2001) analysed PIMA data to characterize clay minerals in alteration assemblages associated with sandstone-hosted unconformity-type U deposits. Comparison with XRD and other lab spectroscopic technologies showed that precision and accuracy of the SWIR data are comparable to the XRD results, with detection limits of single clay species in mixtures down to ~5 per cent relative abundance. Argillic mineral assemblages were detected by Yang *et al.* (2005) in the porphyry related Cu–Au deposit at Tuwu, Xinjang, northwest China, using a PIMA that records SWIR reflectance spectra. Silicification and phyllic alteration, covered by a blanket of argillic alteration, are spatially associated with the Cu–Au mineralization at this deposit.

6.5.4.1 Drill Core Scanning Technologies

During the first decade of the twenty-first century, the field of hyperspectral drill core scanning devices has seen major developments, including routine line profile scanners and drill core imaging spectrometers operating in the VNIR, SWIR and TIR. CSIRO's HyLogging system, for example, collects spectra at 0.8 × 0.8 cm spatial resolution as a central profile series of measurements along the drill core. This system is currently employed by the iron ore industry for routine detection of clay minerals in iron ores. The highly detailed spatial data were also used to (1) characterize the mineralogy of ore zones and contaminants (e.g. clays) in CID (Haest *et al.*, 2012a, 2012b;) and (2) map lithological boundaries and hydrothermal alteration patterns in mafic–ultramafic rock packages in Archean Au systems (Cudahy *et al.*, 2009; Laukamp *et al.*, 2012). Haest *et al.* (2012b) used HyLogging data from diamond and RC drill holes to model the distribution of kaolinite, Al-smectites and Fe-smectites associated with the Rocklea Dome CID, Western Australia (Fig. 6.11). Results showed that highly crystalline kaolinite dominates the channel margins and base of the CID, whereas poorly crystalline kaolinite forms layers within the deposit. Al-smectites

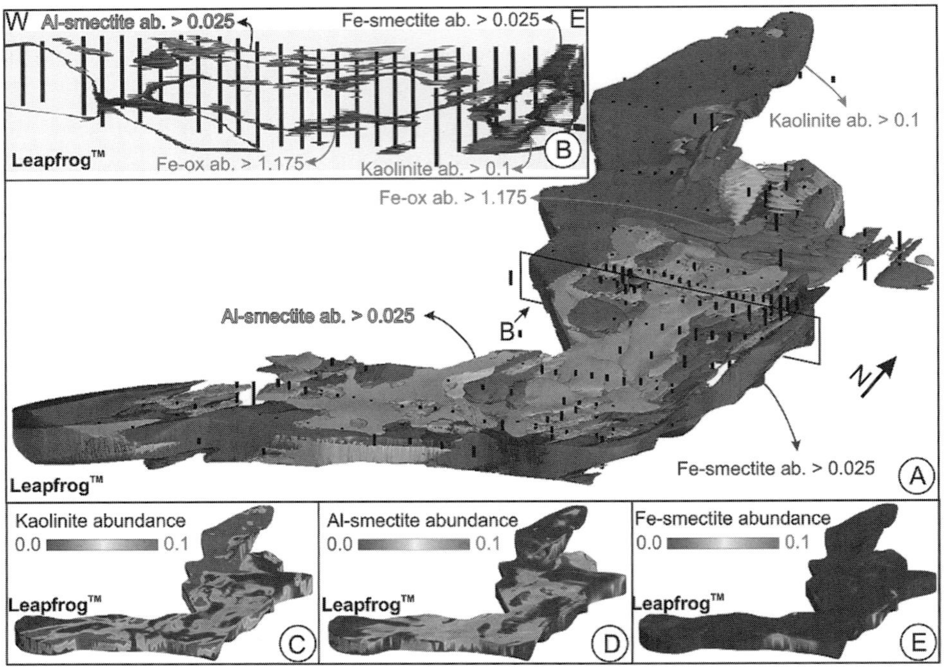

Figure 6.11. Three-dimensional representation of the Rocklea Dome CID (Western Australia) showing relative mineral abundances derived from HyLogging data using a multiple-feature extraction method modified after Haest *et al.* (2012a). (A) Shells showing the Fe (oxyhydr-)oxide (Kruse, 1988), kaolinite (green), Al-smectite (yellow) and Fe-smectite (purple) abundances. (B) East–west profile through the centre of the CID (view from south-west), showing the distribution of the clay shells, relative to the zone with spectrally derived Fe (oxyhydr-)oxide abundances. (C–E) Three-dimensional volume outlining all RC drill cores scanned at Rocklea Dome, evaluated as a function of the different spectrally defined clay types including kaolinite (C), Al-smectite (D) and Fe-smectite (E). Colour stretches showing increasing relative abundance of clay species from blue to red (i.e. 0.0 to 0.1). For setup of applied algorithms and thresholds see Haest *et al.* (2012a, 2012b). A black and white version of this figure will appear in some formats. For the colour version, please refer to the plate section

cover the CID at the surface, whereas Fe-smectites are developed in parts along the southern and eastern margin of the Tertiary channel. The distribution of the various clay mineral phases allows improved resource delineation and enhances our understanding about which clay minerals are formed in association with this important resource type, which can find applications in remote sensing for mineral exploration of other iron ore deposits.

6.5.5 Impediments/Challenges and Future Developments

Visible and infrared reflectance and emission spectroscopy is increasingly used for numerous applications, many of them involving the detection and characterization of kaolinite and smectite group minerals. Imaging spectroscopy does face critique in the geoscience community, as (1) it is 'just' a surface technology and (2) inherent aspects of remote sensing, such as atmospheric gases, clouds, vegetation and man-made features, complicate the interpretation of the data. Nonetheless, the announcement of launches of several new geoscience-tuned hyperspectral satellites in the next years, such as EnMAP (Guanter *et al.*, 2015) or HySpiri (Lee *et al.*, 2015), highlights the increasing importance of this technology for earth observation. Many of the new satellites have hyperspectral VNIR and SWIR sensors (Fig. 6.7), which will enable mapping of the abundance and composition of distinct mineral species, compared to the broad mineral group-level information accessible with ASTER. Hyperion is the only hyperspectral satellite to date and has been trialled for the detection of hydrothermal alteration (Gersman *et al.*, 2008). However, the poor signal-to-noise ratio of Hyperion limits the ability to process the satellite data into detailed mineral maps (Kruse *et al.*, 2002). Furthermore, recent publications show that many of the complicating effects such as vegetation can be unmixed from the imaging data, enabling the production of quantitative Al-clay mineral maps (Haest *et al.*, 2013; Rodger and Cudahy, 2009). The future will see major improvements in remote and proximal hyperspectral sensing technologies, providing increasingly reliable measures of abundance, composition and crystallinity of clays and associated minerals.

6.6 Dating of Clays

6.6.1 Introduction

Over the last approximately 40 years a number of dating methods have found application for characterizing the weathered mantle of ancient palaeocontinental surfaces and in helping establish the chronology of past geodynamic and climatic events as recorded in palaeoweathering surfaces and materials (Ricordel-Prognon *et al.*, 2010). Clauer and Chaudhuri (1995) and Savin and Hsieh (1998) provide a detailed treatment of isotopic dating techniques as applied to clay minerals, whereas Pillans (1998) provides a more practical guide to a wide variety of numerical dating techniques, principally aimed towards dating of the Australian regolith but which may be applied to regolith materials and weathering profiles in general. These are not all treated here in detail, but the following discussion provides examples of the application of some of these techniques and of the type of information such as regolith evolution that may be obtained.

Of the methods directly applicable to elucidating age-related weathering events recorded in clay-rich (i.e. kaolinite or smectite) environments, K–Ar dating and, in particular, stable isotope analysis have the most relevance. Palaeomagnetic dating techniques have also been applied to ferruginous weathering profiles and Fe-rich clay materials (Pillans and Bourman, 1996;Ricordel-Prognon *et al.*, 2010; Schmidt and Ollier, 1988; Schmidt *et al.*, 1983) to record changes in the geomagnetic field and recover the palaeomagnetic pole position, which when compared to the local apparent polar wander path (APWP) provides an age for the weathering surface (Ricordel-Prognon *et al.*, 2010). This technique records the palaeogeomagnetic remanence as a chemical remnant magnetization (CRM) in neoformed and secondary iron oxides. However, as the CRM is carried by iron oxides, such as goethite and hematite, rather than clays, this method of dating is not discussed further.

6.6.2 Stable Oxygen ($^{18}O/^{16}O$) and Hydrogen (D/H) Isotope Analysis

The oxygen (i.e. $^{18}O/^{16}O$) and hydrogen (D/H) isotopic composition of clay minerals are related to the conditions under which the mineral formed, such as the temperature of formation and the isotopic composition of the fluid (i.e. water) from which the mineral precipitated (Chivas, 1993; Savin and Hsieh, 1998). At or close to the Earth's surface the groundwater (i.e. meteoric) is in isotopic equilibrium with the atmosphere, and its isotopic signature is a reflection of the climate (Savin and Hsieh, 1998). In near-surface weathering environments clays are considered to form in isotopic equilibration with meteoric water, and so their $^{18}O/^{16}O$ and D/H ratios record the temperature of their formation (Savin and Hsieh, 1998). To infer something of the palaeoclimatic conditions at the time of formation, the clay mineral must retain its original isotopic composition (Savin and Hsieh, 1998), and not have been subjected to post-formational changes, through chemical alteration (e.g. dissolution and recrystallization) or have experienced elevated temperatures (i.e. >100–200 °C) that may have resulted in isotopic exchange (Savin and Hsieh, 1998).

It is generally considered that under near-surface ambient conditions, once formed exchange between structural oxygen in kaolinite (and illite) and oxygen in water can be considered negligible in the absence of any physicochemical change (Clauer and Chaudhuri, 1995). However, hydrogen may be more susceptible to isotopic exchange as suggested by Bird and Chivas (1989), where kaolinite from Queensland and South Australian mid-Mesozoic to post-mid-Tertiary regolith profiles was shown to have experienced hydrogen-isotope exchange post-formation, rendering hydrogen-isotope analysis in this case unsuitable for dating purposes. Similar findings were reported for kaolinite in North American profiles and interpreted as indicating post-formational H isotopic exchange had occurred (Lawrence and Meaux, 1993; Savin and Hsieh, 1998).

Compared to kaolinite, there appears to be a lack of consensus regarding the oxygen and hydrogen isotopic exchange of smectite. For example, findings cited by Clauer and Chaudhuri (1995) showed that the O and H isotope composition of low-temperature smectite can vary widely and that at elevated temperatures (e.g. >≈100 °C) isotopic exchange occurs more rapidly in smectite than for illite or kaolinite. However, the findings reported by Savin and Hsieh (1998) showed that smectite demonstrated no O isotopic exchange at ambient temperatures over periods of 25×10^6 years. It is perhaps partly for these reasons and the fact that kaolinite is widespread in weathering environments and is comparatively easy to prepare in samples pure enough for analysis, that most isotopic research has focused on using kaolinite to derive information on palaeoclimatic conditions.

Applying O-isotopic analysis to regolith evolution in Australia, Bird and Chivas (1988) demonstrated that $\delta^{18}O$ values for kaolinite retained their original isotopic composition for an extended period up to 240 Ma (i.e. early Triassic). Coupled with the known

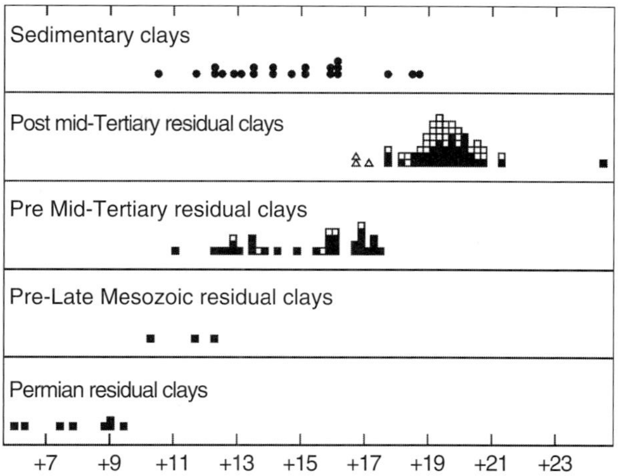

Figure 6.12. Inferred weathering periods for the mid-Tertiary to pre-Late Mesozoic based on $\delta^{18}O$ values for kaolinite in Australian weathered and regolith profiles as modified from Bird and Chivas (1993). Residual clays were collected from deeply weathered profiles. Refer to Bird and Chivas (1989) for a more detailed discussion.

northward migration of the Australian plate during this time, and the expected increase in $\delta^{18}O$ values of meteoric water since the mid-Cretaceous with decreasing latitude, an interpolated age of weathering was deduced for a range of regolith profiles across the Australian continent. These findings (Bird and Chivas, 1988, 1989, 1993) indicated that the deeply weathered regolith of Australia is the result of four weathering events spanning the mid-Tertiary to the Permian (Fig. 6.12), with the low $\delta^{18}O$ values of some pre-Late Mesozoic profiles suggesting weathering may have taken place under moist, cooler conditions rather than under tropical or subtropical conditions as previously considered (Bird and Chivas, 1993).

More recently, oxygen isotopic ratios measured for kaolinite and smectite in palaeosols in the Siwalik Himalayan molasse, northern Pakistan, were shown to be viable palaeoclimate indicators, supporting a significant change in climate (i.e. rainfall) in the region around 8.5–8.6 Ma (Stern *et al.*, 1997). In lateritic profiles north of Manaus (Amazon Basin, Brazil), small changes in the $\delta^{18}O$ values of kaolinite were related to variations in particle size and reflected seasonal climatic variations at the local scale with relatively higher $\delta^{18}O$ values for coarse-grained kaolinite forming during the wet season, and finer particles either forming throughout the year or under slightly drier conditions (Giral *et al.*, 1993a, 1993b). Similarly, variations in $\delta^{18}O$ values reported for kaolinite from a soil profile developed over basement gneiss in eastern Cameroon did not suggest a complete or progressive isotopic equilibration between kaolinite and meteoric waters (Clauer and Chaudhuri, 1995;Clauer *et al.*, 1989) but were interpreted to indicate formation of different generations of kaolinite at slightly different temperatures or under different microchemical conditions (Clauer and Chaudhuri, 1995).

6.6.3 K–Ar Dating: Illite/Smectite (I/S) Interlayer Clay

Dating using the K–Ar method is based on the natural, radioactive decay of ^{40}K to ^{40}Ar through β emission and electronic capture (Clauer and Chaudhuri, 1995; Pillans, 1998). Conventional K–Ar analysis is suitable for a wide range of materials and involves separate measurement of K and Ar, with bulk K analysis determined for one sample aliquot and the amount of ^{40}K determined from the essentially constant $^{40}K/K$ ratio for terrestrial materials (Pillans, 1998). Determination of ^{40}Ar is made on a separate aliquot by isotopic dilution (Pillans, 1998). In a related technique, the ^{40}Ar–^{39}Ar method, both Ar and K are measured on the same sample aliquot by isotopic analysis following neutron irradiation in a reactor which converts a proportion of ^{39}K to ^{39}Ar (Clauer and Chaudhuri, 1995). As the $^{40}K/^{39}Ar$ ratio is essentially constant in nature, the $^{40}Ar/Ar$ ratio, and therefore the age, may be calculated (Pillans, 1998).

However, despite its advantages over the conventional K–Ar method, the ^{40}Ar–^{39}Ar method is not without its drawbacks, as described by Clauer and Chaudhuri (1995), with the most significant being recoil effects that can result in the ejection of isotopes from the clay structure and, thus, affect age determinations. Hence, the majority of K–Ar-based dating studies use the conventional K–Ar method. A major caveat to the application of the method is the requirement of a closed-system condition, particularly in the case of Ar, which being a noble gas is not strongly retained with increasing temperature and so may be lost through diffusion (Clauer and Chaudhuri, 1995).

Use of the K–Ar method has found extensive application to investigations concerned with the diagenetic evolution of sedimentary basins, and in particular to the timing of formation of K-bearing phases, such as illite and, in particular, interlayered illite/smectite (I/S) (see Chapter 2 for the definition). I/S is of particular importance as it serves as an index of the thermal maturity of organic matter in sedimentary basins (Moore and Reynolds, 1997; Pollastro, 1993) and as illite formation occurs over the same temperature range as oil formation, K–Ar dating of I/S may be used as an independent geothermometer to model the timing of heating events and associated burial history of sedimentary basins (Pevear, 1999; Pollastro, 1993; Środoń, 1999, 2002). In similar applications K–Ar dating of (I/S) interlayer mixtures may provide a means of distinguishing different hydrothermal alteration stages superimposed at a site (Inoue, 1995; WoldeGabriel and Goff, 1989). For example, K–Ar dating of illitic-rich clay from the Mercur gold district, north-central Utah, indicated that Au mineralization, at 152 Ma, was not related to Tertiary igneous activity (Wilson and Parry, 1995).

The exact nature of the transformation of smectite to illite, a process also known as smectite illitization, is not universally agreed, but the transformation is most obviously driven by the combined effects of time, temperature (i.e. a source of heat) and the availability of K (Moore and Reynolds, 1997). It is generally agreed that smectite illitization is a continuous process over a broad temperature range, initiated at 60–70 °C with the onset of burial diagenesis through to completion, with the onset of the epizone at 300 °C (Środoń, 1999, 2002). As a result, K–Ar dating often derives an integral age as the thinnest

I/S particles give older K–Ar dates than thicker particles that typically yield younger K–Ar ages, implying a time-related, crystal growth effect (Clauer *et al.*, 1997).

At the temperatures associated with smectite illitization, diffusion of Ar is now not regarded a problem. Work reviewed by Clauer and Chaudhuri (1995) demonstrated that Ar is retained within smectite and illite under diagenetic and anchimetamorphic conditions, and that any loss of Ar is likely related to recrystallization processes. Clauer and Chaudhuri (1995) also discuss the likely effects on the K–Ar age determinations of various mechanical (e.g. ultrasonic dispersion) and chemical treatments (e.g. acid leaching) used to prepare clay-rich material for isotopic analysis. Of most concern are the specific nature of the illitization process and the problem of contamination from detrital K-bearing phases, which introduces bias into K–Ar age determinations (Środoń *et al.*, 2002). This has resulted in numerous workers applying theoretical unmixing approaches using linear and non-linear extrapolation of pure detrital and diagenetic endmembers to determine I/S K–Ar ages in shales (Clauer and Chaudhuri, 2001; Środoń, 2002; Środoń *et al.*, 2002). However, this approach requires accurate mineral quantification and determination of the proportion of diagenetic vs detrital illitic components (Środoń, 2002). More promising techniques may involve modelling of the illite polytype ratio (e.g. 2M1 or detrital vs 1Md or diagenetic), and the proportion of illite (detrital) vs I/S (diagenetic) using XRD peak fitting software (Środoń, 2002).

6.7 Summary

In exploration, clay minerals can work to our favour – for example, as major constituents of alteration mineral assemblages related to hydrothermal ore deposits. Conversely, clays are the host for mineralization in limited cases, e.g. Ni-laterite, bauxite and REEs (see Chapter 5). As mineral exploration begins first and foremost at the Earth's surface, clay minerals should be considered in almost all exploration campaigns. Many ore deposits are concealed by surficial sediments and for a successful discovery it is important to understand the role of clays – as host, vector mineral, contaminant or sample medium – in the respective geological setting.

Nowadays, exploration geologists have a choice between numerous technologies for their exploration programme. Which tool of the here-presented toolbox, comprising geochemistry, geophysics, spectroscopy and age dating, is chosen may be based on the background of the project team (training and experience in real-world scenarios) and the budget for the exploration programme.

Geochemical analyses can provide detailed information about the distribution of major and trace elements, which then can be used to determine the source and pathways of mineralizing fluids, the chemical reactions during the formation of the ore deposit and to identify indicator elements. Clays play an important role during the decision about the sample medium in this case, and the understanding about which elements are leached and which are retained in the system. However, a direct conversion of the geochemical data into clay mineralogy is extremely difficult without further petrographic work when considering the multiple mineral phases that are based on the same elements.

Geophysical methods are applied in mineral exploration to detect discontinuities in the subsurface (Telford *et al.*, 1990). Clay minerals can impede this approach, for example when they form conductive layers in the regolith profile and obscure the subsurface architecture. On the other hand, when the regolith is of particular interest (e.g. palaeochannel mapping using EM), mapping the subsurface distribution of clay-rich layers becomes increasingly important. Furthermore, in hydrocarbon exploration, the detection of clay minerals can assist in the estimation of porosity (e.g. gamma logs). Radiometrics can for example be applied to approximately map the distribution of potassium-bearing (i.e. illite, illite–smectite) versus potassium-void regolith, which may be important for exploration in regolith-dominated terrains.

Spectroscopic methods can directly identify clay mineral groups or even clay species using remote and/or proximal sensing applications. Although the methods have been applied for decades to characterize clays, including on other planets, this technology is still not utilized to its full potential in mineral exploration, possibly due to a less straightforward interpretation of the raw data. Spectroscopy can be used (1) for quantitative analysis of kaolinite and smectite group minerals; (2) to record the crystallinity of kaolinite group minerals; (3) to estimate the hydration state; and (4) to infer other aspects of the chemical composition of clays. These mineralogical factors can be important for (1) regolith characterization (e.g. transported vs *in situ* material); (2) mapping of hydrothermal alteration patterns (e.g. porphyry or epithermal systems); (3) mapping of diagenesis and low-grade metamorphism (e.g. sedimentary basins, hydrocarbon exploration); and (4) provenance analysis (heavy mineral deposits). It is likely that more applications will be developed in the near future.

Isotope analysis of clays can help to determine the age of clay minerals and therefore the age of the formation of a mineral deposit or the age of a palaeoweathering event. This can support stratigraphic correlations necessary for exploration in sedimentary basins. Furthermore, stable oxygen ($^{18}O/^{16}O$) and hydrogen (D/H) isotope analysis provides information about the conditions under which the clay mineral formed (e.g. temperature, influence of meteoric water). This can help to understand the palaeoclimatic conditions during the formation of the clay and therefore during the formation of a sediment-hosted deposit. Isotope data can also indicate the relative amount of hydrothermal and meteoric waters involved in the development of a hydrothermal ore deposit.

The integration of the various exploration technologies designed according to the respective mineral system may provide the key for successful exploration. Examples are the identification of mineral assemblages associated with distinct layers in the regolith stratigraphy when combining airborne EM with hyperspectral drill core logging data, and the comparison of radiometric with multi-/hyperspectral remote sensing data and surface geochemistry for regolith characterization. Kaolinite and smectite are key components in these scenarios, where the quest for concealed new ore deposits demands advanced approaches and technologies for exploration.

Acknowledgements

The authors would like to acknowledge the support of CSIRO and the various industry sponsors, CSIRO's Minerals Down Under Flagship Program and the WA Government Exploration Incentive Scheme for supporting the Western Australian Centre of Excellence for 3D Mineral Mapping.

References

Abrams, M. J., Ashley, R. P., Rowan, L. C., Goetz, A. F. H. & Kahle, A. B. 1977. Mapping of hydrothermal alteration in the Cuprite mining district, Nevada, using aircraft scanner images for the spectral region 0.46 to 2.36 μm. *Geology*, 5 (12), 713–718.

Amonette, J. E. & Zelazny, L. W. 1994. *Quantitative Methods in Soil Mineralogy*. Madison, WI: Soil Science Society of America.

Anand, R. R. & Paine, M. 2002. Regolith geology of the Yilgarn Craton, Western Australia: Implications for exploration. *Australian Journal of Earth Sciences*, 49 (1), 3–162.

Anand, R. R. & Robertson, I. D. M. 2012. The role of mineralogy and geochemistry in forming anomalies on interfaces and in areas of deep basin cover: Implications for exploration. *Geochemistry: Exploration, Environment, Analysis*, 12 (1), 45–66.

Anand, R. R. 2005. Weathering history, landscape evolution and implications for exploration. In: Anand, R. R. & De Broekert, P. (eds) *Regolith Landscape Evolution Across Australia: A Compilation of Regolith Landscape Case Studies with Regolith Landscape Evolution Models*. Perth: CRC LEME.

Bettinelli, M., Beone, G. M., Spezia, S. & Baffi, C. 2000. Determination of heavy metals in soils and sediments by microwave-assisted digestion and inductively coupled plasma optical emission spectrometry analysis. *Analytica Chimica Acta*, 424 (2), 289–296.

Bird, M. I. & Chivas, A. R. 1988. Oxygen isotope dating of the Australian regolith. *Nature*, 331 (6190), 513–516.

Bird, M. I. & Chivas, A. R. 1989. Stable-isotope geochronology of the Australian regolith. *Geochimica et Cosmochimica Acta*, 53 (12), 3239–3256.

Bird, M. I. & Chivas, A. R. 1993. Geomorphic and palaeoclimatic implications of an oxygen-isotope chronology for Australian deeply weathered profiles. *Australian Journal of Earth Sciences*, 40 (4), 345–358.

Bishop, J., Lane, M., Dyar, M. & Brown, A. 2008. Reflectance and emission spectroscopy study of four groups of phyllosilicates: Smectites, kaolinite-serpentines, chlorites and micas. *Clay Minerals*, 43 (1), 35–54.

Bohor, B. F. & Hughes, R. E. 1971. Scanning electron microscopy of clays and clay minerals. *Clays and Clay Minerals*, 19 (1), 49–54.

Borsaru, M., Rojc, A., Ronaszeki, J. & Smith, C. 2002. The determination of Fe, Si and Al content in iron ore blastholes by nuclear logging. In: Ausimm, T. (ed.) *Iron Ore Conference 2002*. Perth: The Australasian Institute of Mining and Metallurgy, 171–176.

Burger, P. A. 1996. Origins and characteristics of lateritic nickel deposits. In: Ausimm, T. (ed.) *Nickel '96, Mineral to Market, 27–29 November*. Kalgoorlie, WA: The Australasian Institute of Mining and Metallurgy, 179–183.

Burnham, G. F. & Hawke, P. J. 2003. Smart solution to a sticky problem: In-mine clay mapping using high-resolution geophysics. *The Australian Society of Exploration Geophysicists' Extended Abstracts*, 2003 (2), 1–5.

Butt, C. R. M., Lintern, M. J. & Anand, R. R. 2000. Evolution of regoliths and landscapes in deeply weathered terrain: Implications for geochemical exploration. *Ore Geology Reviews*, 16 (3–4), 167–183.

Cameron, E. M., Hamilton, S. M., Leybourne, M. I., Hall, G. E. M. & McClenaghan, M. B. 2004. Finding deeply buried deposits using geochemistry. *Geochemistry: Exploration, Environment, Analysis*, 4 (1), 7–32.

Cariati, F., Erre, L., Micera, G., Piu, P. & Gessa, C. 1983. Effects of layer charge on the near-infrared spectra of water molecules in smectites and vermiculites. *Clays and Clay Minerals*, 31 (6), 447–449.

Chabrillat, S., Goetz, A. F. H., Krosley, L. & Olsen, H. W. 2002. Use of hyperspectral images in the identification and mapping of expansive clay soils and the role of spatial resolution. *Remote Sensing of Environment*, 82 (2), 431–445.

Chang, Z. & Yang, Z. 2012. Evaluation of inter-instrument variations among short wavelength infrared (SWIR) devices *Economic Geology*, 107 (7), 1479–1488.

Chivas, A. R. 1993. Palaeoclimate from Gondwanaland clays. In: Churchman, G. J., Fitzpatrick, R. W. & Eggleton, R. A. (eds) *10th International Clay Conference: Clays Controlling the Environment, July 18–23, 1993*. Adelaide: CSIRO Publishing, 333–338.

Churchman, G. J. & Gilkes, R. J. 1989. Recognition of intermediates in the possible transformation of halloysite to kaolinite in weathering profiles. *Clay Minerals*, 24 (4), 579–590.

Clark, R. N. 1999. Spectroscopy of rocks and minerals, and principles of spectroscopy. In: Rencz, A. (ed.) *Manual of Remote Sensing*. New York: Wiley & Sons.

Clauer, N. & Chaudhuri, S. 1995. *Clays in Crustal Environments: Isotope Dating and Tracing*. Berlin: Springer-Verlag.

Clauer, N. & Chaudhuri, S. 2001. Extracting K–Ar ages from shales: The analytical evidence. *Clay Minerals*, 36 (2), 227–235.

Clauer, N., Muller, J. P. & O'Neil, J. R. 1989. Oxygen isotope signature of several generations of kaolinite in a laterite: Geochemical implications. In: Farmer, V. C. & Tardy, Y. (eds) *9th International Clay Conference: Clay–Organic Interactions Clay Minerals In Soils, August 28–September 2, 1989*. Strasbourg: Université Louis Pasteur, Centre National de la Recherche Scientifique, 1.

Clauer, N., Środoń, J., Francu, J. & Šucha, V. 1997. K–Ar dating of illite fundamental particles separated from illite–smectite. *Clay Minerals*, 32 (2), 181–196.

Clayton, T. & Pearce, R. B. 2007. Rapid chemical analysis of the <2 μm clay fraction using an SEM/EDS technique. *Clay Minerals*, 42 (4), 549–562.

Cocks, T., Jenssen, R., Stewart, A., Wilson, I. & Shields, T. 1998. The Hymap Airborne Hyperspectral Sensor: The system, calibration and performance. In: *1st EARSEL Workshop on Imaging Spectroscopy, October 1998*. Zurich, Switzerland: EARSEL, 7.

Cohen, D. R., Shen, X. C., Dunlop, A. C. & Rutherford, N. F. 1998. A comparison of selective extraction soil geochemistry and biogeochemistry in the Cobar area, New South Wales. *Journal of Geochemical Exploration*, 61 (1–3), 173–189.

Cohen, D. R., Kelley, D. L., Anand, R. & Coker, W. B. 2010. Major advances in exploration geochemistry, 1998–2007. *Geochemistry: Exploration, Environment, Analysis*, 10 (1), 3–16.

Cook, S. E., Corner, R. J., Groves, P. R. & Grealish, G. J. 1996. Use of airborne gamma radiometric data for soil mapping. *Soil Research*, 34 (1), 183–194.

Cooke, D. R., Hollings, P. & Walshe, J. L. 2005. Giant porphyry deposits: Characteristics, distribution, and tectonic controls. *Economic Geology*, 100 (5), 801–818.

Coulter, D., Hauff, P. L. & Kerby, W. L. 2007. Airborne hyperspectral remote sensing. In: Milkereit, B. (ed.) *Exploration 07: Fifth Decennial International Conference in Mineral Exploration, 9–12 September, 2007*. Toronto, ON: Decennial Mineral Exploration Conferences, 357–378.

Craig, M. A. & Anand, R. R. 1993. *Kalgoorlie-Kurnalpi Regolith-Landform Map, 1:250,000 Special edition*. Adelaide: Australian Geological Survey Organisation.

Crosta, A. P., Sabine, C. & Taranik, J. V. 1998. Hydrothermal alteration mapping at Bodie, California, using AVIRIS hyperspectral data. *Remote Sensing of Environment*, 65 (3), 309–319.

Crowley, J. K. & Hook, S. J. 1996. Mapping playa evaporite minerals and associated sediments in Death Valley, California, with multispectral thermal infrared images. *Journal of Geophysical Research: Solid Earth*, 101 (B1), 643–660.

Crowley, J. K. & Vergo, N. 1988. Near-infrared reflectance spectra of mixtures of kaolin group minerals: Use in clay studies. *Clays and Clay Minerals*, 36 (4), 310–316.

Crowley, J. K., Brickey, D. W. & Rowan, L. C. 1989. Airborne imaging spectrometer data of the Ruby Mountains, Montana: Mineral discrimination using relative absorption band-depth images. *Remote Sensing of Environment*, 29 (2), 121–134.

Cudahy, T. 1997. *PIMA-II Spectral Characteristics of Natural Kaolins – AMIRA Project P435*. North Ryde, NSW: CSIRO.

Cudahy, T. J., Caccetta, M., Cornelius, A., *et al.* 2005. *Regolith, Geology and Alteration Mineral Maps from New Generation Airborne and Satellite Remote Sensing Technologies*. Perth: CSIRO Minerals Exploration.

Cudahy, T., Jones, M., Thomas, M., *et al.* 2008. *Next Generation Mineral Mapping: Queensland Airborne HyMap and Satellite ASTER Surveys 2006–2008*. North Ryde, NSW: CSIRO.

Cudahy, T., Hewson, R., Caccetta, M., *et al.* 2009. Drill core logging of plagioclase feldspar composition and other minerals associated with Archean gold mineralization at Kambalda, Western Australia, using a bidirectional thermal infrared reflectance system. In: Bedell, R., Crósta, A. P. & Grunsky, E. (eds) *Reviews in Economic Geology*. Littleton, CO: Society of Economic Geologists, Inc.

Cudahy, T., Caccetta, M., Hewson, R., *et al.* 2012. Satellite ASTER Geoscience Map of Australia. v1. CSIRO.

Cudahy, T., Caccetta, M., Thomas, M., *et al.* (2016): Satellite-derived mineral mapping and monitoring of weathering, deposition and erosion. *Nature Scientific Reports*. DOI: 10.1038/srep23702

Davis, J. L. & Annan, A. P. 1989. Ground-penetrating radar for high-resolution mapping of soils and rock stratigraphy *Geophysical Prospecting*, 37 (5), 531–551.

Delineau, T., Allard, T., Muller, J. P., *et al.* 1994. FTIR reflectance vs. EPR studies of structural iron in kaolinites. *Clays & Clay Minerals*, 42 (3), 308–320.

Dickson, B. L. & Scott, K. M. 1997. Interpretation of aerial gamma-ray surveys: Adding the geochemical factors. *AGSO Journal of Australian Geology and Geophysics*, 17 (2), 187–200.

Ducart, D. F., Crosta, A. P., Souza Filho, C. R. & Coniglio, J. 2006. Alteration mineralogy at the Cerro la Mina epithermal prospect, Patagonia, Argentina: Field mapping, short-wave infrared spectroscopy, and ASTER images. *Economic Geology*, 101 (5), 981–996.

Elias, M., Donaldson, M. J. & Giorgetta, N. 1981. Geology, mineralogy, and chemistry of lateritic nickel–cobalt deposits near Kalgoorlie, Western Australia. *Economic Geology*, 76, 1775–1783.

Emerson, D. W. & Yang, Y. P. 1997. Effects of water salinity and saturation on the electrical resistivity of clays. In: *1997 Conference Handbook: ASEG 12th International Conference and Exhibition*. Sydney: Australian Society of Exploration Geophysicist, 19–24.

Ferbey, T. 2010. *Geochemistry of the Nadina River Map Area (093E/15), West-Central British Columbia*. Vancouver: BC Ministry of Energy Mines and Petroleum Resources.

Fisher, L., Gazley, M. F., Baensch, A., *et al.* 2014. Resolution of geochemical and lithostatigraphic complexity: A workflow for application of portable X-ray fluorescence to mineral exploration. *Geochemistry: Exploration, Environment, Analysis*, 14 (2), 149–159.

Flehmig, W. & Kurze, R. 1973. Die quantitative infrarotspektroskopische Phasenanalyse von Mineralmengen. *Neues Jahrbuch Mineralogischer Abhandlungen*, 119 (1), 101–102.

Frost, R. & Johansson, U. 1998. Combination bands in the infrared spectroscopy of kaolins: A drift spectroscopic study. *Clays and Clay Minerals*, 46 (4), 466–477.

Gaudin, A., Deacrreau, A., Noack, Y. & Grauby, O. 2005. Clay mineralogy of the nickel laterite ore developed from serpentinised peridotites at Murrin Murrin, Western Australia. *Australian Journal of Earth Sciences*, 52 (2), 231–241.

Gersman, R., Ben-Dor, E., Beyth, M., *et al.* 2008. Mapping of hydrothermally altered rocks by the EO-1 Hyperion sensor, Northern Danakil Depression, Eritrea. *International Journal of Remote Sensing*, 29 (13), 3911–3936.

Giral, S., Savin, S. M., Girard, J.-P. & Nahon, D. B. 1993a. The oxygen isotope geochemistry of kaolinites from lateritic profiles: Implications for pedology and paleoclimatology. *Chemical Geology*, 197 (3–4), 237–240.

Giral, S., Savin, S. M., Girard, J.-P. & Nahon, D. B. 1993b. Oxygen isotope geochemistry of kaolinites from two lateritic profiles: Implications for pedology and paleoclimatology. In: Churchman, G. J., Fitzpatrick, R. W. & Eggleton, R. A. (eds) *10th International Clay Conference: Clays controlling the environment, 18–23 July, 1993*. Adelaide: CSIRO Publishing, 65.

Gray, D. J., Wildman, J. E. & Longman, G. D. 1999. Selective and partial extraction analyses of transported overburden for gold exploration in the Yilgarn Craton, Western Australia. *Journal of Geochemical Exploration*, 67 (1–3), 51–66.

Grunsky, E. C. 2010. The interpretation of geochemical survey data. *Geochemistry: Exploration, Environment, Analysis*, 10 (1), 27–74.

Guanter, L., Kaufmann, H., Segl, K., *et al.* 2015. The environmental mapping and analysis program (ENMAP) spaceborne imaging spectroscopy mission for earth observation. *Remote Sensing*, 7 (7), 8830–8857.

Haest, M., Cudahy, T., Laukamp, C. & Gregory, S. 2012a. Quantitative mineralogy from infrared spectroscopic data: I. Validation of mineral abundance and composition scripts at the Rocklea channel iron deposit in Western Australia. *Economic Geology*, 107 (2), 209–228.

Haest, M., Cudahy, T., Laukamp, C. & Gregory, S. 2012b. Quantitative mineralogy from infrared spectroscopic data: II. Three-dimensional mineralogical characterization of the Rocklea channel iron deposit, Western Australia. *Economic Geology*, 107 (2), 229–249.

Haest, M., Cudahy, T., Rodger, A., *et al.* 2013. Unmixing the effects of vegetation in airborne hyperspectral mineral maps over the Rocklea Dome iron-rich palaeochannel system (Western Australia). *Remote Sensing of Environment*, 129, 17–31.

Hamilton, S. M. 1998. Electrochemical mass-transport in overburden: A new model to account for the formation of selective leach geochemical anomalies in glacial terrain. *Journal of Geochemical Exploration*, 63 (3), 155–172.

Hamilton, V. E. & Christensen, P. R. 2000. Determining the modal mineralogy of mafic and ultramafic igneous rocks using thermal emission spectroscopy. *Journal of Geophysical Research*, 105 (E4), 9717–9733.

Hedenquist, J. W., Arribas, A. & Reynolds, T. J. 1998. Evolution of an intrusion-centered hydrothermal system: Far Southeast-Lepanto porphyry and epithermal Cu–Au deposits, Philippines. *Economic Geology*, 93 (4), 373–404.

Hesselbo, S. P. 1996. Spectral gamma-ray logs in relation to clay mineralogy and sequence stratigraphy, Cenozoic of the Atlantic margin, offshore New Jersey. In: Mountain, G. S., Miller, K. G., Blum, P., Poag, C. W. & Twichell, D. C. (eds) *Proceedings of the Ocean Drilling Program, Scientific Results, 150: New Jersey Continental Slope and Rise*. College Station, TX: Ocean Drilling Program, Texas A&M University, The National Science Foundation, Joint Oceanographic Institutions, Inc., 411–422.

Hewson, R. & Cudahy, T. 2011. Issues affecting geological mapping with ASTER data: A case study of the Mt Fitton Area, South Australia. In: Ramachandran, B., Justice, C. O. & Abrams, M. J. (eds) *Land Remote Sensing and Global Environmental Change*. New York: Springer.

Hewson, R. D., Cudahy, T. J., Mizuhiko, S., Ueda, K. & Mauger, A. J. 2005. Seamless geological map generation using ASTER in the Broken Hill-Curnamona province of Australia. *Remote Sensing of Environment*, 99 (1–2), 159–172.

Hunt, G. R. & Salisbury, J. W. 1970. Visible and infrared spectra of minerals and rocks: I. Silicate minerals. *Modern Geology*, 1, 283–300.

IAEA. 2003. Guidelines for radioelement mapping using gamma ray spectrometry data. Technical Document No. 1363.

Inoue, A. 1995. Formation of clay minerals in hydrothermal environments. In: Velde, B. (ed.) *Origin and Mineralogy of Clays*. Berlin: Springer.

King, P. L., Ramsey, M. S., McMillan, P. F. & Swayze, G. 2004. Laboratory Fourier transform infrared spectroscopy methods for geologic samples. In: King, P. L., Ramsey, M. S. & Swayze, G. A. (eds) *Infrared Spectroscopy in Geochemistry, Exploration Geochemistry, and Remote Sensing*. Québec: Mineralogical Association of Canada.

Kogel, J. E. & Lewis, S. A. 2001. Baseline studies of the clay minerals society source clays: Chemical analysis by inductively coupled plasma-mass spectroscopy (ICP-MS). *Clays and Clay Minerals*, 49 (5), 387–392.

Kruse, F. A. 1988. Use of airborne imaging spectrometer data to map minerals associated with hydrothermally altered rocks in the northern grapevine mountains, Nevada, and California. *Remote Sensing of Environment*, 24 (1), 31–51.

Kruse, F. A., Lefkoff, A. B., Boardman, J. W., *et al.* 1993. The spectral image processing system (SIPS): Interactive visualization and analysis of imaging spectrometer data. *Remote Sensing of Environment*, 44 (2–3), 145–163.

Kruse, F. A., Boardman, J. W. & Huntington, J. F. 2002. Comparison of EO-1 Hyperion and airborne hyperspectral remote sensing data for geologic applications. In: Woerner, D. F. (ed.) *Aerospace Conference Proceedings*. Big Sky, MO: Institute of Electrical and Electronics Engineers, Inc., 1501–1513.

Laukamp, C., Cudahy, T., Cleverley, J. S., Oliver, N. H. S. & Hewson, R. 2011a. Airborne hyperspectral imaging of hydrothermal alteration zones in granitoids of the Eastern Fold Belt, Mount Isa Inlier, Australia. *Geochemistry: Exploration, Environment, Analysis*, 11 (1), 3–24.

Laukamp, C., Cudahy, T., Thomas, M., *et al.* 2011b. Hydrothermal mineral alteration patterns in the Mount Isa Inlier revealed by airborne hyperspectral data. *Australian Journal of Earth Sciences*, 58 (8), 917–936.

Laukamp, C., Termin, K. A., Pejcic, B., Haest, M. & Cudahy, T. 2012. Vibrational spectroscopy of calcic amphiboles: Applications for exploration and mining. *European Journal of Mineralogy*, 24 (5), 863–878.

Laukamp, C., Salama, W. & González-Álvarez, I. 2015. Proximal and remote spectroscopic characterisation of regolith in the Albany-Fraser Orogen (Western Australia). *Ore Geology Reviews*, 73 (3), 540–554.

Lawrence, J. R. & Meaux, J. R. 1993. The stable isotopic composition of ancient kaolinites of North America. In: Swart, P. K., Lohmann, K. C., Mckenzie, J. & Savin, S. (eds) *Climate Change in Continental Isotopic Records*. Washington, DC: Wiley & Sons.

Lee, C. M., Cable, M. L., Hook, S. J., *et al.* 2015. An introduction to the NASA Hyperspectral InfraRed Imager (HyspIRI) mission and preparatory activities. *Remote Sensing of Environment*, 167, 6–19.

Loizeau, D., Mangold, N., Poulet, F., *et al.* 2007. Phyllosilicates in the Mawrth Vallis region of Mars. *Journal of Geophysical Research: Planets*, 112 (E08S08), 2156–2202.

Macnae, J. C. 2007. Developments in broadband airborne electromagnetics in the past decade. In: Milkereit, B. (ed.) *Exploration 07: Fifth Decennial International Conference on Mineral Exploration*. Toronto: Prospectors and Developers Association of Canada, 387–398.

Madejova, J. & Komadel, P. 2001. Baseline studies of the Clay Minerals Society source clays: Infrared methods. *Clays and Clay Minerals*, 49 (5), 410–432.

Mann, A. W., Birrell, R. D., Fedikow, M. A. F. & de Souza, H. A. F. 2005. Vertical ionic migration: Mechanisms, soil anomalies, and sampling depth for mineral exploration. *Geochemistry: Exploration, Environment, Analysis*, 5 (3), 201–210.

Mars, J. C. & Rowan, L. C. 2010. Spectral assessment of new ASTER SWIR surface reflectance data products for spectroscopic mapping of rocks and minerals. *Remote Sensing of Environment*, 114 (9), 2011–2025.

Martinez-Alonso, S., Rustad, J. & Goetz, A. 2002. Ab initio quantum mechanical modeling of infrared vibrational frequencies of the OH group in dioctahedral phyllosilicates. Part I: Methods, results and comparison to experimental data. *American Mineralogist*, 87 (8–9), 1215–1223.

McCann, D. M. & Forster, A. 1990. Reconnaissance geophysical methods in landslide investigations. *Engineering Geology*, 29 (1), 59–78.

McNeill, J. D. 1980. *Electromagnetic Terrain Conductivity Measurement at Low Induction Numbers*. Mississauga, ON: Geonics Limited.

Mermut, A. R. & Canot, A. F. 2001. Baseline studies of the Clay Minerals Society source clays: Chemical analyses of major elements. *Clays and Clay Minerals*, 49 (5), 381–386.

Meunier, A. 2005. *Clays*. Berlin: Springer.

Mokhtari, A. R., Cohen, D. R. & Gatehouse, S. G. 2009. Geochemical effects of deeply buried Cu–Au mineralization on transported regolith in an arid terrain. *Geochemistry: Exploration, Environment, Analysis*, 9 (3), 227–236.

Moore, D. M. & Reynolds, R. C. J. 1997. *X-ray Diffraction and the Identification and Analysis of Clay Minerals*. Oxford: Oxford University Press.

Munday, T. 2008. Regolith geophysics. In: Scott, K. & Pain, C. (eds) *Regolith Science*.Berlin: Springer.

Munday, T. J., Macnae, J., Bishop, J. & Sattel, D. 2001. A geological interpretation of observed electrical structures in the regolith: Lawlers, Western Australia. *Exploration Geophysics*, 32 (1), 36–47.

Ninomiya, Y., Fu, B. & Cudahy, T. J. 2005. Detecting lithology with Advanced Spaceborne Thermal Emission and Reflection Radiometer (ASTER) multispectral thermal infrared 'radiance-at-sensor' data. *Remote Sensing of Environment*, 99 (1–2), 127–139.

Noble, R. R. P. & Stanley, C. R. 2009. Traditional and novel geochemical extractions applied to a Cu–Zn soil anomaly: A quantitative comparison of exploration accuracy and precision. *Geochemistry: Exploration, Environment, Analysis*, 9 (2), 159–172.

Paquet, H., Colin, F., Duplay, J., Nahon, K. & Millot, G. 1987. Ni, Mn, Zn, Cr-smectites, early and effective traps for transition elements in supergene ore deposits. In: Rodrigez-Clemente, R. & Tardy, Y. (eds) *Geochemistry and Mineral Formation in the Earth Surface*. Madrid: Centre National de la Récherche Scientifique.

Patterson, J. H. 1993. A review of the effects of minerals in processing of Australian oil shales. *Fuel*, 73 (3), 321–327.

Patterson, J. H. & Henstridge, D. A. 1990. Comparison of the mineralogy and geochemistry of the Kerosene Creek Member, Rundle and Stuart oil shale deposits, Queensland, Australia. *Chemical Geology*, 82, 319–339.

Perry, S. L. 2004. Review of airborne and orbiting remote sensing systems applied to worldwide mineral exploration: Case histories focussed on infrared spectroscopy. In: King, P. L., Ramsey, M., Swayze, G. A. (ed.) *Infrared Spectroscopy in Geochemistry, Exploration Geochemistry, and Remote Sensing*. London, ON: Mineralogical Association of Canada.

Pevear, D. R. 1999. Illite and hydrocarbon exploration. *Proceedings of the National Academy of Science USA*, 96, 3440–3446.

Pillans, B. 1998. *Regolith Dating Methods: A Guide to Numerical Dating Techniques*. Millaa Millaa: Cooperative Research Centre for Landscape Evolution and Mineral Exploration (CRCLEME).

Pillans, B. & Bourman, R. 1996. The Brunhes/Matuyama polarity transition (0.78 Ma) as a chronostratigraphic marker in Australian regolith studies. *AGSO Journal of Australian Geology and Geophysics*, 16 (3), 289–294.

Pollastro, R. M. 1993. Considerations and applications of the illite/smectite geothermometer in hydrocarbon-bearing rocks of Miocene to Mississippian age. *Clays and Clay Minerals*, 41 (2), 119.

Post, J. L. & Noble, P. N. 1993. The near-infrared combination band frequencies of dioctahedral smectites, micas, and illites. *Clays and Clay Minerals*, 41 (6), 639–644.

Poulet, F., Bibring, J. P., Mustard, J. F., *et al.* 2005. Phyllosilicates on Mars and implications for early Martian climate. *Nature*, 438 (7068), 623–627.

Richter, N., Staenz, K. & Kaufmann, H. 2008. Spectral unmixing of airborne hyperspectral data for baseline mapping of mine tailings areas. *International Journal of Remote Sensing*, 29 (13), 3937–3956.

Ricordel-Prognon, C., Lagroix, F., Moreau, M.-G. & Thiry, M. 2010. Lateritic paleoweathering profiles in French Massif Central: Paleomagnetic datings. *Journal of Geophysical Research*, 115 (B10104), 19.

Robb, L. 2005. *Introduction to Ore-Forming Processes*. Oxford: Blackwell Publishing.

Roberts, D. A., Smith, M. O. & Adams, J. B. 1993. Green vegetation, nonphotosynthetic vegetation, and soils in AVIRIS data. *Remote Sensing of Environment*, 44 (2–3), 255–269.

Robertson, I. D. M. 2001. Geochemical exploration around the Harmony gold deposit, Peak Hill, Western Australia. *Geochemistry: Exploration, Environment, Analysis*, 1 (3), 277–288.

Rodger, A. & Cudahy, T. 2009. Vegetation corrected continuum depths at 2.20 μm: An approach for hyperspectral sensors. *Remote Sensing of Environment*, 113 (10), 2243–2257.

Rodriguez, J. M., Ustin, S. L. & Riaño, D. 2011. Contributions of imaging spectroscopy to improve estimates of evapotranspiration. *Hydrological Processes*, 25 (26), 4069–4081.

Rowan, L. C. & Mars, J. C. 2003. Lithologic mapping in the Mountain Pass, California area using Advanced Spaceborne Thermal Emission and Reflection Radiometer (ASTER) data. *Remote Sensing of Environment*, 84 (3), 350–366.

Rowan, L. C., Hook, S. J., Abrams, M. J. & Mars, J. C. 2003. Mapping hydrothermally altered rocks at Cuprite, Nevada, using the Advanced Spaceborne Thermal Emission and Reflection Radiometer (ASTER), a new satellite-imaging system. *Economic Geology*, 98 (5), 1019–1027.

Savin, S. M. & Hsieh, J. C. C. 1998. The hydrogen and oxygen isotope geochemistry of pedogenic clay minerals: Principals and theoretical background. *Geoderma*, 82 (1–3), 227–253.

Schlumberger. 1989. *Log Interpretation Principles/Applications*. Sugarland, TX: Schlumberger.

Schmidt, P. W. & Ollier, C. D. 1988. Palaeomagnetic dating of late Cretaceous to early Tertiary weathering in New England, NSW, Australia. *Earth-Science Reviews*, 25 (5–6), 363–371.

Schmidt, P. W., Prasad, V. & Ramam, P. K. 1983. Magnetic ages of some Indian laterites. *Palaeogeography, Palaeoclimatology, Palaeoecology*, 44 (3–4), 185–202.

Scott, K. 2008. Appendix 2: Regolith geochemistry of elements. In: Scott, K. & Pain, C. (eds) *Regolith Science*. Berlin: Springer.

Short, N. M. 1982. *The Landsat Tutorial Workbook: Basics of Satellite Remote Sensing*. Washington, DC: NASA.

Simmons, S. F., White, N. C. & John, D. A. 2005. Geological characteristics of epithermal precious and base metal deposits. *Economic Geology*, 100th Anniversary (3), 485–522.

Sonibare, O. O., Jacob, D. E., Ward, C. R. & Foley, S. F. 2011. Mineral and trace element composition of the Lokpanta oil shales in the Lower Benue Trough, Nigeria. *Fuel*, 90, 2843–2849.

Sonntag, I., Laukamp, C. & Hagemann, S. G. 2012. Low potassium hydrothermal alteration in low sulfidation epithermal systems as detected by IRS and XRD: An example from the Co–O mine, Eastern Mindanao, Philippines. *Ore Geology Reviews*, 45, 47–60.

Środoń, J. 1999. Nature of mixed-layer clays and mechanisms of their formation and alteration. *Annual Reviews of Earth Planetary Science*, 27, 19–53.

Środoń, J. 2002. Quantitative mineralogy of sedimentary rocks with emphasis on clays and with applications to K–Ar dating. *Mineralogical Magazine*, 66 (5), 677–687.

Środoń, J., Clauer, N. & Eberl, D. D. D. 2002. Interpretation of K–Ar dates of illitic clays from sedimentary rocks aided by modeling. *American Mineralogist*, 87 (11–12), 1528–1535.

Stern, L. A., Chamberlain, C. P., Reynolds, R. C. & Johnson, G. D. 1997. Oxygen isotope evidence of climate change from pedogenic clay minerals in the Himalayan molasse. *Geochimica et Cosmochimica Acta*, 61 (4), 731–744.

Telford, W. M., Geldhart, L. P. & Sheriff, R. E. 1990. *Applied Geophysics*. Cambridge: Cambridge University Press.

Thompson, A. J. B., Hauff, P. L. & Robitaille, A. J. 1999. Alteration mapping in exploration: Application of Short-Wave Infrared (SWIR) Spectroscopy. *SEG Newsletter*, 39 (1), 16–27.

Thomson, S., Fountain, D. & Watts, T. 2007. Airborne geophysics: Evolution and revolution. In: Milkereit, B. (ed.) *Fifth Decennial International Conference on Mineral Exploration*. Toronto: Prospectors and Developers Association of Canada, 19–37.

Tosdal, R. M., Dilles, J. H. & Cooke, D. R. 2009. From source to sinks in auriferous magmatic-hydrothermal porphyry and epithermal deposits. *Elements*, 5 (5), 289–295.
van der Meer, F. D., van der Werff, H. M. A., van Ruitenbeek, F. J. A., *et al.* 2012. Multi- and hyperspectral geologic remote sensing: A review. *International Journal of Applied Earth Observation and Geoinformation*, 13 (1), 112–128.
Vane, G., Green, R. O., Chrien, T. G., *et al.* 1993. The airborne visible/infrared imaging spectrometer (AVIRIS). *Remote Sensing of Environment*, 44 (2–3), 127–143.
Vaughan, R. G., Calvin, W. M. & Taranik, J. V. 2003. SEBASS hyperspectral thermal infrared data: Surface emissivity measurement and mineral mapping. *Remote Sensing of Environment*, 85 (1), 48–63.
Velde, B. B. & Meunier, A. 2008. *The Origin of Clay Minerals in Soils and Weathered Rocks*. Berlin: Springer.
Wells, M. & Ramanaidou, E. 2007. Clay quantification of channel iron deposits (CID), Pilbara, Western Australia. In: Ausimm, T. (ed.) *Iron Ore Conference 2007*. Perth: The Australasian Institute of Mining and Metallurgy, 203–207.
White, A. J. R., Robb, V. M., Robb, L. J. & Waters, D. J. 2010. Portable infrared spectroscopy as a tool for the exploration of gold deposits in tropical terrains: A case study at the Damang Deposit, Ghana. *Economic Geology*, 15, 67–84.
Wilford, J. R., Bierwirth, P. N. & Craig, M. A. 1997. Application of airborne gamma-ray spectrometry in soil/regolith mapping and applied geomorphology. *AGSO Journal of Australian Geology and Geophysics*, 17 (2), 201–216.
Wilson, P. N. & Parry, W. T. 1995. Characterization and dating of argillic alteration in the Mercur gold district, Utah. *Economic Geology*, 90 (5), 1197–1216.
Witt, W. K., Hagemann, S. G., Ojala, J., *et al.* 2013. Multiple methods for regional- to mine-scale targeting, Pataz gold field, northern Peru. *Australian Journal of Earth Sciences*, 61 (1), 43–58.
WoldeGabriel, G. & Goff, F. 1989. Temporal relations of volcanism and hydrothermal systems in two areas of the Jemez volcanic field, New Mexico. *Geology*, 17 (11), 986–989.
Woodcock, R., Simons, B., Duclaux, G. & Cox, S. 2010. AuScope's use of standards to deliver earth resource data In: Malamud, B. D. (ed.) *European Geosciences Union General Assembly 2010, 2–7 May, 2010*. Vienna: Copernicus, 15–56.
Yamaguchi, Y., Fujisada, H., Kudoh, M., *et al.* 1999. ASTER instrument characterization and operation scenario. *Advances in Space Research*, 23 (8), 1415–1424.
Yang, K., Huntington, J. F., Boardman, J. W. & Mason, P. 1999. Mapping hydrothermal alteration in the Comstock mining district, Nevada, using simulated satellite-borne hyperspectral data. *Australian Journal of Earth Sciences*, 46 (6), 915–922.
Yang, K., Lian, C., Huntington, J. F., Peng, Q. & Wang, Q. 2005. Infrared spectral reflectance characterization of the hydrothermal alteration at the Tuwu Cu–Au deposit, Xinjiang, China. *Mineralium Deposita*, 40 (3), 324–336.
Yang, K., Huntington, J. F., Gemmell, J. B. & Scott, K. M. 2011. Variations in composition and abundance of white mica in the hydrothermal alteration system at Hellyer, Tasmania, as revealed by infrared reflectance spectroscopy. *Journal of Geochemical Exploration*, 108 (2), 143–156.
Zhang, G., Wasyliuk, K. & Pan, Y. 2001. The characterization and quantitative analysis of clay minerals in the Athabasca Basin, Saskatchewan: Applications of shortwave infrared reflectance. *The Canadian Mineralogist*, 39 (5), 1347–1363.

7

Clay Minerals in Flotation and Comminution Operations

Elizaveta Forbes, Mark Ma and Warren Bruckard

7.1 Introduction

Mineral comminution and froth flotation are key processes in mineral beneficiation and often occur side-by-side in operations or production plants, referred to as concentrators. Comminution can be described simply as the physical breakage of ore via crushing and grinding. Comminution is necessary prior to flotation because most minerals are finely disseminated and intimately associated with the gangue, and so they must be liberated before separation using froth flotation can occur. During comminution the particle size is progressively reduced by crushing and grinding until the valuable particles are sufficiently liberated for effective flotation and are at a size where flotation will be efficient – typically this means in the size range 10–200 microns. For effective, cost-efficient comminution, crushing and grinding are normally carried out in stages with particle classification (sorting or sizing) between the stages. Circuits are typically closed, whereby oversize material from the classifier is directed back to the primary comminution unit and undersize product flows forward to the next unit process. This ensures minimal over-crushing or over-grinding and allows each comminution device to operate efficiently with a regulated and more closely sized feed. Crushing is the primary comminution step and is normally a dry process, where the classification devices are typically single- or multi-decked vibrating screens. Grinding occurs after crushing and is more often a wet process providing a slurry feed for flotation. In grinding circuits the classification devices are more typically hydrocyclones.

Flotation, which follows grinding, is a versatile, well-established and widely used separation unit process designed to treat complex sulphide, oxide and non-metallic ores. Separation is achieved when differences in the physio-chemical surface properties of the various minerals in the flotation feed are exploited with the aid of added chemicals or reagents, which help to impart hydrophobicity on the particles (normally the valuable mineral particles) to be floated. When air is bubbled into the ground-conditioned slurry in the flotation cells the hydrophobic particles attach to the rising air bubbles and float to the top of the pulp where the so-called froth can be removed by skimming the surface of the slurry. This fraction is termed the flotation concentrate while, the unfloated portion is called the flotation tailing. Froth flotation is a continuous process and like comminution also occurs in stages since high grade and high recovery cannot be achieved in one operation. Cleaning

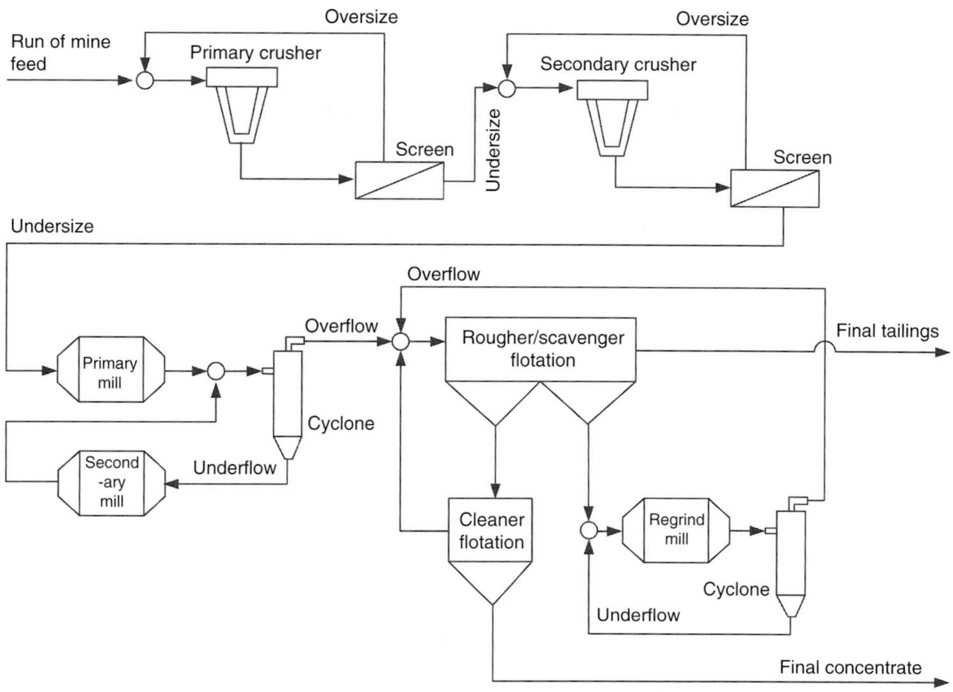

Figure 7.1. Example of a consolidated crushing–grinding–flotation circuit encompassing two-stage closed-circuit dry crushing, two-stage closed-circuit wet grinding, and rougher–scavenger–cleaner flotation, incorporating regrinding.

stages are needed to reject free gangue that has been entrained in the primary or rougher concentrates and scavenger stages are typically required to deal with middling particles, which may be low-grade composites, coarse or difficult to float particles. These particles generally need to be reground and returned to the flotation circuit. Flotation circuits are varied and may be simple or complex, often containing a regrind circuit, but all are ore-specific and their design and operation can only be established after careful laboratory and pilot-scale comminution and flotation testwork. An example of a consolidated crushing–grinding–flotation circuit is shown in Fig. 7.1 – here, two-stage closed-circuit crushing is followed by two-stage closed-circuit grinding to produce feed material for a rougher–scavenger–cleaner flotation circuit incorporating a regrind step.

Generally there are only two outlets in a flotation circuit: a flotation concentrate and tailings. The enriched concentrate is normally further processed by smelting and refining. Concentrators receive their feed material directly from the mine site, so unlike downstream processes they cannot impose a penalty on high proportions of deleterious mineral content within the feed. They simply must deal with this through appropriate circuit design and operation. However, since they produce a product which is further treated downstream, mineral

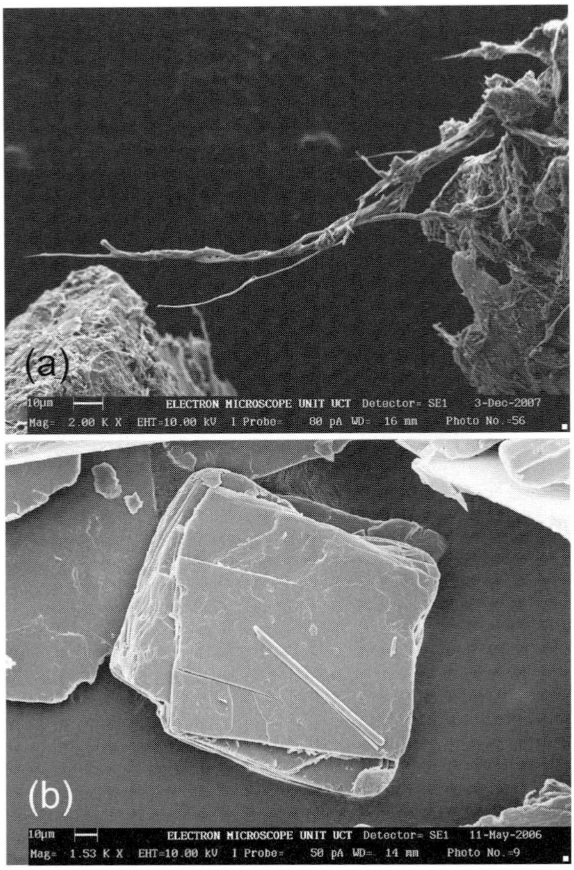

Figure 7.2. SEM images of phyllosilicate minerals: (a) chrysotile and (b) talc in industrial ores.

concentrators themselves are often subject to stringent penalties if their product, the flotation concentrate, contains high quantities of unwanted or undesirable elements or minerals.

One such type of unwanted mineral is clays. Not only are high proportions of clay minerals in concentrates penalized by smelters, but the presence of these minerals in flotation and comminution pulps causes a wide array of problems for the operations. This chapter outlines the types of problems that clay minerals can cause both in flotation and comminution circuits, as well as potential treatments that can alleviate these problems. As clay minerals are also important commodities in their own right, it also provides a brief description of the application of flotation to upgrade clay minerals.

7.2 Clay Structure and Classification

The mineral structure of clays, as well as the classification of different clay types, has been described in detail in Chapters 1, 2 and 4. Clay minerals fall within the phyllosilicate mineral family. Phyllosilicate minerals are so named because they typically display a platy/leafy habit, *fyllo* meaning leaf in Greek. The habit exhibited is simply a reflection of the structure. There are a large number of different types of phyllosilicates in existence, with a variety of different properties. With specific minerals like chrysotile, palygorskite and sepiolite, phyllosilicate particles can also have a fibrous nature, as illustrated in Fig. 7.2 (see Section 4.2.1). The faces and edges of phyllosilicates carry different electrical charges, making these particles electrostatically anisotropic. The faces (or basal planes) tend to be negatively charged, while the edges carry a charge that alters from positive to negative as a function of pH (Van Olphen, 1951). This is of particular relevance to both rheology and coagulation/flocculation behaviour; the charge anisotropy is detailed in Chapters 2 and 8.

Many classification systems include a specific clay subgroup, which includes minerals such as kaolinite, smectite and illite (Deer *et al.*, 1978). However, it would be erroneous to assume that the deleterious effects often encountered in industrial practice and attributed to clays are limited to the minerals within this subgroup. In fact, in mineral processing literature presented in this review, minerals such as talc and serpentine often get clustered together with kaolinite and montmorillonite, and generically referred to as clay minerals. The disparity in the terminology used to describe deleterious minerals often leads to confusion with regards to what constitutes a clay mineral. In terms of rheology, as noted previously in Chapter 2, the clay mineral that exhibits the most extreme behaviour (chrysotile) belongs to the serpentine group, but in this chapter it will be treated as a clay.

7.3 Effect of Clay Minerals on Suspension Rheology

To discuss the effect of clay minerals on mineral beneficiation in water-based slurries it is necessary to understand the effect that these minerals have on the rheology (study of flow and deformation of matter) of mineral suspensions. Slurry rheological properties affect a host of different mineral processing operations, including comminution and flotation (Boger, 2000). In the context of beneficiation this will be considered here in terms of a number of applied studies, but for a detailed examination of the fundamental aspects of rheological behaviour, please refer to Chapter 2.

The rheological properties of fluids are normally represented in the form of flow curves or rheograms, where the fluid shear stress is plotted vs the shearing rate as shown in Fig. 7.3. Some of the more common fluid properties that can be used to describe suspension behaviour are its viscosity and yield stress.

Viscosity is defined as the resistance to deformation by a fluid, and is generally given as the ratio between the shear stress and shear rate, at constant temperature. In the case of a Newtonian fluid, the viscosity is constant throughout the entire shear rate range. However, as can be seen from Fig. 7.3, in the case of non-Newtonian fluids (pseudoplastic, shear thickening), this value changes as a function of shear rate. Therefore the viscosity of a non-Newtonian fluid at any point is referred to as apparent viscosity.

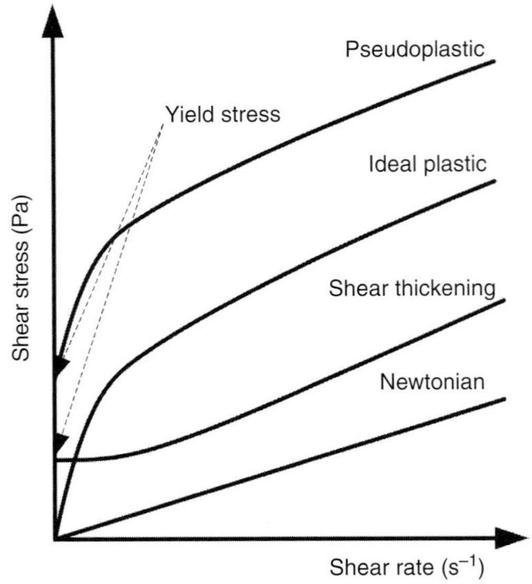

Figure 7.3. Rheograms of various types of fluids.

The yield stress of a suspension is the minimum force required for deformation to take place and is a characteristic property of pseudoplastic fluids. Yield stress is a difficult property to measure directly, as it requires the measurement of the shear stress of the fluid at shear rates approaching zero. A direct measurement method has been developed by Nguyen and Boger, which makes use of a vane rotating within a fluid at a constant shear rate $\ll 1\ s^{-1}$ (Nguyen and Boger, 1983, 1985).

Rheology of suspensions is strongly affected by a large number of factors, including particle size and concentration (Barnes *et al.*, 1989; Bhattacharya *et al.*, 1998; Farris, 1968; Shi and Napier-Munn, 1996; Subbanna *et al.*, 1998). Shi and Napier-Munn (1996) have studied the results of 45 surveys of industrial comminution equipment and have produced the most comprehensive description of rheological behaviour of ground minerals slurries with respect to these two properties. They demonstrated a complex relationship that has been reproduced and discussed in detail in a number of review publications since (Farrokhpay, 2012; He *et al.*, 2004), so only a brief summary will be presented here.

It was found that at low solids concentrations and a low proportion of fine particles (<38 μm), mineral slurries behave like Newtonian fluids, with no yield stress and a constant viscosity. However, as both solids content and fines proportion increase, slurries exhibit increasingly pseudoplastic behaviour, with increasingly large yield stress and an apparent viscosity that decreases with increasing shear rate. In the context of mineral processing operations, the majority of process slurry streams behave as pseudoplastic fluids, and exhibit a yield stress (Boger, 2000).

However, other factors such as surface chemistry and mineralogical properties of suspended particles have also been shown to play a strong role. The change in the surface charge of minerals brings about strong variations in the inter-particle forces, in accordance with classical DLVO theory (Derjaguin and Landau, 1941; Verwey and Overbeek, 1948). These forces have a significant effect on the rheological properties of mineral suspensions, whereby an increased state of dispersion is characterized by lower yield stress and the opposite is true for an increased state of aggregation (Burdukova *et al.*, 2007; Johnson *et al.*, 1999; Scales *et al.*, 2000). It follows that for oxide minerals (such as phyllosilicates/clays) suspension rheology is strongly influenced by solution pH as well as the presence of other dispersing and flocculating reagents.

It has also been demonstrated that the rheological behaviour of different mineral ores differs significantly based on the proportion and type of clay minerals present as gangue species, independently of particle size (Burdukova *et al.*, 2008), as shown in Fig. 7.4. Slurry rheological properties (apparent viscosity and yield stress) were also shown to depend strongly on the surface charge of anisotropic phyllosilicate particles, their shape (plates vs fibres) and swelling capacity (Ndlovu *et al.*, 2010, 2011).

This is consistent with reports that the presence of clay minerals causes problems in the treatment of certain ore types due to their tendency to give rise to high pulp viscosities (Bulatovic *et al.*, 1998; Senior and Thomas, 2005). In these cases, the elevated viscosities could be caused by a combination of clay mineral properties as well as their fine size upon comminution.

7.4 Clay Gangue in Mineral Beneficiation

The presence of clays in mineral beneficiation circuits can cause a wide range of different problems. The soft nature of these minerals induces the production of a large amount of slimes. The presence of slimes has a negative impact on slurry rheology, with detrimental effects on both flotation and comminution. The presence of fine hydrophilic slimes also causes slime coatings and high amounts of entrainment in flotation (the concept of entrainment will be dealt in more detail in the sections that follow). This section describes and discusses these effects in more detail and provides examples of some mitigation techniques.

In mineral processing parlance it is common to refer to the slimes resulting from ores that contain large amounts of clay minerals and produce large amounts of very fine particles. This is typically attributed to the soft nature of the clay minerals, where they get preferentially ground compared to harder mineral particles. However, there are only a few studies present in the literature that provide quantitative information with regard to size-by-size mineralogical analysis.

Senior and Thomas (2005) reported a size-by-size breakdown of the milling products of Mt Keith nickel ore (Western Australia). They showed that over 30 wt% of feed material ended up in the <10 μm size fraction, with serpentine minerals comprising over 30 per cent of the fines. Similarly, Fornasiero and Ralston (2006) noted a large portion of Mt Keith ore was ground finer than 20 μm, with about 28 wt% of the fines comprising magnesium oxide material (serpentines, talc and chlorite). Visser *et al.* (1994) described the results of a grinding pilot study of Rooihuiskraal mine ore in South Africa. They showed that a large

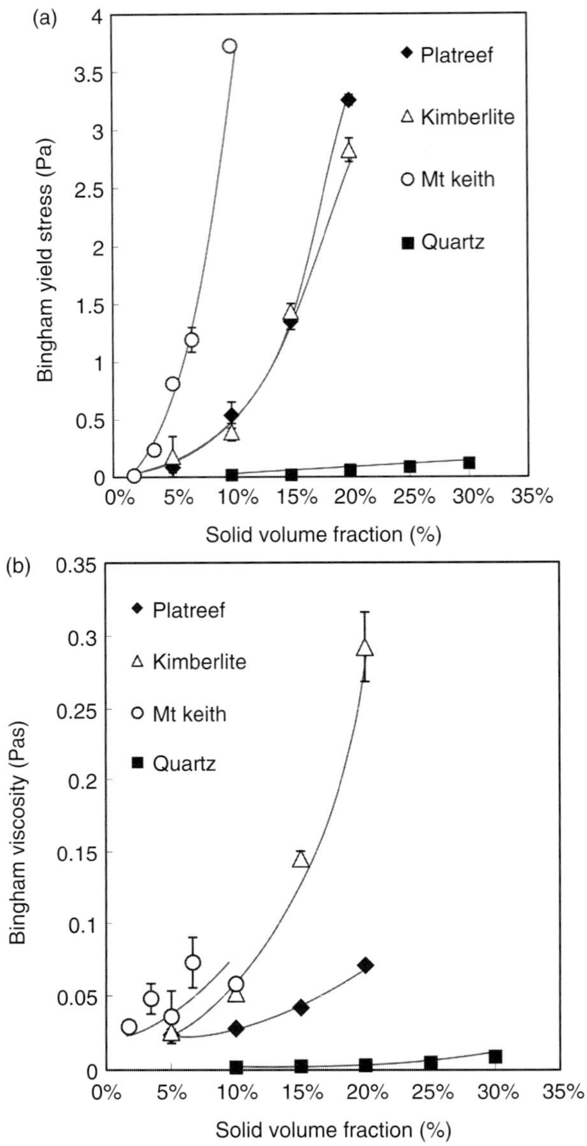

Figure 7.4. (a) Bingham yield stress and (b) viscosity of three different ores: Platreef – rich in non-swelling phyllosilicates; Kimberlite – rich in swelling phyllosilicates; Mt. Keith – rich in fibrous phyllosilicates; compared to quartz (modified from Burdukova *et al.* 2008).

proportion of the mill product was in the <11 μm size fraction (27 per cent for an AG mill and 30 per cent for a SAG mill),[1] with the majority of the fines comprising talc, chlorite and serpentine minerals.

Less specifically, Bulatovic *et al.* (1998) surveyed the mineralogy of four Chilean porphyry deposits (El Salvador, Disputada, Chuquicamata and El Teniente). They showed that these ores contain large quantities of phyllosilicate minerals, mainly chlorite, illite and kaolinite. While size-by-size mineralogy data are not provided, they do ascribe the majority of the difficulties in treating these ores to the presence of clay minerals. Specific reference is made to the tendency of clay minerals to increase pulp viscosities, which is typically associated with fine particle sizes (Boger, 2000). Furthermore, Rosario *et al.* (2011) state that the analysis of a large number of porphyry ores contain ~30 per cent very soft to moderately soft material, which is likely to report ultrafine size fractions (>38 μm) upon comminution. Although not directly stated, this soft material is likely to be phyllosilicates.

7.4.1 Clay Minerals and Comminution

7.4.1.1 Mineral Hardness

The hardness of composite ores, in terms of the crushing response, is generally determined by drop weight tests (Napier-Munn *et al.*, 1996), while the grinding response is determined by the Bond Work Index (Bond, 1952). While some authors have undertaken studies to establish a closer link between breakage characteristics and mineralogical content (Genç and Benzer, 2009; Pourghahramani, 2012), very little information is to be found with specific regard to clay minerals.

In geology, the relative hardness of minerals is measured on a Mohs hardness scale. A Mohs index of 1 is assigned to talc, which can be scratched using a fingernail, while diamond, the hardest known substance, is assigned an index of 10. While this index is insufficient to describe overall hardness of a composite ore, it gives a qualitative idea as to the likely difference between breakages of clay minerals as compared to other commonly encountered mineral types. Table 7.1 summarizes the Mohs hardness indices of four mineral types commonly found in industrial ores: phyllosilicates (clays), metal sulphides, metal oxides and silicates.

7.4.1.2 Clay Particle Breakage

The breakage characteristics of pure clay particles are well described in the literature pertaining to ceramic and coatings applications. Breakage of platy phyllosilicate particles can occur in two ways: delamination and fracture (Aglietti *et al.*, 1986; Makó *et al.*, 2001; Mekhamer, 2011; Vdović *et al.*, 2010).

[1] AG – autogenous grinding; SAG – semiautogenous grinding.

Table 7.1. *Mohs hardness of common minerals, including clays (Mindat, 2012).*

Phyllosilicates/clays		Metal sulphides	
Talc	1	Chalcopyrite	3.5–4
Kaolinite	1.5–2	Pentlandite	3.5–4
Vermiculite	1.5–2	Pyrrhotite	3.5–4
Lizardite	2.5	Sphalerite	3.5–4
Silicates		**Metal oxides**	
Quartz	7	Goethite	5–5.5
Olivine	6.5–7	Hematite	5.5
Plagioclase	6–6.5	Ilmenite	5.5
Augite (pyroxene)	5–6.5	Magnetite	5.5

Delamination occurs as a result of cleavage along the particle basal planes, breaking weaker ionic and van der Waals forces within the particle interlayers. This results in the formation of thinner and thinner platelets, with a higher number of exposed basal planes. Delamination is the dominant breakage mechanism within the first few minutes of grinding, and once the plates have been rendered sufficiently thin, the process of particle fracture becomes more dominant (Baudet *et al.*, 1999).

Upon prolonged grinding, the particles begin to break up by fracturing in a direction perpendicular to the basal planes. In this process, the stronger covalent bonds holding crystalline plates together are broken, exposing more particle edges. Further grinding (2–4 h in duration) causes substantial alteration to the morphology of clay particles, eventually rendering them amorphous. This effect has been observed for a number of different clay minerals, including kaolinite (Makó *et al.*, 2001) and bentonite (Dellisanti and Valdré, 2005).

Closed-circuit comminution equipment present in mineral concentrators such as flotation plants tends to have relatively low mean residence times, between 1 and 5 min (Gardner *et al.*, 1982; Makokha *et al.*, 2011; Rogers *et al.*, 1982). It is therefore likely that the dominant clay breakage mechanism for the clay minerals contained in the ore bodies typically treated by flotation is that of delamination, with a smaller amount of basal plane fracture. On the basis of what is known it is therefore unlikely that clay-rich ores spend sufficient time within the grinding circuit to create amorphous clay particles.

The presence of an increased number of thin plates greatly increases the number of highly charged basal plane surfaces present in the mineral slurry. This has a pronounced effect on slurry rheology through increased particle–particle interactions and an increase in the effective volume fraction of the suspension. These effects have been described in detail in Chapter 2.

However, it is very important to stress that the comminution behaviour of pure minerals differs substantially from that of composite ore particles; in the presence of grinding media it would represent an approximation of an extreme case. Therefore, the trends observed in pure clay comminution systems are unlikely to be fully transferrable to composite ores that are part of typical mining operations in which clays form part of the gangue species. For this reason the information given above must be treated with some caution. Overall, there is a shortage of specific information in public-domain literature pertaining to the effect of the presence of clay minerals on the performance of comminution equipment in terms of their grindability.

Nonetheless, one can *speculate* that even in composite mineral particles, short grinding times (the kind existing in comminution circuits) result in extensive delamination and more limited basal plane fracture of clay particles. This would result in the creation of fine particles that still exhibit a laminar structure. The presence of these platelets can be evidenced in SEM micrographs of industrial ores containing clay minerals, as shown in Fig. 7.2.

7.4.1.3 Slurry Rheology

The effect of slurry rheology on grinding performance has been relatively well documented; therefore the effect of clays on grinding can rather be discussed in rheological terms. A detailed review of previous work on this subject was published by Yue and Klein (2004) as part of their own research on the performance of stirred mills. The results of individual studies of the effect of slurry rheology on grinding performance reported by various researchers are often conflicting, due to the complex nature of the phenomenon, but trends can still be identified.

Several groups conclude that slurry viscosity can have both beneficial and detrimental effects on grinding performance, depending on its magnitude (Gao and Forssberg, 1993; Kawatra and Eisele, 1988; Klimpel, 1982; Shi and Napier-Munn, 2002; Wang and Forssberg, 1997). However, Yue and Klein (2004) have demonstrated that the effect of viscosity is insignificant compared to the effect of yield stress.

An increase in the slurry yield stress has the effect of significantly increasing the grinding power consumption, while decreasing the particle breakage rate. This indicates that the particle size distribution of the milled product can be manipulated by controlling the factors that influence yield stress, namely feed particle size distribution and slurry density, in a manner illustrated in Fig. 7.5.

For a fixed feed size, an increase in the slurry yield stress obtained by increasing slurry percentage by weight of the solid's density dramatically reduced the amount of <10 μm material present in the milling product. If, on the other hand, the feed slurry solid's density was kept constant, the increase in the slurry yield stress led to the narrowing of the product size distribution (Yue and Klein, 2004).

Another significant aspect of the effect of slurry rheology on grinding performance is the effect of slurry chemistry. It has been demonstrated that the addition of dispersing agents and alteration of slurry pH can significantly improve mill performance, through the modification of slurry rheology (Fuerstenau *et al.*, 1990; Klimpel and Hansen, 1989; Wang and Forssberg, 1997). Though the majority of researchers only report the effects on pulp viscosity, slurry yield stress is also likely to be strongly influenced by chemistry, through the alteration in the nature of the particle–particle interactions (see Chapter 2 for details).

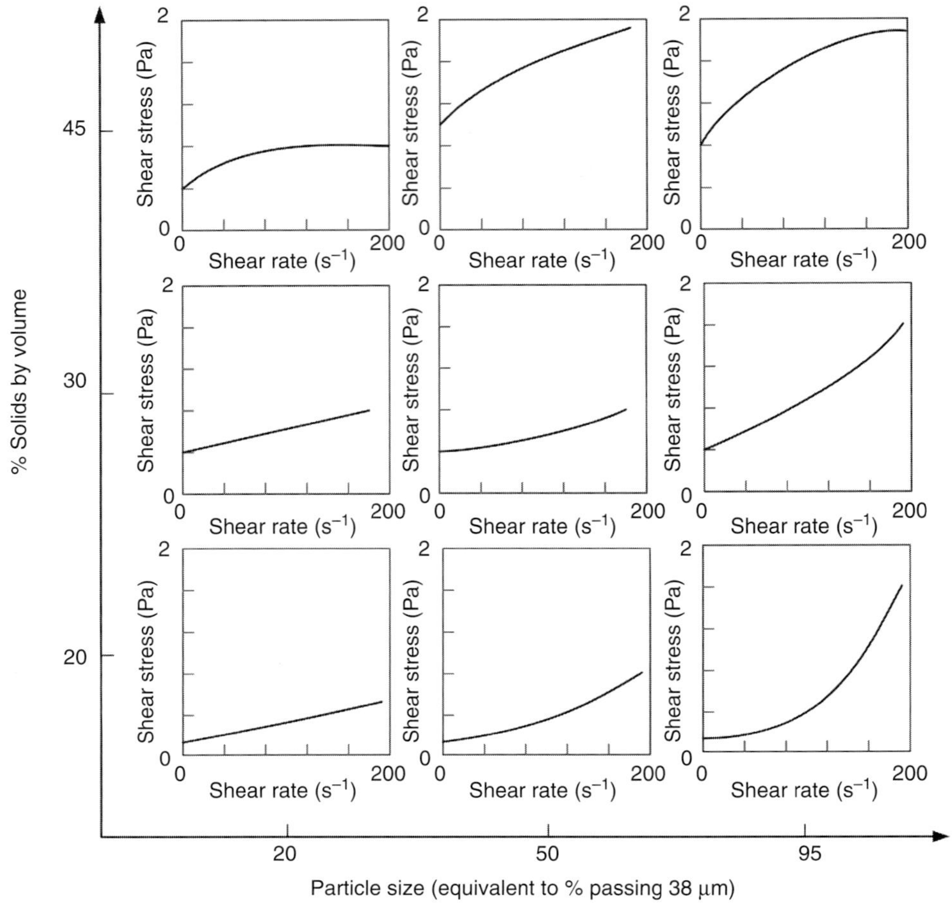

Figure 7.5. Effect of particle size distribution and percentage solids fraction on slurry rheology, adapted from Shi and Napier-Munn (1996) with permission.

7.4.2 Effects of Clay Gangue on Flotation

7.4.2.1 Entrainment

The most obvious consequence of the presence of very fine slime particles in flotation is entrainment. Entrainment has been described as a non-selective process whereby mineral particles report to the froth phase by being upswept in the wake of rising bubbles, while not actually being attached to them (Gaudin, 1957). The effect of particle size in entrainment has been very widely studied and it has been shown that the entrainability of mineral particles increases dramatically at particle sizes below approximately 20 μm (Neethling and Cilliers, 2002; Savassi *et al.*, 1998; Yianatos and Contreras, 2010).

With the majority of clay particles being present in the <20 μm fraction (Fornasiero and Ralston, 2006; Senior and Thomas, 2005; Visser *et al.*, 1994), the presence of clay minerals understandably contributes to the decrease in the concentrate grade.

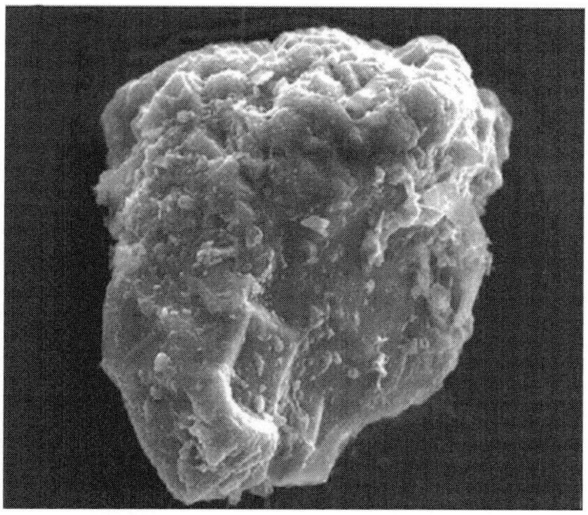

Figure 7.6. Example of a chalcopyrite particle coated in clay slimes, adapted from Bradshaw *et al.* (2011) with permission.

7.4.2.2 Slime Coatings

Another detrimental and well-known effect of slimes on flotation is referred to as slime coatings. In this phenomenon, relatively coarse valuable particles become either completely or partially coated by a layer of hydrophilic slimes, thereby rendering the valuables hydrophilic and inhibiting collector adsorption (Arnold and Aplan, 1986b; Del Guidice, 1934; Edwards *et al.*, 1980; Sutherland and Wark, 1955; Trahar, 1981). An example of this is illustrated in Fig. 7.6, which shows a slime-coated chalcopyrite particle from the tailings of the Kennecott Utah Copper Concentrator (Bradshaw *et al.*, 2011).

The mechanism of coating by slimes has been widely studied. Earlier work postulated that the slime coatings attached to mineral particles through chemisorption, resulting from the ion exchange between slime particles and the surrounding solution (Del Guidice, 1934). However, the current consensus is that slime coatings arise due to electrostatic forces between oppositely charged mineral particles (Arnold and Aplan, 1986b; Attia and Deason, 1989; Fuerstenau *et al.*, 1958; Iwasaki *et al.*, 1962; Xu *et al.*, 2003). A well-studied example of this phenomenon occurs in nickel flotation in Western Australia, where negatively charged pentlandite particles become completely coated with positively charged lizardite particles. This coating renders the pentlandite surfaces hydrophilic, preventing flotation (Peng and Bradshaw, 2012; Peng and Seaman, 2011). Another study that showed a similar effect was that by Forbes *et al.* (2014). The study inferred that fine kaolinite particles were forming hydrophilic coatings on chalcopyrite surfaces at low pH values, where chalcopyrite becomes less negatively charged, thus significantly retarding flotation.

A number of studies have been conducted on slime coatings with specific reference to clay minerals, and a range of different treatment options were investigated. Bremmell *et al.*

(2005) and Fornasiero and Ralston (2006) proposed using negatively charged carboxymethyl cellulose polymers. The adsorption of these reagents onto clay surfaces induces steric repulsion between clay and metal sulphide species. Attia and Deason (1989) concluded that slime coatings can be controlled by selective flocculation of fine clay particles using flocculants such as polyethylene glycol. Arnold and Aplan (1986a) suggested that clay slimes could be mitigated by the presence of ions in flotation pulp. In this process, multivalent ions adsorb onto clay surfaces, masking their surface charge. Alternatively, Senior and Thomas (2005) proposed solving this problem by completely re-designing a flotation circuit such that the coarse and slime streams are treated separately.

7.4.2.3 Froth Stability

The stability of flotation froths is an important property that has a strong effect on overall flotation performance through processes such as froth carrying capacity, entrainment and drainage (Rubenstein and Melik-Gaikazyan, 1998; Subrahmanyam and Forssberg, 1988). Froth *stability* is difficult to define, but can be generally described as the measure of a rate at which bubbles coalesce within the froth structure. On the other hand, froth *mobility* can be described as a rate at which the froth overflows the cell weir.

Stable froths are generally characterized by reduced mobility (Moolman *et al.*, 1996). Low mobility prevents effective froth overflow into the collection launders, with detrimental effects on flotation recovery. High froth stability also tends to retard drainage of entrained material, thus decreasing the grade of the froth product. In most cases, optimum froth stability is desired, as opposed to a highly stable or unstable regime to balance the loss of valuable particles through drop-pack with the removal of entrained material though drainage.

One of the major factors influencing the stability of flotation froths is the presence of mineral particles and their properties (Dippenaar, 1978, 1982; Johansson and Pugh, 1992; Klassen and Makrousov, 1963; Livshits and Dudenkov, 1965; Lovell, 1976; Tao *et al.*, 2000). It is generally agreed that the size of particles and their hydrophobicity/hydrophilicity play the key role in determining the degree to which the particles stabilize froths (although there are still disagreements with regards to the exact magnitudes of these properties and their effects) (Barbian *et al.*, 2005).

Lovell (1976) noted that flotation froths tended to become highly stable and unmanageable in the presence of large quantities of fine hydrophobic particles. Furthermore, both Klassen and Makrousov (1963) and Livshits and Dudenkov (1965) noted that the particle stabilizing effect was far more prevalent in the presence of laminar (plate-like) particles rather than spherical ones. This further reinforces the notion that platy, hydrophobic phyllosilicate particles are likely to contribute to the formation of highly stable flotation froths, and hence have a detrimental effect on drainage and froth recovery. This description correlates well with the type of slimes generated in the presence of clay minerals such as hydrophobic talc.

A good example of froth stabilization by talc was observed at Lac des Iles PGM Concentrator, in Ontario. In this operation, the stability of froth was significantly reduced by the use of carboxymethyl cellulose as a talc depressant, to the point where the froth became too unstable. The instability was controlled by *reducing* the amount of depressant to ensure that there was enough talc present in the froth to ensure optimum stability (Martin *et al.*, 2002).

7.4.2.4 Pulp Rheology

As in the case of ore comminution, the presence of fine clay particles in flotation operations causes changes in the rheology of the flotation pulp, resulting in pseudoplastic behaviour (Burdukova *et al.*, 2008). Pulp is the generic flotation term for the suspended solid's slurry. The influence of pulp rheology on flotation processes has been recently reviewed in detail by Farrokhpay (2012), and only a brief overview will be presented here.

As rheology describes flow properties of fluids it has a significant impact on flotation cell hydrodynamics. Schubert (2008) argues that low pulp viscosity causes a reduction in turbulence damping (i.e. higher overall turbulence). While this aids flotation kinetics by increasing the probability of particle/bubble attachment, it also has a negative effect through increasing the probability of bubble/particle detachment. This was verified by Xu *et al.* (2011), who found that increased pulp viscosity (and hence lower cell turbulence) resulted in an increased stability of the bubble/particle aggregates, thus decreasing detachment probability. On the other hand, Forbes *et al.* (2014) demonstrated that the increase in flotation pulp viscosity can also serve to significantly deter flotation performance, an effect that was attributed to a decrease in particle/bubble contact through turbulence damping.

The effect of increased pulp viscosity was also shown to have a detrimental effect on gas dispersion, through limiting the turbulent region of flotation cells to a small turbulent cavern surrounding the impeller (Bakker *et al.*, 2009; Shabalala *et al.*, 2011).

It has also been suggested that the rheological properties of the flotation pulp have an effect on the rheological properties of the froth phase. Froth rheology is still poorly understood, with a very limited number of studies dedicated to it (Rubenstein and Melik-Gaikazyan, 1998; Shi and Zheng, 2003). While it is clear that froth rheological properties bear a strong influence on factors such as froth mobility and stability, the exact mechanisms involved are still unknown.

7.4.3 Clay Rejection Techniques

In order to counteract and overcome the detrimental effects of clay minerals in flotation and comminution circuits, a number of different methods have been investigated. These techniques include mechanical size classification to remove slimes prior to flotation, as well as the use of chemical reagents, such as selective flocculants, to mitigate the presence of clay slimes in flotation pulps.

Hydrocyclones have been widely used in the mining industry to reject clays prior to the beneficiation processes. For example, in the iron ore industry, hydrocyclones are used as a desliming tool to reject ultrafine mineral particles prior to flotation. Desliming significantly improves the performance of the subsequent froth flotation process, by mitigating the deleterious processes described in the previous sections (Ma, 2011b).

The major advantages of hydrocyclones are (1) mechanical simplicity, (2) high throughput of mineral suspensions, (3) small installation footprint and (4) low capital cost in comparison to other mechanical classifiers (Mohanty and Das, 2010). The disadvantage of using hydrocyclones is that the fine particles of the valuable minerals are also rejected to the overflow, together with the clays and other slime minerals. Mohanty and Das (2010) report that the level of alumina-containing clay minerals could be reduced to 3.5 per cent with the use of hydrocyclones. Srivastava *et al.* (2001) find that, with the use of a hydrocyclone followed by a spiral concentrator, they could reduce the content of alumina-containing clay minerals in an iron ore slime sample from 2.93 per cent to 1.17 per cent with 37 per cent iron recovery.

Another desliming tool is the Floatex Density Separator (FDS). FDS operates as a hindered settling-based hydraulic classifier, incorporating counter-current and autogenous teetered-bed separators, and is used in the minerals and coal industries (Kumar *et al.*, 2010; Mohanty and Das, 2010). Raghukumar *et al.* (2012) studied an iron ore sample with a high content of alumina-containing clays from Noamundi, India, using an FDS. They reduced the content of alumina-containing clays from 5.89 per cent to 2.10 per cent, but lost approximately 75 per cent of the iron value.

Another process for dealing with clay minerals is selective flocculation. This technique has been well-studied at laboratory level for a number of mineral systems (Filippov *et al.*, 2012; Liu *et al.*, 2012; Ma, 2011a, 2011b, 2012a, 2012b, 2012c; Mark, 2011). This process involves the effective dispersion of clay minerals and other gangue minerals such as quartz, and the flocculation of valuable minerals. The dispersed clays and other gangue minerals stay in the supernatant, and the flocculated valuable minerals settle to the bottom and thus are separated from the gangue minerals. In this process, the effective dispersion of clays is of critical importance. While the traditional dispersant (sodium silicate) was found ineffective in the dispersion of kaolinite (Ma, 2011b; Mark, 2011), polyacrylic acid was found to be an effective dispersant of the clay mineral under iron ore flotation conditions, which has the potential to significantly improve the efficiency of this process (Ma, 2011a, 2011b).

Reverse froth flotation can also be employed for the rejection of clay minerals (Ma, 2012b; Ma *et al.*, 2012). Ma *et al.* (2009) investigated the flotation mechanism of kaolinite and found that the pH dependence of kaolinite flotation was opposite to that of oxide minerals, with lower flotation recovery obtained at higher pH when using amines as collectors. The special flotation behaviour of kaolinite was related to its unique edge/plane structure, described in detail in Chapters 2 and 8.

7.5 Clay Purification Techniques

While clays and clay minerals within the mineral values process chain, and across most commodities, represent an unwanted gangue species, they are also an important commodity in the manufacture of paints, coatings, ceramics, cosmetics and pharmaceutical products and will often be beneficiated to produce a purified clay product. For example, kaolinite is widely used as paper filler and a coating pigment, as an extender in ink and water-based paints, and as a functional additive in polymers, as well as its traditional role as a major feed component of ceramics (Braggs, 1993). Increasingly demanding requirements on clay purity has driven improved beneficiation processes (Prasad *et al.*, 1991). In the case of kaolinite, methods have targeted the removal of impurities that reduce its brightness, where brightness is a measure of the blue reflectance in pigments and coatings. These methods include: size classification; magnetic separation; selective flocculation; and froth flotation (Mathur, 2002;Matis *et al.*, 1993). Of these, froth flotation is generally considered to be the most effective beneficiation method where inorganic gangue is the problem and it has been extensively investigated for clay purification over the last five decades (Mathur, 2002; Matis *et al.*, 1993).

7.5.1 Gravity Separation

Gravity separation is a process based on the differential settling rates of particles in a fluid medium. With kaolinite in dilute suspensions the particles settle freely, i.e. minimal hindrance between the particles (Wilson and Middleton, 1937). The main parameter determining the extent to which the minerals can be separated in gravity concentration is the concentration criterion (Wills and Finch, 2016), as defined in Eq. (7.1), where D_h and D_l are the densities of the heavier and lighter minerals, and D_m is the density of the medium (which is normally water, except in heavy-medium separation):

$$\frac{D_\mathrm{h} - D_\mathrm{m}}{D_\mathrm{l} - D_\mathrm{m}} \tag{7.1}$$

The use of gravity concentration for minerals liberated at a coarse particle size can significantly reduce the mass and volume load in subsequent beneficiation processes and is inexpensive and environmentally friendly. Simple gravity concentration works most efficiently with coarser particle size ranges and a concentration criterion of at least 1.25 (Wills and Finch, 2016).

7.5.2 Magnetic Separation

Magnetic separation is a process for extracting magnetically susceptible mineral particles, such as magnetite, from ores using a non-homogeneous magnetic field (Svoboda and Fujita, 2003). Its selectivity depends on the different responses of the susceptible mineral particles and gangue components to magnetic forces (Wasmuth and Unkelbach, 1991). The important impurities in kaolinite clay, e.g. hematite, anatase, rutile and pyrite, tend to be weakly magnetic. When placed in a magnetic field they become magnetized in such a way as to either aid or oppose the applied field. The amount of magnetization induced in the particles depends on the mass and the magnetic susceptibility of the those particles and the intensity of the applied magnetic field (Notebaart and Meer, 1992). The process becomes more effective at higher magnetic gradients.

A technical breakthrough in high gradient magnetic separator (HGMS) technology came with the first industrial application of superconducting HGMS. Using superconducting technology, extremely high magnetic fields may be achieved with virtually zero power consumed by the magnet (Notebaart and Meer, 1992). Superconducting HGMS units are capable of generating magnetic fields up to 50 kG, in comparison to the maximum magnetic fields of 20 kG generated conventionally and requiring only a fraction of the energy consumed by conventional magnetic separators (Prasad *et al.*, 1991). Ohara reviews the history of HGMS for dilute suspension processing (Ohara, 2014)

7.5.3 Flotation

Kaolinite purification using froth flotation was uniquely the first application of flotation to submicron ultrafine mineral particles (Mathur, 2002). It had been generally accepted that only mineral particles larger than 10 μm could be effectively beneficiated in flotation processes (Attia and Deason, 1989; Bandini *et al.*, 2001; Edwards *et al.*, 1980) as flotation generally loses effectiveness at submicron sizes for a number of reasons. First, there are mechanical factors such as entrainment, entrapment, poor drainage of the extremely fine mineral particles and reduced probabilities of collision and attachment. Added to this is the very high specific surface area of fine particles, causing extremely high consumption of flotation reagents, thereby increasing production cost. The very fine mineral particles also form a layer of slime coatings on coarser mineral particles and consequently reduce the selectivity of the froth flotation process (Arnold and Aplan, 1986b; Attia and Deason, 1989; Fuerstenau *et al.*, 1958; Iwasaki *et al.*, 1962; Xu *et al.*, 2003).

7.5.3.1 Carrier Flotation

The generally ultrafine nature of kaolinite particles mean they are amenable to treatment by carrier flotation. This involves the utilization of coarse mineral particles, usually larger than 150 μm in size, to act as coagulation sites. The ultrafines heterocoagulate with the coarse mineral particles and are carried or ‘piggy-backed’ into the froth product. Carrier flotation has been widely studied in the literature (Matis *et al.*, 1993; Pindred and Meech, 1984; Trahar, 1981). In the case of kaolinite purification the carrier flotation involves the use of tall oil (by-product of wood pulp manufacture containing resin and fatty acids) as a collector and calcium carbonate as the carrier of very fine particle-size impurity minerals (Basilio *et al.*, 2000; Matis *et al.*, 1993). While tall oil macromolecules chemically adsorb on the surfaces of calcium carbonate by forming chemical bonds with the surface calcium sites, the impurity minerals such as anatase require the use of activators (dissolved calcium ions) to provide adsorption sites (Luz *et al.*, 2000). After conditioning with tall oil the surfaces of the impurity minerals and the carrier mineral become net negatively charged via the hydrophobic surfactant layer. Since the negative charge on the impurity minerals and the carrier mineral generate repulsive electrostatic forces between the particles, the impurity mineral particles attach on the surfaces of calcium carbonate through hydrophobic forces, which is the so-called 'piggy-back' phenomenon (Mathur *et al.*, 2000; Matis *et al.*, 1993).

The traditional carrier flotation process employed in the kaolinite industry was significantly improved by Freeport Kaolinite with the titanium removal and extraction process (TREP) for the removal of impurity minerals (Young *et al.*, 1985). In TREP, the slurry is heated up to 93 °C during conditioning. A dispersant is then added to the slurry, followed by flotation using column flotation cells, rather than mechanical flotation cells (Bacon, 1984). The high temperature during conditioning increases the solubility of oleic acid, which is used as a collector of the impurity minerals. Although the pH of the slurry during conditioning is as low as 6.1 to 6.3, oleic acid can still adsorb onto the surfaces of anatase particles (which are again activated with calcium ions) because of the high temperature of the slurry (Mathur *et al.*, 2000). The significant improvement with TREP is the high solids concentration at the conditioning stage (greater than 25 per cent solids) that results in high throughput and lower dewatering costs (Bacon, 1984).

7.5.3.2 Carrierless Flotation

In addition to carrier flotation, a carrierless flotation process has also been employed in clay purification. The first was developed by English China Clay International (ECCI) and was designed to remove titaniferous impurity minerals from clays (Cundy, 1969). Here, the clay suspensions, typically controlled at 40–60 per cent solids, are scrubbed with high-speed agitation. This serves to facilitate the liberation of the impurity minerals from the kaolinite, as well as induce coagulation between the very fine anatase particles (i.e. shear flocculation). The resultant increased coagulated particle size then facilitates the anatase flotation (Mathur *et al.*, 2000). After the scrubbing process, the kaolinite suspensions are conditioned with calcium salts to activate the surfaces of the impurity minerals, and oleic acid is employed as an impurity collector under alkaline conditions. For flotation the concentration of solids is diluted to 15–20 per cent.

7.5.3.3 Collector Issues

The use of tall oil type collectors in the purification of clays requires the use of divalent activator cations. Excessive use of such activator ions can cause the coagulation of the clay particles and may also induce the flotation of the clay together with the impurity minerals, leading to poor flotation performance. Coagulation can be countered, and overall flotation performance improved, when centrifuging is employed to separate the coagulated calcium carbonate–titania particles from the very fine kaolinite particles entrained in the froth, and a flotation collector is used to agglomerate the impurity (Gantt *et al.*, 1994).

High-speed conditioning is also reported to be beneficial for improved selectivity and flotation performance. Though tall oil adsorption on kaolinite does occur, it is not as strong as on calcite and anatase. The high-speed (and hence high-shear) agitation preferentially strips the tall oil coating from the kaolinite surfaces, whereas the tall oil remains strongly adsorbed on the anatase (Mathur *et al.*, 2000).

Hydroxamates (R-CONHO.M where R is a general hydrocarbon group and M is a metal) are reported to be a better collector choice than tall oil. As hydroxamates can chemically adsorb on the metallic cation sites on the surfaces of the impurity minerals, they can be used at alkaline pH without requiring the use of activators (Yoon *et al.*, 1992). Cytec developed a hydroxamates-based collector, reagent S6493, which has been successfully used by Thiele Kaolinite. The slurry pH for S6493 is typically controlled in the alkaline range because in acidic slurries the efficiency of the collector understandably decreases (Wang *et al.*, 2005) due to kaolinite self-flocculation.

7.6 Summary

Overall, it is evident that the presence of phyllosilicate clay minerals in mineral processing streams has a strong effect on the performance of both comminution and flotation circuits. In the case of grinding, this takes the form of slime production, where soft phyllosilicate minerals are preferentially ground and report to the ultrafine size fractions. The fine sizes as well as the mineralogical properties of these clay particles results in complex slurry rheology with elevated apparent viscosities and yield stresses.

It has been demonstrated that the rheology of mineral slurries has a profound effect on grinding performance. However, the precise mechanism of this influence is still not clearly elucidated. There is no fundamental model that predicts the extent of the influence of rheology on milling parameters.

In flotation circuits, the presence of fine phyllosilicate particles is known to result in the following detrimental effects:

(1) Entrainment of fine clay particles into the froth phase, causing the loss of selectivity.
(2) Slime coatings, which prevent particle/bubble attachment and collector adsorption on valuable coarse particles.
(3) Froth stabilization resulting in retarded froth drainage, with further loss of selectivity.
(4) Poor gas dispersion through increased pulp viscosity, resulting from poor pulp hydrodynamics.
(5) Poor bubble/particle contact resulting from decreased cell turbulence with increased pulp viscosity, which is balanced by an increased probability of particle/bubble detachment.

Several techniques have been developed to counteract the detrimental effects of clay minerals in flotation, such as hydrocyclones and density separators, as well as selective flocculation methodologies.

Clay minerals require beneficiation to produce a purified clay product for a wide variety of applications. The demanding requirements for improving the purity of clay products have been driving researchers and clay manufacturers to seek improved beneficiation processes.

Froth flotation is generally considered to be the most effective beneficiation method and also offers further potential for technical improvements in this area. Therefore, froth flotation, in particular carrier flotation and carrierless flotation, has been widely investigated for improved process development in clay purification.

References

Aglietti, E. F., Porto Lopez, J. M. & Pereira, E. 1986. Mechanochemical effects in kaolinite grinding: II. Structural aspects. *International Journal of Mineral Processing*, 16 (1–2), 135–146.

Arnold, B. J. & Aplan, F. F. 1986a. The effect of clay slimes on coal flotation, Part II: The role of water quality. *International Journal of Mineral Processing*, 17 (3–4), 243–260.

Arnold, B. J. & Aplan, F. F. 1986b. The effect of clay slimes on coal flotation, Part I: The nature of the clay. *International Journal of Mineral Processing*, 17 (3–4), 225–242.

Attia, Y. A. & Deason, D. M. 1989. Control of slimes coating in mineral suspensions. *Colloids and Surfaces*, 39 (1), 227–238.

Bacon, F. C. 1984. Froth flotation apparatus and process. US Patent US4472271.

Bakker, C. W., Meyer, C. J. & Deglon, D. A. 2009. Numerical modelling of non-Newtonian slurry in a mechanical flotation cell. *Minerals Engineering*, 22 (11), 944–950.

Bandini, P., Prestidge, C. A. & Ralston, J. 2001. Colloidal iron oxide slime coatings and galena particle flotation. *Minerals Engineering*, 14 (5), 487–497.

Barbian, N., Hadler, K., Ventura-Medina, E. & Cilliers, J. J. 2005. The froth stability column: Linking froth stability and flotation performance. *Minerals Engineering*, 18 (3), 317–324.

Barnes, H. A., Hutton, J. F. & Walters, K. 1989. *Introduction to Rheology*. Amsterdam: Elsevier.

Basilio, C., Lowe, R. A., Gorken, A., Magliocco, L., & Hagy, R. 2000. Modified hydroxamate collectors for kaolin flotation. *Developments in Mineral Processing*, 13, C8b-51–C8b-55.

Baudet, G., Perrotel, V., Seron, A. & Stellatelli, M. 1999. Two dimensions comminution of kaolinite clay particles. *Powder Technology*, 105 (1–3), 125–134.

Bhattacharya, I. N., Panda, D. & Bandopadhyay, P. 1998. Rheological behaviour of nickel laterite suspensions. *International Journal of Mineral Processing*, 53 (4), 251–263.

Boger, D. V. 2000. Rheology and the minerals industry. *Mineral Processing and Extractive Metallurgy Review*, 20 (1), 1–25.

Bond, F. C. 1952. Third theory of comminution. *Transactions of the American Institute of Mining, Metallurgical, and Petroleum Engineers*, 193, 484.

Bradshaw, D. J., Triffett, B. & Kashuba, D. 2011. The role of process mineralogy in identifying the cause of low recovery of chalcopyrite at KUCC. In: Broekmans, M. (ed.) *10th International Congress for Applied Mineralogy*. Trondheim: Springer, 73–80.

Braggs, B. R. 1993. The controlled modification and characterisation of the kaolinite surface. PhD Thesis, University of South Australia, Mawson Lakes, SA.

Bremmell, K. E., Fornasiero, D. & Ralston, J. 2005. Pentlandite–lizardite interactions and implications for their separation by flotation. *Colloids and Surfaces A: Physicochemical and Engineering Aspects*, 252 (2–3), 207–212.

Bulatovic, S. M., Wyslouzil, D. M. & Kant, C. 1998. Operating practices in the beneficiation of major porphyry copper/molybdenum plants from Chile: Innovation technology and opportunities, a review. *Minerals Engineering*, 11 (4), 313–331.

Burdukova, E., Bradshaw, D. J. & Laskowski, J. S. 2007. Effect of CMC and pH on the rheology of suspensions of isotropic and anisotropic minerals. *Canadian Metallurgical Quarterly*, 46 (3), 273–278.

Burdukova, E., Becker, M., Ndlovu, B., Mokgethi, B. & Deglon, D. A. 2008. Relationship between slurry rheology and its mineralogical content. In: Wang, D. D., Xao, S. C., Wang, F. L., Cheng, Z. U. & Long, H. (eds) *24th International Minerals Processing Congress*. Beijing: China Scientific Book Service Co. Ltd, 2169–2178.

Cundy, E. G. 1969. *Processing of Clay*. US Patent US3450257.

Deer, W. A., Howie, R. A. & Zussman, J. 1978. *Rock Forming Minerals, Volume 3. Sheet Silicates*.London: Longmans.

Del Guidice, G. R. M. 1934. A study of slime coatings in flotation. *Transactions of the American Institute of Mining, Metallurgical, and Petroleum Engineers*, 112, 398–409.

Dellisanti, F. & Valdré, G. 2005. Study of structural properties of ion treated and mechanically deformed commercial bentonite. *Applied Clay Science*, 28 (1–4), 233–244.

Derjaguin, B. V. & Landau, L. 1941. Theory of stability of strongly charged lyophobic sols and the adhesion of strongly charged particles in solution of electrolytes. *Acta Physiochimica*, 14, 633–662.

Dippenaar, A. 1978. *The Effects of Particles on the Stability of Flotation Froths*. Randburg, RSA: National Institute of Metallurgy.

Dippenaar, A. 1982. The destabilization of froth by solids: I. The mechanism of film rupture. *International Journal of Mineral Processing*, 9 (1), 1–14.

Edwards, C. R., Kipkie, W. B. & Agar, G. E. 1980. The effect of slime coatings of the serpentine minerals, chrysotile and lizardite, on pentlandite flotation. *International Journal of Mineral Processing*, 7 (1), 33–42.

Farris, R. J. 1968. Prediction of the viscosity of multimodal suspensions from unimodal viscosity data. *Transactions of the Society of Rheology*, 12 (2), 281–302.

Farrokhpay, S. 2012. The importance of rheology in mineral flotation: A review. *Minerals Engineering*, 36–38, 272–278.

Filippov, L. O., Duverger, A., Filippova, I. V., Kasaini, H. & Thiry, J. 2012. Selective flotation of silicates and Ca-bearing minerals: The role of non-ionic reagent on cationic flotation. *Minerals Engineering*, 36–38, 314–323.

Forbes, E., Davey, K. J. & Smith, L. 2014. Decoupling rheology and slime coatings effect on the natural flotability of chalcopyrite in a clay-rich flotation pulp. *Minerals Engineering*, 56, 136–144.

Fornasiero, D. & Ralston, J. 2006. Effect of MgO minerals on pentlandite flotation. In: Onal, G., Acarkan, N., Celik, M. S., *et al.* (eds) *23rd International Mineral Processing Congress, 2006*. Istanbul: Promed, 750–755.

Fuerstenau, D. W., Gaudin, A. M. & Miaw, H. L. 1958. Iron oxide slime coatings in flotation. *Transactions of the American Institute of Mining, Metallurgical, and Petroleum Engineers*, 211, 792–793.

Fuerstenau, D. W., Kapur, P. C. & Velamakanni, B. 1990. A multi-torque model for the effects of dispersants and slurry viscosity on ball milling. *International Journal of Mineral Processing*, 28 (1–2), 81–98.

Gantt, G. E., Adkisson, T. J., Gladin, P. B. & Ussery, R. E. 1994. Selective separation of finely-divided minerals by addition of selective collector reagent and centrifugation. US Patent US5358120.

Gao, M. & Forssberg, E. 1993. The influence of slurry rheology on ultra-fine grinding in a stirred ball mill. In: Ausimm, T. (ed.) *18th International Mineral Processing Congress, 23–28 May 1993*. Sydney: Australasian Institute of Mining and Metallurgy, 237–244.

Gardner, R. P., Aissa, M. & Verghese, K. 1982. Determination of ball mill residence time distributions from tracer data taken in closed-circuit operation. *Powder Technology*, 32 (2), 253–266.

Gaudin, A. M. 1957. *Flotation*. New York: McGraw-Hill.

Genç, Ö. & Benzer, A. H. 2009. Single particle impact breakage characteristics of clinkers related to mineral composition and grindability. *Minerals Engineering*, 22 (13), 1160–1165.

He, M., Wang, Y. & Forssberg, E. 2004. Slurry rheology in wet ultrafine grinding of industrial minerals: A review. *Powder Technology*, 147 (1–3), 94–112.

Iwasaki, I., Cooke, S. R. B., Haraway, D. H. & Choi, H. S. 1962. Iron wash slimes: Some mineralogical characteristics. *Transactions of the American Institute of Mining, Metallurgical, and Petroleum Engineers*, 223, 97–108.

Johansson, G. & Pugh, R. J. 1992. The influence of particle size and hydrophobicity on the stability of mineralized froths. *International Journal of Mineral Processing*, 34 (1–2), 1–21.

Johnson, S. B., Franks, G. V., Scales, P. J., Boger, D. V. & Healy, T. W. 1999. Surface chemistry: Rheology relationships in concentrated mineral suspensions. *International Journal of Minerals Processing*, 58 (1–4), 267–304.

Kawatra, S. K. & Eisele, T. C. 1988. Rheological effects in grinding circuits. *International Journal of Mineral Processing*, 22 (1–4), 251–259.

Klassen, V. I. & Makrousov, V. A. 1963. *Introduction to the Theory of Flotation*.London: Butterworth.

Klimpel, R. R. 1982. Slurry rheology influence on the performance of mineral/coal grinding circuits. *Mining Engineering*, 34 (12), 1665–1668.

Klimpel, R. R. & Hansen, R. D. 1989. Chemistry of mineral slurry rheology control grinding aids. *Minerals and Metallurgical Processing*, 6 (1), 35–43.

Kumar, C. R., Tripathy, S. K., Mohanan, S., Venugopalan, T., Suresh, N. 2010. Evaluation of Floatex density separator performance using silica sand. In: Singh, R., Das, A., Banerjee, P. K., Bhattachryya, K. K. & Goswami, N. G. (eds) *XI International Seminar on Mineral Processing Technology, 15–17 December, 2010*. Jamshedpur: Allied Publishers Pvt. Ltd, 701–706.

Liu, W., Sun, W. & Hu, Y. 2012. Effects of water hardness on selective flocculation of diasporic bauxite. *Transactions of Nonferrous Metals Society of China*, 22 (9), 2248–2254.

Livshits, A. K. & Dudenkov, S. V. 1965. Some factors in flotation froth stability. In: Arbiter, N. (ed.) *7th International Minerals Processing Congress*. New York: Gordon and Beam.

Lovell, V. M. 1976. *Froth Characteristics in Phosphate Flotation*. Baltimore, MD: Port City Press.

Luz, P. A. B., Yildirim, I., Yoon, R. H. 2000. Purification of Brazilian kaolin clay by flotation. *Developments in Mineral Processing*, 13, C8b-79–C8b-83.

Ma, M. 2011a. Enhancement of hematite flocculation in the hematite–starch–(low-molecular-weight) poly(acrylic acid) system. *Industrial & Engineering Chemistry Research*, 50 (21), 11950–11953.

Ma, M. 2011b. The dispersive effect of sodium silicate on kaolinite particles in process water: Implications for iron-ore processing. *Clays and Clay Minerals*, 59 (3), 233–239.

Ma, M. 2012a. The dispersive effect of sodium hexametaphosphate on kaolinite in saline water. *Clays and Clay Minerals*, 60 (4), 405–410.

Ma, M. 2012b. Froth flotation of iron ores. *International Journal of Mining Engineering and Mineral Processing*, 1 (2), 56–61.

Ma, M. 2012c. The significance of dosing sequence in the flocculation of hematite. *Chemical Engineering Science*, 73, 51–54.

Ma, X., Bruckard, W. J. & Holmes, R. 2009. Effect of collector, pH and ionic strength on the cationic flotation of kaolinite. *International Journal of Mineral Processing*, 93 (1), 54–58.

Ma, M., Bruckard, W. J. & McCall, D. W. 2012. Role of water structure-making/breaking ions in the cationic flotation of kaolinite: Implications for iron ore processing. *International Journal of Mining Engineering and Mineral Processing*, 1 (1), 17–20.

Makó, É., Frost, R. L., Kristóf, J. & Horváth, E. 2001. The effect of quartz content on the mechanochemical activation of kaolinite. *Journal of Colloid and Interface Science*, 244 (2), 359–364.

Makokha, A. B., Moys, M. H. & Bwalya, M. M. 2011. Modeling the RTD of an industrial overflow ball mill as a function of load volume and slurry concentration. *Minerals Engineering*, 24 (3–4), 335–340.

Mark, M. 2011. The dispersion of kaolinite. In: Ausimm, T. (ed.) *Iron Ore 2011, 11–13 July, 2011*. Perth: Australasian Institute of Mining and Metallurgy, 471–474.

Martin, C., Ohrling, T., Olsen, T. & Taggart, P. 2002. From tribulation to triumph: Flotation improvements at the new Lac des Iles concentrator. SGS Mineral Services – Technical Bulletin 2002-10.

Mathur, S., Singh, P. & Moudgil, B. M. 2000. Advances in selective flocculation technology for solid–solid separations. *International Journal of Mineral Processing*, 58 (1–4), 201–222.

Mathur, S. 2002. Kaolin flotation. *Journal of Colloid and Interface Science*, 256 (1), 153–158.

Matis, K. A., Gallios, G. P. & Kydros, K. A. 1993. Separation of fines by flotation techniques. *Separations Technology*, 3 (2), 76–90.

Mekhamer, W. K. 2011. Stability changes of Saudi bentonite suspension due to mechanical grinding. *Journal of Saudi Chemical Society*, 15 (4), 361–366.

Mindat. 2012. Mineral and Locality Database [Online]. Available: www.mindat.org (accessed 22 June 2012).

Mohanty, S. & Das, B. 2010. Optimization studies of hydrocyclone for beneficiation of iron ore slimes. *Mineral Processing and Extractive Metallurgy Review*, 31 (2), 86–96.

Moolman, D. W., Eksteen, J. J., Aldrich, C. & van Deventer, J. S. J. 1996. The significance of flotation froth appearance for machine vision control. *International Journal of Mineral Processing*, 48 (3–4), 135–158.

Napier-Munn, T. J., Morrell, S., Morrison, R. D. & Kojovic, T. 1996. *Mineral Comminution Circuits, Their Operation and Optimization*.Indooroopilly, QLD: Julius Kruttschnitt Mineral Research Centre, University of Queensland.

Ndlovu, B., Burdukova, E., Becker, M., *et al.* 2010. An investigation on the effect of chrysotile mineralogy and surface chemistry on slurry rheology. In: Ausimm, T. (ed.) *25th International Minerals Processing Congress*. Brisbane, QLD: Australasian Institute of Mining and Metallurgy, 367–376.

Ndlovu, B. N., Forbes, E., Becker, M., *et al.* 2011. The effects of chrysotile mineralogical properties on the rheology of chrysotile suspensions. *Minerals Engineering*, 24 (9), 1004–1009.

Neethling, S. J. & Cilliers, J. J. 2002. The entrainment of gangue into a flotation froth. *International Journal of Mineral Processing*, 64 (2–3), 123–134.

Nguyen, D. Q. & Boger, D. V. 1983. Yield stress measurement for concentrated suspensions. *Journal of Rheology*, 27 (4), 321–349

Nguyen, D. Q. & Boger, D. V. 1985. Direct yield stress measurement with the vane rheometer. *Journal of Rheology*, 29 (3), 335–347.

Notebaart, C. W. & Meer, F. P. 1992. High selectivity in HGMS by capture on the downstream side of matrix elements. *Minerals Engineering*, 5 (10–12), 1135–1145.
Ohara, T. 2014. Progress in magnetic separation technology for processing large quantities of dilute suspensions. *Journal of the Japan Society of Powder and Powder Metallurgy*, 61 (S1), S139–S144.
Peng, Y. & Bradshaw, D. 2012. Mechanisms for the improved flotation of ultrafine pentlandite and its separation from lizardite in saline water. *Minerals Engineering*, 36–38, 284–290.
Peng, Y. & Seaman, D. 2011. The flotation of slime–fine fractions of Mt. Keith pentlandite ore in de-ionised and saline water. *Minerals Engineering*, 24 (5), 479–481.
Pindred, A. & Meech, J. A., 1984. Interparticular phenomena in the flotation of hematite fines. *International Journal of Mineral Processing*, 12 (1–3), 193–212.
Pourghahramani, P. 2012. Effects of ore characteristics on product shape properties and breakage mechanisms in industrial SAG mills. *Minerals Engineering*, 32, 30–37.
Prasad, M. S., Reid, K. J. & Murray, H. H. 1991. Kaolin: Processing, properties and applications. *Applied Clay Science*, 6 (2), 87–119.
Raghukumar, C., Tripathy, S. K. & Monahan, S. 2012. Beneficiation of Indian high alumina iron ore fines: A case study. *International Journal of Mining Engineering and Mineral Processing*, 1 (2), 94–100.
Rogers, R. S. C., Bell, D. G. & Hukki, A. M. 1982. A short-lived radioactive tracer method for the measurement of closed circuit ball mill residence time distributions. *Powder Technology*, 32 (2), 245–252.
Rosario, P., Hall, R., Grundy, M. & Klein, B. 2011. A preliminary investigation into the feasibility of a novel HPGR-based circuit for hard, weathered ores containing clayish material. *Minerals Engineering*, 24 (3–4), 290–302.
Rubenstein, J. B. & Melik-Gaikazyan, V. I. 1998. Characterisation of flotation froths. In: Laskowski, J. S. & Woodburn, E. T. (eds) *Frothing in Flotation II*. Amsterdam: Gordon & Breach.
Savassi, O. N., Alexander, D. J., Franzidis, J. P. & Manlapig, E. V. 1998. An empirical model for entrainment in industrial flotation plants. *Minerals Engineering*, 11 (3), 243–256.
Scales, P. J., Johnson, S. B. & Kapur, P. C. 2000. The influence of surface chemistry on the rheology and flow of flocculated particulate suspensions. *Mineral Processing and Extractive Metallurgy Review*, 20 (1), 27–40.
Schubert, H. 2008. On the optimization of hydrodynamics in fine particle flotation. *Minerals Engineering*, 21 (12–14), 930–936.
Senior, G. D. & Thomas, S. A. 2005. Development and implementation of a new flowsheet for the flotation of a low grade nickel ore. *International Journal of Mineral Processing*, 78 (1), 49–61.
Shabalala, N. Z. P., Harris, M., Leal Filho, L. S. & Deglon, D. A. 2011. Effect of slurry rheology on gas dispersion in a pilot-scale mechanical flotation cell. *Minerals Engineering*, 24 (13), 1448–1453.
Shi, F. N. & Napier-Munn, T. J. 1996. A model for slurry rheology. *International Journal of Mineral Processing*, 47 (1–2), 103–123.
Shi, F. N. & Napier-Munn, T. J. 2002. Effects of slurry rheology on industrial grinding performance. *International Journal of Mineral Processing*, 65 (3–4), 125–140.
Shi, F. N. & Zheng, X. F. 2003. The rheology of flotation froths. *International Journal of Minerals Processing*, 69 (1–4), 115–128.
Srivastava, M. P., Pan, S. K., Prasad, N., & Mishra, B. K., 2001. Characterization and processing of iron ore fines of Kiriburu deposit of India. *International Journal of Mineral Processing*, 61 (2), 93–107.

Subbanna, M., Pradip & Malghan, S. G. 1998. Shear yield stress of flocculated alumina–zirconia mixed suspensions: Effect of solid loading, composition and particle size distribution. *Chemical Engineering Science*, 53 (17), 3073–3079.

Subrahmanyam, T. V. & Forssberg, E. 1988. Froth stability, particle entrainment and drainage in flotation: A review. *International Journal of Mineral Processing*, 23 (1–2), 33–53.

Sutherland, K. L. & Wark, I. W. 1955. *Principles of Flotation*. Melbourne, VIC: Australian Institute of Mining and Metallurgy.

Svoboda, J. & Fujita, T. 2003. Recent developments in magnetic methods of material separation. *Minerals Engineering*, 16 (9), 785–792.

Tao, D., Luttrell, G. H. & Yoon, R. H. 2000. A parametric study of froth stability and its effect on column flotation of fine particles. *International Journal of Mineral Processing*, 59 (1), 25–43.

Trahar, W. J. 1981. A rational interpretation of the role of particle size in flotation. *International Journal of Mineral Processing*, 8 (4), 289–327.

Van Olphen, H. 1951. Rheological phenomena of clay sols in connection with the charge distribution on the micelles. *Discussions of the Faraday Society*, 11, 83–96.

Vdović, N., Jurina, I., Škapin, S. D. & Sondi, I. 2010. The surface properties of clay minerals modified by intensive dry milling: Revisited. *Applied Clay Science*, 48 (4), 575–580.

Verwey, E. J. W. & Overbeek, J. T. G. 1948. *Theory of Stability of Lyophobic Solids*.Amsterdam: Elsevier.

Visser, P. R., Coetzee, M. L. & Kendall, S. G. 1994. Flotation of autogenously milled material. *Minerals Engineering*, 7 (2–3), 357–369.

Wang, J., Somasundaran, P. & Nagaraj, D. R. 2005. Adsorption mechanism of guar gum at solid–liquid interfaces. *Minerals Engineering*, 18 (1), 77–81.

Wang, Y. & Forssberg, E. 1997. Ultra-fine grinding and classification of minerals. In: Kowatra, S. (ed.) *Comminution Practice*. Littleton, CO: Society for Mining, Metallurgy and Exploration, 203–214.

Wasmuth, H. D. & Unkelbach, K. H., 1991. Recent developments in magnetic separation of feebly magnetic minerals. *Minerals Engineering*, 4 (7–11), 825–837.

Wills, B. A. & Finch, J. A. 2016. Gravity concentration. In: *Wills' Mineral Processing Technology*, 8th edition. Boston, MA: Butterworth-Heinemann.

Wilson, H. & Middleton, G. J. 1937. Purification of kaolin: Some data on the classification of quartz and muscovite. *Journal of American Ceramic Society*, 20 (1–12), 126–136.

Xu, D., Ametov, I. & Grano, S. R. 2011. Detachment of coarse particles from oscillating bubbles: The effect of particle contact angle, shape and medium viscosity. *International Journal of Mineral Processing*, 101 (1–4), 50–57.

Xu, Z., Liu, J., Choung, J. W. & Zhou, Z. 2003. Electrokinetic study of clay interactions with coal in flotation. *International Journal of Mineral Processing*, 68 (1–4), 183–196.

Yianatos, J. & Contreras, F. 2010. Particle entrainment model for industrial flotation cells. *Powder Technology*, 197 (3), 260–267.

Yoon, R. H., Nagaraj, D. R., Wang, S. S. & Hildebrand, T. M. 1992. Beneficiation of kaolin clay by froth flotation using hydroxamate collectors. *Minerals Engineering*, 5 (3–5), 457–467.

Young, R. H., Morris, H. H., Brooks, R. L. 1985. *Method of Treating Clay to Improve its Whiteness*. US Patent US4492628.

Yue, J. & Klein, B. 2004. Influence of rheology on the performance of horizontal stirred mills. *Minerals Engineering*, 17 (11–12), 1169–1177.

8

Solid–Liquid Separation of Clay Tailings

Phillip Fawell

8.1 Solid–Liquid Separation

8.1.1 Techniques Used and the Relationship to Particle Size

Hydrometallurgical processing of minerals invariably involves at least one stage of solid–liquid separation. This can be for a variety of purposes, ranging from the clarification of liquors through to high-solids slurry dewatering, usually with the needs of downstream processes in mind. Particle size is always a primary consideration when selecting the appropriate unit operation for concentrating or dewatering slurries. Fig. 8.1 shows the general approaches that can be applied, depending on the size range. Note the high degree of overlap, particularly in the range from 10^{-6} to 10^{-4} m (or 1–100 μm, generally the preferred unit when describing fine particles in these applications).

What Fig. 8.1 fails to capture are the additional aspects of throughput and stability (i.e. consistency) of the feed volumes. Gravity thickeners are typically employed to treat dilute suspensions when high volumetric throughputs are expected. There are few situations in which screens, hydrocyclones, filtration and centrifugation are viable alternatives, but their number is increasing as new equipment technologies become available. Screens and hydrocyclones will always struggle to process particles smaller than 20–30 μm, but they increasingly represent practical pre-treatment options when feeds contain broad distributions, e.g. coarse sand can be removed before thickening a mainly clay-based feed. While the throughputs are limited for filtration and centrifugation, both can be used successfully in batch operations. Importantly, thickening at any significant scale requires the implementation of continuous operation.

The high volumetric throughputs and low feed solids concentrations typical within the minerals industry leads to solid–liquid separation being achieved predominantly via gravity thickening. The main features of a thickener are represented in Fig. 8.2. Low-solids feed streams are delivered to the feedwell either though pipes or open launders, generally with a high velocity. The solids are then allowed to settle and consolidate in the main body of the thickener, while clarified liquor is removed at the overflow. An important function of the feedwell is to dissipate much of the input stream's energy. Rakes may be present in the cone section in order to assist the movement of solids to the underflow and potentially

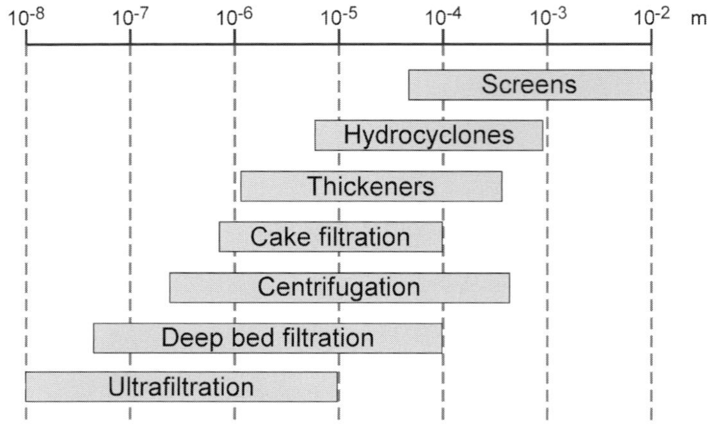

Figure 8.1. Typical size ranges for different dewatering techniques.

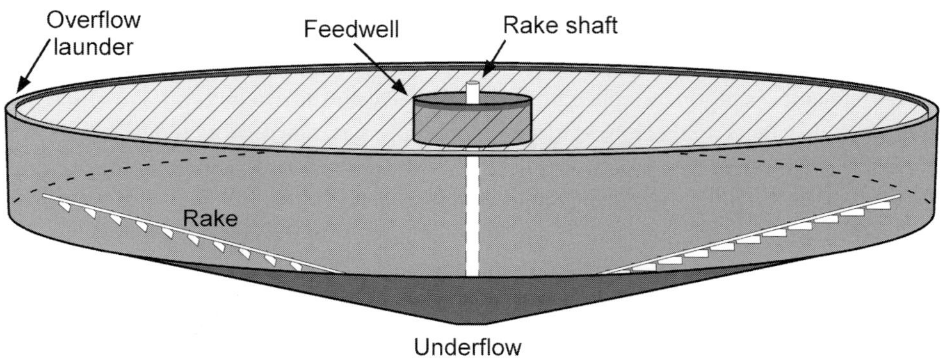

Figure 8.2. Schematic representation of a thickener.

provide additional consolidation forces. The underflow rate may represent 10–20 per cent of the original volumetric feed rate.

It is worthwhile clarifying some basic terminology for readers unfamiliar with solid–liquid separation. ‘Settling rate’ refers to the initial linear portion of the mud-line settling curve, in which the visible interface of the settling solids is monitored as a function of time, as schematically illustrated in Fig. 8.3a. The equilibrium (‘final’) mud-line height can be used to provide an indication of settled bed density, although it can be unreliable and is of questionable value in anything other than raked tests. More detailed information can be achieved from the complete analysis of settling curves (e.g. Grassia *et al.*, 2011), but that is beyond the scope of this review.

As particles settle in a bed, the solids concentration (ideally expressed as the volume of the solid phase divided by total volume of all phases, termed the volume fraction, ϕ) reaches a value called the gel point (ϕ_{gel}), where particles ‘just touch’ and form a continuous network. The gel point is taken as the lowest concentration giving a detectable yield

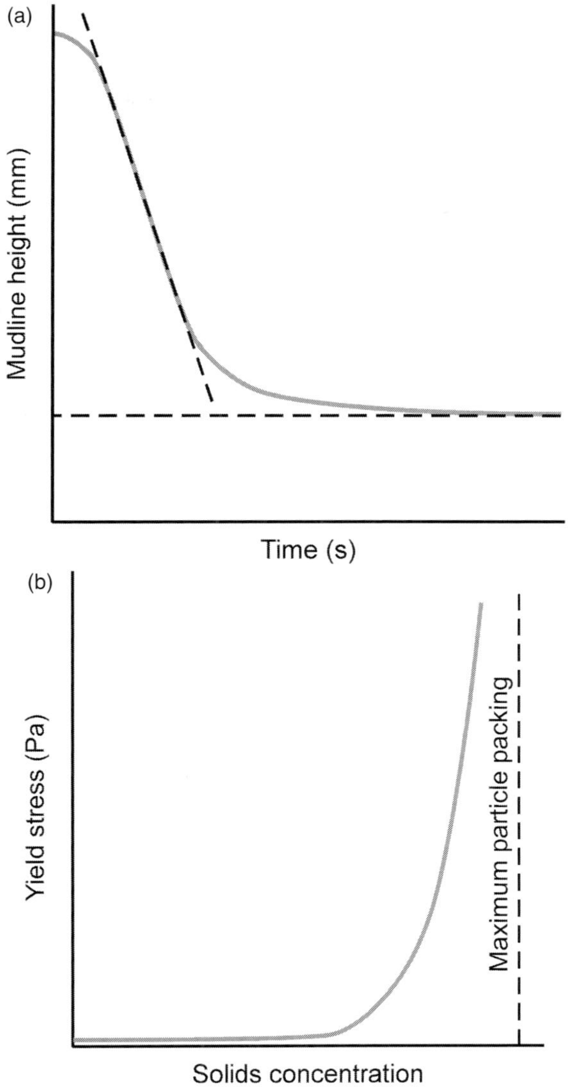

Figure 8.3. Schematic representations of (a) mud-line settling curve and (b) compressive yield stress response.

stress, $P_y(\phi)$, and is the concentration at the very top of a sediment at steady-state (i.e. after several days of batch settling). The compressive yield stress is a measure of network strength – Fig. 8.3b shows a typical yield stress response as a function of solids concentration. For $\phi < \phi_{gel}$, $P_y(\phi)$ is 0. When $\phi \geq \phi_{gel}$ it can be represented by Eq. (8.1):

$$P_y(\phi) = k\left(\left(\frac{\phi}{\phi_{gel}}\right)^n - 1\right) \tag{8.1}$$

where k and n are constants. The expected range for n is from 4 to 30, with clays being at the lower end, given their likelihood to form more compressible beds. The vertical dashed line in Fig. 8.3b is the maximum particle packing concentration, which may be approached through filtration but is unlikely to ever be practically achieved.

8.1.2 Need for Aggregation

An important parameter in any thickening application is the rate of liquor upflow (often referred to as the nominal thickener rise rate or velocity), determined by dividing the volumetric flow to the overflow by the available horizontal cross-sectional settling area (hatched area in Fig. 8.2). This assumes perfectly symmetrical flows throughout the vessel, which are rarely achieved, but the calculation provides a practical guide to the minimum settling rates required (industry targets range from 1.4 to 2 times the rise rate).

Fine particle slurries have very low settling rates and thickening is often only practical through aggregation to enhance sedimentation. This is especially the case for clay slurries, which will all be <45 μm and may contain significant submicron fractions (note that clay mineralogists would consider clays to all be <2 μm, but clays are rarely fully dispersed in tailings). Thickeners over 80 m in diameter still exist, operating at throughputs that produce nominal rise rates of <0.3 m/h. Under such conditions even fine clay slimes in moderate ionic strength liquors may settle without additional treatment. Such designs are now the exception, as the need for high throughputs was mostly met with the development of synthetic flocculants. Via flocculation, aggregate sizes up to several millimetres can form from particles in the micron range, thereby raising settling rates by several orders of magnitude.

Aggregation also favours filtration methods, with the open structures leading to cakes of enhanced permeability. Several studies have tested flocculants to improve solid–liquid separation in hydrocyclones (Franks *et al.*, 2005; Visman and Hamza, 1973; Yalcin, 1996), although the high shear would lead to significant aggregate degradation and high dosages. The direct application of flocculants to high-solids clay slurries prior to deposition is also becoming an established process to achieve even greater dewatering (Adkins, 2008; Wells *et al.*, 2011, 2015). All of these related processes require the properties of the fragile aggregates to be optimized, and while they have similarities, their specific needs differ significantly. Given that thickening represents the most widespread application of flocculation in the minerals industry, it will be used as the template to explore the many intricacies of the processes influencing flocculation. The other applications will then be contrasted against this background.

8.2 The Flocculation Process as It Occurs in Mineral Processing

Thickeners and other unit operations mentioned above are applicable across a wide range of commodities, and are often discussed in generic terms. The flocculation process itself often benefits from being considered in this way, and clay slurries have been utilized as

Table 8.1. *Comparison of aggregation by both coagulation and flocculation.*

Property/consideration	Coagulation	Flocculation
Aggregate strength	Low	Medium
Response to application of brief duration of high shear	Reversible breakage; aggregates will reform under low shear	Irreversible breakage; ruptured bridges have reduced bridging capacity
Optimal primary particle size range	<2 μm	1–50 μm
Outside optimal size range	Surface charge effects ineffective at holding larger particles together	Dosages required to contact all particles are prohibitively high at small sizes; larger sizes will settle rapidly and do not require aggregation, but will increase settling rates if captured in aggregates of fines
Typical settling rates achieved	1–2 m/h, but often lower; higher only from subsequent flocculant addition	5–50 m/h; much higher is possible, but rarely required

the substrate in numerous studies of flocculant performance or aggregate properties. This is both a boon and a curse; while the body of literature dealing with clay flocculation is immense, much of this is done without a proper appreciation of flocculation issues specific to minerals processing. Consequently, much of the literature is either irrelevant or misleading. The flocculation of clay slurries is also particularly affected by the distinct surface properties for different particle faces, a significant point of differentiation from other common substrates.

It is necessary to first define terminology related to the aggregation process that is frequently used either incorrectly or ambiguously:

(1) *Coagulation* represents aggregation achieved solely through the manipulation of the surface charge to enable particles to come together, either through the presence of areas or patches of opposite charge (through salts or low molecular weight polymers) or the reduction of the effective charge magnitude by increasing ionic strengths or adjusting pH.
(2) *Flocculation* occurs through the formation of bridges between particles, generally by high molecular weight synthetic polyelectrolytes (predominantly acrylamide/acrylate copolymers).

This distinction is very important, as indicated by the summary in Table 8.1. Flocculated aggregates will be stronger, and under most conditions the bridging process will lead to larger aggregates than those formed by coagulation. While adsorption of individual flocculant chain functionalities is weak, their large number in an adsorbed segment (or 'train') leads to an effectively irreversible bond (Lyklema, 1988). When these aggregates are broken under shear, it is through chain rupture rather than functional detachment (Gregory, 1988). As the remaining 'tail' has reduced capacity for bridging, the aggregates will not reform thereafter. This is a critical aspect of optimizing flocculation in mineral processing in order to achieve the highest throughput.

The coagulation process is the domain of classical colloid theory dealing with the impact of electrolytes and adsorbed cations on double-layer structures. The relevance of such theories to aggregation is strong with submicron particles, but diminishes as their size increases and shear conditions become more intense. Bridging flocculation can be applied over a broad particle size range, but is not viable as a stand-alone process for submicron solids. Hogg (1999) gave a statistical analysis of the probability of a flocculant (molecular weight 20×10^6) at a moderate dosage (25 g/t) adsorbing on different sized particles. For a median size of 1 μm, only ~0.2 of a polymer molecule would be expected to adsorb per particle; at 0.5 μm this dropped well below 0.1, i.e. there would be more than ten particles for every polymer chain present. At small particle sizes, the dosages that would then be required to ensure all particles have some probability of contacting flocculant become prohibitively high.

Such estimates also assume adsorption to be fast and a homogeneous distribution of flocculant throughout the slurry. Given that flocculant volumes are typically small relative to those of the slurries to which they are applied, an even distribution is highly unlikely. These factors mean that flocculation by itself is not a practical option when treating a slurry with a significant fraction of submicron particles. Pre-treatment with a coagulant (either a salt or a much lower molecular weight polyelectrolyte) to initially produce a low degree of aggregation will significantly reduce the effective surface area of the suspension, making subsequent flocculant addition much more effective at realistic dosages (Chaplain *et al.*, 1995). This approach is relevant across all substrates, particularly so for the treatment of clay-based tailings. The reduction of effective surface area available to a flocculant is the key idea behind the enhancement of flocculation (or reduction in required dosages) from the split or staged addition of flocculant. The factors influencing the success of such dosing in mineral systems has been discussed elsewhere (Owen *et al.*, 2008).

The majority of flocculants used in the thickening of mineral systems are copolymers of acrylamide and sodium acrylate, with products varying in terms of molecular weight, charge density and physical form (solid powders vs. emulsion). Cationic flocculants are frequently discussed in the literature, particularly for clays, but this is often for low-throughput ceramic manufacture. These flocculants are typically of lower molecular weight, although still giving some bridging between particles. They are rarely applied in high-throughput mineral processing applications.

In general, the main purpose of the anionic functionality of the flocculant is to provide chain extension of the polymer in solution (Fig. 8.4), reducing the applied dosages by enhancing the probability of bridging adsorption. Actual surface adsorption is usually through the amide functionality by hydrogen bonding to neutral surface groups on the

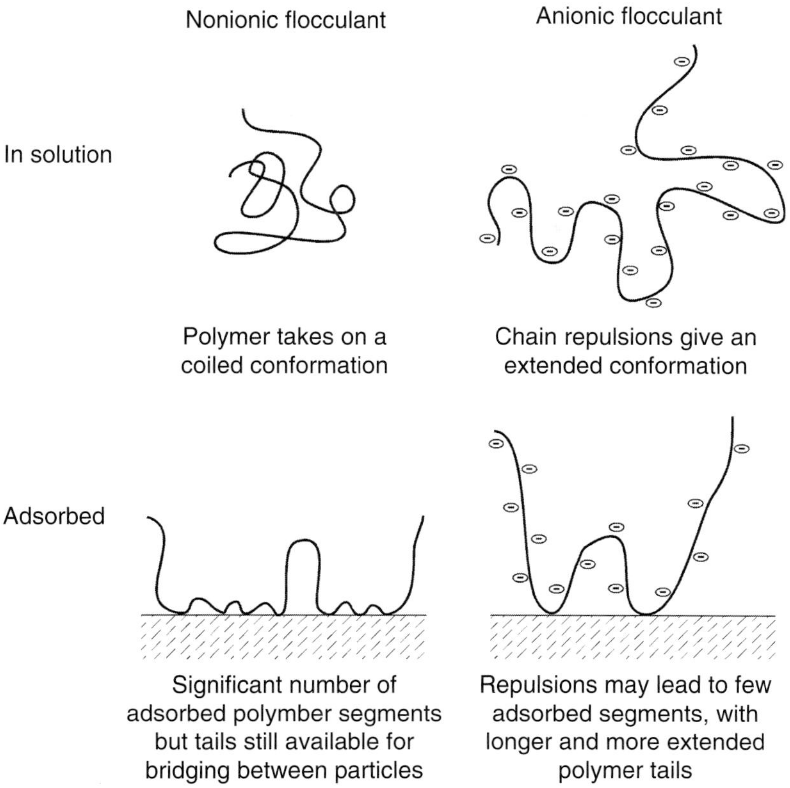

Figure 8.4. How flocculant solution and adsorbed conformation influences polymer-bridging capacity. Note that chain repulsions are reduced at high ionic strength, leading to anionic polymer behaviour more like their nonionic equivalents.

particle substrate (surface adsorbed cations can enable adsorption through anionic functionalities, allowing polyacrylate homopolymers to induce aggregation). Adsorption will be favoured near the point of zero charge (pzc) for a particular substrate slurry, but it is important to note that even at pH values several units either side of the pzc, a distribution of charges will exist across the particle surface, including a proportion of neutral sites.

Surface structures become even more important when clay flocculation is being considered, given the distinction between edge and basal face structures, as shown in Fig. 8.5 and discussed in earlier chapters. Consequently, any measurement of bulk pzc values is of little value for clay slurries, as they cannot be used reliably to predict coagulation or adsorption responses. The basal faces are dominated by silanol groups, which are negatively charged over most pH values and effectively control the bulk pzc as they represent the major part of the total surface area. In contrast, the edge faces have a higher proportion of aluminol groups that have a greater prospect of providing neutral sites for flocculant adsorption. Flocculants will therefore almost exclusively adsorb on the edges, which thereby determine

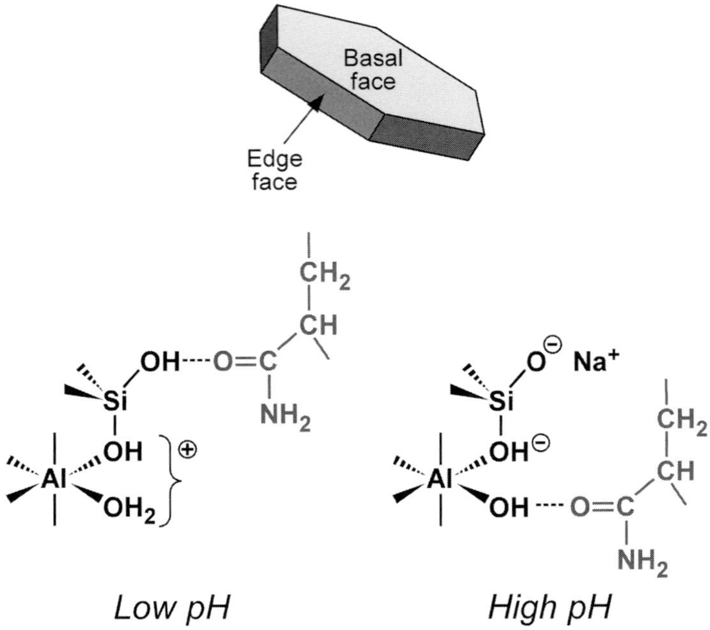

Figure 8.5. Adsorption of polyacrylamide flocculant amide functionality onto clay surfaces at low and high pH.

eventual aggregate structures. While neutral sites are expected to be most prevalent at pH values between 5 and 8, they may still be present at quite high pH values, as represented in Fig. 8.5 for adsorption of the amide functionality. Subsequent aggregation can then still be expected, although both the kinetics and extent of the process may be affected, particularly as the pH reaches 10 or above.

The aggregates formed by coagulation or flocculation are both highly porous, with this porosity critically impacting upon aggregate growth, sediment dewatering and rheology. Aggregate porosity is usually described through fractal geometry, thereby accounting for aggregates becoming increasingly porous as size increases. The total enclosed volume of an aggregate (solid particles plus void volume) is treated as a sphere and taken to increase with the cube of a characteristic length (L, either diameter or radius). The mass (m) increases to a lower power, using the relationship:

$$m \propto L^{D_f} \tag{8.2}$$

where D_f is the mass-length fractal dimension and can take a value between 1 and 3, with the latter representing a solid sphere. Typically, coagulated aggregates will give lower values of D_f, frequently quoted in the range 1.5–2, while flocculated aggregates will be greater than 1.8. In reality, a single value of D_f may be insufficient to fully capture a system's behaviour, with multiple or variable fractal models also proposed (Maggi *et al.*, 2007; Vahedi and Gorczyca, 2012). However, for most practical purposes the use of a single value is effective at representing the impact of aggregate porosity on performance measures. In the case of clays, the porosity is further influenced by particle shape and surface effects; their impact on flocculation kinetics is discussed below.

8.3 The Kinetics of Aggregate Growth in Flocculation Processes

The majority of published studies of clay flocculation can be grouped into two areas, being of either low or intermediate solids concentrations studies. Defining the boundaries for these regions is very much dependent upon the particle size and mineralogy, but for this discussion 'low' will be taken as well below 10 kg/m^3 (note that kg/m^3 is equivalent to g L^{-1}), with 'intermediate' roughly as 10–150 kg/m^3, and more specifically as the concentrations encountered for clay tailings feeds to gravity thickening. Actual studies of high-solids flocculation are comparatively rare and few of these can be considered robust or detailed, although an area of increasing interest, mainly through oil sands tailings applications (Section 8.9). Before high solids can be discussed, the contrasting behaviour at low and intermediate solids concentrations must first be considered.

8.3.1 Low Solids

Aggregation, whether by coagulation or flocculation, is predicated on achieving collisions between particles that lead to an increase in the effective size. This can be particle–particle, particle–aggregate or aggregate–aggregate, and for most practical systems is achieved in turbulent fluid flow. Saffman and Turner (1956) proposed that the aggregation kernel, β_{ij}, between particles i and j can be described as:

$$\beta_{ij} = 1.294\alpha G\left(a_i + a_j\right)^3 \tag{8.3}$$

$$G = \sqrt{\frac{\varepsilon}{\upsilon}} \tag{8.4}$$

where a_i and a_j are the particle radii (m)
G is the mean shear rate (s^{-1})
α is the capture (or collision) efficiency [0,1]
ε is the energy dissipation rate (m^2/s^3)
v is the kinematic viscosity (m^2/s)

The term α is added in recognition of the fact that not every collision is effective in leading to aggregation. Hydrodynamic and electrostatic repulsions both influence the collision efficiency, the latter reflected in the surface sites available for flocculant adsorption previously discussed. The adsorption of bridging polymers in itself influences α, with La Mer and Healy (1963) proposing a relationship with the flocculant fractional surface coverage (θ) of:

$$\alpha = \theta(1 - \theta) \tag{8.5}$$

This relationship implies α (and therefore aggregation rates) will increase with flocculant dosage up to 50 per cent surface coverage, but at higher coverages it decreases as a consequence of steric hindrance. There have been numerous studies that have refined this relationship (Berlin and Kislenko, 1995; Hsu *et al.*, 1995; Olsen *et al.*, 2006), but the original version suffices in the context of this discussion.

The aggregation rate is second order with respect to particle concentration, and at low solids concentrations both the aggregation rate and aggregate size will increase as concentration is increased. Higher shear will also increase the aggregation rate, although the impact on aggregate size will be dependent upon a number of factors, including the flocculant dosage. Increasing aggregation rate through higher collision rates is an important aspect of most low-solids water treatment applications, where fines capture is critical for clarification.

With the lower collision rates at low solids, the values for θ tend to be higher, and certainly values approaching or exceeding 0.5 may be realized, i.e. there is the danger of flocculant overdosing. This is further complicated by the difficulty in efficiently dispersing flocculants throughout slurries in such situations. Very high dilution of the flocculant solutions are required (i.e. to 0.005 wt/wt% or even lower) to achieve any sort of efficiency, and without such dilution, applied dosages can be inflated. As large aggregate sizes are rarely sought, lower molecular weight flocculants, which provide a higher number of polymer chains per mass unit, are preferred.

8.3.2 Intermediate Solids

While further increases in solids concentrations will continue to raise the aggregation rate, at some point the effect of aggregate breakage rate (S_i) becomes a significant contributing factor (Curtis and Hocking, 1970; Heath *et al.*, 2006a; Oles, 1992).

$$S_i \propto \left(\frac{\varepsilon\rho}{\mu}\right)^n \qquad (8.6)$$

where ε is the energy dissipation rate, μ is the suspension viscosity (N s m^{-2}), ρ the suspension density (kg/m^3) and n an exponent higher than that in the aggregation kernel. Aggregation is typically taken as proportional to $\varepsilon^{0.5}$, with n therefore in the range 0.5 to 1.0. In other words, both aggregation and breakage rates are increased by higher shear rates, but breakage dominates and the aggregate sizes that can be achieved will reduce.

For simple coagulation reactions, the aggregate size will achieve a plateau value under a constant applied mean shear rate, and there is the prospect that lowering the shear will lead to an increase in the plateau size. This reflects that aggregation and breakage reach an effective equilibrium. No such equilibrium exists in polymer-bridging flocculation; as already noted, aggregates broken under shear will not reform in the absence of fresh flocculant. A peak in aggregate size is therefore observed, with the decline in size at longer reaction times being a reflection of the applied shear, the aggregate structure, the flocculant dosage and (potentially) flocculant type.

Hence, in contrast to the low-solids situation, aggregate size does not continue to rise with increasing solids concentration at a fixed flocculant dosage (calculated per unit mass of solid), but reaches an optimum, above which the size then declines. This decline can be attributed in part to a slightly reduced particle collision rate at a lower mean shear rate (a consequence of higher viscosity), but more so to an increased breakage rate due to the increase in viscosity and energy dissipation rate at higher solids (Heath *et al.*, 2006a). The increase in viscosity relative to the unflocculated solids is heightened by the very low density of the flocculated aggregates, which greatly increases the effective solids volume fraction (ϕ_{eff}), as shown in Eq. (8.7) (Heath *et al.*, 2006b; Potanin and Uriev, 1991; Richardson and Zaki, 1954):

$$\phi_{eff} = \phi_s \left(\frac{d_{agg}}{d_p} \right)^{3-D_f} \tag{8.7}$$

where ϕ_s is the solids volume fraction, d_{agg} and d_p are the aggregate and particle diameters, respectively, and D_f the fractal dimension. Note that from Eq. (8.7) a low aggregate density (low D_f) will elevate ϕ_{eff} and so increase viscosity, thereby lowering the optimum solids concentration.

Understanding and quantifying a particle system's flocculation response to changes in solids concentration is almost always critical to predicting or resolving performance issues, and yet is often neglected or poorly controlled in published studies. Flocculation at too high a solids concentration is done at the risk of reducing sensitivity to (or even completely masking) many of the effects of flocculant selection/dosage or flocculation conditions. Dosage responses will also vary substantially at different solids loadings. Given that flocculant selection can also impact upon aggregate structures, the best-performing flocculant at one solids concentration may not be the best at another concentration.

As noted, higher mean shear rates accelerate both aggregate growth and breakage, with the higher exponent in the breakage response limiting the aggregate size that can be attained. In kinetic experiments where constant mean shear rates are applied across the full reaction volume, whenever the shear rate is increased, the peak aggregate size that is attained will substantially drop, and the degree of aggregate breakage at reaction times beyond this peak will increase (i.e. there is a steeper decrease in size at these longer reaction times). The advantage of 'tapered shear', in which the initial shear is high but is then decreased, can then be seen. It favours the initial mixing of flocculant and aggregate formation, but then provides milder conditions to maximize aggregate growth. The concept is well understood in water treatment applications (Nan and He, 2012; Yukselen *et al.*, 2006) and while not widely recognized as such in the minerals industry, actually forms the basis for optimal flocculation in well-designed and -operated thickener feedwells (Fawell *et al.*, 2009a; Owen *et al.*, 2008).

Given that flocculation will take place in regions of low and high shear, it is surprising the number of studies that attempt to quantify flocculation kinetics within impeller-stirred vessels, often using the mean shear rate for the immediate zone surrounding the impeller to characterize the applied mixing. This ignores the lower shear conditions within the bulk of the vessel volume. Such geometries may provide useful information for slow coagulation reactions at very low solids concentrations, for which reaction times are measured in minutes, but much shorter times are required in polymer-bridging flocculation processes, e.g. peak aggregate sizes may be achieved within 3–20 s. Under the broad shear rate distributions within impeller-stirred vessels, the measured aggregation response (aggregate size, settling rate) after flocculant addition is averaged across the full volume, almost certainly masking the true peak in performance, which is often achieved over a narrow range of reaction times.

The most detailed quantification of flocculation kinetics in relation to mineral systems has been with the pipe reactor system introduced by Swift *et al.* (2004), in which flocculant and slurry are mixed under turbulent pipe flow and the reaction time controlled precisely by the pipe length to the monitoring position. Essential to the success of this approach are focused beam reflectance measurement (FBRM; Lasentec, now Mettler-Toledo) probes that provide real-time monitoring of aggregate chord length distributions at all solids concentrations of interest (i.e. no need for sample dilution). In this method, profiles of aggregate size and settling rate as a function of reaction time are collected for a range of mean shear rates, solids concentrations and flocculant dosages. Heath and co-workers (Heath *et al.*, 2006a, 2006c) utilized such data to develop the first population balance (PB) model for flocculation to capture the inevitable irreversible aggregate breakage at longer reaction times for polymer-bridging flocculation (the majority of PB models claiming to describe flocculation actually deal with aggregation by reversible coagulation). This PB has been applied to more than 20 mineral systems, including clay slurries. Others in the same team have also used the pipe reactor to achieve new insights on the flocculation process and the interpretation of FBRM chord length distributions (Fawell *et al.*, 2009a; Grabsch *et al.*, 2013; Owen *et al.*, 2008).

8.3.3 Implications for Clay Systems

The flocculation issues discussed above are essentially generic across all particulate systems considered in mineral processing applications involving sedimentation, with the possible exception of ultra-low solids precipitation products from waste streams. The issues are all relevant to the flocculation of clay suspensions. Indeed, their importance is heightened by the platelet shape of clay particles and the commensurate lower proportion of the total surface area available for flocculant adsorption at typical process pH values (see Fig. 8.4). With polymer-bridging only through particle edge faces, capture efficiencies will be low. The edge–edge bridging within the aggregates leads to higher porosity (lower aggregate density), thereby substantially reducing both the optimum solids concentration for flocculation and the concentrations at which breakage rates and suspension viscosity are affected.

The PB for polymer-bridging flocculation developed by Heath *et al.* (2006a) provides an excellent tool to demonstrate the above effects on flocculation kinetics. In Fig. 8.6 the parameter set estimated for calcite flocculation (which included a value for D_f of 2.41) is

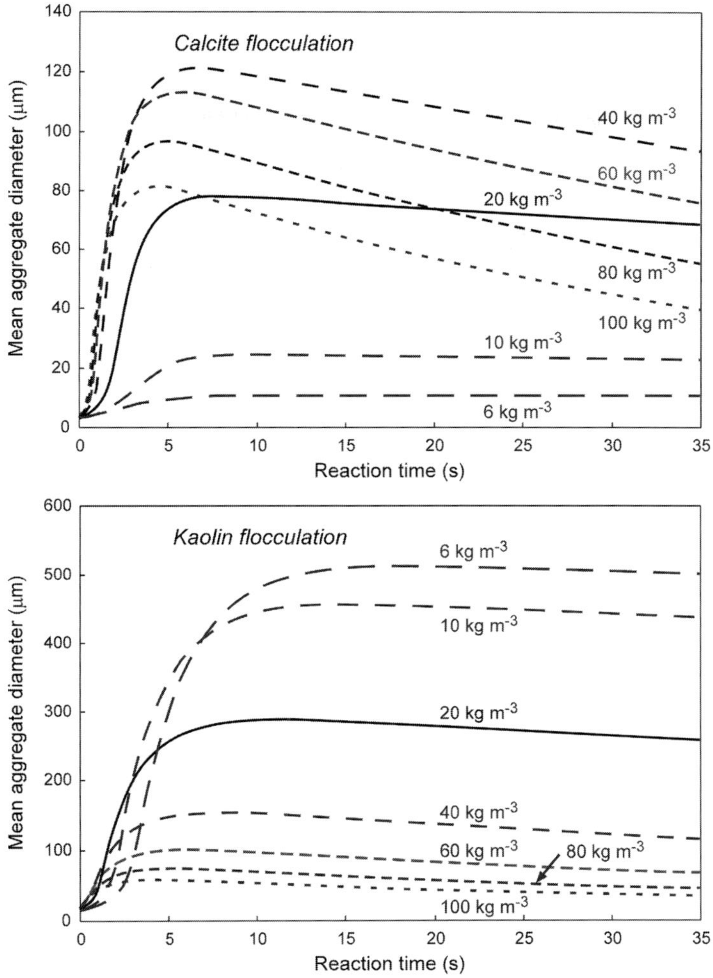

Figure 8.6. Population balance (PB) predictions for the effect of solids concentration on the flocculation of standard calcite and kaolinite slurries (pH ~6.8, low ionic strength) in turbulent pipe flow (25.4 mm ID) at a fixed mean shear rate (934 s^{-1}) and flocculant dosage (20 g/t, acrylamide/acrylate copolymer).

used to predict the effect of solids concentration for a fine calcite (d_{50} 2.8 μm) at a fixed flocculant dosage. Aggregation is clearly poor at the lowest concentration of 6 kg/m^3; collision rates and fractional surface coverage (θ) are low, leading to poor collision efficiencies (α) and slow aggregate growth. As the reaction time increases, flocculant adsorbed to the surface of a particle will lose bridging capacity as polymer tails reconform and adsorb to a greater extent on the original particle. Aggregation is thereby curtailed and only a low plateau size is achieved. Increasing the solids concentration to 10 and then 20 kg/m^3 produces higher values for α, elevating both the rate and extent of aggregation.

The impact of irreversible aggregate breakage is evident at 20 kg/m, the decline in sizes beyond a reaction time of 5 s being consistent with an increased breakage rate and no available active flocculant to promote aggregation. The optimum solids concentration in terms of aggregate size is 40 kg/m^3; above this concentration, the effective aggregate volume fraction becomes a major factor, with the effect of suspension viscosity limiting the size of aggregates that can be achieved.

The calcite solids for such purposes are considered as spherical or 'blocky' particles, with the available sites for flocculant adsorption distributed evenly across the surface. In contrast, the platelet shape of clay particles and the exclusive adsorption of flocculant onto the minor edge faces of the particles under most slurry conditions imposes an added degree of porosity to the aggregates that are formed. Heath (unpublished results) used a modification of Eq. (8.7) to account for the impact on the effective aggregate volume fraction, inserting a packing factor, K_p:

$$\phi_{eff} = \frac{\phi_s}{K_p}\left(\frac{d_{agg}}{d_p}\right)^{3-D_f} \tag{8.8}$$

Parameter estimation then applied to pipe reactor results for flocculation of a standard kaolinite slurry with an anionic flocculant gave a parameter set not greatly dissimilar to that for calcite, but the parameter setting the dosage response was 30 per cent lower (i.e. less flocculant required). Critically, the fit could only be achieved with the use of both a lower D_f (2.20) and a K_p of 0.333. These parameters were then applied to the same particle size distribution and flocculation as used for calcite, with the effect of solids concentration on the reaction profiles also shown in Fig. 8.6.

In contrast to calcite, no clearly optimum solids concentration was evident in the range 6–100 kg/m^3, with aggregate sizes decreasing as solids concentration increased. It is likely that the optimum in terms of aggregate size is near or slightly below 6 kg/m^3, well below the 40 kg/m^3 predicted for the fine calcite feed.

Note that the peak aggregate size predicted is much higher for kaolinite flocculation. This reflects: (1) the more open structure and lower fractal dimension of the kaolinite aggregates; and (2) the higher fractional adsorption onto the surface area available to flocculant, affecting collision efficiencies and the dosage response. Such large, porous aggregates are only possible because of the very low solids concentration, which means the effective aggregate volume fraction also remains low. This volume fraction becomes significant enough to impact upon the aggregate size at 10 and 20 kg/m^3, while at 40 kg/m^3 the sizes have deteriorated to be comparable to calcite at the same solids concentration. At higher solids concentrations, the flocculated calcite aggregate sizes were predicted to be larger at all reaction times.

8.3.4 Sedimentation

Figure 8.6 encapsulates perhaps the most significant problem with clay flocculation – the formation of large and fast-settling aggregates is often readily achieved, but only at very

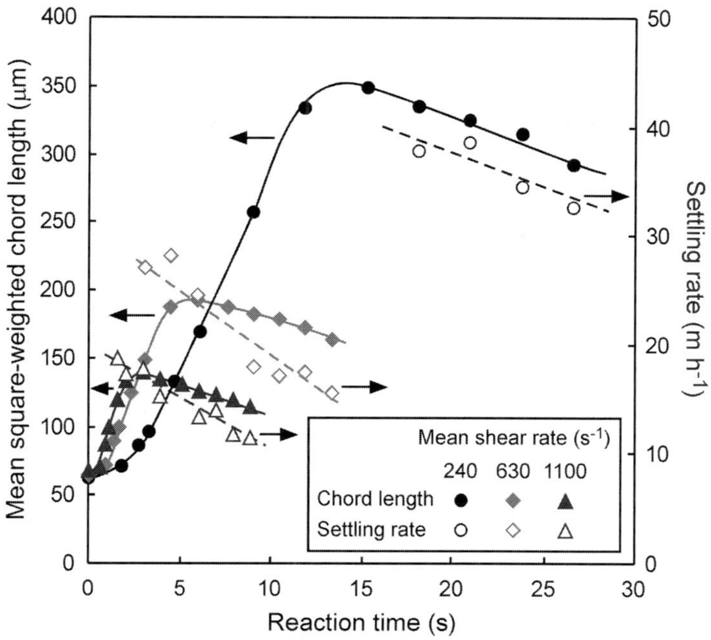

Figure 8.7. Effect of mean shear rate on aggregate size as a function of reaction time for 21 kg/m^3 kaolinite flocculated with a high molecular weight acrylamide/acrylate copolymer at a dosage of 60 g/t (adapted from Fawell *et al.*, 2009b; copyright (2009) Australasian Institute of Mining and Metallurgy).

low solids concentrations. There are some systems in mineral processing in which the thickener feeds are dilute clay slurries of below 1 wt/wt% solids, but most are in the range 3–14 wt/wt% (ignoring the presence of coarse sand fractions). Such slurries generally cannot be settled effectively without solids dilution *prior* to flocculation (dilution after flocculation will give some noticeable enhancement of mud-line settling rates, but to a far smaller degree than from actual flocculation to a larger aggregate size through solids dilution before dosing).

The actual situation is more extreme than indicated in Fig. 8.6, which only predicts aggregate sizes under the applied shear conditions. Sedimentation will take place under much milder (if not quiescent) conditions that raise the prospect of additional aggregate growth, thereby increasing settling rates beyond what would be predicted from the sizes achieved in pipe reactor studies. However, this enhancement is almost always much greater following short flocculation reaction times under applied shear than for longer times, as dramatically illustrated in Fig. 8.7. This reflects the residual activity of surface-adsorbed flocculant at short reaction times, thus allowing further growth under subsequent reduced shear (Fawell *et al.*, 2009a). Experiments with staged shear (i.e. from high to low in different pipe diameters) have been used to confirm such behaviour, including the absence of further growth at reaction times beyond the peak (Owen *et al.*, 2008). However, it should be noted that the larger settling rates at short reaction times also tend to coincide with poor supernatant clarity, with poor fines capture expected from the short duration of mixing.

Figure 8.7 also shows how the decline in settling rates at extended reaction times is accelerated under more intense mixing. In many cases (assuming the solids concentration is appropriate) this decline can be compensated for with higher flocculant dosages. While an increased dosage of an expensive consumable to treat waste tailings is hardly desirable, the cost of flocculant is also rarely the primary process driver. However, excessive flocculant usage will lead to an increase in network strength, resulting in the yield stress response to solids concentration (see Fig. 8.3b) shifting to the left. The consequences can be higher rake torque, lower underflow densities or a higher rheology for the early stages of underflow pumping.

8.3.5 Problematic Clays: Issues with Flocculation or Physical Limits?

The terms 'problematic clays' or 'poor settling clays' are frequently used in discussions on the decline in flocculation and thickening performance resulting from shifts in mineralogy, i.e. an increased proportion of one of these types of clay (e.g. an increased presence of smectite relative to kaolinite). Making this conclusion from such observations may be inappropriate for two reasons:

(1) Flocculation performance will be predominantly determined by particle size (perhaps more precisely, particle number), shape and the proportion of surface area available to flocculant adsorption.
(2) A decline in thickener performance does not necessarily imply poor flocculation. Many so-called problematic feeds actually flocculate very well *at the correct solids concentration for their particle size*, but expectations on what can be achieved must be revised.

With regards to the first point, this is not to say that mineralogy is irrelevant, as it will clearly influence the surface charge distribution and particle shape. Depending on the pH, ionic strength and adsorbed cations, the dosage response and flocculation behaviour will then shift. The distinction between swelling and non-swelling clays is significant, as is the extensive work done on the effective dispersion of clays by suitable multivalent cations. However, in most mineral processing applications, clay tailings slurries are received in a state and in liquors that are determined by upstream (and sometimes downstream) requirements, with minimal scope for adjustment. The large volumes to be considered rarely make pH or liquor composition adjustment viable, although some examples are known. One factor that is potentially controllable and which will be discussed is the level of mixing/shear applied to the slurry.

To illustrate the second point, an increase in smectite content within an otherwise well-behaved clay feed will in all likelihood lead to very poor thickener performance at the solids concentration and flocculant dosage considered optimal for the original feed (i.e. that which gives the desired settling rate to exceed the thickener rise rate). Throughput (and from this, thickener size) is determined from the settling flux, calculated from the product of the settling rate with the solids concentration, having units of mass per square metre of horizontal settling area. The 'poor' feed may require dilution to (say) one-half the previous optimum solids concentration to achieve the target settling rate, but that would then halve the throughput. Either the settling area would need to be doubled (a major capital cost) or the settling rate doubled. In a conventional thickener with a very low rise rate, this may be possible, but for a more modern high rate or high density/paste thickener with a high rise rate, higher settling rates may be difficult to consistently achieve, even with higher flocculant dosages. This would be inevitable for finer particle sizes, imposing physical limits on the throughputs that can be achieved.

8.4 Practical Aggregate Characterization

There are numerous reviews that provide excellent detail on the techniques available to characterize aggregate structures (Bushell *et al.*, 2002; Jarvis *et al.*, 2005; Klimpel and Hogg, 1991; Liao *et al.*, 2006). There is no intention to further review that work here, other than to point out that the majority of such techniques are beyond the resources of most industry hydrometallurgists attempting to obtain information of direct relevance to full-scale flocculation processes. Towards this end, the level of characterization can usefully be restricted to (1) flocculation performance assessment and (2) quantifying feed properties as a way of predicting potential performance issues.

8.4.1 Flocculation Performance

The primary tool in flocculation studies has always been and will likely remain the simple cylinder or jar tests, the latter being stirred vessel tests intended for very low solids in wastewater clarification studies (Garcia *et al.*, 2014). Cylinder tests wherein mixing is achieved by inversions, or the action of a plunger, are preferred in most mineral processing studies. It is therefore disturbing that such a core test is often misused or done with a poor selection of conditions. Given the sensitivities to such conditions already highlighted, it raises concerns (from slight to very serious) over some results and their interpretation within the published literature. Examples of the mistakes or omissions made just within clay-related studies include the following.

8.4.1.1 Inadequate Flocculant Dilution Prior to Dosing

The flocculant will always represent a minor volume in relation to the slurry. Its efficient distribution through the slurry is critical, as adsorption for most feeds at practical dosages is rapid. Dilution of the flocculant aids the distribution process by both increasing the dosed volume and reducing the flocculant solution's viscosity. The latter is critical, as viscous solutions have been shown to mix with difficulty even in turbulent pipe flow. Mpofu and co-workers (Mpofu *et al.*, 2003, 2004, 2005) applied flocculants at 0.1 wt/wt% to clay slurries with fairly mild mixing and for high dosages; the mixing may have been appropriate at a lower flocculant concentration, possibly reducing the dosage required. A diluted flocculant concentration of 0.02 per cent would be far more suitable, with 0.01 per cent potentially better. In low-solids applications or when supernatant clarity is being examined, even further dilution can be an option.

8.4.1.2 Comparison of Flocculant Products at a Single Dosage

Comparing the performance of different flocculant products at a single fixed dosage (Caskey and Primus, 1986; McFarlane *et al.*, 2008) may possibly be appropriate when flocculation is only being done as a preparative stage for evaluation of properties much further downstream (i.e. rheology, filtration, deposition), but it is difficult to envisage when this can be justified for detailed studies of the flocculation process. Dosage response curves are never linear, and their shapes will often vary for different flocculant products, which means trends observed at one dosage may change at higher or lower dosages, particularly for distinct flocculant molecular weights and chemistries. Few flocculants will have the same molecular weight, so comparisons purely on a grams per tonne basis are only relevant when flocculant cost is a key issue. Single dosage comparisons also suffer as flocculant products will have different viscosities, which impacts on flocculation performance (as discussed above) unless done at high flocculant dilutions. Proper dosage curves are always preferable, from the onset of aggregation through to (where possible) high settling rates well above what may be required; in some cases, this may only represent a narrow range of dosages.

8.4.1.3 Excessive Flocculant Dosages

Few thickeners with clay feeds would seek settling rates much greater than 30 m/h. Few experimentalists can reliably measure settling rates of 40 m/h, given that represents a mud-line moving at 11 mm/s, assuming the mud-line is distinct. Even for those that can capture these settling rates, e.g. through recording images of the settling mud-line, as described by Zhu *et al.* (2000), at such large aggregate sizes the sensitivity to small variations in dosage and mixing conditions will often cause substantial shifts, making the values obtained highly questionable for comparisons. Yet there are cases where settling rates of 100 m/h are quoted and values near 50 m/h are contrasted (Chen, 1998).

8.4.1.4 Comparisons Made at a Single Solids Concentration

The comparison of flocculants or other reagents in a series of tests at just one fixed solids concentration is a very common practice and is acceptable if done after a number of concentrations have been screened. However, a fixed concentration that is too high or low raises concerns, especially when different flocculants are being compared. At too high a concentration, the sensitivity to differences between products may be reduced and even masked by mixing effects (Besra *et al.*, 2002; Dollimore and Horridge, 1973; Sabah and Cengiz, 2004). If one flocculant is able to produce a denser aggregate than another, then the relative response may be quite different, depending on the solids concentration, e.g.

a denser aggregate should result in higher settling rates from more concentrated slurries, while offering no advantage for dilute slurries (Grabsch *et al.*, 2013).

If the solids concentration is too low, then prolonged or more intense mixing is required to give the necessary aggregate-forming collisions, with the danger that aggregate size will be limited by the deactivation of the adsorbed polymer (i.e. reconformation to reduced bridging potential) as reaction time increases. It therefore becomes difficult to know if results then truly reflect the flocculant's properties or are too strongly impacted by the applied conditions.

8.4.1.5 Excessive or Ill-Defined Mixing of Feed and Flocculant Prior to Settling

Many groups inappropriately apply techniques from water treatment science to thickening-related studies, despite them being intended only for low-solids applications and coagulation reactions (Ofori *et al.*, 2011; Sabah and Erkan, 2006). Reaction times of several minutes in stirred beakers are appropriate at low solids, but rarely at high solids. The consequences may be deciding on higher dosages than are actually required and the masking (or even reversal) of performance trends.

8.4.1.6 Failure to Maintain a Constant Solids Concentration

With clay flocculation being very sensitive to this variable, the best test procedures involve a known volume of slurry being mixed with the required flocculant volume and then an appropriate diluant to make up to a set volume; in this way, dosage can be varied while maintaining solids. However, many studies involve simply adding variable flocculant volumes to the slurry (Mpofu *et al.*, 2004). When considering a range of dosages, small changes in solids concentration can have an impact; the use of high flocculant concentrations may negate that effect, but as already discussed, adds another complexity.

8.4.1.7 Assuming That Different Flocculants Have the Same Mixing Requirements

The level of mixing required will vary not only with solids concentration and flocculant solution concentration, but also between different flocculant products. A low molecular weight product that produces more bridges will tolerate more mixing during flocculation, which can favour fines capture, but higher molecular weight products may be far more sensitive. The impact of this is demonstrated in results obtained by Clark *et al.* (1990), who used a standard kaolinite slurry to quantify the effects of both molecular weight and charge density on settling rates (see Fig. 8.8). In an otherwise meticulous study, they applied 20 cylinder inversions in each test, a protocol identified in earlier work with lower molecular weight cationic flocculants. This was likely to be excessive for anionic flocculants (Gill and Herrington, 1987), and certainly for the higher molecular weight products. Over-shearing during flocculation would contribute to the plateaus observed in the dosage response (Fig. 8.8a). That plateaus are more evident for higher charge densities (Fig. 8.8b) may be evidence of a change in aggregate density and strength, such that solids concentration effects play a role in limiting settling rates.

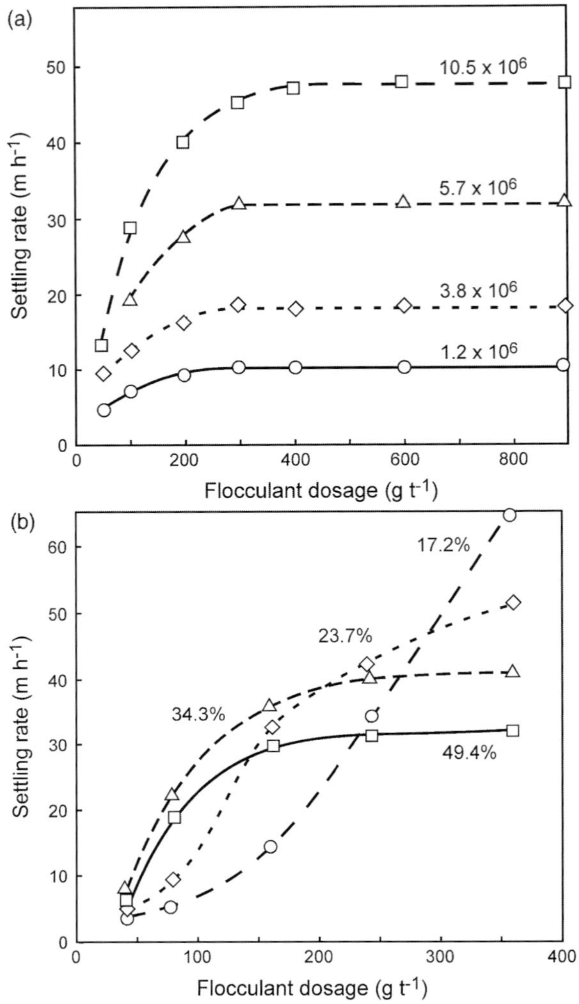

Figure 8.8. Settling rates as a function of flocculant dosage after addition to 20 kg/m^3 kaolinite and 20 cylinder inversions for (a) 21–24 per cent anionic flocculants of varying mean molecular weight, and (b) 13–15 × 10^6 molecular weight flocculants of varying anionic character (adapted from Clark *et al.*, 1990; copyright (1990) with permission from Elsevier).

8.5 Characterization Techniques

8.5.1 Performance Characterization

8.5.1.1 Continuous Flocculation

There have been several devices proposed for practical continuous flocculation on a small scale, most notably the CSIRO Shear Vessel (Farrow and Swift, 1996; Owen *et al.*, 2002) and the Ultrafloc tester (Rulyov *et al.*, 2006, 2011). Both rely on flocculation in a Couette cell, i.e. a concentric cylinder arrangement in which mixing is applied through rotation of the inner cylinder. The Ultrafloc tester is a smaller device and appears to apply higher shear and shorter residence times, conditions that may not be directly relevant to feedwell flocculation. Flocculant and slurry are contacted within the Couette section of the Shear Vessel, whereas they come together in the feedpipe of the Ultrafloc tester, potentially giving some level of flocculation prior to the application of the set shear.

8.5.1.2 Measuring Aggregate Size

Numerous studies have attempted to measure aggregate sizes with commercial instruments either during or immediately after flocculation. Low-angle laser light scattering/laser diffraction instruments are often applied, of which the Malvern Mastersizer is most frequently cited, but a range of comparable instruments exist. Perhaps the earliest detailed application to clays was by Gill and Herrington in examining cationic products as kaolinite flocculants. They considered the impact of molecular weight (Gill and Herrington, 1988) and then extended this to the effect of solids concentration (Gill and Herrington, 1989); in the latter they observed a wide spread of small aggregates at 2 kg/m^3 but larger sizes and similar distributions at higher concentrations up to 50 kg/m^3, indicating the flocculants were ineffective at low solids. Their conclusion may be correct, but the measured distributions had so few channels that any interpretation beyond the trivial has to be questioned.

Jarvis *et al.* (2004) used a Malvern Mastersizer to follow kaolinite coagulation and the breakage of aggregates in jar testing, but this was at the very low solids concentration of 80 mg/L. Given the prolonged stirring required to induce aggregation, the shear associated with the instrument's measurement cell may not be detrimental, but this is certainly an issue with fragile polymer flocculated systems, along with the high dilution required. Rattanakawin and Hogg (2001) achieved distributions from flocculated alumina with a Microtrac laser scattering instrument by a single pass through the measurement cell (as opposed to the normal recirculation), and while successful, the aggregate sizes achieved were mostly below 100 μm, which reduces the sensitivity to shear. Alam *et al.* (2011) examined aggregate size distributions from coal tailings flocculated in a stirred vessel by sampling, dilution and re-suspension prior to introduction to the measurement cell of a Malvern Mastersizer. They saw distinctions between dosages and the use of cationic and anionic flocculants, but again overdosing and overmixing make this study of limited use.

The same authors used the Malvern Mastersizer to extract estimates of aggregate D_f from the scattering as a function of angle, an approach that has been successful with colloidal latexes (Burns *et al.*, 1997), but is strictly not appropriate for primary particles in the micron range and above (Liao *et al.*, 2006). Ofori *et al.* (2011) also used this approach with polymer flocculated coal tailings. While such numbers may be acceptable for relative comparisons, the concerns with larger particles that may be present and the sample handling required prevent the approach from being widely practical.

Sengupta *et al.* (1997) were among the first to use FBRM in flocculation studies, examining the effects of mixing speed, reaction time, staged addition, pH and solids concentration on cationic flocculant addition to a kaolinite. They only gave a relative size term from the ratio of the mean aggregate size to the primary particle size. More detailed studies of silica flocculation in stirred baffled beakers provided strong evidence that staged cationic flocculant addition gave larger sizes than a single dose (Peng and Williams, 1993; Williams *et al.*, 1992). Hecker *et al.* (1999) inserted an FBRM probe below the settling column of the CSIRO Shear Vessel to study kaolinite flocculation. They established an approximate relationship between FBRM aggregate size and hindered settling rate, which is not only important for flocculation studies but also may have implications for control. The relationship was greatly refined by Heath *et al.* (2006b), and FBRM is now routinely used in Shear Vessel flocculation studies, providing more detailed analysis of flocculation extent and efficiency than can be obtained from settling rates (Owen *et al.*, 2007). Newer-generation FBRM probes now offer the ability to detect larger, low-density aggregates as a single object (Fawell *et al.*, 2015; Senaputra *et al.*, 2014).

8.5.1.3 Turbidity

Supernatant turbidity is often used as a guide to flocculation efficiency and likely process clarity, with high turbidity/suspended solids after settling considered evidence of poor (or excessive) mixing, poor fines capture or the failure to aggregate a minor phase. It can be convenient to use turbidity meters (or nephelometer probes) that provide an arbitrary measure of turbidity. At the simplest level, supernatant can be placed in a 'clarity wedge'; a clarity category is then determined by observing the highest value readable through the liquid. These arbitrary values can be difficult to usefully interpret when high clarity is being sought. While some degree of calibration can be achieved this way, actual quantification of suspended solids should be the preferred approach.

Measurement of supernatant turbidity is of greatest value in low-solids applications where the particles are low in both size and density. However, at the higher solids concentrations in most mineral systems, a distinct mud-line is formed in batch testing and free fines can be dragged down with the bulk solids. This is particularly so with flocculated clays, often leading to low turbidity values that do not reflect operating process results. There can be the added complication of selective flocculation, e.g. clays may be readily flocculated but other phases, such as fine quartz, may not be well captured in aggregates, either preferentially reporting to the supernatant or as a top layer on the settled bed. Batch testing is also notoriously unreliable for predicting clarity from continuous flocculation/sedimentation operations that are strongly influenced by currents, short-circuiting streams and bed scouring.

Turbidity fluctuations have also been used to monitor aggregation (Gregory and Nelson, 1986), which has led to a commercial instrument widely applied in low-solids water treatment studies, the Photometric Dispersion Analyzer or PDA (Gregory, 1988). A photodiode converts transmitted light into a voltage with a large dc component and a smaller ac component. The root-mean-square value from the fluctuating signal divided by the dc component gives a ratio that increases with aggregation and is much more sensitive than the raw signal. Sizes cannot be extracted and it can only be used at low solids but has found application to monitor jar tests (Barany *et al.*, 2011; Hopkins and Ducoste, 2003). The Turbiscan has a back-scattered detection head that moves up and down along a flat-bottomed cylindrical cell, and has considerable potential for following dispersion and aggregation processes (Mengual *et al.*, 1999). However, its value is limited by constraints imposed from the fixed geometry.

8.5.2 Feed Characterization

As techniques for the characterization of clay mineralogy and many other physical properties are discussed in detail within other chapters, this section will only consider techniques related to particle sizing.

The issues with excessive mixing that make distributions from the Malvern Mastersizer (and other laser diffraction/scattering instruments) questionable for aggregate sizing are less of a concern when characterizing feed slurries. However, feed slurries with significant clay contents are rarely at a high state of dispersion, and can therefore be highly shear sensitive (Fawell *et al.*, 2015). When conducting particle size distribution measurements, it is therefore necessary to decide whether to measure the slurry ‘as is’ or to apply additional dispersion – doing both can often be highly informative.

There is a tendency to only quote mean or median values from the particle size distributions, which is misleading when these distributions exhibit significant bimodal character. It must be remembered that such distributions are volume-weighted, and thereby emphasize the contributions of the larger particles, whereas the flocculation process can be dominated by the fines within the slurry.

This issue is well illustrated in the work of Du *et al.* (2010), who conducted perhaps the most detailed comparisons of clay identity effects on flocculation. Figure 8.9a shows Malvern Mastersizer distributions for two kaolinites they studied. They appear very similar, each having distinct fines peaks in the submicron region of comparable magnitude. However, the two kaolinites produced very different dosage response curves from flocculation tests (Fig. 8.9b).

The similarity of the distributions may in part be an artefact of them being presented in terms of volume percentage per size channel to provide a degree of normalization. The Snobrite kaolinite has a significant proportion of >50 μm, which would be expected to contribute little surface area during flocculation. However, the minor peaks in the submicron region may dominate the flocculation response, as they represent the majority of the slurry’s particles. Despite their similarity in Fig. 8.9a, it is likely that the Malvern Mastersizer measurements are not sensitive enough to what could be significantly different total particle numbers.

Automated particle counting can be achieved by allowing particles and aggregates to pass singly through an electrical sensing zone, their presence then detected by a suitable sensor (Gregory, 2009). This pulse counting is highly accurate; the Coulter Counter is perhaps the best known of such instruments. While such instruments have been tested to monitor the extent of aggregation, generally with little success, there are no published studies of them being used to actually identify flocculation issues, i.e. high counts indicating likely poor

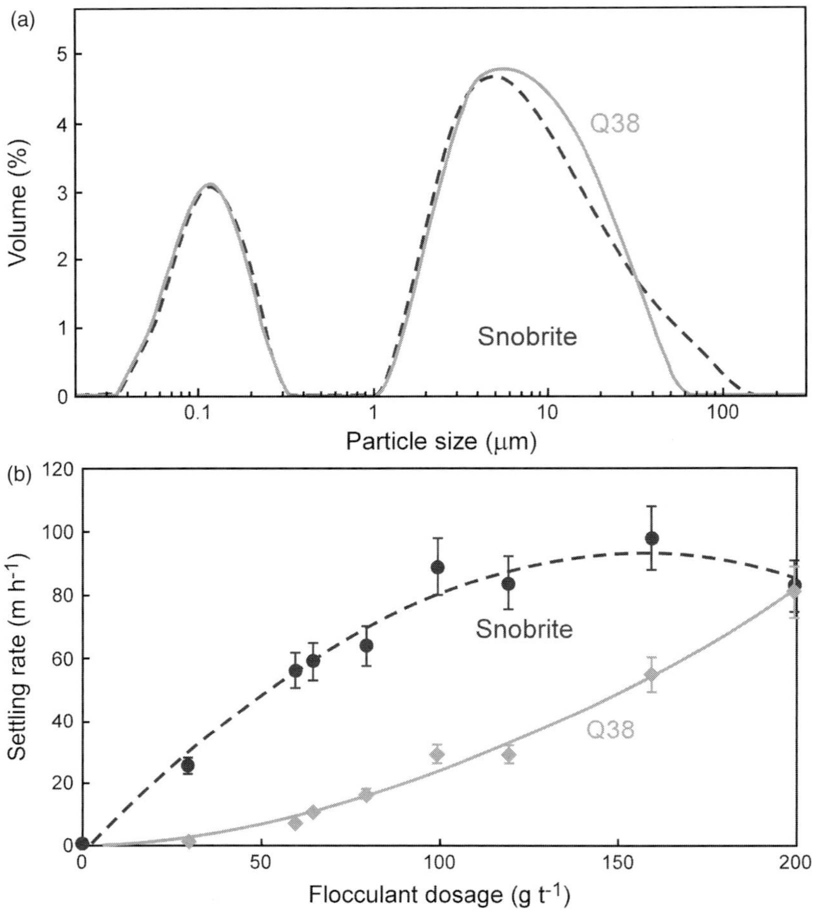

Figure 8.9. (a) Particle size distributions for two different kaolinites as measured with a Malvern Mastersizer. (b) The corresponding dosage response curves after cylinder tests – 0.025 wt% anionic flocculant, 2 wt% solids, four cylinder plunges for mixing (adapted with permission from Du *et al.*, 2010; copyright (2010) American Chemical Society).

flocculation properties/low settling fluxes. Grabsch *et al.* (unpublished results) recently obtained results suggesting the Coulter Counter may offer some enhanced sensitivity relative to Malvern measurements when high particle numbers are a concern, particularly when fines are being generated during stirring. However, Coulter measurements require very high (and therefore very accurate) dilution, and the lower size limit of around 1–2 μm may not be appropriate for some clays. FBRM can detect to just below 1 μm without the need for any solids dilution, but while the unweighted chord length measurements are particle number sensitive, the technique does not give number counts. Heath *et al.* (2002) found that the non-linear response of FBRM to solids concentration limits its ability to detect changes in particle number. The G-series FBRM probes (released in 2012) appear to offer much greater sensitivity to particles in the 1–10 μm range and a more linear response to solids concentration, increasing the potential to provide real-time monitoring of feed particle properties (Fawell *et al.*, 2015).

8.6 Optimizing Operational Flocculation

8.6.1 Flocculant Selection

8.6.1.1 Synthetic Bridging Flocculants

The majority of flocculants used within the minerals industry are copolymers of acrylamide and sodium acrylate, with varying charge density, molecular weight and physical form. As a general rule, nonionic flocculants would be preferred at very low pH (<3), when the anionic carboxylate group is expected to be protonated. As the pH is increased, the level of anionic character is also increased (Connelly *et al.*, 1986). The majority of anionic flocculants applied to clay tailings in mineral processing applications contain 5–50 per cent anionic character, with 20–30 per cent being the most common.

High pH becomes an issue when flocculating tailings from the Canadian oil sands industry that contain significant fractions of <2 μm clays. Bitumen extraction is favoured by caustic treatment of the solids, which elevates the pH to over 9. Xu *et al.* (2004) found that high molecular weight, medium anionic charge density flocculants were the most effective, giving good settling rates and clarity in the absence of caustic treatment, but for caustic treated tailings a 'small fraction of clays' (representing at least ~1 wt% solids) remained in overflows. Lower water quality demands in this process are reflected in 1 wt% overflow solids being considered low. High bicarbonate levels after caustic treatment were assumed to act as a dispersant for fine clays, but any elevation in pH would also contribute to a reduction in surface sites available to flocculant adsorption. The presence of calcium through lime addition definitely enhanced the performance of anionic flocculants (Hamza *et al.*, 1996).

Higher molecular weight flocculants generally favour higher settling rates, but it is a mistake to assume this always applies or is in fact always required. While the performance of nonionic flocculants tends to give a clear relationship with molecular weight, as Fig. 8.4 shows, anionic flocculant chains are much more extended in solution, increasing their bridging capacity. In a study of flocculant solution sensitivity to shear applied prior to dosing, Scott *et al.* (1996) found that turbulent shear that reduced molecular weight of anionic flocculants actually increased the settling rates obtained at low dosages with diamond mine tailings slurries. In this case, the greater number of polymer chains available after the shearing process reduced the dosages required to achieve moderate settling rates. It was only when high settling rates were sought that the sheared solutions gave inferior performance.

In general, variations in the flocculant properties can impact noticeably upon settling, but to a lesser degree on aggregate density. While there have been claims of flocculant products that lead to higher aggregate density, thereby offering superior settling rates at higher feed concentrations and increased underflow densities, there are surprisingly few published studies that conclusively demonstrate such effects. Alternative products typically come at a greater cost, and often the performance enhancements obtained are too small to justify the expense (products successfully developed and applied for bauxite residue settling are specific to the alumina industry only, and therefore will not be considered here). Plant trials of products over short periods of operation can suffer from feed variations or operational inadequacies that place doubt on the results.

One of the few commercial products available for general applications that has become the subject of published work is Rheomax® DR 1050 (BASF). Laboratory and plant trials of the flocculant have been described (Adkins, 2008; Vietti *et al.*, 2008), but without providing details of the product's composition. The flocculant is believed to be based on a polyacrylamide backbone, with an alternative anionic monomer replacing the acrylate monomer used in conventional products. Improved aggregate density is certainly implied from these studies on clay-based systems, although at this time the understanding of why Rheomax® DR 1050 offers greater performance advantages in some instances remains incomplete. Flocculation kinetics measurements and PB modelling with a calcite feed for Rheomax® DR 1050 and a conventional acrylamide/acrylate copolymer has confirmed a significantly higher fractal dimension after flocculation with Rheomax® DR 1050, and therefore a much higher optimal solids concentration for flocculation (Grabsch *et al.*, 2013). Significantly, the benefit of this higher aggregate density is only realized at higher feed solids concentrations and flocculant dosages.

8.6.1.2 Polyethylene Oxide

Polyethylene oxide (PEO) has long been known as a potential flocculant for mineral processing applications, although the reasons behind its success or failure were often not understood. Rubio (1981) concluded PEO was successful as a flocculant with hydrophobic substrates (talc, graphite, chalcopyrite, covellite, phosphate slimes and many clays) and less so with hydrophilic substrates (rutile, quartz, calcite). Scheiner and Wilemon (1987) found that quartz flocculation by PEO was successful when the slurry represents a mixture of minerals. They also felt that the CH_2–CH_2– segments on the PEO chain were able to adsorb on the hydrophobic sites on clay particles; certainly they noticed a distinction for clays flocculated with PEO, from which water is readily 'squeezed' to form a strong, stable aggregate, presumably favouring edge–face or face–face interactions. Oxides that are strongly hydrated (hematite, alumina) are poorly flocculated by PEO (Koksal *et al.*, 1990). Mathur and Moudgil (1997) concluded that PEO adsorbs onto oxides through specific hydrogen bonding in an acid–base reaction between the ether oxygen (a Lewis base) and strong Brønsted acid sites on the surface, and that interface charge was not a significant factor.

PEO as a kaolinite flocculant has also been considered in the context of consolidation, both with and without 'shearing' (in this instance, several minutes of high-rate impeller stirring). McFarlane *et al.* (2005) compared settling rates, consolidation, rheology, pressure filtration and SEM imaging for kaolinite slurries treated with a conventional anionic

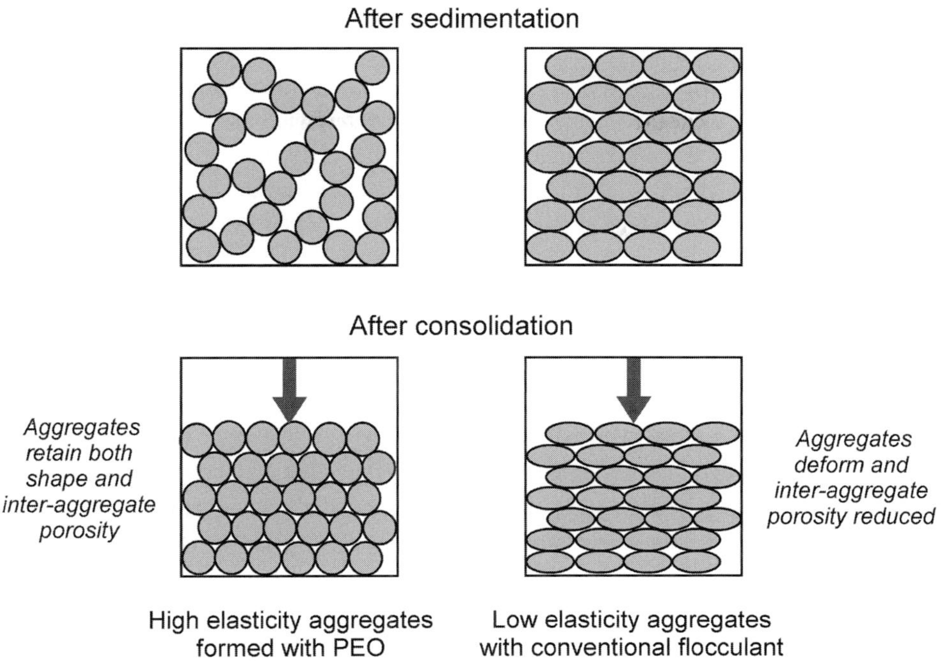

Figure 8.10. Schematic representation of how aggregates (represented by the shaded shapes) formed with PEO and a conventional flocculant may dewater differently (adapted with permission from McFarlane *et al.*, 2005; copyright (2005) Korean Society of Rheology).

flocculant and nonionic PEO. Their most significant observation was that PEO-flocculated aggregates survived the shearing process to a much greater extent, but also dewatered to a higher volume fraction, noting an improved sediment cohesiveness similar to that described by Scheiner and Wilemon (1987). The authors claimed that for the same solids volume fraction, aggregates formed with PEO were denser and retained their structure under shear or consolidation (Fig. 8.10). By contrast, with conventional flocculants the aggregates were readily deformed, making the removal of inter-aggregate liquor more difficult. Retaining internal structure with PEO may have a negative impact on dewatering at high volume fractions, but such values are not expected by thickening. Whether or not this mechanism is correct, a more detailed study of PEO flocculation with greater control over applied mixing conditions would be highly desirable (although the practical use of PEO as a flocculant may be complicated by some of its physical properties).

8.6.1.3 Cationics

Cationic products dominate municipal wastewater sedimentation applications, where aggregation is mainly through charge-neutralization. They have also been extensively studied as clay flocculants, although this is mainly for low solids clarification, with the main mineral process application being in coal tailings thickening (Henderson and Wheatley,

1987). Even then, the lower molecular weight for cationic flocculants leads to lower settling rates than those obtained with anionic products. Nasser and James (2006, 2007) re-examined cationic flocculants for kaolinite flocculation, focusing more on sediment consolidation. They concluded that cationic flocculants produce compact, stronger aggregates compared to the more open, easily deformed structures obtained with anionic flocculants, and actually proposed a mechanism for consolidation identical to that shown in Fig. 8.10. Their arguments about strength resulting from different adsorption properties are clouded by the higher dosages required for cationic flocculants, which themselves impart a degree of strength. In contrast to McFarlane *et al.* (2005), weaker, open aggregates were thought beneficial for thickening, while stronger compact aggregates formed with cationics benefited drainage during filtration.

8.6.1.4 Stimulant-Sensitive Flocculants

An approach proposed for achieving higher underflow densities is stimulant-sensitive flocculation, in which a polymer is initially able to produce aggregates under feedwell conditions but then in response to a change in conditions (e.g. pH or temperature), the polymer's ability to maintain adsorption on surfaces is diminished and the rapid elimination of bridged structures reduces any barrier to particles consolidating further.

Temperature change is the stimulant in the case of poly (*N*-isopropylacrylamide), which is hydrophobic and functions as a flocculant at high temperatures (~50 °C), but becomes hydrophilic and loses any attraction to surfaces when cooled below the critical solution temperature of 32 °C (Franks *et al.*, 2009). This approach has been demonstrated with quartz in a continuous pilot column (Franks *et al.*, 2014), but the actual flocculation achieved did not appear efficient and dosages were high, and few applications will offer the required temperature differential. Long *et al.* (2011) did examine this reagent with oil sands tailings, claiming both excellent initial flocculation and then good final consolidation; however, the applied dosages were alarmingly high (200 ppm) and the short duration of their settling tests (never more than 120 minutes) make it very difficult to truly assess the level of consolidation relative to the untreated slurries. Perhaps most significantly, they noted the potential for residual polymer to have negative downstream impacts, and at such high dosages this appears a realistic concern.

Berger *et al.* (2011, 2013) also considered the stimulant approach, with modification of 'post gel point underflow characteristics' for clay-containing feeds leading to a change in the yield stress of the consolidating bed. Their approach has been tested successfully with plant feeds, but to date there is only limited published detail on full-scale applications (Berger *et al.*, 2012).

8.6.1.5 Natural Products

Flocculants derived from polysaccharides and other natural polymers are commonly used in wastewater treatment, where there is typically a greater demand for environmentally friendly materials to be applied. They include tannins, chitosan, cellulose, and sodium alginate. These reagents are biodegradable, produce no secondary pollutants, their solutions are largely shear-insensitive, and they can be reproducibly obtained from agricultural sources (Bolto and Gregory, 2007). However, they typically only offer moderate efficiency in inducing aggregation, which leads to high dosage requirements and usually limited settling rates, and shelf lives may be short. Such limitations are not considered major in wastewater treatment, but rule out their use in most tailings applications. While tailor-made grafted flocculants, in which synthetic flocculants are grafted onto natural polymer backbones, are gaining increasing interest as a way to combine the best properties of both (Lee *et al.*, 2014), it is unlikely that they would replace synthetic polymers for flocculation of clay tailings.

8.6.1.6 Multi-Stage Flocculant Use

The use of two flocculants in sequence (anionic and cationic) has been proposed in laboratory studies of alumina flocculation, with the two products together giving fast settling rates, whereas when used alone they performed very poorly (Yu and Somasundaran, 1993; 1996). The enhancement was attributed to complexation between the flocculants, although pre-mixing the flocculants led to low activity, indicating complexation needs to occur after pre-adsorption of one polymer. The concept has been applied to sludge dewatering, with Lee and Liu (2001) concluding dual polymer conditioning (cationic and nonionic) gave better aggregate strength and dewatering. While this did not correlate with measured fractal dimensions, their measurements of the latter may be questionable. For both applications, the order of addition appeared to favour the higher molecular weight product being added last to maximize final aggregate size.

Dual reagent addition is very well established in situations when the feed particle size distribution features a substantial submicron fraction and flocculant usage alone will lead to prohibitively high dosages, as was discussed in Section 8.2. A reagent may be added to promote aggregation of that fraction by coagulation, e.g. through a simple electrolyte or pH modification, or a low molecular weight cationic or anionic polymer, with flocculants then much more successful in forming large aggregates. Adding coagulating reagents after flocculants will usually be less effective (Rattanakawin and Hogg, 2005), although it can still impact on fines capture.

Dual flocculants have been applied in dewatering of coal tailings, with McColl and Flanagan (1999) describing a novel approach towards efficient dispersion of the second flocculant by encapsulating it within the primary product. This encapsulation breaks during the flocculation process, releasing the second flocculant, leading to stronger, denser aggregates and possibly even removing unadsorbed primary flocculant. It should be noted that dual flocculant addition (hydroxamate and polyacrylate) is common in primary settling of bauxite residue. Hydroxamate alone typically gives better fines capture and adequate settling rates, but is combined with polyacrylate to achieve better underflow density. To our understanding, no actual synergy between the flocculants is being claimed, but rather a balance between the performance of the two products (and perhaps a lower consumption of the former, which is more expensive).

The most successful application of dual flocculants to clays has been in oil sands tailings flocculation (Vedoy and Soares, 2015). Watson *et al.* (2011) describe how use of very low and high molecular weight flocculants in sequence gave enhanced performance, although the dosages of the former were extremely high. Yuan and Shaw (2007) and Yuan (2011) found that flocculation–coagulation–flocculation (FCF) as three distinct steps gave better settling for such tailings than simple flocculation or when combined with coagulation in any order. It remains unclear if this is just realizing the benefits of multi-stage addition in reducing dosage requirements or if any special synergy results from FCF. The extended mixing applied during flocculation stages would minimize the impact of flocculant molecular weight on settling rates. It should be noted that the flocculants used most frequently at plant level with oil sands tailings are believed to be blends of high and low molecular weight products. In part their success comes from their lower viscosity relative to high molecular weight flocculants, as they can be made-up and dosed at higher concentrations, quite important in applications requiring high dosages.

While the advantages of multi-stage or split addition of flocculant in terms of reduced dosages has been well established (Owen *et al.*, 2008), Glasgow and Liu (1995a) have also shown that multiple flocculant addition to very dilute kaolinite slurries offers the potential for higher aggregate density and strength. Such impacts have not as yet been considered in any detail for flocculation in thickening applications.

8.6.2 Flocculation Hydrodynamics

Poor feedwell flocculation performance is often blamed on the flocculant. In reality, while there may be scope for incremental improvements via flocculant selection, in the majority of cases, poor performance is the result of suboptimal feedwell design or operation. When a feedwell does not provide appropriate conditions for flocculation, whether that arises from insufficient contact between feed and flocculant or from excessive shear, the actual flocculant becomes almost irrelevant. The common response of increasing dosages exacerbates the inefficient use of flocculant and creates additional problems further down within the bed.

Issues with feedwell flocculation are often heightened in the case of clay flocculation by the need to achieve a suitable degree of solids dilution prior to flocculant dosing. Some 'open' feedwells (i.e. with no constrictions at the exit or discharge plane, allowing some dilution liquor to rise into the feedwell) are able to achieve a two-fold dilution. However, this is presuming that the flows are stable, the flocculant is dosed in a position that can fully utilize this dilution, and the residence time and shear are conducive to flocculation, conditions that are rarely all met. Solids dilution will therefore often be achieved external to the feedwell, either through direct addition of streams to the feed or through self-dilution devices that utilize the feed velocity to draw in dilution from the clarification zone in a Venturi arrangement (Bedell *et al.*, 2015).

Solids dilution is often important in thickening tailings from mineral sands operations, but such applications are complicated by the variability of the feed's mineralogy (or PSD) and solids concentration. Both can lead to excessive solids dilution prior to flocculant addition, which can then be highly detrimental to flocculation performance. Mineralogy issues, when understood, can be dealt with through careful blending of the feeds, but sudden changes in solids concentration from spiral circuits can be harder to control. While feed tanks prior to the thickener can be an effective buffer, this can lead to air entrainment and over-shearing of clay solids to generate additional fines, both of which can be counter to good flocculation.

With access to computational fluid dynamics (CFD) modelling that incorporates PB models of flocculation, a more detailed understanding of how to optimize feedwell design and operation is now appearing (Fawell *et al.*, 2009b; Nguyen *et al.*, 2012). Many modern feedwells are capable of very high performance, yet older designs remain in wide use and (depending on performance expectations) can still perform adequately. A discussion on feedwells is outside this review's scope, but the following points are of general relevance to clay applications:

(1) Feedwells that function well are best considered as a two-compartment design, with the first (upper) compartment providing mixing of feed and flocculant, as well as some dissipation of the feed momentum/energy, while the second (lower) compartment provides lower shear for aggregate growth. This gives the correct shear profile for flocculation, and situations that apply higher shear to aggregates late in the process should be avoided.
(2) Simple open feedwells (i.e. cylinders with no internal shelves) are still common, but can only function well under a narrow range of conditions that are rarely achieved or maintained. A horizontal shelf of width ~10 per cent of the feedwell diameter located just below the feed entry generates more appropriate hydrodynamics, provided that feed entry is tangential, i.e. creating a swirling flow. Ideally, 50 per cent of the side-wall height should be below the shelf to create a suitable stable zone for flocculation.
(3) The inlet velocity is crucial to creating the correct hydrodynamics. If the velocity as the feed enters the feedwell is below 1 m/s, it is unlikely that flocculation will be effective, as the feed stream will sink rapidly. Velocities higher than 2.5 m/s are a concern in smaller feedwells (diameter <6 m), while in larger feedwells velocities higher than 3 m/s may lead to excessive shear. In such cases, splitting the feed into two streams that enter the feedwell travelling in opposite directions is a common strategy (particularly in many large-diameter feedwells), but this does create difficulties with asymmetric discharge.
(4) As discussed above, split or staged flocculant addition can be beneficial. Adding some flocculant to the feed prior to the feedwell is common, but the importance of solids dilution should not be forgotten. If the feedwell is adequately mixed, then the major fraction of the flocculant dose should be added there. If the feedwell has a high inlet velocity, the feedwell dosing position should be moved downstream of the feed entry point. If there are concerns that the feed stream is sinking rapidly, dosing at the entry point (or within the feedpipe/launder) is required.

8.6.3 Action of Rakes, Pickets and Other Dewatering Devices

Rakes feature within most thickeners, with their slow rotation (usually much less than 1 rpm) allowing the angled blades (Fig. 8.2) to move solids towards the underflow discharge point without any major disruption to the bed that could see settled solids re-entrained and carried to the overflow. The angle of attack may vary from 20 to 60°, but is typically 30° in mineral applications. CFD has been used to examine rake blade design and configuration (Frost *et al.*, 1993; Sutalo *et al.*, 2003), with solids transport also considered from tracer studies at pilot scale (Rudman *et al.*, 2008, 2010).

In addition to solids transport, raking also potentially assists in solids dewatering. Rakes and associated structures (e.g. vertical pickets) produce channels for water to rise (Pearse, 1980), an important aspect given that bed permeability reduces under consolidation. Farrow and Swift (1991) characterized the size and settling rate of free settling aggregates at high dilution to demonstrate that samples taken from the underflow of a diamond mine tailings thickener were smaller but denser than when formed in the feedwell. This work was later extended in sampling of flocculated kaolinite from a continuous raked pilot column (Farrow *et al.*, 2000), confirming aggregate densification after raking that was well above that seen from compressive dewatering in the bed region above the rake.

Gladman *et al.* (2005) examined the impact of low shear on aggregates during sedimentation (i.e. post-feedwell) and observed the potential for enhanced sedimentation, which was attributed to densification of the highly porous structures under extended shear conditions low enough to minimize breakage. This may be similar to the kaolinite aggregate 'ageing' described by Glasgow and Liu (1995b) following gentle stirring to give re-suspension; they observed a decrease in size due to erosion, but settling rates were maintained, thanks to higher densities.

A practical approach towards applying low shear to settling aggregates within a thickener is needed to take advantage of any potential densification effect. Enhanced sedimentation over inclined walls is well known. Aggregate densification occurs as they 'roll' down the surface, experiencing extra shear (Usher and Scales, 2009). This concept was utilized within the ultra-high rate thickeners (E-cat, Ultrasep), through the use of dewatering cones. Lamella thickeners, in which settling is through a number of inclined plates stacked vertically, are well known in low-solids sludge clarification – the theoretical prediction of performance enhancement from such units is described by Rushton *et al.* (1996).

Rake structures that protrude high up within the settling zone could potentially assist in this way, although the difficulty is in ensuring contact with a significant fraction of the settling material. Vertical pickets or fences (a set of pickets with low-profile horizontal supports) are normally thought to assist consolidation by providing channels for rising liquor. In a study of fence design in sludge thickening, Hoyland and Day (1986) claim a partial contribution to faster consolidation from the enhancement of flocculation, likely to be caused by the gentle mixing action of the fence (presumably near the top of the bed), although their published results gave insufficient evidence to support this.

This potential was recognized in the deep cone design developed by the National Coal Board (later British Coal) in which the upper stirring section 'enhances flocculation'

(Svarovsky, 2000a). Abbott (1979) states that a preferred design has 'four radial arms to carry the blades … in the upper section of the cone the blades are vertical, about six inches wide and inclined at 45° to the radial arms to induce an inward flow of material. Blades are set at different radial distances on each arm so that good coverage of the whole volume occurs.'

In highlighting the importance of this low-speed stirring, Abbott (1979) noted that highly flocculated solids would otherwise produce aggregates with high voidage, with water not released and very fluid underflow discharges. Other than postulating that such stirring may be less successful on incompressible materials, no further insights into the process are given.

Loan and Triglavcanin (2009) have described an interesting refinement of the Abbott (1979) approach to achieve such enhancements in sedimentation. The radial arm fixed higher up on the rake shaft had a series of 'shearing elements' distributed along its length, and angled with respect to the vertical plane. Aggregates passing through this zone are expected to experience 'uniform cumulative shear' from contact with the elements and from the turbulence the elements produce during their passage. Increasing the spacing of the shearing elements with radial distances is stated as critical towards achieving uniform shear.

The term 'aggregate densification' is frequently used, and should be taken as the effective removal of intra-aggregate liquor (i.e. liquor from within the aggregate's structure) from the flocculated suspension. This is complicated by the fact that aggregates are fractal structures, having higher density at smaller sizes, and that aggregate erosion can leave a denser core without any expression of liquor. True densification should be through contraction of the aggregate structure releasing intra-aggregate liquor without significant loss of the particles within. Presumably this can only occur through rearrangement and reconformation of the bridging polymer flocculants, possibly with additional internal bridges being formed. Alternatively, densification can be achieved through aggregate breakage, but this will only be practical if there is no disruptive effect on the sedimentation process. In other words, a process to break aggregates should preferentially occur within the already consolidated bed.

A significant insight into the densification process has been provided by Du *et al.* (2009), who sampled flocculated kaolinite aggregates before and after 'raking' (although what they describe is akin to a vane or paddle), froze the slurry under liquid nitrogen and examined the structures by field emission SEM. The plate-like kaolinite particles would be expected to favour flocculant adsorption through their edges in preference to the basal faces, the latter representing the major portion of the total particle surface area. Flocculated aggregates are expected to form through edge–edge interactions, giving open structures, and this is supported by the images obtained. They also observed restructuring on raking that led to some face–face interactions and overall a much more closed structure.

While the claim of 'direct imaging' is questionable, given the sampling and pre-treatment applied, the images provided by Du *et al.* (2009) are still highly useful. The published flocculation and raking procedures are crude, the contribution of compressive effects are not adequately accounted for, and it is possible that the images represent considerable over-raking. As such, it seems unlikely that this can be related to the effects described by Gladman *et al.* (2005) and Loan and Triglavcanin (2009). The conclusions are also specific to aggregates of platelet particles, and so are most applicable to clay systems.

8.7 Hydrocyclones

The use of hydrocyclones for the removal of coarse solids from feed streams prior to thickening is well established (Abd-El Rahman, 2000; Knight *et al.*, 2012; van Ryssen and Steenkamp, 2012). This can greatly decrease the mass of solids going to tailings disposal areas, and the coarse phase removed is often highly useful for building impoundment walls. There may be some advantages to subsequent thickening stages, although it should be pointed out that the presence of some sand within the thickener feed is not detrimental to flocculation processes, as co-flocculation of sand with clay fines can lead to enhanced settling rates (or the option of lower dosages), as well as the possibility of higher underflow density.

Hydrocyclones as normally used can remove solids greater than 20 μm in size, so in most applications any clays will report to the overflow. Visman and Hamza (1973) lowered the cut-size to ~10 um for an iron oxide feed by adding flocculant to the hydrocyclone. Presumably this would be less successful with lower-density clay particles. They recognized that the fragility of the aggregates formed would certainly limit the sizes that can be achieved, given the shear experienced within the units.

In studying flocculant addition to a copper–nickel mill tailings, Yalcin (1996) found that feed pulp density is a critical factor, and tailings with pulp densities of greater than ~15 per cent require a two-stage hydrocyclone operation where the flocculant-assisted hydrocyclone was employed at the second stage to process the lower solids overflow of a conventional hydrocyclone. This idea was extended to an operating plant treating iron ore tailings, with test work conducted on cyclone overflow product which becomes the feed to thickeners (Yalcin and Brunet, 2004). Interestingly, they initially screened flocculants in settling tests, and then claimed the best hydrocyclone performance to be with a 'shear resistant flocculant', which was actually a cationic of lower molecular weight than the anionic products studied. The shear-resistance is likely to come from the formation of a smaller aggregate size and perhaps the greater number of chains available from a lower molecular weight product. The relative viscosity of the flocculant solutions must also be a factor, but was not considered.

In examining flocculant effects on hydrocyclone performance for two coal tailings slurries, Franks *et al.* (2005) found that 90 per cent of the aggregates formed from the submicron tailings were greater than 38 μm after shearing, sufficiently large to be suitable for hydrocyclone dewatering. With density being the most important hydrodynamic effect, Woodfield and Bickert (2004) state that reduced aggregate density compared to primary particles is the main reason for limitations on the improvement flocculation could achieve in hydrocyclone separation and not, as often claimed, aggregate break-up.

Wallace *et al.* (1980) found that addition of a polyacrylamide flocculant to a 2 wt% water–clay suspension resulted in significant changes to centrifugal separation efficiency of a 10 mm hydrocyclone operating at entrance Reynolds numbers $>10^4$ (either positive or negative depending on the history of the polymer solution). As conical hydrocyclones have several flow structures that can be significantly influenced by small amounts of a high molecular weight polymer, the observed physical phenomena may be related to the underlying viscoelastic behaviour of dilute polymer solutions. It should be noted that the polymer concentration studied (264 ppm) is more than an order of magnitude higher than would ever be applied in feedwell flocculation.

Williams and Roldan-Villasana (1991) reported improved efficiency when flocculated kaolinite was treated in a set of six 10 mm diameter hydrocyclones arranged in parallel and fixed with a 2 mm spigot and a 2.6 mm vortex finder. In the same arrangement, titania ceramic pulps (~0.1 μm) were coagulated to 4–5 μm diameter by pH change, with the aggregates exhibiting marked tolerance to multiple hydrocyclone passes. The same authors (Roldan-Villasana and Williams, 1999) developed a model of aggregate formation and breakage in this environment to support their conclusion that small diameter (10–50 mm) hydrocyclones can allow classification and clarification of slurries containing fine particles (down to <2 μm in size). At present, full-scale applications of this approach to clay slimes treatment are not known, but it is possible this could be practical in some lower volume situations.

8.8 Filtration/Centrifugation

Reagent addition to clay slurries with the intention of then proceeding directly to filtration or centrifugation requires specific consideration. In the case of filtration, this is typically done at higher solids concentrations than feedwell flocculation. For the purposes of this review, reagent addition to high-solids streams prior to deposition but normally without filtration is discussed separately in Section 8.9.

The kaolin industry, in which the clay is purified for ceramic applications, favours filtration over sedimentation for solid–liquid separation, which is not surprising given the need to eventually generate a dry product. Rotary vacuum or drum filters dominate for this purpose (Whatley, 1986), with the coagulated feed solids concentrations varying from 18 to 35 wt%, depending on particle size. Pressure filtration is also known to be used (Johns, 1991; Mackowski, 1989) and belt press filters combining pressure with gravity drainage have been applied at larger scales (e.g. coal tailings) as a reflection of their lower capital cost (Mathewson *et al.*, 2006). Clarifiers or lamella thickeners may be used to treat the low-solids filtrates.

Flocculants will enhance filtration rates, but this often comes at the cost of higher cake moisture, and overdosing will increase cake viscosity (Bowman *et al.*, 1993). Besra *et al.* (2006) observed that cationic flocculants were better for kaolinite filtration than anionics, with the latter more likely to be present in an unadsorbed state and thereby being detrimental to filtration. Given the ability of anionic flocculants to adsorb almost quantitatively on clays, it is more likely that overdosing and the high viscosity of anionic flocculant solutions will combine to give poor mixing, a major issue at high solids. Pre-dosing with surfactant prior to flocculation was observed to improve filtration properties (Besra *et al.*, 2004).

Within the minerals industry, clay slimes filtration is difficult, as fine solids build-up in cakes leads to a high resistance to flow, slowing dewatering and limiting throughputs. Crossflow filtration can keep the cake to a limited thickness through shear from a slurry flow tangential to the filter medium. This was the basis of the Kalgoorlie Filter Pipe, in which a flexible hose of a woven medium provided dewatering of slurries under pressure (Yan *et al.*, 2003).

An additional concern with reagent addition in filtration applications is the potential for the polymers to block or 'blind' pores on filter membranes and cloths, thereby reducing throughputs. While this is observed with cationic polymer reagents, it is thought to be worse with anionics (Gregory and De Moor, 1984), almost certainly a reflection of the higher molecular weight (and thereby viscosity) for the latter. A cationic product applied after careful addition of an anionic product is thought to reduce issues with filter blinding.

An innovation that has some potential to make filtration an option at realistic throughputs for the minerals industry is the Z-Filter (Graham and Graham, 2007). It uses a tubular loop of filter fabric ('Sock') which opens to accept the material, closes to hold it throughout processing, and then opens again to discharge it. This sock allows the Z-Filter to hold the required material in a suspended form long enough for gravity and the action of rollers (comparable to a peristaltic pump) to provide dewatering. The level of dewatering achieved may not be any higher than with other techniques, but it should be attained at a much faster rate. It could, for example, be readily applied to underflows from conventional thickening of clay slimes to give a stackable/conveyable product.

Centrifugation is becoming an increasingly practical option. Fig. 8.11 shows the suitability of different centrifuge types in terms of volumetric throughput and particle size (Svarovsky, 2000b). Of these, the scroll (or decanter) centrifuge offers perhaps the best potential for continuous operation at practical solids concentrations, certainly when clarification is the goal. The total removal of solids down to 1–2 μm is readily achieved, although this cut-off will rise at higher throughputs and feed solids concentrations above 10 per cent by volume.

Decanter centrifugation has been identified as an appropriate technology for producing ultrapure bentonite clays with high fines recovery (Boylu *et al.*, 2012). They are now applied extensively as part of the cleaning and recycling of clays used as drilling fluids (Merkl and Steiger, 2012; Reif *et al.*, 1990).

8.9 Enhanced Dewatering by Reagent Addition at High Solids

8.9.1 Dewatering Beyond Underflow Concentrations

The typical curve for the yield stress response to solids concentration (Fig. 8.3b) suggests that the underflow density gain on going from conventional to paste thickening may not

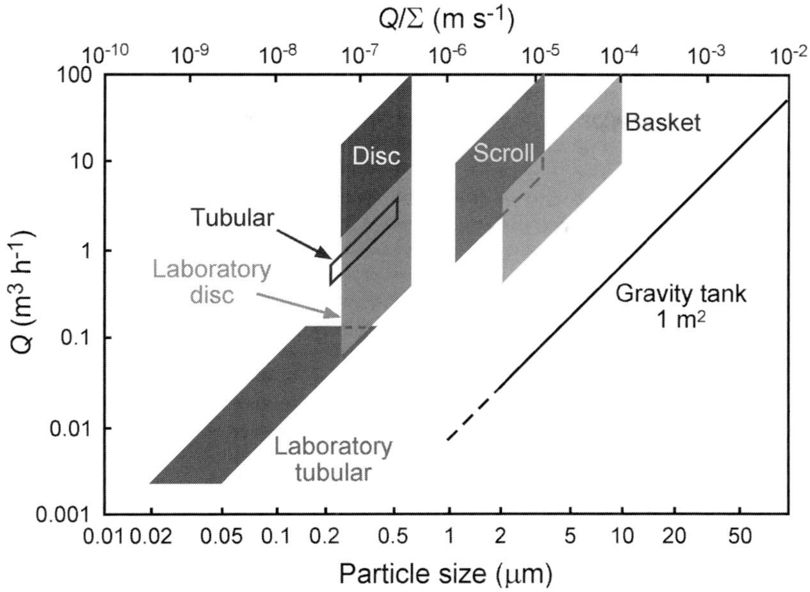

Figure 8.11. Performance of different centrifuge systems, where Q is the volumetric flow rate and Σ is an index of the centrifuge size (adapted from Svarovsky, 2000b; copyright (2000) with permission from Elsevier).

always be large. For example, an increase in washed bauxite residue solids concentration from 50 to 55 wt% may increase the yield stress from 20–30 Pa to well over 100 Pa. For many feeds, the gains from paste thickening from additional short-term water return may be small, but the reduction in moisture content for the tailings, combined with effective deposition practices, leads to a quicker drying time in order to form a load-bearing state.

Such advantages, however, may not be fully achieved for some fine clay-based tailings. The plate-like structure of clays, and the fact that flocculants (under most conditions) only adsorb to functionalities on the edges, mean that very open (low-density) aggregate structures are formed. Even when paste behaviour is achieved, solids concentrations may still be less than 25 wt%, meaning that the paste is still mainly water. Such pastes can be deposited, but liquefaction (where shocks such as earthquakes lead to the paste becoming fluid) can then be a concern. In the case of clay-based tailings from the Canadian oil sands industry, a combination of residual bitumen, high pH and clay mineralogy results in slurries that are difficult to thicken conventionally, but that also do not consolidate well over time. As a consequence, there is a huge legacy issue in tailings that now require further treatment prior to any rehabilitation.

The addition of high dosages of flocculant to high-solids tailings streams has been shown to generate greatly enhanced dewatering relative to what can be achieved from thickening alone, and is the subject of several recent patents, with several Canadian oil sands companies having moved to full-scale applications, followed by deposition or centrifugation.

A number of names have been used to describe this process, few of which are properly descriptive and most are associated with a particular variation on the theme. As an example, the review by Wells *et al.* (2015) uses the term 'inline flocculation' as a reflection of the polymer reagents being dosed directly into pipe flow prior to deposition. For the purposes of this review, it will be referred to simply as 'condensing aggregation', in recognition that the use of extremely high dosages leads in some way to an effective densification of aggregate structures. The discussion will be speculative, given the low level of detail in most of the available publications.

McColl *et al.* (2004) describe a process that takes existing thickened tailings and through the addition of a suitable reagent at high dosages, achieves rapid dewatering such that upon deposition the water runs off freely. Gaillard and Poncet (2010) published a patent application in a similar area.

The McColl *et al.* (2004) process involves adding a polymeric flocculant (alone or in combination with a natural/semi-natural flocculant) at high concentration to a pipeline containing thickener underflow. The tailings and flocculant mix to form 'super-aggregates' – essentially a continuous network rather than individual aggregates. On deposition, the tailings exhibit a lava-like flow, channels quickly appear and clear water is rapidly released. The approach is comparable to the Suncor TRO process and 'thin-lift drying' as proposed by Syncrude and Shell Canada (discussed in more detail below in relation to oil sands tailings), but without relying on tailings being spread out in thin layers over a large area.

Flocculant and tailings are mixed in a conduit such as pipe or launder leading from an area in which thickened tailings is produced. Optimization of the mixing is not described, but it is stressed that those advantages described cannot be achieved by adapting the flocculation step in a thickener, as the material produced is not conducive to pumping (no evidence is presented to support this). The use of thickened tailings material as the feed is advocated, reflecting the need for a high solids concentration.

8.9.2 Oil Sand Tailings Dewatering

The dewatering of tailings derived from the treatment of oil sands is discussed as a specific example, given that this is a growing industry, as well as being without doubt the area in which high-solids reagent addition has the most immediate and practical implications. There are currently a large number of technologies being used by the oil sands industry to deal with both current fluid fine tailings (FFT) inventory and future production. The industry acknowledges that there will most likely not be one single solution to tackle the problem, but rather a number of complimentary techniques, some of which are outlined below.

The most common technique applied for consolidation of FFT within existing operations was the consolidated tailings (CT) process, also referred to as non-segregating tailings (NST). In the 1990s, Suncor pioneered CT technology to help speed up the consolidation of FFT into a soil-like deposit that can be re-vegetated and reclaimed. The technique involved mixing FFT, coarse sand and gypsum to produce an engineered, non-segregating trafficable deposit. CT was not without its own issues, despite being the most common approach until recent years. Suncor's own CT studies have shown that a mixture of deposit types can result. Some exhibit sand-like behaviour and readily consolidate, quickly releasing water, while others consolidate slowly and need further dewatering applied. Sand use in CT also competes with the need for sand to build containment structures for the tailings areas, and ultimately the volumes of FFT greatly exceed that of sand available for use in CT. Soluble calcium within returned water is potentially detrimental to the bitumen extraction process.

Suncor developed a new technique called tailings reduction operations (TRO) (Revington *et al.*, 2011a) that promises significant improvements. TRO claims to more rapidly convert FFT into a solid landscape suitable for reclamation. FFT at high solids is mixed with a polymer flocculant, and then deposited in thin layers over sand beaches with shallow slopes. The resulting product is a dry material capable of being reclaimed in place or moved to another location for final reclamation. This drying process occurs over a matter of weeks; options that reduce this time further and use less land area are being actively examined.

Syncrude has investigated a number of options for tailings management that include FFT water capping, composite tails, FFT centrifugation and accelerated dewatering. Water capping involves placing water over an existing FFT deposit, which eventually forms a lake. Syncrude claims that over time the lakes evolve into natural ecosystems capable of eventually supporting plants, animals and fish. FFT centrifugation involves adding FFT and a flocculant to a horizontal solid bowl centrifuge and applying up to 2000 G, aiming to produce a relatively particle-free centrate and a cake that can be rapidly dried and become trafficable. It is believed this technology may be implemented at full scale despite its high cost, although the need for high throughputs may make it difficult to achieve centrate solids concentrations of below 10 kg/m^3.

Accelerated dewatering (a process pioneered in the Florida phosphate mining industry) involves pumping highly flocculated tailings into a pit and then draining the water out. Syncrude is also trailing a modified version of this ('thin-lift drying'), whereby the tailings are spread out in 10 cm layers allowing rapid drying to occur. They claim that the layer is dried sufficiently within a week to allow another layer to be added on top.

Shell Canada has taken the 'thin-lift drying' technology further by establishing a 300 hectare demonstration plant at its Muskeg River Mine to produce high-density trafficable tailings. They term their process Atmospheric Fines Drying (AFD); it has similarities to both the TRO and thin-lift processes. Shell is also believed to be looking at altering its composite tailings process by adding alum rather than gypsum as a chemical thickening agent.

Shell has been the main oil sands producer to use thickeners within its extraction operations. In doing this it has been able to improve energy efficiency by recovering and immediately re-using warm water in the bitumen extraction process, although it should be noted that the thickener underflow does not achieve high solids concentrations, and the thickeners at the Muskeg River Mine are not designed for this purpose. Shell included thickeners at its more recent Jackpine mine as well, although again, they are believed to be designed more for throughput. Syncrude is researching the use of thickened paste technology. The main question over the use of thickeners is the limit on underflow density and that further consolidation beyond that may not be achieved without additional treatment (e.g. TRO).

Canadian Natural Resources (CNRL) currently uses CO_2 injection for tailings management at its newly commissioned Horizon mine. Waste CO_2 is injected into the tailings slurry line prior to the tailings entering a containment pond. The CO_2 reacts to form carbonic acid that changes the pH of the tailings, allowing fine clays, silt and sand to settle quickly through an aggregation process and leaving behind clear water that can immediately be re-used in the extraction process. During the next phase of the Horizon mine, CNRL plans to implement cyclones and thickeners into the process. Cyclones will remove water from the coarse sand and thickeners will remove water from the fines and silt component. The two dewatered streams will be combined and CO_2 injected. An advantage of this process is that CO_2 is a waste product from the plant and its use produces a water chemistry that is beneficial to the extraction process, so there is no concern over adding too much to the tailings streams.

8.9.3 Assessment of Known Information

The Suncor TRO patent application (Revington *et al.*, 2011a) and the recent review by Wells *et al.* (2015) provide the most practical detail on how enhanced dewatering of oil sands tailings is achieved from direct addition of high flocculant dosages. The key aspect is the method of flocculant injection via an orifice plate (a co-annular injector – this is the subject of another patent application (Revington *et al.*, 2011b), which itself is not conceptually different to other forms of injection). The expectation is that this will apply a high degree of shear across a wider area than simpler mixing approaches, with a greater prospect of achieving good distribution. However, Wells *et al.* (2015) do note that the wide range for tailings and flocculant rheologies in field applications may present conditions that prevent efficient mixing, and hence impellers and pre-mixing vessels may be required, along with more complex control systems.

In discussing the TRO process, Wells *et al.* (2011) emphasized the importance of adding flocculant under high shear, with subsequent tapered shear to give the optimum discharge properties. Capillary suction time (CST) measurements are extensively used in their studies. CST is common in wastewater sludge studies and is clearly a simple test that can give good qualitative comparisons in this application. However, given the time-based changes in the treated solids and the variable clay contents, it is highly questionable whether it can be relied upon for quantifying flocculation effects.

Wells *et al.* (2011) not only highlight the need for the process to be done at high solids, but also the importance of the correct applied shear regime. Figure 8.12 represents the impact of mixing time on the main performance measures and the main process stages. The stages (as for conventional flocculation) are: (1) the mixing and adsorption of the flocculant, with the start of aggregate growth; (2) 'floc rearrangement', during which it is claimed aggregate growth and breakage is in balance; and finally (3) 'floc breakdown', which sees a reduction in yield stress and the greatest water release. Extending mixing too

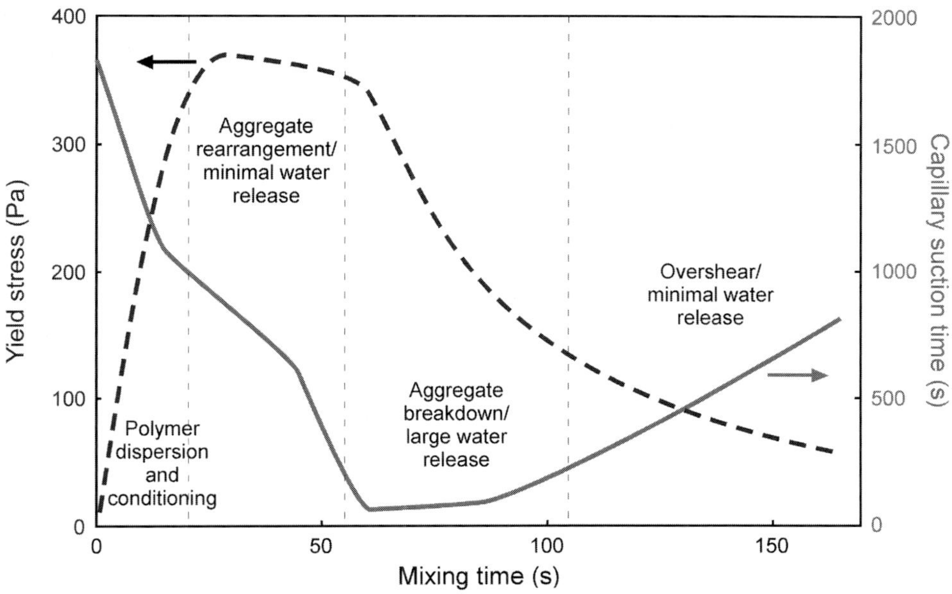

Figure 8.12. Key stages in Suncor's TRO process for dewatering FFT (adapted from Wells *et al.*, 2011; image reproduced with kind permission of the Australian Centre for Geomechanics. 14th International Paste and Thickened Tailings Seminar, Perth, Australia, 2011; www.acg.uwa.edu.au).

long leads to 'overshear', and while the yield stress continues to reduce, little additional water is released.

An important implication to be drawn from Fig. 8.12 is that while the yield stress may appear high, the corresponding numbers for material flocculated traditionally at feedwell solids and then settled/thickened to the same solids concentrations as in the TRO process would in theory be substantially higher (although filtration would probably be necessary to attain such concentrations). An inescapable conclusion is that the aggregates as formed are at a higher density than is achieved under normal feedwell flocculation conditions. However, this cannot be the only contributing factor, as alone it would not explain the enhanced dewatering.

A potential trap is to think of the processes involved in 'condensing aggregation' as being closely related to those now well understood for polymer-bridging flocculation at typical feedwell solids concentrations. Possible points of differentiation are summarized in Table 8.2, and Fig. 8.13 provides a schematic representation (not to scale) of how the two aggregation processes may differ. Flocculant chains are also not represented in order to simplify the images.

While there is potential for unadsorbed flocculant to be discharged when applying such extreme dosages, it is more probable that the solids will scavenge any residual polymer prior to discharge if effectively mixed. The flocculant is expected to persist for much longer within the system as a consequence of the overdosing, raising the prospect of quite distinct aggregation processes.

Table 8.2. *Contrasting conventional feedwell flocculation vs condensing aggregation.*

Aspect	Polymer-bridging flocculation	Condensing aggregation
Flocculant mixing and adsorption	Adsorption is rapid, and effective surface coverage is almost always <20 per cent.	Initial adsorption is rapid, but at high dosages surface coverages may be >50 per cent, slowing adsorption. Some unadsorbed?
Initial aggregation	High proportion of available surface makes bridging collisions effective; higher dosages increase collision efficiency. Lower solids concentrations reduce growth rates, but lower viscosity and shear-induced breakage.	High surface coverage will lower collision efficiency (fewer sites for adsorption). High solids will increase collision rate, but also increases shear-induced breakage. High viscosity also slows aggregation rate.
Adsorbed flocculant conformation	Flocculant adsorbs irreversibly through a series of 'trains'; bridging through tails, ~30–50 per cent of polymer mass (Fig. 8.3).	Surface crowding limits extent of trains, especially after initial adsorption and also bridging links. Will some adsorption be reversible?
	Non-bridging tails rapidly reconform to attach to surface, losing bridging capacity.	Possible that tails are less likely to reconform flat, retaining some activity.
Aggregate size	Limited by applied shear and dosage. Large sizes form, but there is an optimum solids concentration, above which size decreases.	Determined by solid concentration/viscosity, with limited impacts of applied shear (in terms of maximum size) and dosage.
Aggregate breakage	Breakage is irreversible – bridging chains are generally halved, and with shorter tails available for adsorption.	Sheared chains may retain bridging capacity at close aggregate proximity. Weak adsorption may allow chain detachment. Both reflected in a degree of reversible breakage (like coagulation), supported by Fig. 8.13.
Aggregate structure	Initial fractal aggregates further flocculated in a clustering mechanism. Multiple bridges may exist. Large aggregates are weak, but small core structures persist to underflow.	Initial aggregates small and dense. Clusters have many bridges, and this may be where rearrangement can further densify. High-level clusters are broken, but core persists.

Table 8.2 (*cont.*)

Aspect	Polymer-bridging flocculation	Condensing aggregation
Applied shear	Mean shear rate through feedwell 40–80 s^{-1}, residence times 10–40 s; higher shear favourable during initial mixing (i.e. 2–5 s).	Mean shear rate 20–40 s^{-1} (?), residence times 20–100 s.
Aggregate response to shear	Extended shear during flocculation favours breakage over densification. Low shear post-flocculation may give densification.	Initial 'rearrangement' followed by breakage releases water. Initial step may involve substantial internal densification.

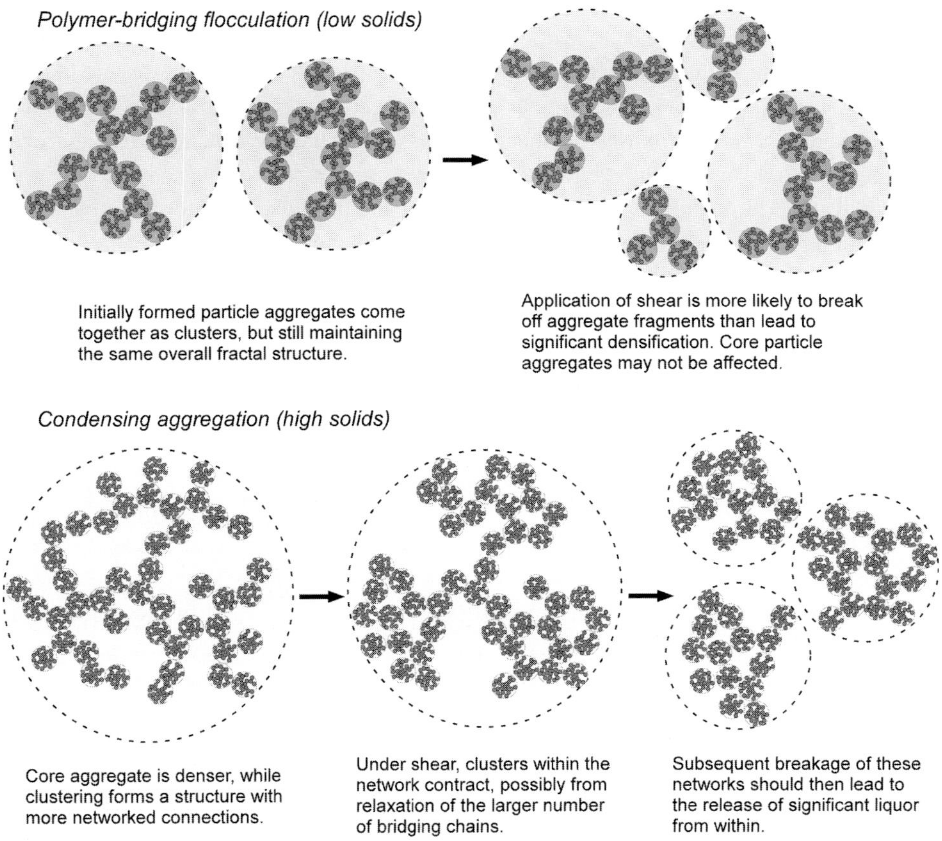

Figure 8.13. Schematic representation of possible aggregate structures for both polymer-bridging and condensing aggregation processes, and the influence of shear forces.

In polymer-bridging flocculation, self-similar fractal structures are maintained as aggregates grow from initial particle aggregates through subsequent clustering. The aggregates are weak, reflecting the open structures and limited binding points, and so the main response to shear applied after formation is breaking or rupture to fragments, with little actual densification. Some water is inevitably released, but much is retained within the structure.

With the condensing process, the initial core aggregate will be denser, due to closer proximity of the particles at the higher solids volume fraction. Clustering of these aggregates will also be influenced by the higher flocculant dosages, with additional bridges giving some further networking within a cluster. The second step shown in Fig. 8.12 is more difficult to illustrate, but is intended to represent the rearrangement of internal clusters, with the possibility of relaxation/contraction of multiple bridging chains leading to densification. When the overall cluster is broken down to these internal clusters, a greater proportion of the originally encapsulated liquor is released. Any subsequent breakage of these clusters will release less liquor.

Visual observation of laboratory simulations and plant applications shows water 'ejected' or 'squeezed' out of the network structure that forms. This could be the result of the fact that there is intense force/pressure inside the network structure as newly formed aggregates are effectively competing against the water for limited space. As the aggregates are likely to be held in position by a very large number of bridging structures (combination of high flocculant dosage and close proximity of the aggregates to each other causing cross-aggregate bridging), it would make sense that the easiest way to create more space is to eject the unbound water. The 'pressure' could also be responsible for the aggregates effectively 'collapsing' into each other and densifying, creating the beaching effect on deposition.Adkins *et al.* (2014) demonstrated that the network structures formed were an important differentiation from conventional flocculation.

Long mixing times are needed (>30 s) to achieve the sought-after continuous networked structure. This is unlike 'conventional' flocculation, where the solids content is reasonably low and short residence times (<5 s) give optimal flocculation. The reasons for this are likely to be numerous, but may include:

(1) The inevitably high viscosity of the flocculant solution, a result of the high concentration at which the flocculant needs to be added to achieve the required dosages, means transport through the slurry is slow.
(2) The very high solids content of the slurry creates 'crowding' and further inhibits dispersion of the flocculant in a short time frame.

It is suspected that the shear regime needed in the mixing zone should be laminar in nature and not fully turbulent, as the aggregates that form are already exposed to intense breakage events due to the very high number of particles in the slurry. Having a turbulent mixing zone will only exacerbate aggregate breakage. The high prospect of aggregate breakage is probably the main reason that the current full-scale operators add very high dosages of flocculant, effectively using the flocculant as a blunt instrument. It seems unlikely that the process as described within the patents and patent application is properly optimized in terms of application or reagent selection. If issues associated with mixing can be resolved, there is then scope to properly identify which reagents are best suited to the process. The possibility of using chaotic advection in such applications is being considered (Costine *et al.*, 2014).

Limited information exists on the most appropriate choice of flocculant, complicated by the fact that high dosages necessitate the use of a high flocculant concentration, and hence the dosed flocculant solution's rheology becomes an issue. The flocculant chemistries appear to be in general the same as those applied in feedwell applications, with the main difference being in the molecular weights selected – not surprisingly, lower molecular weights can favour mixing, but it is believed that the presence of some higher molecular weight component can be beneficial.

The major weakness with all work in this area has been the lack of detailed and appropriate dewatering performance characterization. While suitable (and often irrefutable) evidence can be shown simply from the solids concentration of the final slurry, the majority of other quantification measures are limited to vane yield stress, capillary suction times and slump tests, all of which are of questionable sensitivity in this application, and also are not relevant to examination of intermediate material states. Vane and slump measures are of value when considering deposition behaviour, but not to flow conditions. Some limited pressure drop measurements have been described, and these are of greater value in regards to transitory stages, but are still an indirect measure of the material state. The application of seepage-induced consolidation (Abu-Hejleh *et al.*, 1996) and small-scale pressure filtration (Usher *et al.*, 2001) for permeability and compressibility characterization is known, but little has been openly published as yet. In addition, many of the issues discussed in Section 8.4.1 in association with flocculation performance testing are further heightened in high-solids tailings applications, and contribute to why the fundamental aspects remain poorly understood.

8.10 Summary

This chapter has not sought to provide a comprehensive review of all clay flocculation studies, but rather to use the literature to illustrate a number of key issues that are either overlooked or sometimes not well understood when undertaking flocculation of mineral systems. Flocculation is a generic process that can be applied across a wide variety of minerals, and while it is undoubtedly true that there are specific complications associated with the study of clays, in the majority of cases the flocculation response can be attributed to:

(1) physical properties, such as particle size/number, available surface area, fractal dimension and effective solids volume fraction; and
(2) the applied conditions, such as the mean shear rate, shear rate distribution, reaction time, solids concentration, flocculant solution properties and dosing strategy.

Flocculation can occur across a wide range of conditions and performance can vary substantially, particularly in response to the applied shear intensity and duration. It is therefore important to understand that even for the tightest control of flocculation conditions in a test, the results are directly relevant to the conditions in that test, but not necessarily to the conditions within full-scale applications. Having at least a qualitative understanding of the applied shear conditions is important for better tailoring of tests to an application.

A clear distinction must be drawn between polymer-bridging flocculation as it occurs in feedwells and feedpipes/launders in thickeners and aggregation achieved from adding reagents at high dosages to high-solids suspensions for enhanced dewatering prior to tailings deposition. The latter process was given the rather unsatisfactory name of 'condensing aggregation' in this review, without any expectation that name will survive, but rather to emphasize that it may be highly misleading to think of the process as flocculation. Condensing aggregation is almost certain to involve different mechanisms, structures and responses to applied conditions from conventional polymer-bridging flocculation, and the understanding of these processes represents a major scientific challenge.

Acknowledgements

Alton Grabsch is thanked for the generation of PB model predictions of flocculation reaction profiles; Andrew Chryss and Allan Costine are thanked for advice in relation to rheology and high-solids applications, respectively.

References

Abbott, J. 1979. Control systems and operational features of the deep cone thickener. *Filtration & Separation*, 16, 376–380, 393.

Abd-El Rahman, M. K. 2000. Dewatering of phosphatic clay waste by flocculation. *Chemical Engineering & Technology*, 23 (5), 457–461.

Abu-Hejleh, A. N., Znidarcic, D. & Barnes, B. L. 1996. Consolidation characteristics of phosphatic clays. *Journal of Geotechnical Engineering*, 122 (4), 295–301.

Adkins, S. J. 2008. Underflow rheology optimisation to enhance thickener performance. In: Fourie, A. B., Jewel, R. J., Paterson, A. & Slatter, P. (eds) *Proceedings of the Eleventh International Seminar on Paste and Thickened Tailings*. Perth: Australian Centre for Geomechanics, 83–95.

Adkins, S. J., Bellwood, J., Beveridge, A., Edgar, M. & Flanagan, I. A. 2014. Difference between secondary flocculation and rigidification via Rheomax®-enhanced tailings disposal technology. In: Jewell, R. J., Fourie, A. B., Wells, P. S. & Van Zyl, D. (eds) *Proceedings of the 17th International Seminar on Paste and Thickened Tailings*. Vancouver: Infomine, 187–197.

Alam, N., Ozdemir, O., Hampton, M. A. & Nguyen, A. V. 2011. Dewatering of coal plant tailings: Flocculation followed by filtration. *Fuel*, 90 (1), 26–35.

Barany, S., Meszaros, R., Marcinova, L. & Skvarla, J. 2011. Effect of polyelectrolyte mixtures on the electrokinetic potential and kinetics of flocculation of clay mineral particles. *Colloids and Surfaces A: Physicochemical and Engineering Aspects*, 383 (1–3), 48–55.

Bedell, D., Fawell, P., Slottee, S. & Schoenbrunn, F. 2015. Thickening. In: Jewell, R. J. & Fourie, A. B. (eds) *Paste and Thickened Tailings: A Guide*, 3rd edition. Nedlands, WA: Australian Centre for Geomechanics, 113–137.

Berger, A., Adkins, S., Hess, S., Flanagan, I. & Stocks, P. 2011. Step change improvements in underflow rheology. In: Jewell, R. J. & Fourie, A. B. (eds) *Paste 2011: Proceedings of the 14th International Seminar on Paste and Thickened Tailings*. Nedlands, WA: Australian Centre for Geomechanics, 135–141.

Berger, A., Muhor, J., Adkins, S., Flanagan, I. & Brodt, G. 2012. BASF Novel Flocculant Technology (NFT): 'Thickener performance beyond the threshold'. In: Pradip & Mishra, B. K. (eds) *XXVI International Mineral Processing Congress: IMPC 2012, September 24–28, 2012*. New Delhi: Technowrites Pvt. Ltd.

Berger, A., Muhor, J. & Adkins, S. 2013. BASF novel flocculant technology: Benefits for counter current decantation circuits based on pilot-scale trials. In: Jewell, R. J., Fourie, A. B., Caldwell, J. & Pimenta, J. (eds) *Paste 2013: Proceedings of the 16th International Seminar on Paste and Thickened Tailings*. Nedlands, WA: Australian Centre for Geomechanics, 161–174.

Berlin, A. A. & Kislenko, V. N. 1995. Kinetic-models of suspension flocculation by polymers. *Colloids and Surfaces A: Physicochemical and Engineering Aspects*, 104 (1), 67–72.

Besra, L., Sengupta, D. K., Roy, S. K. & Ay, P. 2002. Studies on flocculation and dewatering of kaolin suspensions by anionic polyacrylamide flocculant in the presence of some surfactants. *International Journal of Mineral Processing*, 66 (1–4), 1–28.

Besra, L., Sengupta, D. K., Roy, S. K. & Ay, P. 2004. Influence of polymer adsorption and conformation on flocculation and dewatering of kaolin suspension. *Separation and Purification Technology*, 37 (3), 231–246.

Besra, L., Sengupta, D. K. & Roy, S. K. 2006. Influence of unadsorbed and weakly adsorbed flocculants on separation properties of kaolin suspensions. *International Journal of Mineral Processing*, 78 (2), 101–109.

Bolto, B. & Gregory, J. 2007. Organic polyelectrolytes in water treatment. *Water Research*, 41 (11), 2301–2324.

Bowman, L. E., Downey, C. E., Pillai, K. J. & Roe, W. J. 1993. Kaolin filter aids: Key to improved plant productivity. *Mining Engineering*, 45 (6), 600–602.

Boylu, F., Hojiyev, R., Ersever, G., Ulcay, Y. & Celik, M. S. 2012. Production of ultrapure bentonite clays through centrifugation techniques. *Separation Science and Technology*, 47 (6), 842–849.

Burns, J. L., Yan, Y. D., Jameson, G. J. & Biggs, S. 1997. A light scattering study of the fractal aggregation behavior of a model colloidal system. *Langmuir*, 13 (24), 6413–6420.

Bushell, G. C., Yan, Y. D., Woodfield, D., Raper, J. & Amal, R. 2002. On techniques for the measurement of the mass fractal dimension of aggregates. *Advances in Colloid and Interface Science*, 95 (1), 1–50.

Caskey, J. A. & Primus, R. J. 1986. The effect of anionic polyacrylamide molecular-conformation and configuration on flocculation effectiveness. *Environmental Progress*, 5 (2), 98–103.

Chaplain, V., Janex, M. L., Lafuma, F., Graillat, C. & Audebert, R. 1995. Coupling between polymer adsorption and colloidal particle aggregation. *Colloid and Polymer Science*, 273 (10), 984–993.

Chen, W. J. 1998. Effects of surface charge and shear during orthokinetic flocculation on the adsorption and sedimentation of kaolin suspensions in polyelectrolyte solutions. *Separation Science and Technology*, 33 (4), 569–590.

Clark, A. Q., Herrington, T. M. & Petzold, J. C. 1990. The flocculation of kaolin suspensions with anionic polyacrylamides of varying molar mass and anionic character. *Colloids and Surfaces*, 44, 247–261.

Connelly, L. J., Owen, D. O. & Richardson, P. F. 1986. Synthetic flocculant technology in the Bayer process. *Light Metals*, 2, 61–68.

Costine, A., Lester, D., Fawell, P. & Chryss, A. 2014. Shear isn't mixing: How to build larger aggregates using chaotic advection for accelerated dewatering. In: Jewell, R. J., Fourie, A. B., Wells, P. S. & Van Zyl, D. (eds) *Paste 2014: 17th International Seminar on Paste and Thickened Tailings*. Vancouver: Infomine.

Curtis, A. S. G. & Hocking, L. M. 1970. Collision efficiency of equal spherical particles in a shear flow: Influence of London-van der Waals forces. *Transactions of the Faraday Society*, 66 (570), 1381–1390.

Dollimore, D. & Horridge, T. A. 1973. Dependence of flocculation behavior on China clay–polyacrylamide suspensions on suspension pH. *Journal of Colloid and Interface Science*, 42 (3), 581–588.

Du, J. H., Pushkarova, R. A. & Smart, R. S. C. 2009. A cryo-SEM study of aggregate and floc structure changes during clay settling and raking processes. *International Journal of Mineral Processing*, 93 (1), 66–72.

Du, J. H., Morris, G., Pushkarova, R. A. & Smart, R. S. 2010. Effect of surface structure of kaolinite on aggregation, settling rate, and bed density. *Langmuir*, 26 (16), 13227–13235.

Farrow, J. B. & Swift, J. D. 1991. Improving thickener technology. *Fifth AUSIMM Extractive Metallurgy Conference*. Carlton, VIC: Australasian Institute of Mining and Metallurgy, 227–232.

Farrow, J. B. & Swift, J. D. 1996. A new procedure for assessing the performance of flocculants. *International Journal of Mineral Processing*, 46 (3–4), 263–275.

Farrow, J. B., Johnston, R. R. M., Simic, K. & Swift, J. D. 2000. Consolidation and aggregate densification during gravity thickening. *Chemical Engineering Journal*, 80 (1–3), 141–148.

Fawell, P. D., Owen, A. T., Grabsch, A. F., *et al.* 2009a. Factors affecting flocculation within gravity thickeners. *Water in Mining 2009, Proceedings*. Carlton, VIC: Australasian Institute of Mining and Metallurgy, 71–76.

Fawell, P. D., Farrow, J. B., Heath, A. R., *et al.* 2009b. 20 years of AMIRA P266 'Improving Thickener Technology': How has it changed the understanding of thickener performance? In: Jewel, R., Fourie, A., Barrera, S. & Wiertz, J. (eds) *Paste 2009: Proceedings of the 12th International Seminar on Paste and Thickened Tailings*. Nedlands, WA: Australian Centre for Geomechanics, 59–68.

Fawell, P. D., Costine, A. D. & Grabsch, A. F. 2015. Why small-scale testing of reagents goes wrong. In: Jewell, R. & Fourie, A. (eds) *Paste 2015: 18th International Seminar on Paste and Thickened Tailings*. Nedlands, WA: Australian Centre for Geomechanics, 153–165.

Franks, G. V., Yates, P. D., Lambert, N. W. A. & Jameson, G. J. 2005. Aggregate size and density after shearing, implications for dewatering fine tailings with hydrocyclones. *International Journal of Mineral Processing*, 77 (1), 46–52.

Franks, G. V., Li, H., O'Shea, J.-P. & Qiao, G. G. 2009. Temperature responsive polymers as multiple function reagents in mineral processing. *Advanced Powder Technology*, 20, 273–279.

Franks, G. V., O'Shea, J. P. & Forbes, E. 2014. Controlling thickener underflow rheology using a temperature responsive flocculant. *AIChE Journal*, 60, 2940–2948.

Frost, R. C., Halliday, J. & Dee, A. S. 1993. Continuous consolidation of sludge in large-scale gravity thickeners. *Water Science and Technology*, 28 (1), 77–86.

Gaillard, N. & Poncet, F. 2010. Treating sludge from mining or mineral industry before spreading out into soil, comprises contacting the sludge with a branched flocculant such as a water soluble organic polymer having a specified anionicity for a specified

duration. Patent FR2937635-A1; US2010105976-A1; CA2682542-A1.
Garcia, R. A., Riner, S. A. & Piazza, G. J. 2014. Design of a laboratory method for rapid evaluation of experimental flocculants. *Industrial & Engineering Chemistry Research*, 53, 880–886.
Gill, R. I. S. & Herrington, T. M. 1987. The flocculation of kaolin suspensions with cationic polyacrylamides of varying molar mass but the same cationic character. *Colloids and Surfaces*, 22 (1), 51–76.
Gill, R. I. S. & Herrington, T. M. 1988. Floc size studies on kaolin suspensions flocculated with cationic polyacrylamides. *Colloids and Surfaces*, 32 (3–4), 331–344.
Gill, R. I. S. & Herrington, T. M. 1989. The effect of colloid concentration and pH on kaolin suspensions flocculated with cationic polyacrylamides of high molar mass. *Colloids and Surfaces*, 42 (1–2), 23–37.
Gladman, B., de Kretser, R. G., Rudman, M. & Scales, P. J. 2005. Effect of shear on particulate suspension dewatering. *Chemical Engineering Research & Design*, 83 (A7), 933–936.
Glasgow, L. A. & Liu, S. X. 1995a. Effects of dosing regimen and agitation profile upon floc characteristics. *Chemical Engineering Communications*, 132, 223–237.
Glasgow, L. A. & Liu, S. X. 1995b. Effects of macromolecular conformation upon solid–liquid separation and water-treatment plant residuals. *Environmental Technology*, 16 (10), 915–927.
Grabsch, A. F., Fawell, P. D., Adkins, S. J. & Beveridge, A. 2013. The impact of achieving a higher aggregate density on polymer-bridging flocculation. *International Journal of Mineral Processing*, 124, 83–94.
Graham, N. D. B. & Graham, A. D. B. 2007. Materials handling and treatment. Patent WO2007143780-A1.
Grassia, P., Usher, S. P. & Scales, P. J. 2011. Closed-form solutions for batch settling height from model settling flux functions. *Chemical Engineering Science*, 66 (5), 964–972.
Gregory, J. 1988. Polymer adsorption and flocculation in sheared suspensions. *Colloids and Surfaces*, 31, 231–253.
Gregory, J. 2009. Monitoring particle aggregation processes. *Advances in Colloid and Interface Science*, 147–148, 109–123.
Gregory, J. & De Moor, A. E. L. 1984. Filterability of polymer-flocculated suspensions. In: Goddard, E. D. & Vincent, B. (eds) *Polymer Adsorption and Dispersion Stability*. Washington, DC: ACS, 445–458.
Gregory, J. & Nelson, D. W. 1986. Monitoring of aggregates in flowing suspensions. *Colloids and Surfaces*, 18 (2–4), 175–188.
Hamza, H. A., Stanonik, D. J. & Kessick, M. A. 1996. Flocculation of lime-treated oil sands tailings. *Fuel*, 75 (3), 280–284.
Heath, A. R., Fawell, P. D., Bahri, P. A. & Swift, J. D. 2002. Estimating average particle size by focused beam reflectance measurement (FBRM). *Particle & Particle Systems Characterization*, 19 (2), 84–95.
Heath, A. R., Bahri, P. A., Fawell, P. D. & Farrow, J. B. 2006a. Polymer flocculation of calcite: Population balance model. *AIChE Journal*, 52 (5), 1641–1653.
Heath, A. R., Bahri, P. A., Fawell, P. D. & Farrow, J. B. 2006b. Polymer flocculation of calcite: Relating the aggregate size to the settling rate. *AIChE Journal*, 52 (6), 1987–1994.
Heath, A. R., Bahri, P. A., Fawell, P. D. & Farrow, J. B. 2006c. Polymer flocculation of calcite: Experimental results from turbulent pipe flow. *AIChE Journal*, 52 (4), 1284–1293.
Hecker, R., Kirwan, L., Jefferson, A., *et al.* 1999. Focused beam reflectance measurement for the continuous assessment of flocculant performance. In: Laskowski, J. S. (ed.) *Polymers in Mineral Processing, Aug 17–19, 1999*. Quebec City: Canadian Institute of Mining, Metallurgy & Petroleum, 91–105.

Henderson, J. M. & Wheatley, A. D. 1987. Factors effecting a loss of flocculation activity of polyacrylamide solutions: Shear degradation, cation complexation, and solution aging. *Journal of Applied Polymer Science*, 33 (2), 669–684.

Hogg, R. 1999. Polymer adsorption and flocculation. In: Laskowski, J. S. (ed.) *Polymers in Mineral Processing, Aug 17–19, 1999*. Quebec City: Canadian Institute of Mining, Metallurgy & Petroleum, 3–17.

Hopkins, D. C. & Ducoste, J. J. 2003. Characterizing flocculation under heterogeneous turbulence. *Journal of Colloid and Interface Science*, 264 (1), 184–194.

Hoyland, G. & Day, M. 1986. An evaluation of picket fences for assisting the consolidation of sewage sludges. *Water Pollution Control*, 85 (3), 291–303.

Hsu, J. P., Lin, D. P. & Tseng, S. 1995. The sticking probability of colloidal particles in polymer-induced flocculation. *Colloid and Polymer Science*, 273 (3), 271–278.

Jarvis, P., Jefferson, B. & Parsons, S. 2004. The duplicity of floc strength. *Water Science and Technology*, 50 (12), 63–70.

Jarvis, P., Jefferson, B. & Parsons, S. A. 2005. Measuring floc structural characteristics. *Reviews in Environmental Science and Bio/Technology*, 4, 1–18.

Johns, F. E. 1991. High-pressure filtration of fine particle slurries. *Proceedings of the Sixteenth International Conference on Coal & Slurry Technologies, Apr 22–25, 1991*. Washington, DC: Coal & Slurry Technology Association, 297–308.

Klimpel, R. C. & Hogg, R. 1991. Evaluation of floc structures. *Colloids and Surfaces*, 55, 279–288.

Knight, M. A., Wates, J. A. & Du Plessis, I. 2012. Application of hydrocyclone technology to tailings storage facilities to reduce water consumption. In: Jewel, R. J., Fourie, A. B. & Paterson, A. (eds) *Paste 2012: Proceedings of the 15th International Seminar on Paste and Thickened Tailings*. Nedlands, WA: Australian Centre for Geomechanics, 233–241.

Koksal, E., Ramachandran, R., Somasundaran, P. & Maltesh, C. 1990. Flocculation of oxides using polyethylene oxide. *Powder Technology*, 62 (3), 253–259.

La Mer, V. K. & Healy, T. W. 1963. Adsorption–flocculation reactions of macromolecules at the solid–liquid interface. *Reviews of Pure and Applied Chemistry*, 13, 112–133.

Lee, C. H. & Liu, J. C. 2001. Sludge dewaterability and floc structure in dual polymer conditioning. *Advances in Environmental Research*, 5 (2), 129–136.

Lee, C. S., Robinson, J. & Chong, M. F. 2014. A review on application of flocculants in wastewater treatment. *Process Safety and Environmental Protection* 92, 489–508.

Liao, J. Y. H., Selomulya, C., Bushell, G., Bickert, G. & Amal, R. 2006. On different approaches to estimate the mass fractal dimension of coal aggregates. *Particle & Particle Systems Characterization*, 22 (5), 299–309.

Loan, C. & Triglavcanin, R. 2009. A quantum change in the thickening process. In: Amelunxen, P., Kracht, W. & Kuyvenhoven, R. (eds) *PROCEMIN 2009: Proceedings of the VI International Mineral Processing Seminar, 2–4 December 2009*. Santiago: GECAMIN, 503–512.

Long, J., Li, H., Xu, Z. & Masliyah, J. H. 2011. Improving oil sands processability using a temperature-sensitive polymer. *Energy and Fuels*, 25, 701–707.

Lyklema, J. 1988. The colloidal background of flocculation and dewatering. In: Moudgil, B. M. & Scheiner, B. J. (eds) *Flocculation and Dewatering*. New York: Engineering Foundation, 1–20.

Mackowski, S. J. 1989. Pressure filtration of Weipa kaolin. *Dewatering Practice & Technology*. Brisbane: AusIMM, 137–140.

Maggi, F., Mietta, F. & Winterwerp, J. C. 2007. Effect of variable fractal dimension on the floc size distribution of suspended cohesive sediment. *Journal of Hydrology*, 343, 43–55.

Mathewson, D., Norris, R. & Dunne, M. 2006. Tailings dewatering dry screening and water clarification for reduced water usage. *Proceedings: Water in Mining 2006 – Multiple Values of Water*. Parkville, VIC: Australasian Institute of Mining and Metallurgy, 315–322.

Mathur, S. & Moudgil, B. M. 1997. Adsorption mechanism(s) of poly(ethylene oxide) on oxide surfaces. *Journal of Colloid and Interface Science*, 196, 92–98.

McColl, P. & Flanagan, I. J. 1999. The Twintec advanced dewatering system for the solid–liquid separation of coal tailings. In: Laskowski, J. S. (ed.) *Polymers in Mineral Processing, Aug 17–19, 1999*. Quebec City: Canadian Institute of Mining, Metallurgy & Petroleum, 579–589.

McColl, P., Scammell, S., Philip, M. & Stephen, S. 2004. Treatment of mineral material, especially waste mineral slurries transferring material with dispersed particulate solids as fluid to deposition area by combining with material aqueous solution of water-soluble polymer. Patent WO2004060819-A1 WOEP000042 07 Jan 2004 and numerous country patents.

McFarlane, A. J., Addai-Mensah, J. & Bremmell, K. 2005. Rheology of flocculated kaolinite dispersions. *Korea-Australia Rheology Journal*, 17 (4), 181–190.

McFarlane, A., Yeap, K. Y., Bremmell, K. & Addai-Mensah, J. 2008. The influence of flocculant adsorption kinetics on the dewaterability of kaolinite and smectite clay mineral dispersions. *Colloids and Surfaces A: Physicochemical and Engineering Aspects*, 317 (1–3), 39–48.

Mengual, O., Meunier, G., Cayre, I., Puech, K. & Snabre, P. 1999. Characterisation of instability of concentrated dispersions by a new optical analyser: The TURBISCAN MA 1000. *Colloids and Surfaces A: Physicochemical and Engineering Aspects*, 152 (1–2), 111–123.

Merkl, R. & Steiger, W. 2012. Properties of decanter centrifuges in the mining industry. *Minerals & Metallurgical Processing*, 29 (1), 6–12.

Mpofu, P., Addai-Mensah, J. & Ralston, J. 2003. Investigation of the effect of polymer structure type on flocculation, rheology and dewatering behaviour of kaolinite dispersions. *International Journal of Mineral Processing*, 71 (1–4), 247–268.

Mpofu, P., Addai-Mensah, J. & Ralston, J. 2004. Flocculation and dewatering behaviour of smectite dispersions: Effect of polymer structure type. *Minerals Engineering*, 17 (3), 411–423.

Mpofu, P., Addai-Mensah, J. & Ralston, J. 2005. Interfacial chemistry, particle interactions and improved dewatering behaviour of smectite clay dispersions. *International Journal of Mineral Processing*, 75 (3–4), 155–171.

Nan, J. & He, W. P. 2012. Characteristic analysis on morphological evolution of suspended particles in water during dynamic flocculation process. *Desalination and Water Treatment*, 41 (1–3), 35–44.

Nasser, M. S. & James, A. E. 2006. The effect of polyacrylamide charge density and molecular weight on the flocculation and sedimentation behaviour of kaolinite suspensions. *Separation and Purification Technology*, 52 (2), 241–252.

Nasser, M. S. & James, A. E. 2007. Effect of polyacrylamide polymers on floc size and rheological behaviour of kaolinite suspensions. *Colloids and Surfaces A: Physicochemical and Engineering Aspects*, 301 (1–3), 311–322.

Nguyen, T. V., Farrow, J. B., Smith, J. & Fawell, P. D. 2012. Design and development of a novel thickener feedwell using computational fluid dynamics. *The Journal of the Southern African Institute of Mining and Metallurgy*, 112, 939–948.

Ofori, P., Nguyen, A. V., Firth, B., McNally, C. & Ozdemir, O. 2011. Shear-induced floc structure changes for enhanced dewatering of coal preparation plant tailings. *Chemical Engineering Journal*, 172 (2–3), 914–923.

Oles, V. 1992. Shear-induced aggregation and breakup of polystyrene latex-particles. *Journal of Colloid and Interface Science*, 154 (2), 351–358.

Olsen, A., Franks, G., Biggs, S. & Jameson, G. J. 2006. An improved collision efficiency model for particle aggregation. *Journal of Chemical Physics*, 125 (18).

Owen, A. T., Fawell, P. D., Swift, J. D. & Farrow, J. B. 2002. The impact of polyacrylamide flocculant solution age on flocculation performance. *International Journal of Mineral Processing*, 67 (1–4), 123–144.

Owen, A. T., Fawell, P. D. & Swift, J. D. 2007. The preparation and ageing of acrylamide/acrylate copolymer flocculant solutions. *International Journal of Mineral Processing*, 84 (1–4), 3–14.

Owen, A. T., Fawell, P. D., Swift, J. D., *et al.* 2008. Using turbulent pipe flow to study the factors affecting polymer-bridging flocculation of mineral systems. *International Journal of Mineral Processing*, 87 (3–4), 90–99.

Pearse, M. 1980. Factors affecting the laboratory sizing of thickeners. In: Somasundaran, P. (ed.) *Fine Particle Processing, Proceedings of the International Symposium*. New York: AIME, 1619–1642.

Peng, S. J. & Williams, R. A. 1993. Control and optimization of mineral flocculation and transport processes using online particle-size analysis. *Minerals Engineering*, 6 (2), 133–153.

Potanin, A. A. & Uriev, N. B. 1991. Microrheological models of aggregated suspensions in shear-flow. *Journal of Colloid and Interface Science*, 142 (2), 385–395.

Rattanakawin, C. & Hogg, R. 2001. Aggregate size distributions in flocculation. *Colloids and Surfaces A: Physicochemical and Engineering Aspects*, 177 (2–3), 87–98.

Rattanakawin, C. & Hogg, R. 2005. Reagent combinations in flocculation. *Minerals & Metallurgical Processing*, 22 (4), 211–218.

Reif, F., Stahl, W. & Langeloh, T. 1990. Optimising decanter centrifuges. *Filtration & Separation*, 27 (6), 408–410.

Revington, A. P., Omotoso, O., Wells, P. S., *et al.* 2011a. Process for flocculating and dewatering oil sand mature fine tailings. Patent WO2011/032258 A1.

Revington, A. P., Martin, W. M. & Eastwood, J. 2011b. Method and device for in-line injection of flocculant agent into a fluid flow of mature fine tailings. Canada patent application WO2011/032258 A1.

Richardson, J. F. & Zaki, W. N. 1954. Sedimentation and fluidisation: Part 1. *Transactions of the Institution of Chemical Engineers*, 32, 35–53.

Roldan-Villasana, E. J. & Williams, R. A. 1999. Classification and breakage of flocs in hydrocyclones. *Minerals Engineering*, 12 (10), 1225–1243.

Rubio, J. 1981. The flocculation properties of poly (ethylene-oxide). *Colloids and Surfaces*, 3 (1), 79–95.

Rudman, M., Simic, K., Paterson, D. A., *et al.* 2008. Raking in gravity thickeners. *International Journal of Mineral Processing*, 86 (1–4), 114–130.

Rudman, M., Paterson, D. A. & Simic, K. 2010. Efficiency of raking in gravity thickeners. *International Journal of Mineral Processing*, 95 (1–4), 30–39.

Rulyov, N. N., Korolyov, B. Y. & Kovalchuk, N. M. 2006. Application of ultra-flocculation for improving fine coal concentrate dewatering. *Coal Preparation*, 26 (1), 17–32.

Rulyov, N. N., Laskowski, J. S. & Concha, F. 2011. The use of ultra-flocculation in optimization of the experimental flocculation procedures. *Physicochemical Problems of Mineral Processing*, (47), 5–16.

Rushton, A., Ward, A. S. & Holdich, R. G. 1996. *Solid–Liquid Filtration and Separation Technology*. Weinheim, VCH: Wiley.

Sabah, E. & Cengiz, I. 2004. An evaluation procedure for flocculation of coal preparation plant tailings. *Water Research*, 38 (6), 1542–1549.

Sabah, E. & Erkan, Z. E. 2006. Interaction mechanism of flocculants with coal waste slurry. *Fuel*, 85 (3), 350–359.

Saffman, P. G. & Turner, J. S. 1956. On the collision of drops in turbulent clouds. *Journal of Fluid Mechanics*, 1 (1), 16–30.

Scheiner, B. J. & Wilemon, G. M. 1987. Applied flocculation efficiency: A comparison of polyethylene oxide and polyacrylamides. In: Attia, Y. A. (ed.) *Flocculation in Biotechnology and Separation Systems*. Amsterdam: Elsevier, 175–185.

Scott, J. P., Fawell, P. D., Ralph, D. E. & Farrow, J. B. 1996. The shear degradation of high-molecular-weight flocculant solutions. *Journal of Applied Polymer Science*, 62 (12), 2097–2106.

Sengupta, D. K., Kan, J., Al Taweel, A. M. & Hamza, H. A. 1997. Dependence of separation properties on flocculation dynamics of kaolinite suspension. *International Journal of Mineral Processing*, 49 (1–2), 73–85.

Senaputra, A., Jones, F., Fawell, P. D. & Smith, P. G. 2014. Focused beam reflectance measurement for monitoring the extent and efficiency of flocculation in mineral systems. *AIChE Journal*, 60, 251–265.

Sutalo, I. D., Paterson, D. A. & Rudman, M. 2003. Flow visualisation and computational prediction in thickener rake models. *Minerals Engineering*, 16 (2), 93–102.

Svarovsky, L. 2000a. Gravity clarification and thickening. In: Svarovsky, L. (ed.) *Solid–Liquid Separation*, 4th edition. Oxford: Butterworth-Heinemann, 166–190.

Svarovsky, L. 2000b. Separation by centrifugal sedimentation. In: Svarovsky, L. (ed.) *Solid–Liquid Separation*, 4th edition. Oxford: Butterworth-Heinemann, 246–280.

Swift, J. D., Simic, K., Johnston, R. R. M., Fawell, P. D. & Farrow, J. B. 2004. A study of the polymer flocculation reaction in a linear pipe with a focused beam reflectance measurement probe. *International Journal of Mineral Processing*, 73 (2–4), 103–118.

Usher, S. & Scales, P. 2009. Predicting settler/clarifier behaviour: The role of shear effects. *Filtration*, 9 (4), 308–314.

Usher, S. P., de Kretser, R. G. & Scales, P. J. 2001. Validation of a new filtration technique for dewaterability characterization. *AIChE Journal*, 47 (2), 1561–1570.

Vahedi, A. & Gorczyca, B. 2012. Predicting the settling velocity of flocs formed in water treatment using multiple fractal dimensions. *Water Research*, 46, 4188–4194.

van Ryssen, G. J. & Steenkamp, P. L. 2012. Thickening of tailings using hydrocyclones operating under vacuum conditions. In: Jewel, R. J., Fourie, A. B. & Paterson, A. (eds) *Paste 2012: Proceedings of the 15th International Seminar on Paste and Thickened Tailings*. Nedlands, WA: Australian Centre for Geomechanics, 225–231.

Vedoy, D. R. L. & Soares, J. B. P. 2015. Water-soluble polymers for oil sands tailing treatment: A review. *The Canadian Journal of Chemical Engineering*, 93, 888–904.

Vietti, A. J., Cooke, R., Ntshabele, K., Cooks, M. & Busani, B. 2008. Rheomax™ and water conservation at Orapa Mine. In: Fourie, A. B., Jewel, R. J., Paterson, A. & Slatter, P. (eds) *Paste 2008: Proceedings of the Eleventh International Seminar on Paste and Thickened Tailings*. Perth: Australian Centre for Geomechanics, 353–364.

Visman, J. & Hamza, H. A. 1973. Application of flocculants in hydrocyclone separation. *CIM Bulletin*, 66 (730), 78–85.

Wallace, L. B., Dabir, B. & Petty, C. A. 1980. Preliminary findings on the effect of polyacrylamide on particle–liquid separation in hydrocyclones. *Chemical Engineering Communications*, 7 (1–3), 27–36.

Watson, P., Farinato, R., Fenderson, T., *et al.* 2011. Novel polymeric additives to improve oil sands tailings consolidation. *Proceedings – SPE International Symposium on Oilfield Chemistry*. Houston, TX: Society of Petroleum Engineers, 703–709.

Wells, P. S., Revington, A. & Omotoso, O. 2011. Mature fine tailings drying: Technology update. In: Jewell, R. & Fourie, A. (eds) *Proceedings of the 14th International Seminar on Paste and Thickened Tailings*. Perth: Australian Centre for Geomechanics, 155–166.

Wells, P. S., Charlebois, L., Diep, J., *et al.* 2015. Inline flocculation. In: Jewell, R. J. & Fourie, A. B. (eds) *Paste and Thickened Tailings: A Guide*, 3rd edition. Nedlands, WA: Australian Centre for Geomechanics, 231–241.

Whatley, J. S. 1986. Filtration and solid–liquid separation in the kaolin industry. *SME Annual Meeting, March 2–6, 1986*. New Orleans, LA: SME, Littleton.

Williams, R. A. & Roldan-Villasana, E. J. 1991. Opportunities for dewatering aggregated colloidal materials using small diameter hydrocyclones. In: Williams, R. A. & De Jaeger, N. C. (eds) *Advances in Measurement and Control of Colloidal Processes*.Oxford: Butterworth-Heinemann, 18–34.

Williams, R. A., Peng, S. J. & Naylor, A. 1992. In situ measurement of particle aggregation and breakage kinetics in a concentrated suspension. *Powder Technology*, 73 (1), 75–83.

Woodfield, D. & Bickert, G. 2004. Separation of flocs in hydrocyclones: Significance of floc breakage and floc hydrodynamics. *International Journal of Mineral Processing*, 73 (2–4), 239–249.

Xu, Y., Hamza, H. & Matthews, J. 2004. Enlargement of fine particles in the oil sands industry: Flocculation and thickening. In: Laskowski, J. S. (ed.) *Particle Size Enlargement in Mineral Processing*. Montreal: Canadian Institute of Mining, Metallurgy & Petroleum, 271–288.

Yalcin, T. 1996. Tailings disposal using flocculant-assisted hydrocyclones. *CIM Bulletin*, 89 (1005), 98–100.

Yalcin, T. & Brunet, S. 2004. Testing of flocculant-assisted hydrocyclone on Quebec Cartier Mining Company's iron ore tailings. In: Laskowski, J. S. (ed.) *Particle Size Enlargement in Mineral Processing*. Montreal: Canadian Institute of Mining, Metallurgy & Petroleum, 289–301.

Yan, D., Parker, T. & Ryan, S. 2003. Dewatering of fine slurries by the Kalgoorlie Filter Pipe. *Minerals Engineering*, 16 (3), 283–289.

Yu, X. A. & Somasundaran, P. 1993. Enhanced flocculation with double flocculants. *Colloids and Surfaces A: Physicochemical and Engineering Aspects*, 81, 17–23.

Yu, X. & Somasundaran, P. 1996. Role of polymer conformation in interparticle-bridging dominated flocculation. *Journal of Colloid and Interface Science*, 177 (2), 283–287.

Yuan, X. S. 2011. Ore types impact on flocculation and the treatment strategies for different types of oil sands tailings. In: Jewell, R. & Fourie, A. (eds) *Proceedings of the 14th International Seminar on Paste and Thickened Tailings*. Perth: Australian Centre for Geomechanics, 143–153.

Yuan, X. S. & Shaw, W. 2007. Novel processes for treatment of Syncrude fine transition and marine ore tailings. *Canadian Metallurgical Quarterly*, 46 (3), 265–272.

Yukselen, M. A., O'Halloran, K. R. & Gregory, J. 2006. Effect of tapering on the break-up and reformation of flocs formed using hydrolyzing coagulants. In: Ng, H. Y. & Hu, J. (eds) *Drinking Water Treatment, Supply and Management in Asia*. London: IWA Publishing, 139–145.

Zhu, Y., Wu, J., Shepherd, I. S., *et al.* 2000. An automated measurement technique for slurry settling tests. *Minerals Engineering*, 13 (7), 765–772.

9

Effects of Clay in Tailings Handling and Storage

Andrew Chryss

9.1 Introduction

Tailings are the waste stream generated during the recovery of mineral commodities from ore. They are a mixture of clays, rock flour and rocks, dispersed in an effluent containing process chemicals – e.g. thiocyanates and dissolved metals. The storage of tailings in a stable, low-risk environment is a key design criterion for tailings management. Leaching of acidic water (acid mine drainage (AMD)) and metalliferous drainage are potential environmental hazards from tailings disposal, along with safety risks from catastrophic releases of water and slurry during containment failure. These events can be mitigated or avoided by using more recent tailings disposal technologies based around increasing the solids concentration of the deposited material. This approach can have the additional benefit of an improved water balance. The ability to employ these techniques is largely dependent on topography and the components of the tailings stream; however, the desire to implement them is often damped by the increased cost of infrastructure. The effects of clays on tailings management processes are varied and generally negative, with a few exceptions.

The volumes of tailings produced annually are proportional to the rate of production of mined commodities, and tailings storage facilities (TSF) are therefore a large and significant component of their environment (Fig. 9.1). Considering the example of a clay waste from the mining and beneficiation used to produce phosphate (Pittman and Sweeney, 1983), a typical company producing 3×10^6 tonnes/year mines about 1.6 km^2 of land, removing 11×10^6 m^3 of overburden and producing 7×10^6 m^3 of matrix. Depending on the nature of the phosphate matrix, for each tonne of phosphate produced, approximately 1 tonne of sand tailings and 1 tonne of clays are sent for disposal. Oil from tar sands production in Alberta, Canada produces 200×10^3 m^3 of partially consolidated fine tailings per day (Mikula *et al.*, 2008). The volumes from base metal production are larger still.

The major effects that clay has on tailings (handling and storage) are in concentrating, transport, settling, dewatering and water recovery processes. Clay affects aggregate density (see Chapter 8), so high concentration thickened tailings are harder to achieve. The swelling and gelation properties of certain clays (e.g. smectites) alter the tailings rheology, which can change the rate of particle settling, cause particle size segregation with depth and increase pumping energy requirements. The high surface area of clays in tailings

Figure 9.1. End of the line. Estimates of mine tailings production range up to 7 billion tonnes per annum globally. The thickening, transport and deposition of this non-Newtonian slurry are strongly influenced by its clay content. Photograph courtesy of Andreas Mönch.

can increase chemical consumption to uneconomic levels and affect the clarity of water returned to the refining process. Clays can affect beaching angle for tailings deposited above ground and alter the beach concavity that is often present. Importantly, high clay content in a deposited tailings beach will reduce its shear strength, allowing for possible collapse.

9.2 Physical Characteristics of Mine Tailings

The physical characteristics of mine tailings vary greatly. This variation depends on location, ore type, milling and deposition methods. From the processing side, tailings may be classified as coarse-grained (>20 μm) or fine-grained (<2 μm), or a mixture of the two. During processing or after discharge, the coarse and fine fractions separate at different rates depending on the type of tailings storage facility used. A distribution in particle size can occur laterally or vertically with coarser (sand) particles deposited close to the discharge point and settling faster. If the coarse fraction is extracted for embankment and retaining wall construction, only fine tailings fractions will remain in the disposal area.

The clay types present in tailings streams are diverse. The type of clay depends less on the metal ore involved than the region it occurs in. As an example, the copper tailings from

Table 9.1. *Physical characteristics of tailings (adapted from Vick (1990) with amendments as shown).*

Category	Characteristics
Soft rock tailings	
Fine coal refuse Trona insols (from soda ash production) Potash (high-salt effluent, slow sedimentation (Coates and Yu, 1977)	Contain both sand and slime fractions. The slimes fraction may dominate overall properties because of presence of clay
Hard rock tailings	
Lead/zinc Au/Ag, e.g. Abitibi, Canada; talc, chlorite and mica (Benzaazoua *et al.*, 2004) Molybdenum Nickel sulphide ores, e.g. Sudbury, Canada; chlorite and biotite (Dimitrova and Yanful, 2012) Nickel laterite ores, e.g. Moa Bay, Cuba; serpentine (Dimitrova and Yanful, 2012)	May contain both sand and slime fractions. Slimes are usually of low plasticity to non-plastic. Sands usually control overall properties for engineering purposes; clays are therefore less problematic for these conditions
Fine tailings	
Phosphatic clays Bauxite red muds (contains no clay) Slimes from tar sands tails Fines from base metal production, e.g. Val d'Or, Canada; muscovite and chlorite (Ouangrawa *et al.*, 2009)	The sand fraction is generally small or undetectable. The rheological and sedimentation–consolidation characteristics of the material is dominated by silt or clay-sized particles and may pose disposal volume problems
Coarse tailings	
Oil sands tailings Uranium tailings (clay content varies greatly depending on ore source and processing, e.g. acid leaching may change mineralogical character of clay fraction when present) Gypsum tailings Coarse iron tailings (in the absence of fine grinding for magnetic separation) Phosphate sand	Contain either principally sands or non-plastic silt-sized particles exhibiting sand-like behaviour. These are low in their effect on rheology and show generally favourable engineering characteristics.

Minera Escondida (Chile) contain clay-like minerals such as: muscovite, illite, kaolinite, pyrophyllite, and chlorite. Comparison can be made to other Chilean copper deposits such as La Andina (alpine climate) chlorite, El Teniente (Mediterranean climate) chlorite, biotite, kaolinite and monmorillonite, and El Salvador (hyper-arid climate) biotite (Dold and Fontboté, 2001). It should be noted that recent climatic conditions are not necessarily a

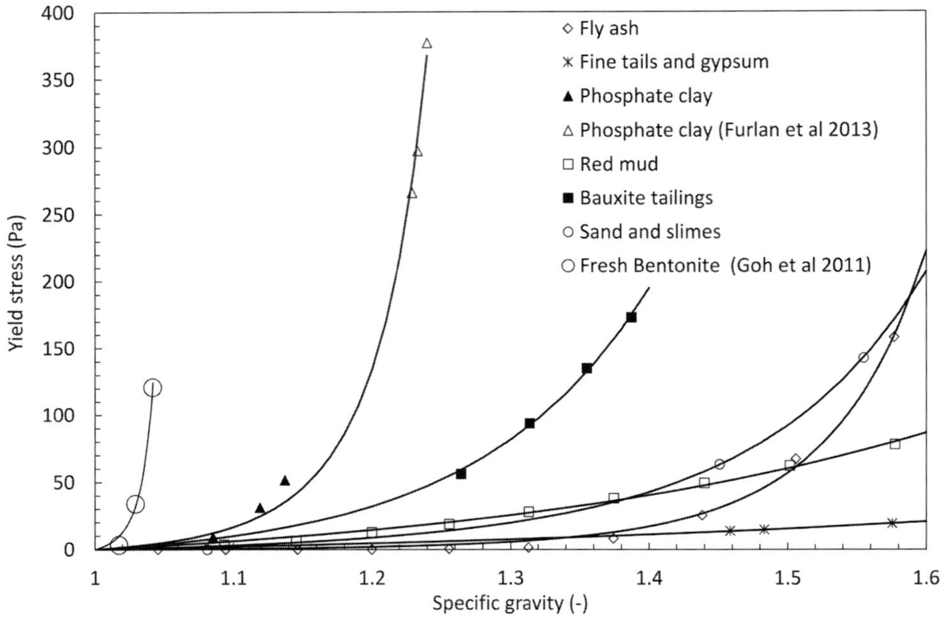

Figure 9.2. Yield stress variation with concentration/density (adapted from Addie *et al.* (2005) except where otherwise shown).

strong guide to the weathering conditions present when clays are formed. The amount of clay in tailings is influenced by the method of mining employed; open pit mining will incorporate greater volumes and fractions of clay than subsurface mining does. Other generalized tailing characteristics are shown in Table 9.1.

Variation of clay concentration will of course occur between sites, but once deposited significant variations in clay content with depth have also been recorded. For a fine nickel laterite tailings deposit, Wijewickreme *et al.* (2005) found 35.0 wt% clay at 5.3 m depth (with 65.0 wt% silt) and 2.0 wt% clay at 8.0 m depth (with 91.0 wt% silt). This variation would have an effect on the strength and stability under cyclic load (e.g. traffic or seismic events) for the deposited bed. Tin tailing slurry from tropical regions is a stable mineral suspension. Frequently this is returned to worked-out pits forming long-term pond storage. These ponds are often environmental hazards subject to overflow of slime from poor reclamation techniques and embankment collapse during heavy rain. The fine solid fraction composed of kaolinite, halloysite or montmorillonite (Chan *et al.*, 1994) do not settle to form a compact sediment, due to its stable colloidal dispersion. Only the larger sand fractions eventually form a low-solids concentration slime at the bottom of the pond.

A wider discussion of the many contributors to multi-phase suspension rheology is given in Chapter 2. The contribution of clay fractions is shown in Figure 9.2. The yield stress is used here as an analogue for the wider rheological behaviour of each slurry. It can be seen that the more rapid rates of yield stress increase as solids concentration occurs for the tailings with the highest clay content, with no clay in red mud, and mixed phases in sand and slimes. The fine tails/gypsum are similar in size range to the clays but are substantially less rheologically active. This behaviour is most apparent in beaching angle (Section 9.5) and pumping (Section 9.8).

9.3 Tailings Definitions

There is no universally accepted nomenclature for describing tailings. The classification used in Section 9.2 is based on ore type and source, and is less relevant to those designing the tailings transport and placement scheme. The system outlined in Jewell (2006) for characterizing thickened tailings is logically consistent and will be used in Section 9.3.2. For the purpose of designing aspects of a TSF, tailings are usually classified by concentration or density, but this has limitations. Tailings slurries are best described on the basis of their 'strength' (often substituted by yield stress) as no consistent description is achieved by using density or concentration. This results from rheological and settling behaviour being a function of reduced volume fraction ($\frac{v}{c_{vm}}$ from Chapter 2) and the maximum packing fraction, c_{vm}, value is highly variable for different tailings conditions. The broad categories of tailings types are as discussed below.

9.3.1 Dilute

Low-strength/low-concentration tailings have been the default approach for tailings disposal in large dams for the greater period of modern mineral exploitation. They are generally segregating after discharge, i.e. the solid phase will settle differentially based mainly on particle size. As they are rapidly settling they are transported from the plant to the TSF as highly turbulent flows in pipes or in open channels. These low-density slurries are generally unthickened, readily produce large quantities of supernatant water and are not greatly influenced by the presence of clays, other than for off-shore disposal (see Section 9.3.5).

9.3.2 Thickened

Thickened tailings are one of the means of reducing area of the TSF and improving mine water balances. The simplest definition of thickened tailings is those tailings that have been processed in some type of gravity thickener, centrifuge, etc. This is unhelpful for management purposes. Jewell (2006) applies the term 'thickened tailings slurry' for one that will flow a sufficient distance from the point of discharge for practicable, above-ground surface deposition. This implies that the solids concentration is sufficient for particle interactions to affect the slurry rheology. At such a solids concentration, clays will strongly influence slurry rheology, assuming they are present. Further division into medium- (low yield stress), high- to very high-density slurries (high yield stress, minimal segregation but still flow under gravity) are possible, but are of value only for making relative distinctions in behaviour.

9.3.3 Paste

Paste tailings are usually those that have effectively no drainage of water and segregation of particles after deposition. Jewell (2006) applies the term for a tailings slurry thickened to an 'ultra-high density', or one with a yield stress >200 ± 25 Pa. Manufacture of paste tailings goes further towards achieving the goals of thickened tailings. For base and precious metal ores solids, concentrations by volume may exceed 45 per cent, depending on the particle size distribution (tailings with at least 15 per cent passing 20 μm will exhibit paste properties). Pastes can be made at lower solids content when clay-containing tailings are handled (Addie *et al.*, 2005). The presence of clays makes the higher densities harder to achieve in solid–liquid separation processes (see Chapter 8), but the high yield stress values are more readily obtained due to the effect of clays on suspension rheology (see Chapter 2).

9.3.4 Cake

Tailings cake is a semi-solid usually derived from a filtration process in series with other forms of thickening. It is generally not easy to pump and is transported by conveyor or trucking.

9.4 Types of Tailings Storage Facilities

Tailings storage facility is the current enveloping phrase for what were previously described as dams, dumps, waste piles, etc. It is more than a hollow euphemism, however, as in many cases old tailings sites are reworked with modern methods for the small concentrations of useful metal still retained. Gold recovery at sites around Johannesburg (South Africa) by hydro-mining is one example. With some of the more intractable tailings such as those derived from mineral sands or tar sands, multiple stages are required and tailings may be transferred from a low-concentration dam to another for final storage (Kaminsky *et al.*, 2009). The ultimate aim of all TSF is the stable, long-term confinement of post-process mine solids without excess transmission of metalliferous and acid drainage or other damaging components, with the added desire to return the environment to something similar to its native state.

TSF are broadly divided into above-ground storage and below-ground storage; the latter are usually returned to worked-out pits. These are further divided in the sections below.

9.4.1 Low-Concentration Dams

Unthickened or low-concentration tailings are typically sent to a water retention type dam or a raised embankment impoundment (Fig. 9.3) (Vick, 1990). Globally this is the most

Figure 9.3. Embankment of an arid region tailings dam (non-discharge end). Note the proximity to the local water supply (left).

important method of tailings storage either in terms of tonnes stored or number of facilities. A common arrangement is to have supernatant water and rainfall collected as surface water and recovered for reuse in the process. The clay content affects clarity of the recycled water and may require treatment before return to the process streams.

Dam or retaining wall construction requires large quantities of solid material, e.g. overburden or coarse tailings fractions. The use of finer grinds for improved recovery, usually associated with declining ore grades, or the presence of clays reduces the amount of material available for construction. Clays and rock flour reduce the shear strength of the constructed wall and are avoided, necessitating the search for transferred material (material dug for use at another, nearby, location). This increases TSF costs substantially. Seismic stability of dams built on soft clays (i.e. as a substrate) is generally unsatisfactory (Masengo, 2004). Further discussion of dam construction is outside the scope of this chapter and the reader is referred to Vick (1990).

9.4.2 Thickened Tailings Disposal

Tailings with a higher underflow density or concentration, the medium- to very high-concentration slurries described above, up to paste consistency may be disposed of above ground with minimal earthworks via a centrally thickened discharge (CTD) or down-valley discharge (DVD) scheme. The deposited tailings form a cone for a CTD (Fig. 9.4) or a wedge (DVD) built by successive thin layers (see Al and Blowes, 1999). The beach slope (or beaching angle) is the angle formed between the top layer of tailings and the solid

Figure 9.4. Centrally discharged thickened tailings. Note the shallow gradient of the beach. Primary discharge is from near the peak, with a minor test spigot discharging to the left. The strength of the recently deposited tailings is shown by the footprints (right).

substrate. Beach slope and slope stability are the two key design criteria for implementing a CTD or DVD facility.

A tailings beach can be considered to have two distinct flow regimes. A turbulent slurry stream in a narrow, self-formed channel transports the tailings to the toe of the beach, where the sheet flow takes over and settling of particles occurs (Williams, 2001). The flow in this latter region is laminar or transitional-turbulent and does not support coarse solids. The channel flow behaviour and consequently the beaching angle is highly dependent on the tailings rheology, which is in turn influenced by clay content. This is discussed further in Section 9.5.

9.4.3 Dry Stacking

Dry stacking of filter cake has advantages for tailings management in areas that are water constrained. The operating costs of the filtering and placement process are generally higher than other management options (Davies, 2011), and it is only under circumstances in which water balance, climate, topography or legislation dictates that filtration is used.

Efficient filtration processes are dependent on the permeability of the developing filter cake. Clays generally have a low permeability, between 10^{-8} and 10^{-13} ms^{-1} (Mesri and Olsen, 1971). It can be seen in Fig. 9.5 that permeability decreases by orders of magnitude in the

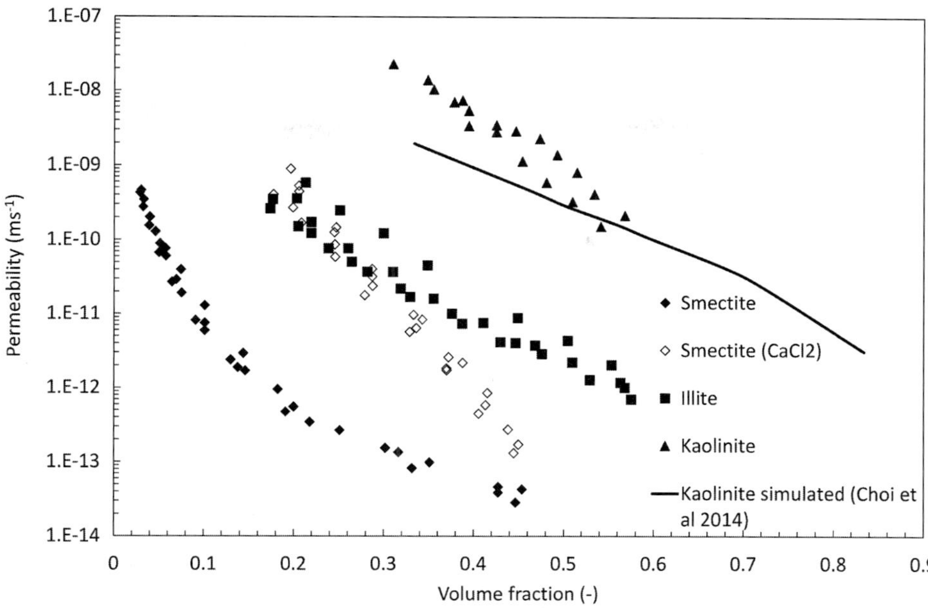

Figure 9.5. Permeability of clays in NaCl solution (unless otherwise indicated). Data from Mesri and Olsen (1971), molecular dynamics simulation from Choi *et al.* (2014).

order: kaolinite, illite, smectite. The effect of exchangeable ions can be seen to improve permeability as well. In samples of coal tailings from different seams, McKee and O'Brien (1989) showed differences in clay mineralogy could be correlated with varying degrees of filter performance. Sodium montmorillonite-rich coal tailings displayed poor filter performance. They were able to improve performance with gypsum for cation exchange and cationic flocculants, in line with the work of de Kretser *et al.* (1997) showing an excess Ca^{2+} electrolyte concentration as necessary to provide full cation exchange. Ion exchangeable clay filter feeds were also more sensitive to the application of pressure.

Despite the imperatives listed above, only with low clay concentrations or with a high permeability clay does filtration become available for consideration in a tailings management scheme.

9.4.4 Mine Back Fill

The simple concept that mine waste can be returned to the hole it was removed from is complicated by two points: tailings are generally of a larger volume than the ore-body due to water addition; and there is a need to keep tailings separate from the current work area. One above-ground example is mineral sand mining. This is a source of ilmenite and rutile for titanium production and is commonly extracted via dredging operations. This method uses large floating dredges in a purpose-built pond to pump the slurried ore. Gravity spiral concentrators are used to recover heavy minerals from the sand and clay. As mining of the ore-body progresses, overburden and low mineral content sands are returned to fill the excavated area and the slime fraction (clays and rock flour) is pumped to solar drying cells for later rehabilitation. In more arid regions water recovery is a pressing issue, and water lost in the drying cells may be recovered by co-depositing slimes with the sand fraction as it is returned to the actively worked pond. The ratio of clay slimes returned needs to be carefully determined as the clay is slow to settle and will contaminate freshly mined material, effectively lowering the ore grade.

Tailings in paste form has been returned to worked-out pits and has been sent underground for stabilizing worked areas (Brackebusch, 1994). The latter often has cement added to increase its strength. The presence of clay in backfilling operations will decrease cemented shear strength (Horpibulsuk *et al.*, 2012) and can be improved by de-sliming (Kesimal *et al.*, 2003).

9.4.5 Offshore Disposal

Offshore disposal of tailings is generally of two types: estuarine disposal (which leads to tailings ejected to the ocean at estuaries); and deep-sea tailings placement (DSTP). In DSTP practice, tailings are transported by pipeline to a point below the lowest mixed-layer depth and deposited on the sea bed. They are distinct from the TSF approaches discussed above as the mechanism is not to contain the tailings and effluent in a defined area (most usually within the mine site boundaries), but to disperse the tailings over a marginally defined area outside the control of the mine operator. Some advantages of the method are obvious, such as minimal capital and operating costs (apart from pipelining and environment monitoring). The method is viewed most seriously in regions of high positive water balance, seismic instability, steep topology and (not surprisingly) near the coast. Under these circumstances, DSTP may be the only economically viable and safe option, compared with land TSF. The potential effect on the benthos (seabed flora and fauna) in both the long and short term, with consequent effects on local health and economic enterprise, leave this method as problematic for social rather than technical reasons, and it is prohibited in many regions. Examples of DSTP exist in British Columbia, the Philippines, Central America, Indonesia, Norway, Papua New Guinea and historically elsewhere. A review of the ongoing Norwegian experience is given by Ramirez-Llodra *et al.* (2015).

Major problems with DSTP are water turbidity, spreading beyond intended placement zones and slumping of the deposit. This is exacerbated by the presence of fine or clay particles in a saline environment. The engineered solution to turbidity is to place the tailings outfall below the mixed-layer depth (e.g. greater than 120 m in the tropical western pacific (Lukas and Lindstrom, 1991)), or below 3000 m in submarine canyons (Batu Hijau copper/gold mine in Sumbawa, Indonesia). Clay and silt deposited in this manner will eventually form a settled layer that will damage the benthic layer (Edinger, 2012), although recovery can be rapid after the end of deposition if deposits are stable and slumping does not occur (Burd *et al.*, 2000). Environmental management objectives will consequently be to keep deposition away from highly productive ecosystem areas such as coral reefs or fisheries and to use areas of low productivity that can later be re-colonized (Gwyther *et al.*, 2009). Longer-term effects (>5 years), however, are still poorly understood and more longitudinal studies such as that by Angel *et al.* (2013), where sites are revisited over a decade, are required.

9.5 Clay Effects on Beach Slope and Slope Stability

For a CTD or DVD discharge scheme the mass of tailings solids per area depends on the final beaching angle; for a conical CTD the relationship is linear. Any increase in beaching angle will improve the feasibility of using thickened tailings for tailings management providing the final structure is stable. The precise modelling and prediction of tailings beach slopes is yet to be achieved on a regular and reliable basis (Jewell, 2012). This is largely due to many variables that have an impact on beach slope. The three most descriptive prediction methods in the tailings literature are (Simms *et al.*, 2011):

(1) Tailings in self-formed channels (Fitton *et al.*, 2006;Pirouz *et al.*, 2005; Williams, 2001) adapts the hydrodynamic theories of channel flow to the more complex non-Newtonian tailings flow, allowing a minimum solids conveying angle to be estimated, and equating that to the beach slope.
(2) Stream power (McPhail and Blight, 1998) again considers channelized tailings, but predicts a variable beaching slope via the maximum information entropy principle. Curiously, this method applies several Newtonian viscosity assumptions.
(3) Lubrication theory (Simms, 2007), which applies lubrication type flow to describe thin layer deposition, again using several Newtonian simplifying assumptions.

Despite their substantial differences, all of the above modelling methods have a significant viscosity or rheological component. In general, they predict an increase in viscosity will give rise to a steeper beaching angle. This can be envisioned as a higher viscosity reducing the settling rate and segregation of suspended solids. For discharged tailings with sufficient clay content flow may almost be as a pseudo-homogeneous viscous fluid without settling out of particles for a greater length of the beach, also influencing the radial variation in beach slope. Working with paste-like coal slurries Wennberg *et al.* (2008), found that as the clay content, and therefore viscosity, increased a larger slope for a given flow rate would develop.

From what was seen in Chapter 2 and Section 9.1, it could be inferred that the presence of clays is likely to have a beneficial effect during deposition by increasing the beaching angle. This possible advantage needs to be considered along with the requirements for slope stability. Dimitrova and Yanful (2012) examined the effect of bentonite and kaolinite clay addition to artificial mine tailings and created beds with different composition, thickness and age through sedimentation. After shearing in drained and partially drained conditions they found that clay addition to mine tailings caused a decrease in the frictional strength of the settled tailings bed. The decrease was greater for bentonite addition than it was for kaolinite. Ageing of the bentonite tailings slurry led to an increase in its shear strength, which reflects an increase in solids content after its initial poor dewatering performance.

From the above description of thickened tailings discharge (Section 9.3.2) it can be seen that tailings flowing across a beach for several hundred metres may experience prolonged shearing, which for a thixotropic suspension would lead to time-dependent rheological changes (see Section 2.6.1). Significant reductions in yield stress have been reported by Charlebois (2012) for high-clay thickened tailings under highly turbulent flow conditions. This must be considered relative to the generally higher rate of shearing that is imposed during pipeline transportation of tailings. It may then be reasonably expected to see a slow *rise* in viscosity due to the rebuilding of structure at the lower rate of shear as occurs in channel flow. The effect on settling of coarse particles of this changing viscosity is important to the beach profile.

9.6 Dewatering

Tailings streams will often undergo several stages of dewatering. Apart from the thickening stage that occurs prior to transport to the TSF or centrifuging just prior to deposition, both of which are discussed in Chapter 8, dewatering will occur on the TSF as well. Strictly this is only the case for thickened tailings as dilute tailings in a low solids concentration dam undergo settling. Water runoff on a CTD-TSF is recovered at the toe of the stack and returned to the process. This assists in the water balance and lowers the phreatic surface, increasing the shear strength of the tailings. High shear strength allows additional layers of tailings to be deposited, improves trafficability and allows remediation to occur at the end of mine life. Kaolinite and smectite pose a significant problem to dewatering due to floc structures retaining water. The high levels retained after flocculant-assisted gravity thickening (typically >88 per cent water (Hogg, 2000)) are not easily removed by further consolidation in the tailings dam. This may persist after several months due to the slow compaction rate of the clay pulp (McFarlane *et al.*, 2005).

Many of the same influences and effects of clay on dewatering as described in Sections 8.3 and 8.8 apply to the dewatering on a TSF, and will not be repeated here. The main differences are: the solids concentration on a TSF is higher (making reagent addition more complex); timescales for consolidation are longer; and any unrecovered water may still drain away to less saturated layers below or evaporate. Apart from the effects on water recovery, the presence of swelling clays can be expected to lower beach shear strength by decreasing permeability and maintaining higher moisture content in the deposited tailings, limiting the potential volume of solids stored.

An example that has received much attention is that of oil sands. Hydrocarbons are extracted from North American tar sands via the hot water extraction process. In this process, bitumen is floated off the top, sand is settled out and a middling stream contains suspended fines and residual bitumen. The fine tailings are allowed to settle in a holding pond to produce more concentrated mature fine tailings (MFT). The slow rate of settling that occurs is a major bottleneck in tailings processing and consequently large areas are required for tailings ponds. At Fort McMurray, Alberta, the clay mineralogy of the formation and the tailings sludge is mainly kaolinite and illite with traces of smectite, chlorite, vermiculite and mixed-layer clays (Scott *et al.*, 1985). The clay is predominantly non-swelling, largely due to the small proportion of smectite and vermiculite, which would otherwise not cause the poor consolidation observed. It has been suggested variously that the cause is due to residual bitumen (which is highly viscous) bridging between particles and creating a tortuous path for water (Scott *et al.*, 1985), or that the enrichment of the swelling clays in the middling stream creates sufficient concentration for slow settling (Kaminsky *et al.*, 2009). The search for remedies to the problems with MFT is ongoing (Wells *et al.*, 2011).

9.6.1 Polymer Addition for Property Modification

As with solid–liquid separation within the refining stages, various chemical treatments have been applied to tailings to improve their dewatering, rheological and other properties. Polymers have been added to MFT from oil sands to improve consolidation and also to improve rheology (i.e. increase yield stress) prior to deposition for colliery wastes and chromium tailings (Brumby *et al.*, 2008; Daubermann and Földvári, 2011), but those results are still waiting independent verification. The technique has its origins in the alumina industry for bauxite residue disposal (Kaiser *et al.*, 2006) and has been slow to be adopted elsewhere, largely due to economics.

Apart from polymer addition, dewatering in the presence of montmorillonite has been improved by the controlled dispersion of Ca^{2+} ions followed by shear for high-clay colliery waste (de Kretser *et al.*, 1997), bentonite (Lim *et al.*, 2009) and smectite/kaolinite (McFarlane *et al.*, 2005). This will prevent swelling and help consolidation; however, application to TSF would be less effective as shear ceases after deposition.

The mixing of post-thickener, or end-of-pipe, added polymers is complicated by the poor mixing that occurs when high- and low-viscosity materials are brought together. Various approaches have been trialled, from simple injection at the wall to more elaborate ring distributors (Wells *et al.*, 2011).

9.7 Effect on the Environment of the Tailings Storage Facility

Sorption onto clay minerals can reduce metalliferous drainage. In a study on coal waste containing ion-exchangeable clay, Dreher and Finkelman (1992) found selenium loss to the surroundings was decreased with increased clay concentration. Addition of bentonite can reduce AMD (Gitari *et al.*, 2011) in laboratory tests. The large quantities of clay required to produce effective change would appear to limit this as a useful technique. Galan et *al.* (1999) examined the effect of AMD on the heavily polluted Río Tinto (river) in Spain. Of the clay species present, illite and kaolinite did not show appreciable variation but chlorite underwent a loss of Fe. Smectite and calcite were rapidly and fully dissolved, causing the water's low pH to be partly neutralized. AMD is often addressed by 'capping' tailings with a low-permeability clay layer (Zhou *et al.*, 1993).

Many TSF are designed to have low water retention. This presents a problem at the end of a mine's life and remediation is required (or before if progressive remediation is employed). Phytoremediation, the use of vegetation to help stabilize tailings, requires a soil type that is quite different to a tailings beach, and the presence of additional material, such as clays, to improve the water-holding capacity and nutrient uptake is valuable (Mendez and Maier, 2008). The management of a TSF can reflect this if a change in the content of deposited tailings is made as a region approaches readiness for remediation.

9.8 Pipeline Transport

Pipeline transport of tailings is the most common method of delivering tailings slurries from a plant or thickener to the TSF, the exceptions being open channels for dilute tailings, open channels for thickened tailings in steep terrain or conveyors for filter cake. The design of a slurry pipeline is dependent on the rheology of the slurry and the settling characteristics of any non-colloidal particles. The slurry viscosity has a direct effect on pressure drop in the pipeline, and as was seen in Chapter 2, the presence, concentration and type of clay have a major impact on the rheology.

The higher viscosity of a clay-containing slurry over a coarse-solids-in-water slurry (see Fig. 9.2) can be advantageous for the transport of coarse solids in a co-disposal scheme. Non-settling slurries consist of fine solid particles and water where the solid particles settle slowly, or not at all in the case of very fine clay–water slurries. Laminar flow may be used to transport slurries of this type. This can be compared to a silt slurry with a size range below 75 μm. The silt slurry is settling in nature and can shut down a pipeline in a short time by building up a settled bed, unless the flow is turbulent (Addie *et al.*, 2005). A pseudo, non-settling slurry may contain clay particles and a coarser fraction. The higher viscosity of the clay–water component allows the larger particles to settle slowly or, if the yield stress is sufficient, not at all. This stable suspension effect is lost during transport as the non-Newtonian slurry viscosity decreases and the coarse particles settle (Thomas *et al.*, 2004). Settling in laminar pipe flow is inevitable, even if bench testing of unsheared systems indicates stability. Transport of sand (coarse) fractions will result in bed formation. The presence of clay in the carrier fluid has, however, been linked to reduced settled bed concentration in simulated sand/slime transport (Gillies *et al.*, 2007). The lower energy penalty of laminar flow over turbulent makes these pseudo-stable suspensions advantageous for short pipe runs.

A lower energy requirement and easier dewatering in the presence of swelling clays may be achieved with the controlled dispersion of exchangeable ions (Nguyen and Boger, 1998). This would create a lower-viscosity slurry, among other things. As was seen in the paragraph above, this method would achieve the opposite effect in terms of settling behaviour, and would be less suitable for tailings containing coarse fractions.

9.8.1 Pumps and Pump Deration

The pumping of a tailings slurry to the TSF requires pumps capable of moving the slurry with minimal settling. This necessitates high speeds and develops high-pressure drops for turbulent flows, or lower speeds and potentially higher pressure drops for the pumping of a highly viscous slurry. A higher pressure drop often requires using a more expensive (for the same throughput) positive displacement pump in place of the near-ubiquitous centrifugal pump. For thickened tailings slurries, which are capable of being moved by centrifugal pumps, the pressure head loss is enhanced over that of a homogeneous Newtonian fluid. A dramatic reduction in available head at low flow rates is characteristic of non-Newtonian slurries such as a clay suspension (see Sery and Slatter, 2002), but the presence of coarse particles can have a still greater effect (Pullum *et al.*, 2007; Wennberg and Sellgren, 2007). The presence of clay in tailings will therefore increase pressure drop due to increased viscosity, reduce available pump head due to its non-Newtonian behaviour and make prediction of pump deration more involved (Pullum *et al.*, 2007).

9.9 Summary

As may be seen from the above, tailings management is a multidisciplinary endeavour with a wide variety of options and conditions under which it can operate. The presence, concentration, type and condition of clay in tailings is one of several components that is of greater or lesser significance, depending on where and what is being sent to a TSF. Improvements that may occur by increasing beach slope in the presence of clay are offset by the decrease in slope stability and greater water retention. The ability to transport coarse particles at lower flow rates without settling comes at the price of higher pumping energy. At present the fundamental science behind many aspects of tailings management is still under-developed, and is certainly not sufficiently refined to make robust predictions about how a clay may influence a TSF amid all the 'noise' arising from changing grind sizes, mineralogy, throughputs, climate, etc. What has been presented here is not so much a 'rough guide' as a series of discussion points on what effect clays may have on tailings handling, and where research may need to be directed. There is much to be done.

References

Addie, G. R., Carstens, M. R., Sellgren, A., Visintainer, R. & Whitlock, L. 2005. Pipeline design characteristics of some industrial paste-like slurries. In: Ausimm, T. (ed.) *First Extractive Metallurgy Operators' Conference, 7 – 8 November, 2005*. Brisbane, QLD: Australasian Institute of Mining & Metallurgy, 147–152.

Al, T. A. & Blowes, D. W. 1999. The hydrogeology of a tailings impoundment formed by central discharge of thickened tailings: Implications for tailings management. *Journal of Contaminant Hydrology*, 38 (4), 489–505.

Angel, B. M., Simpson, S. L., Jarolimek, C. V., *et al.* 2013. Trace metals associated with deep-sea tailings placement at the Batu Hijau copper–gold mine, Sumbawa, Indonesia. *Marine Pollution Bulletin*, 73 (1), 306–313.

Benzaazoua, M., Marion, P., Picquet, I. & Bussière, B. 2004. The use of pastefill as a solidification and stabilization process for the control of acid mine drainage. *Minerals Engineering*, 17 (2), 233–243.

Brackebusch, F. W. 1994. Basics of paste backfill systems *Mining Engineering*, 46 (1), 1175–1178.

Brumby, M., Brough, J. & Flanagan, I. 2008. Rheomax™ Modified Tailings Disposal at Welbeck Colliery. In: Fourie, A. B., Jewell, R. J., Paterson, A. & Slatter, P. (eds) *11th International Seminar on Paste and Thickened Tailings, May 5–9, 2008*. Kasane, Botswana: 199–210.

Burd, B., Macdonald, R. & Boyd, J. 2000. Punctuated recovery of sediments and benthic infauna: A 19-year study of tailings deposition in a British Columbia fjord. *Marine Environmental Research*, 49 (2), 145–175.

Chan, Y. K., Ho, C. C., Lee, K. C. & Yeap, E. B. 1994. Physicochemical characterization of tin tailings slurries. *Applied Clay Science*, 9 (2), 83–96.

Charlebois, L. E. 2012. Pahoehoe to paste: Rheological field methods for characterizing overland tailings flow behavior. In: Overton, D. (ed.) *Tailings & Mine Waste 2012. October 14–17, 2012*. Keystone, CO: A.A. Balkeema: 187–198.

Choi, J., Chae, B., Kawamura, K. & Ichikawa, Y. 2014. Calculation of permeability of clay mineral in natural slope by using numerical analysis. In: Al, K. S. E. (ed.) *World Landslide Forum 3, 2–6/6/2014*. Beijing: Springer International Publishing Switzerland, 21–25.

Coates, D. F. & Yu, Y. S. 1977. *Pit Slope Manual*. Ottawa, ON: Canada Centre for Mineral and Energy Technology.

Daubermann, R. R. & Földvári, Z. 2011. Ciba® Rheomax™ ETD technology evaluation on chrome thickener underflow for enhanced tailings management. In: Fourie, A. B., Jewell, R. J., Paterson, A. & Slatter, P. (eds) *11th International Seminar on Paste and Thickened Tailings, May 5–9, 2008*. Kasane, Botswana: 377–382.

Davies, M. 2011. Filtered dry stacked tailings – the fundamentals. In: *Proceedings Tailings and Mine Waste 2011, November 6 to 9*. Vancouver, BC: Norman B. Keevil Institute of Mining Engineering, University of British Columbia, 1–9.

de Kretser, R., Scales, P. J. & Boger, D. V. 1997. Improving clay-based tailings disposal: Case study on coal tailings. *American Institute of Chemical Engineers Journal*, 43 (7), 1894–1903.

Dimitrova, R. S. & Yanful, E. K. 2012. Factors affecting the shear strength of mine tailings/clay mixtures with varying clay content and clay mineralogy. *Engineering Geology*, 125, 11–25.

Dold, B. & Fontboté, L. 2001. Element cycling and secondary mineralogy in porphyry copper tailings as a function of climate, primary mineralogy, and mineral processing. *Journal of Geochemical Exploration*, 74 (1–3), 3–55.

Dreher, G. B. & Finkelman, R. B. 1992. Selenium mobilization in a surface coal mine, Powder River Basin, Wyoming, U.S.A. *Environmental Geology and Water Sciences*, 19 (3), 155–167.

Edinger, E. 2012. Gold mining and submarine tailings disposal: Review and case study. *Oceanography*, 25 (2), 184–199.

Fitton, T. G., Chryss, A. G. & Bhattacharya, S. N. 2006. Tailings beach slope prediction: A new rheological method. *International Journal of Mining, Reclamation and Environment*, 20 (3), 181–202.

Galan, E., Carretero, M. I. & Fernandez-Caliani, J. C. 1999. Effect of acid mine drainage on clay minerals suspended in the Tinto River (Río Tinto, Spain): An experimental approach. *Clay Minerals*, 34 (1), 99–108.

Gillies, R. G., Sun, R., Sanders, R. S. & Schaan, J. 2007. Lowered expectations: The impact of yield stress on sand transport in laminar, non-Newtonian flows. *Journal of the South African Institute of Mining and Metallurgy*, 107 (6), 351–357.

Gitari, W. M., Kaseke, C. & Nkuzani, B. B. 2011. Passive remediation of acid mine drainage using bentonite clay: A laboratory batch experimental study. In: Rüde, T. R., Freund, A. & Wolkersdorfer, C. (eds) *International Mineral Water Association Congress*. Aachen: International Mine Water Association, 325–329.

Gwyther, D., Batterham, G. J., Waworuntu, J., *et al.* 2009. Recolonisation of mine tailing by meiofauna in mesocosm and microcosm experiments. *Marine Pollution Bulletin*, 58 (6), 841–850.

Hogg, R. 2000. Flocculation and dewatering. *International Journal of Mineral Processing*, 58 (1–4), 223–236.

Horpibulsuk, S., Phojan, W., Suddeepong, A., Chinkulkijniwat, A. & Liu, M. D. 2012. Strength development in blended cement admixed saline clay. *Applied Clay Science*, 55, 44–52.

Jewell, R. 2006. Introduction. In: Jewell, R. J. & Fourie, A. B. (eds) *Paste and Thickened Tailings: A Guide*, 3rd edition. Perth: Australian Centre for Geomechanics.

Jewell, R. J. 2012. Putting beach slope prediction into perspective. *Journal of the Southern African Institute of Mining and Metallurgy*, 112 (11), 927–932.

Kaiser, L. K., Lamperd, J. W., Loan, C. & Cooling, D. J. 2006. A new approach to bauxite residue dry stacking utilising Ciba Rheomax ETD technology. In: Jewell, R. J., Lawson, S. & Newman, P. (eds) *Ninth International Seminar on Paste and Thickened Tailings (Paste06), 3–7 April. Limerick, Ireland.* Perth: Australian Centre for Geomechanics, 57–67.

Kaminsky, H. A. W., Etsell, T. H., Ivey, D. G. & Omotoso, O. 2009. Distribution of clay minerals in the process streams produced by the extraction of bitumen from Athabasca oil sands. *The Canadian Journal of Chemical Engineering*, 87 (1), 85–93.

Kesimal, A., Ercikdi, B. & Yilmaz, E. 2003. The effect of desliming by sedimentation on paste backfill performance. *Minerals Engineering*, 16 (10), 1009–1011.

Lim, J., de Kretser, R. G. & Scales, P. J. 2009. Investigating the influence of total electrolyte concentration and sodium–calcium ion competition on controlled dispersion of swelling clays. *International Journal of Mineral Processing*, 93 (2), 95–102.

Lukas, R. & Lindstrom, E. 1991. The mixed layer of the western equatorial Pacific Ocean. *Journal of Geophysical Research: Oceans*, 96 (S01), 3343–3357.

Masengo, E. 2004. Seismic performance of tailings dams built on soft clay. In: *Tailings and Mine Waste '04*. Boca Raton, FL: CRC Press.

McFarlane, A. J., Bremmell, K. E. & Addai-Mensah, J. 2005. Optimising the dewatering behaviour of clay tailings through interfacial chemistry, orthokinetic flocculation and controlled shear. *Powder Technology*, 160 (1), 27–34.

McKee, W. M. & O'Brien, D. R. 1989. Dewatering of bentonitic rich coal tailings using belt press filter. In: Ausimm, T. (ed.) *Dewatering Technology and Practice Conference, 9–11 October, 1989*. Brisbane, QLD: Australasian Institute of Mining and Metallurgy, 5–10.

McPhail, G. I. & Blight, G. E. 1998. Predicting tailings beach profiles using energy and entropy. In: Nelson, J. D. (ed.) *Tailings and Mine Waste '98, 26–28 January*. Fort Collins, CO: Balkema, 19–26.

Mendez, M. O. & Maier, R. M. 2008. Phytostabilization of mine tailings in arid and semi-arid environments: An emerging remediation technology. *Environmental Health Perspectives*, 116 (3), 278–283.

Mesri, G. & Olsen, R. 1971. Mechanisms controlling the permeability of clays. *Clays and Clay Minerals*, 19, 151–158.

Mikula, R. J., Munoz, V. A. & Omotoso, O. 2008. Water use in bitumen production: Tailings management in surface mined oil sands. In: *World Heavy Oil Congress, 10–12 March 2008*. Edmonton, AB: Curran Associates, Inc., 780–787.

Nguyen, Q. D. & Boger, D. V. 1998. Application of rheology to solving tailings disposal problems. *International Journal of Mineral Processing*, 54 (3–4), 217–233.

Ouangrawa, M., Molson, J., Aubertin, M., Bussière, B. & Zagury, G. J. 2009. Reactive transport modelling of mine tailings columns with capillarity-induced high water saturation for preventing sulfide oxidation. *Applied Geochemistry*, 24 (7), 1312–1323.

Pirouz, B., Kavianpour, M. & Williams, M. 2005. Thickened tailings beach deposition: Field observations and full-scale flume testing. In: Jewell, R. & Barrera, S. (eds) *8th International Seminar on Paste and Thickened Tailings (Paste 2005), 20–22 April 2005*. Santiago: Australian Centre for Geomechanics, 53–72.

Pittman, W. E. & Sweeney, J. W. 1983. *State-of-the-art of Phosphatic Clay Dewatering Technology and Disposal Techniques 1. A Review of Phosphatic Clay Dewatering Research*. Bartow, FL: Florida Industrial and Phosphate Research Institute.

Pullum, L., Graham, L. & Rudman, M. 2007. Centrifugal pump performance calculation for homogeneous and complex heterogeneous suspensions. *Journal of the South African Institute of Mining and Metallurgy*, 107 (6), 373–380.

Ramirez-Llodra, E., Trannum, H. C., Evenset, A., *et al.* 2015. Submarine and deep-sea mine tailing placements: A review of current practices, environmental issues, natural analogs and knowledge gaps in Norway and internationally. *Marine Pollution Bulletin*, 97 (1–2), 13–35.

Scott, J. D., Dusseault, M. B. & David Carrier III, W. 1985. Behaviour of the clay/bitumen/water sludge system from oil sands extraction plants. *Applied Clay Science*, 1 (1–2), 207–218.

Sery, G. & Slatter, P. T. 2002. Centrifugal pump derating for non-Newtonian slurries. *15th International Conference on Hydraulic Transport of Solids: Hydrotransport 15, 3–5 June, 2002*. Cranfield: BHR Group Limited, 679–692.

Simms, P. 2007. On the relation between laboratory flume tests and deposition angles of high density tailings. In: Fourie, A., Jewell, R. & Fourie, A. (eds) *10th International Seminar on Paste and Thickened Tailings (Paste 2007)*. Perth: Australian Centre for Geomechanics, 329–338.

Simms, P., Williams, M. P. A., Fitton, T. G. & McPhail, G. 2011. Beaching angles and evolution of stack geometry for thickened tailings: A review. In: Jewell, R. J. & Fourie, A. B. (eds) *14th International Seminar on Paste and Thickened Tailings (Paste 2011), 5–7 April*. Perth: Australian Centre for Geomechanics, 323–338.

Thomas, A. D., Pullum, L. & Wilson, K. C. 2004. Stabilised laminar slurry flow: review, trends and prognosis. In: *16th International Conference on Hydraulic Transport of Solids: Hydrotransport 16*. Cranfield: BHR Group, 701–716.

Vick, S. G. 1990. *Planning, Design, and Analysis of Tailings Dams*. Richmond, BC: BiTech.

Wells, P. S., Revington, A. & Omotoso, O. 2011. Mature fine tailings drying: Technology update. In: Jewell, R. J. & Fourie, A. B. (eds) *14th International Seminar on Paste and Thickened Tailings (Paste 2011), 5–7 April*. Perth: Australian Centre for Geomechanics, 155–166.

Wennberg, T. & Sellgren, A. 2007. Pumping evaluations with paste tailings thickened close to the surface disposal area. In: *Hydrotransport 17: The 17th International Conference on the Hydraulic Transport of Solids*. Cape Town: BHR Group Limited, 561–574.

Wennberg, T., Sellgren, A. & Goldkuhl, I. 2008. Rheological and depositional characterisation of paste-like tailings slurries. *Minerals Engineering Conference, Feb 3–4, 2008*. Luleå: Luleå University of Technology, 1–10.

Wijewickreme, D., Sanin, M. V. & Greenaway, G. R. 2005. Cyclic shear response of fine-grained mine tailings. *Canadian Geotechnical Journal*, 42 (5), 1408–1421

Williams, M. P. A. 2001. Tailings beach slope forecasting: A review. *High Density and Paste Tailings 2001 Seminar*. Pilanesberg: University of Witwatersrand.

Zhou, Z., Wiwchar, B., Gunter, W. D. & Gao, D. 1993. Polymer-modified clay as impermeable barriers for acid mining tailings. MEND Report 6.2. Vancouver, BC: Mine Environment Neutral Drainage (MEND).

10

Knowledge and Practice Gaps and Research Priorities for Clay-Related Phenomena in the Minerals Processing Value Chain

Markus Gräfe, Angus McFarlane and Craig Klauber
on behalf of the author teams

10.1 Introduction

This final chapter addresses priority research needs for either creating new knowledge of how clays affect one or more unit processes in the mineral processing value chain (MPVC) or to improve utilization of existing knowledge/techniques to assess the impacts or characteristics of clays through improved best practice. These are referred to in this chapter as knowledge and practice gaps, respectively. Knowledge gaps in general were synthesized by the respective chapter authors and the editors subsequently refined the distinction between knowledge and practice gaps. The knowledge and practice gaps are succinctly summarized in Table 10.1. All author teams have provided a conceptual view as to how the knowledge/practice gap could be addressed as well as what the perceived research outcomes would be. In total, the authors identified 15 knowledge/practice gaps and research needs based on the public domain literature surveyed.

10.1.1 Review Statistics

This review has compiled and processed information from more than 1100 references available in the public domain, not counting data compiled for the preg-robbing (74) and characterization (~ 400) tables in Chapter 1. The range of citations used per chapter was between 52 and 212 for a page range of 20 to 72; on average each chapter used 128 ± 49 citations or 3 ± 1 citations per page.

10.2 Practice Gaps

A principal outcome of this book has been a collation of established knowledge regarding clays and their impact upon mineral processing. Since clays affect a range of unit processes across commodities, site-specific studies can remain cloistered and not readily useful to new situations. While such site-specific information could provide useful hypotheses for clay behaviour at other mines (with different geology and processing operations),

Table 10.1. *Knowledge and practice gaps, research priorities and projected research outcomes by chapter.*

Chapter	
	Knowledge gaps
1, 3	*Knowledge gap: Metal sorption (preg-robbing) under industrial-relevant conditions.* The extent of public knowledge for metal–clay interactions is largely restricted to environmental and not industrial conditions. The extent and reversibility of metal sorption onto clays at high temperatures, pressure and ionic strength (among others) at relevant pH is unknown. *Research priority:* Conduct detailed thermodynamic and kinetic studies on the temperature- and pressure-dependence of metal ion sorption under industrially relevant conditions. *Research Outcome:* The research would provide pre-competitive information about rate and thermodynamic constants for metal sorption across a range of common clays (kaolinite, smectite, vermiculite, micas, illites, etc.) that can be used in modelling programs (e.g. PHREQC, GWB) to predict commodity metal sorption and release under a wide range of industrial conditions (pH, T, pressure, I).
2	*Knowledge gap: Modelling particle–particle interactions.* Current theories and models (e.g. DLVO) for predicting and explaining particle–particle interactions are insufficient to reproduce empirical results. A critical lack is perceived in predicting particle orientations and on the nature of surface charge distributions of anisotropic clays. *Research priority:* In order to build a substantially improved predictive model for clay particle–particle interactions, two important techniques need to be engaged to provide mechanistic information. (1) Cryogenic scanning electron microscopy can be used to determine the nature of particle orientations *in situ*. (2) Atomic force microscopy can help shed light on the nature of the surface charge distributions of anisotropic clay particles as well as the surface forces that exist between the different particle planes. *Research outcome:* A significant improvement in the ability to model clay particle–particle interactions in aqueous suspension based on mineralogical analyses. Such a modelling improvement would impact on unit processes in the MPVC such as liquid–solid separation, flotation, and tailings dewatering strategies, among others.
3	*Knowledge gap: The mechanism of gold–cyanide and gold–thiourea adsorption on clays.* Although transmission electron microscopy (TEM) investigation has provided direct evidence to show submicrometre gold appearing on the edges of kaolinite and illite particles from a natural ore sample (e.g. Hanlie *et al.*, 1999), the surface reaction mechanism(s) remain elusive. *Research priority:* Investigate the bonding mechanism of industrially relevant gold complexes on model clays and in real industrial samples. X-ray absorption spectroscopy and X-ray fluorescence spectromicroscopy are key tools for direct empirical evidence of local bonding environments (<4 Å) in homogeneous and mixed environments, while ATR-FTIR techniques can identify which surface sites are involved in the bonding mechanism as well as changes in the symmetry of the gold ligands (as far as infrared active).

Table 10.1 (*cont.*)

Chapter	
	Research outcome: This research would provide direct evidence of the types (inner-sphere/outer-sphere) of bonding mechanisms of gold–cyanide/–thiourea at the clay–water interface of standard clay materials as well as actual industrial samples. The information gleaned would provide a basis for identifying interfering agents to prevent sorption on clays and thus increase gold recovery.
3	*Knowledge gap: Loss of valuable metals during the Al/Si (re-)polymerization*
	The fate of valuable metals during the re-precipitation of Al and/or Si in acidic leach circuits is not well understood, but is likely a significant preg-robbing mechanism in hydrometallurgical operations where clay dissolution is extensive.
	Research priority: Conduct a series of dissolution studies of 1:1 and 2:1 type clays in the presence of different commodity metals (e.g. Cu, Ni or U) under hydrometallurgical relevant conditions.
	Research outcome: A well-structured study will quickly be able to differentiate effects stemming clay sources, co- and counter-ions in solution, and processing conditions (e.g. T) on the dissolution reaction of the clay and the loss of commodity metal from pregnant liquor solutions. This in turn will identify factors in different unit processes where adjustments can be made to minimize such losses.
3	*Knowledge gap: Selective dissolution of host minerals*
	Selective dissolution of thermodynamically highly stable host minerals for commodity metals (e.g. nickeliferous goethite) in a complex gangue matrix is a prospective cost-saving mechanism to attack only the mineral phase of interest. If successful it would avoid wholesale dissolution of the associated gangue, e.g. clays, which is likely mitigating the concentration of target metal in the pregnant liquor solution and causes significant problems in the downstream unit processes, e.g. clarification, thickening, tailings disposal.
	Research priority: (1) Review the existing extensive literature on organic acid interactions with minerals common to a particular ore type (e.g. laterites), (2) identify knowledge gaps with respect to hydrometallurgical processing conditions, (3) conduct small proof-of-concept studies under hydrometallurgically relevant conditions.
	Research outcome: Develop a strategic understanding of the existing knowledge base for further development in hydrometallurgical processing. The existing body of knowledge stems largely from soil and environmental sciences, and it is extensive. For example, plants and microbes are known to employ siderophores and citrate to complex, reduce and transport Fe from insoluble Fe oxides back to the organism. Oxalate at pH 3 is known to reductively dissolve iron oxides. However, none of these processes have been evaluated in an industrial context.
7	*Knowledge gap: Effects of clays on froth rheology*
	While it is clear that froth rheological properties have a strong influence on factors such as froth mobility and stability, the exact mechanisms involved are still unknown. This book has demonstrated that the presence of clay minerals in flotation pulps affect a number of flotation subprocesses, e.g. entrainment, froth stabilization, pulp rheology, slime coating, etc.). However, the relative magnitude of impact that each of these subprocesses has on the final flotation outcome is not known.

Table 10.1 (*cont.*)

Chapter	
	Research priority: A concerted research study is required to determine which of the subprocesses affected by clays has the greatest effect on the outcomes of the flotation process.
	Research outcome: This research would enable direct targeting of one or more flotation subprocesses for major improvement of the flotation process and performance.
8	*Knowledge gap: Substantially increasing aggregate density for flocculated clays*
	In the majority of mineral processing applications, flocculants will adsorb only onto the edge faces, leading to fractal aggregates having an even lower effective density. Achieving a higher aggregate density during flocculation under practical operating conditions would be far preferable to any post-flocculation treatment of beds or thickened slurries.
	Research priority: It is unlikely that the solution will come from variation of the flocculant functionalities that provide adsorption to the surface, but rather through the introduction of intra-chain interactions that favour coil/bridge contraction during or post-aggregation. The 'condensing aggregation' process that is engendering strong interest for the further dewatering of high solids slurries in effect may potentially be achieving this through the application of extreme reagent dosages, however, the mechanism remains very poorly understood. Establishing the mechanisms at play may provide a pathway to similar effects under more widely applicable conditions.
	Research Outcome: Success in this area would lead to a reduced need for solids dilution prior to feedwell flocculation and lower settling areas for thickening. Fine and high-aspect-ratio clays would be more amenable to treatment, with much reduced hold-up of liquor in beds. Benefits may flow through to underflow and paste properties.
	Practice gaps
2	*Practice gap: Rheometry of multi-phase, non-Newtonian mixtures*
	Current practice and instrumentation for rheological measurement is largely based on techniques developed for homogeneous fluids. As clays are often present as multi-phase mixtures that display non-Newtonian characteristics, measurements for research or design purposes are marred by artefacts that distort the true results.
	Research priority: To improve modelling of rheological behaviour, instrumentation and techniques need to be developed that can allow for wall slip phenomena and settling of solid phases in multi-phase mixtures including clay slurries.
	Research outcome: Techniques that can correctly measure the properties of clay slurries and other multi-phase mixtures will lead to a better understanding of the effects of chemical change on clay rheology and improve design outcomes for processes requiring knowledge of the transport properties of clays.
4	*Practice gap: Improvement in accurate quantification of disordered, interstratified clays and poorly crystalline materials. Identification and quantification of non-crystalline materials.*
	Research priority: There are two key areas of priority research: (1) characterization of the properties, in particular cation exchange capacity, of clays in a mineral assemblage; and (2) development of robust techniques for the identification of clays and associated minerals.

Table 10.1 (*cont.*)

Chapter	
	Although there are several standard methods for determining the cation exchange capacity (CEC) of clays, all methods have shortfalls, particularly for clays with high CEC such as smectites. A critical examination of the current techniques for determining CEC is required and techniques that can overcome the current shortfalls need to be developed.
	Several methods for quantifying clay content are available but these methods are, in general, specific to a particular form of clay. Unification of the many clay species-specific quantification methods into a single, comprehensive system for the quantification of clays is therefore desirable. This would need to be accompanied by improvements in the methods used to identify and quantify non-crystalline materials present. Modelling the variation in types of disorder found in clay species is also of critical importance. As part of this method development, it would be necessary to obtain clays in pure form to use as standards for identification and quantification.
	Research outcome: Increased accuracy of quantification for all forms of clays in mineral assemblages plus much better characterization of poorly crystalline and interstratified clays.
4, 5	*Practice gap: High-throughput analysis for identification and quantification of clays and clay assemblages.*
	Reliable and rapid identification and quantification of the clay mineralogy in ore will yield valuable information for sound decisions regarding the economic viability and resource management. For example, accurate identification of clay types and their spatial associations (e.g. interlayered illite–smectite) is important to evaluate the thermal maturity of organic matter in sedimentary basins. Knowledge of the clay type impacts on the extraction and recovery of bitumen from oil sands deposits; or, in a different context, it could provide valuable information regarding the alteration history of porphyry Cu deposits.
	Research priority: Further development of XRD, IR or other techniques to support high throughput, automated or semi-automatic analysis for identification and quantification of clays and clay assemblages.
	Research outcome: In addition to improved geological models, these techniques would provide very useful information regarding the rheological properties of clays, such as smectite, where the nature of water and exchangeable cations at the interlayer site of smectite are important to the viscosity and settling behaviour of clay.
6	*Practice gap: Advanced spectral analysis and data integration*
	Clay minerals play an important role in numerous steps of the mine life cycle, ranging from exploration through mining and processing to mine site remediation: e.g. as host, vector mineral, contaminant, sample medium or slope stability criterion (Chapters 5 and 6). Remote sensing data have become increasingly applied for the detection and characterization of clay minerals (Chapter 6), but more sophisticated methods are needed.

Table 10.1 (*cont.*)

Chapter	
	Research priority: (1) Advanced research in understanding spectral signatures of clay mineral species. (2) Improved methods for integrating surface remote sensing data (e.g. radiometrics and spectroscopy) for clay characterization. (3) Improved inversion of geophysical data by integration with spectroscopic and soil geochemical data (e.g. defining conductive layers in the regolith profile during processing of airborne electromagnetic (AEM) data).
	Research outcome: Routine detection and characterization of clay minerals would benefit the resources sector in the early stages of exploration by characterizing regolith profiles and landforms to map out transported vs *in situ* material using spectroscopic data or for detecting paleochannels using AEM – both are important for planning drill campaigns. The detailed (m-scale) quantification of clay minerals is a crucial step towards modelling a resource (see Haest *et al.*, 2012) on mapping clay-rich layers of channel iron deposits). A better understanding of spectral signatures of clay minerals that could be collected with routine scanning technologies, such as CSIRO's HyLogging system, and the improved interpretation of these data would enhance our ability to model and characterize the clay domains in resources. The achieved knowledge could be transferred to mine face scanning devices to tackle geotechnical challenges in mining environments, such as the mapping of swelling clays that can cause slope failures.
8	*Practice gap: How to process high-aspect-ratio clays*
	The issues associated with low flocculated aggregate density are no more apparent than in the case of clay particles with a high aspect ratio, i.e. needle-like shape. An extreme example of this is palygorskite, for which flocculation may require dilution to below 0.5 wt%, only to then encounter settled densities as low as 1–2 wt%. Thickening is not practical for such feeds, while filtration may give very limited throughputs.
	Research priority: The application of centrifugation or hydrocyclones in association with aggregating reagents may offer some useful scope, although reagents distinct from conventional flocculants may be required. Surface modification (possibly through precipitation of a different mineral phase) could be considered, but may be difficult to achieve at practical scales.
	Research outcome: The potential to dewater slurries previously considered beyond economic treatment (apart from discharge into a water body or long-term storage).
8	*Practice gap: Optimization of shear distribution in slurry storage/leach tanks*
	Prevention is always better than cure, and the best way to enhance solid–liquid separation is to generate fewer fines. However, most processes will have at least one stage in which clay slurries are stirred to maintain a homogeneous suspension. Such tanks are inevitably over-sheared, with a strong likelihood of producing additional fines, detrimentally affecting flocculation and viscosity.

Table 10.1 (*cont.*)

Chapter	
	Research priority: Computational fluid dynamics (CFD) has been used successfully in precipitation tank design for the alumina industry, with impeller design and height adjustments to reduce energy consumption and eliminate zones of high and low shear. Swirl flow technology is just one outcome from this. Applications in other industries have not advanced to the same degree, and this may require detailed quantification of extended shear on feed particle size distributions.
	Research outcome: In thickeners, the potential for reduced flocculant consumption, lower solids dilution requirements and better clarity, as well as greater operational stability. Benefits may also be realized in leach tanks through lower viscosities (or the option to leach at high solids).
9	*Practice gap: Management of clayey tails*
	At present the fundamental science behind many aspects of tailings management is still under-developed, and is certainly not sufficiently refined to make robust predictions about how a clay may influence a tailings storage facility amid all the 'noise' arising from changing grind sizes, mineralogy, throughputs, climate etc.
	Research priority: An integrated research programme examining the rheological/ hydrodynamic interactions during tailings transportation and deposition with a focus on the transport properties of multi-phase mixtures and settling behaviour in sheared suspensions.
	Research outcome: The improved understanding of transport property and settling effects on pipeline flow and tailings deposition will separate the effect of clay type and concentration (via its rheology) and better guide management strategies and TSF planning to mitigate its effects.

any casual application of such knowledge could be counterproductive. Likewise, insights offered by applied research into 'pure' clay systems need to be critically reviewed and tested within given operational contexts. In light of the existing significant volume of clay knowledge, the practical question is how to best utilize this to improve the effectiveness and efficiency of mining, beneficiation, processing and disposal operations.

A conceptual model for an integrated clay characterization programme is illustrated in Fig. 10.1. *At the outset, a sound knowledge and understanding of the clay speciation is essential for any qualitative and/or quantitative evaluation of the spatial distribution of clays and their related processing behaviours.* Of course, clays and clay minerals are usually not the primary interest of mine operators. Rather, it is the impact on processing response (rheology, settling, solids loading) or geological insight (alteration, thermal history, etc.) related to clays that are of greatest importance. Pragmatic testing regimes with sound links to tangible outcomes will invariably compete with the detailed mineralogical analysis required for establishing foundation knowledge. Long-term mining project needs may best be resolved through recognition that both approaches have value.

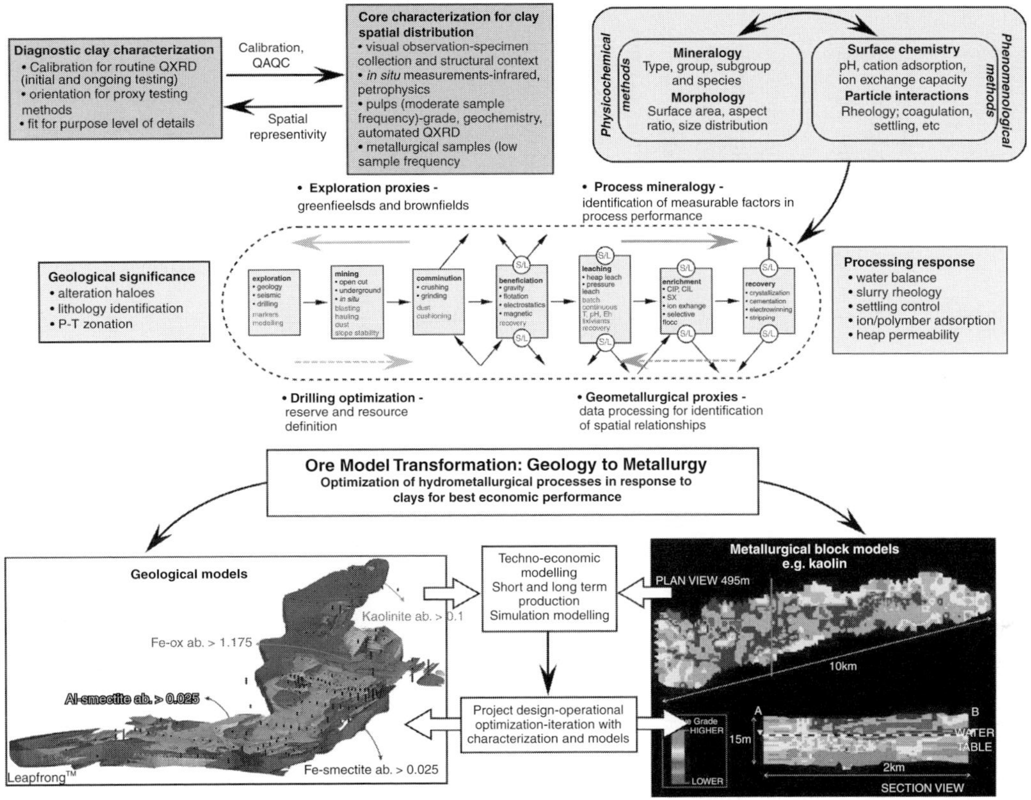

Figure 10.1. Integrated characterization of clays across the minerals value chain. Since clays will inevitably affect different unit processes, thorough diagnostic characterization of the clay mineral species present in the ore is an essential foundation for beginning and sustaining quantitative and qualitative analyses of ores and their processing behaviour. Analysis of drill core sections at various length scales provides the basis for modelling the spatial distribution of ore properties and their processing responses. Ongoing developments in clay-sensitive measurement methods such as XRD and infrared spectrometry can provide cost-effective data at high sampling frequency. Linking clay mineral content of ores to observed processing behaviour is aided by knowledge of the clay species present. As mining and processing operations for the deposit progress, the feedback of understanding gained leads to ongoing efficiency improvements. A black and white version of this figure will appear in some formats. For the colour version, please refer to the plate section.

A rigorous application of proven diagnostic methods for clay content and type can enable 'whole of ore' quantification via increasingly automated high-speed acquisition of characteristic data (e.g. HyLogger) collected using spectral (e.g. IR) or diffraction responses. Such information allows operators to be more confident that any ore samples used for establishing process design are indeed 'representative' of the ore. Such a design approach has two components: (1) a mechanistic approach to explain the average behaviour of the ore based on its physicochemical components (e.g. clay minerals with distinct physicochemical behaviour), and (2) a statistical component that assures that the average behavioural response of the ore represents a certain confidence limit (e.g. 90 per cent, $p < 0.10$). In process variability testing, recourse to high-quality quantitative mineralogical data allows ore classification models to be built which include mechanistic information concerning the unit operation as opposed to empirically driven approaches in which correlations and statistically derived relationships may not be related at all. Likewise, measurements that give process information (proxy or direct, e.g. pH, Eh, viscosity, solids concentration, etc.) can be tailored to provide the most cost-effective target mineral recovery. This may be more intensive clay analyses (e.g. via quantification of clay fractions vs. routine bulk mineralogical quantification) or in other cases, various physical testing options (e.g. slurry yield stress) or a combination of both. Ideally, as knowledge of the ore-body and processing behaviour increases, ongoing routine feedback will take place with minimal costs via high performance data management platforms.

An example of an integrated clay characterization at ore-body scale is the study of smectite and kaolinite spatial distribution at the Yeelirrie Uranium deposit (Lower *et al.*, 2011). Recognizing the presence of clay minerals to be a significant issue for processing and the inadequacy of elemental analysis in estimating clay content, quantification was achieved through XRD-calibrated modelling of spectral data for over 70 000 intervals, i.e. complete XRD characterization of a sample subset with measured spectral response enabled rapid characterization prediction across all intervals. Monitoring the uncertainty in the spectral modelling prompted further sampling for QXRD calibration to ensure the spectral models captured the full extent of clay mineral variability. These clay distributions could then be utilized in mine planning and process design to manage deleterious effects and optimize return.

Ultimately, a cross-disciplinary, integrated clay characterization programme does not lend itself to a 'one size fits all' method and requires discussion between stakeholders regarding decision making and the collaborative use of samples and information.

An integrated characterization approach may allow different stakeholders to improve their use of the data to achieve different project objectives. For example, combining spectral, chemical and geophysical exploration data could help constrain model parameters and reduce uncertainty in inversion models (see Table 10.1). Clayey ores are known to negatively impact upon thickener performance, tailings disposal and associated plant water balance calculations. Estimation of metallurgical process performance requires consideration of a number of variables, e.g. effects of grain and bulk density, particle size distribution, etc. – apparently diverse objectives that ultimately benefit from an integrated

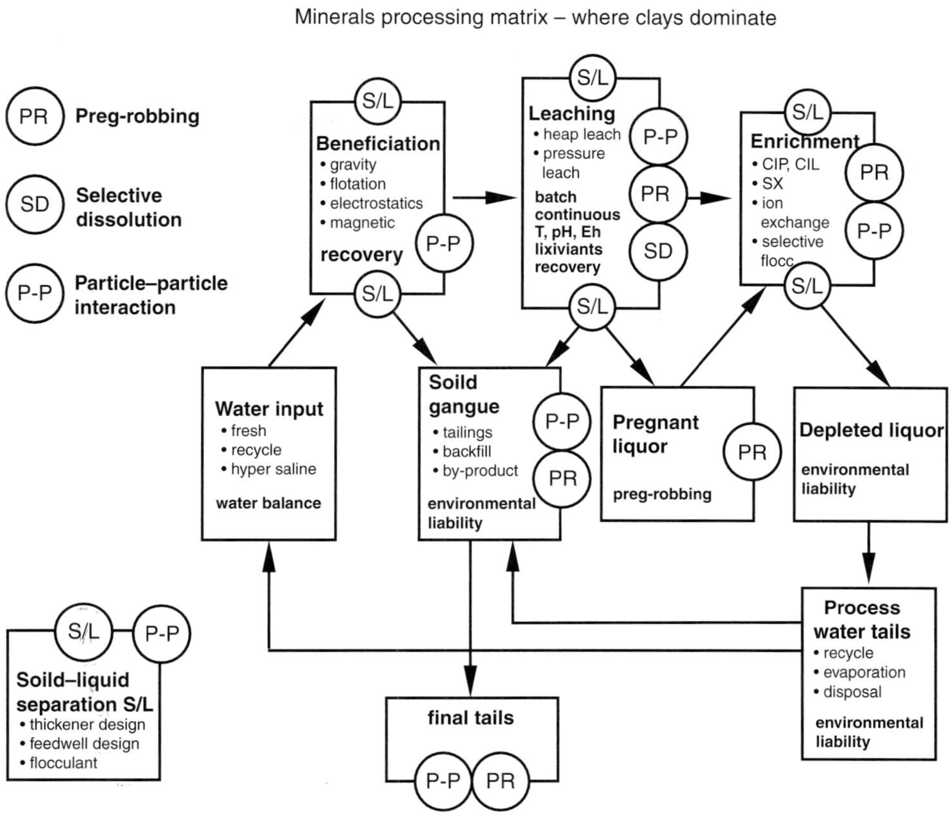

Figure 10.2. Major knowledge gaps in the minerals process value chain.

approach. This may also reveal unexpected benefits – for example, whether routine spectral acquisition contains additional information about clay mineral speciation, or how existing rheological instrumentation might be adapted to overcome artefacts generated by inhomogeneous fluids. Mineralogical characterization and process improvement is also likely to benefit from research that improves the detection and discrimination of various clay species (such as interstratified illite–smectite). The dilemma in an integrated approach is the cost–benefit return when many critical benefits may not be obvious to the primary mining operator.

10.3 Knowledge Gaps

Major knowledge gaps in the minerals process value chain surrounding three major areas: preg-robbing (PR), opportunities for selective dissolution (SD), and particle–particle interaction (PP) (Fig. 10.2).

10.3.1 Chemistry of Clays (Chapters 1 and 3)

10.3.1.1 Metal–Clay Interactions

The extent of public knowledge for metal–clay sorption interactions is largely restricted to environmental and not industrial conditions. The extent and reversibility of metal sorption onto clays at high temperatures, pressure and ionic strength (among others) at relevant pH is unknown, as is the extent of clay stabilities. The mechanisms of preg-robbing, whether through ion exchange, edge-site adsorption and/or re-precipitation with Al and/or Si is not well understood and therefore neither are the means to mitigate such losses. In addition, preg-robbing losses will report as contaminant loading in the residue, and requires costly safeguards with respect to tailings dam design, environmental monitoring and remediation planning.

10.3.1.2 Selective Dissolution

The wholesale dissolution of clays under highly acidic or caustic conditions (usually at elevated temperature) is not only problematic in the context of preg-robbing and reagent consumption, but also for unit processes following extraction, such as clarification, counter-current decantation and for tailings disposal/management due to the creation of fine colloidal, nano-sized materials. Targeted dissolution of thermodynamically highly stable host minerals of commodity metals (e.g. nickeliferous goethite) in a complex, clayey gangue matrix is thus a prospective mechanism to avoid many clay-related problems downstream.

10.3.2 Particle–Particle Interactions (Chapters 2, 7 and 8)

10.3.2.1 The Science Challenge

The basic science of particle–particle interactions has relied on DLVO theory since the 1940s, but it has become clear that DLVO's ability to predict and explain particle–particle interactions is insufficient and often cannot reproduce empirical results. A critical lack is perceived in predicting particle orientations and on the nature of surface charge distributions of anisotropic clays. The lack of further development of DLVO or particle–particle interaction theories in general negatively affects the capacity to predict effects and behaviours of clays in a number of unit processes, e.g. flotation, solid–liquid separation, tailings disposal.

10.3.2.2 Flotation: Effects of Clays on Froth Rheology

While it is clear that froth rheological properties bear a strong influence on factors such as froth mobility and stability, the exact mechanisms involved are still unknown. Clays in flotation pulps affect a number of flotation subprocesses e.g. entrainment, froth stabilization, pulp rheology, slime coating, etc.; however, the relative magnitude of impact that each of these subprocesses has on the final flotation outcome is not known.

10.3.2.3 Condensing Aggregation

In the majority of mineral processing applications, flocculants will adsorb only onto the edge faces, leading to the fractal aggregates having an even lower effective density. Achieving higher aggregate density during flocculation under practical operating conditions would be far preferable to any post-flocculation treatment of beds or thickened slurries. The 'condensing aggregation' process is one mechanism to further dewater slurries; however, the mechanism remains very poorly understood.

References

Haest, M., Cudahy, T., Laukamp, C. & Gregory, S. 2012. Quantitative mineralogy from infrared spectroscopic data: II. Three-dimensional mineralogical characterization of the Rocklea channel iron deposit, Western Australia. *Economic Geology*, 107 (2), 229–249.

Hanlie, H., Qinyan, W., Jianping, C., Shirong, L. & Ruizhong, H. 1999. Occurrence and distribution of invisible gold in the Shewushan supergene gold deposit, southeastern Hubei, China. *The Canadian Mineralogist*, 37 (6), 1525–1531.

Lower, C., Ehrig, K., Macmillan, E., *et al.* 2011. Quantitative clay mineralogy for a calcrete-hosted uranium deposit: Innovative application of existing technology on an unprecedented scale. In: Ausimm, T. (ed.) *The First AusIMM International Geometallurgy Conference, 5–7 September, 2011*. Brisbane, QLD: The Australasian Institute of Mining and Metallurgy, 173–180.

Glossary

absorb/absorption (*chemistry*) – process of absorption of a chemical species to within the pores or other interstices of a solid (as opposed to adsorption, which is attachment to a surface)
acrylamide – $CH_2CHCONH_2$, a monomer in polymer flocculants
acrylate – CH_2CHCOO^-, a monomer in anionic polymer flocculants, generally as a copolymer with acrylamide
activator (*hydrometallurgy*) – a chemical agent that enhances a hydrometallurgical process, e.g. flotation, leaching or dissolution. May act like a catalyst in some circumstances but is not a catalyst
activity (*thermodynamics*) – measure of the 'effective' concentration of a species arising because (especially at high concentrations) those species interact with each other and behave less effectively than at lower concentrations
adamellite (*geology*) – mix of orthoclase and plagioclase feldspars (granite)
adcumulates (*geology*) – rocks formed from solid crystal precipitation from a magma
adsorb/adsorption (*chemistry*) – process of attachment of a chemical species to a surface (as opposed to absorption, which is within pores or other interstices)
adularia – synonym for orthoclase $KAlSi_3O_8$
AEC – anion exchange capacity
aeolian processes (*geology*) – wind erosion
AFM – atomic force microscopy
agglomerate/agglomeration (*hydrometallurgy*) – cluster of particles, act of particles clustering
aggregate/aggregation (*geology*) – composite rock, act of forming a composite
aggregate/aggregation (*hydrometallurgy*) – fine particles in suspension brought together by coagulation or flocculation for a large, faster settling (but fragile) object; act of forming a particle aggregate
albite – $NaAlSi_3O_8$, Na-endmember of the plagioclase (albite-anorthite) series, may have up to 10 per cent Ca-for-Na substitution with concomitant effects on the Si:Al ratio for charge balance.
aliettite – $Ca_{0.2}Mg_6(Si,Al)_8O_{20}(OH)_4{\cdot}4H_2O$, member of interstratified chlorite–smectite group

aliquot (*chemistry*) – a volume of solution, a part thereof, can be any defined volume
alkylammonium – any aliphatic (R) ammonium cation from NRH_3^+ though to NR_4^+
allevardite – non-pure clay mineral, rectorite with ~12.5 per cent dickite
allochthonous (*geology*) – originating in a place other than where found
allophane – a clay-sized mineral of short range order consisting of Al, Si and H_2O in chemical combination and Al:Si ratios ranging from 1 (Al-poor) to 2 (Al-rich), e.g. $Al_2O_3{\cdot}(SiO_2)_{1.3-2}{\cdot}(2.5–3)H_2O$
alluvial (*geology, adj.*) – unconsolidated, terrestrial sediment transported and redeposited from the action of water
alum – any double sulphate salt $A^+M^{3+}(SO_4)_2.12H_2O$ e.g. $KAl(SO_4)_2.12H_2O$
alumina – Al_2O_3
aluminol (*chemistry*) – Al–OH group
alunite – $KAl_3(SO_4)_2(OH)_6$
AMD – acid mine drainage
amesite – $Mg_2Al_2SiO_5(OH)_4$ 1:1 clay mineral, member of kaolinite–serpentine group and tri-octahedral subgroup
amine – organic functional group with a basic N atom, RNH_2, $R_IR_{II}NH$, $R_IR_{II}R_{III}N$
amphibole – general double-chain inosilicate
amphibolite (*geology*) – metamorphic rock consisting mainly of amphibole
amphoteric (*chemistry*) – capable of displaying both acid and base properties
analcime – non-IMA zeolite $NaAlSi_2O_6.2H_2O$
anatase – polymorph of TiO_2
anchimetamorphic (*geology*) – metamorphic grade of sedimentary rock where earlier diagenesis has been overtaken by metamorphism
anhydrite – $CaSO_4$ (anhydrous)
anion (*chemistry*) – an ionic species with one or more formal negative charges
anisotropic – properties that are (usually crystallographically) directionally dependent
ankerite – $CaFe^{2+}_{0.6}Mg_{0.3}Mn^{2+}_{0.1}(CO_3)_2$
anorthic (*crystallography*) – unequal crystal axes intersecting at oblique angles
anoxic – without oxygen
antigorite – $Mg_{2.25}Fe^{2+}_{0.75}(Si_2O_5)(OH)_4$
apatite – phosphate mineral $Ca_5(PO_4)_3(F,Cl,OH)$ which can be dominated by either F^-, Cl^- or OH^{--}
API – American Petroleum Institute
aragonite – $CaCO_3$ polymorph of calcite
Archean eon (*geology*) – before 2500 mya
arenite – sedimentary sandstone
argillic alteration (*geology*) – hydrothermal alteration of rock into clay minerals
argillitic (*geology*) – coming from an agrillic alteration
aridic (*geology*) – a dry moisture regime
arkose (*geology*) – a type of sandstone with at least 25 per cent feldspar

Arrhenius activation energy (*chemistry*) – a simple empirical activation energy for a chemical reaction determined by measuring the reaction rate at several temperatures and then applying the Arrhenius equation

asphaltenes – general group of saturated hydrocarbons found in crude oil

attapulgite – $(Mg,Al)_2Si_4O_{10}(OH)\cdot 4(H_2O)$, less common name for palygorskite

augite – single-chain inosilicate of general composition $(Ca,Na)(Mg,Fe,Al,Ti)(Si,Al)_2O_6$

aurocyanide – $Au(CN)_2^-$

AuScope – national earth science infrastructure programme, see http://portal.auscope.org

autoclave (*mineral processing*) – a unit operation component for heating

autogenous – literally self-produced

baddeleyite – ZrO_2

basal plane (*crystallography*) – crystal plane perpendicular to the principal axis where that axis has higher rotational symmetry than the other crystal axes

basalt (*geology*) – common igneous rock, <20 per cent quartz and >65 per cent feldspar

basins (*geology*) – rock strata depressions, usually approximately circular or elliptical in shape, can be formed a number of ways and may be hundreds of kilometres in diameter

bassanite – $CaSO_4\cdot 0.5H_2O$

bauxite – primary aluminium ore consisting of gibbsite, boehmite, diaspore mixed with iron oxides and clays

beaching angle or slope (*mineral processing*) – angle/slope of deposited, high-density slurry or thickened paste

bedrock (*geology*) – underlying consolidated rock

beidellite – $NaAl_5Si_7O_{20}(OH)_4\cdot 2H_2O$ 2:1 clay mineral, member of smectite group

beneficiate/beneficiation (*mineral processing*) – the process by which the grade of an ore is increased; this may be the entire upgrading process to a saleable commodity or the precursor to further mineral processing

bentonite – rock name and/or commercial product with composition and/or properties dominated by smectite minerals (usually beidellite or montmorillonite; also synonym for both)

BET – Brunauer–Emmett–Teller; theory of physical adsorption of gas molecules on a solid surface; basis of the method of measuring surface area

bidentate (*chemistry*) – where a ligand molecule is coordinated to a central metal through two atoms

bimodal – a distribution with two modes (e.g. peaks in a probability distribution)

biotite – $K(Mg,Fe)_3(AlSi_3O_{10})(F,OH)_2$, 2:1, dominantly tri-octahedral, clay mineral, member of the mica group

birnessite – $(Na_{0.3}Ca_{0.1}K_{0.1})(Mn^{4+},Mn^{3+})_2O_4\cdot 1.5H_2O$

bituminite (*geology*) – one of the three major types of organic matter in oil shale

bituminous – asphalt- or bitumen-like

boehmite – γ-AlOOH

Brownian motion – random motion of particles in a fluid resulting from collisions of the particles with atoms/molecules in the fluid

brucite – $Mg(OH)_2$
BSA – basal surface area in particular reference to clay particles
calc-alkaline (*geology*) – one of two magma compositional series that are found in igneous rocks
calcite – $CaCO_3$
calcrete (*geology*) – hardened deposit of calcium carbonate or a calcium-rich duricrust, also termed caliche
Calgon® – Ca^{2+} complexing agent, originally amorphous $(NaPO_3)_6$, now zeolite and polycarboxylate
calorimetry (*thermodynamics*) – measurement of the heats of chemical reaction or physical phase transformation
Cambrian period (*geology*) – 541–485 mya
carapace (*geology*) – upper part of a fold with a near-horizontal plane
carbonatites (*geology*) – intrusive or extrusive igneous rocks with >50 per cent carbonate mineral content
carbonic acid – H_2CO_3
carboxylate – $RCOO^-$ where R is H or an organic group
carboxymethyl – group – CH_2COOH
carnotite – $K_2(UO_2)_2(VO_4)_2 \cdot 3H_2O$
carrboydite – $(Ni,Cu)_{14}Al_9(SO_4,CO_3)_6(OH)_{43} \cdot 7(H_2O)$
carrierless (*hydrometallurgy*) – froth flotation is difficult with ultrafine particles, one approach is to use a floatable carrier particle on which the ultrafine particles attach, carrierless is the absence of such particles in ultrafine flotation
cassiterite – SnO_2
cation (*chemistry*) – an ionic species with one or more formal positive charges
CEC – cation exchange capacity
celadonite – $K(Mg,Fe^{2+})(Fe^{3+},Al)[Si_4O_{10}](OH)_2$, 2:1 clay mineral, member of the mica group
centi – one-hundredth
centimole – one-hundredth of a mole
centipoise (cP) – unit of dynamic viscosity
chalcedony – intergrowth of quartz and moganite (SiO_2 polymorphs); agate and onyx are varieties of chalcedony
chalcopyrite – $CuFeS_2$
chamosite – $(Fe^{2+},Mg)_5Al(AlSi_3O_{10})(OH)_8$ phyllosilicate, endmember of chlorite
chelation (*chemistry*) – formation of two or more coordinate bonds between a ligand and a central chelated atom; the ligand is usually organic (IUPAC)
chemisorption – the adsorption of an atom, ion or molecule on a surface that occurs through the formation of a chemical bond or bonds; unlike physisorption
chert – microcrystalline quartz
chlorite (*mineral*) – phyllosilicate group with general formula $(Mg,Fe)_3(Si,Al)_4O_{10}(OH)_2 \cdot (Mg,Fe)_3\ (OH)_6$, comprising a 2:1 structure similar to the 2:1 clay minerals and an

additional 'brucite' ($Mg,Fe(OH)_2$) sheet. Has four endmembers where Mg, Fe, Ni and Mn dominate the substitution

chlorite (*chemical*) – ClO_2^-

chromite – nominally $FeCr_2O_4$, can exist as $(Fe,Mg)Cr_2O_4$; belongs to spinel group

chrysotile – $Mg_3(Si_2O_5)(OH)_4$ or white asbestos, exists as one of three serpentine polytypes

circuit (*mineral processing*) – description of the overall mineral processing circuit, usually a succession of different unit processing operations

clarification (*mineral processing*) – the process of removal of suspended solids from a process stream, with the resultant clarity of the stream generally a higher priority than the settling rate of the solids (in contrast to thickening)

classification (*mineral processing*) – the process of separating mixed minerals on the basis of (terminal) velocity. Classifiers are usually based on fluid rising in a column and solids reporting to the overflow or underflow depending on velocity.

classifier (*mineral processing*) – for classification, classifiers are usually based on fluid rising in a column and solids reporting to the overflow or underflow depending on (terminal) velocity

climotoposequence – literally a topographically based climate sequence

clinochlore – $(Mg_5Al)(AlSi_3O_{10})(OH)_8$ phyllosilicate, endmember of chlorite

clinoptilolite – $(Na,K,Ca)_{2-3}\,Al_3(Al,Si)_2Si_{13}O_{36}\cdot 12(H_2O)$, natural zeolite

clinopyroxene (*geology*) – pyroxene crystallized in the monoclinic system

clintonite – $Ca(Mg,Al)_3(Al_3SiO_{10})(OH)_2$, 2:1 clay mineral, member of the mica group

cmol (*chemistry*) – one-hundredth of a mole

coagulant (*hydrometallurgy*) – chemical that enhances a coagulation process

coagulation (*hydrometallurgy*) – process of colloidal particles adhering directly to each other on the basis of their charge

collector (*hydrometallurgy*) – specific molecule used in froth flotation which (normally selectively) attaches to the target mineral to change its surface tension (contact angle), thereby either enhancing or depressing its flotation

colloid (*chemistry*) – small particles (<1 micron in diameter) capable of forming a colloidal suspension

colloidal suspension (*chemistry*) – a specific form of interphase suspension in which small particles (<1 micron in diameter) remain in suspension in another fluid; the actual particles can be conventional particulates e.g. clay particles or non-conventional particles such as hydrosols or emulsions

colluvial (*geology, adj.*) – describing (loose) debris (rocks, rock fragments) transported principally by gravity to the base of a slope, cliff or valley.

colourimetric – colour-based assessment

comminution – reduction in ore fragment size, generally by crushing and grinding

commodity – a generic marketable item, e.g. iron ore, kaolinite

conditioning (*hydrometallurgy*) – usually applied in the context of froth flotation, process of a specific chemical reaction being allowed to occur to improve the outcomes of the next step or process. Usually occurs in separate conditioning tanks

conglomerate (*geology*) – sedimentary rock consisting of rounded fragments cemented together with a finer matrix
Congo red – anionic dye molecule
consolidation (*hydrometallurgy*) – settling of a solids bed, usually in the context of a thickener or a tailings pond or storage facility
cookeite – $LiAl_4(Si_3AlO_{10})(OH)_8$, can be considered a chlorite group member
copolymer (*chemistry*) – a polymer formed from two or more types of monomer
corundum – α-Al_2O_3
Couette flow (*fluid dynamics*) – laminar flow of a fluid between two plates
covellite – CuS
craton (*geology*) – old and stable portion of the continental crust
Cretaceous period (*geology*) – 145–66 mya
cristobalite – quartz polymorphs (α and β)-SiO_2
cryo-SEM – SEM or scanning electron microscopy in which the sample remains frozen
cryptobauxite (*geology*) – gibbsitic layer interbedded between two kaolinitic horizons
cubic – one of the seven lattice point or space groups or crystal systems
cumulate (*geology*) – igneous rocks formed by crystal accumulation from a magma
cuprite – Cu_2O
cyclone (*mineral processing*) – general (continuous) mineral separation concept based on centrifugal forces exerted by means of a cyclonic action; exists in a wide variety of designs and usually relatively low G
deflocculant (*hydrometallurgy*) chemical that reverses a flocculation process
dehydration (*chemistry*) – process of removing water
dehydroxylation (*chemistry*) – process of removing hydroxide groups as water and leaving behind one oxygen atom
densify – to increase the density
depressant (*hydrometallurgy*) – either a specific molecule used in froth flotation that (normally selectively) attaches to the target mineral to change its surface tension (contact angle), to depress its flotation (acting like a collector but in reverse) or a less selective gangue component that acts to generally depress flotation
desertification – to render otherwise fertile land into a desert
desliming (*mineral processing*) – removal of slimes, usually via washing
detrital (*geology*) – particles worn away from solid rock
Devonian period (*geology*) – 419–359 mya
dewater – (*mineral processing*) removal of water from a suspension, e.g. before tailings disposal or to recover needed process water for re-use
diagenesis – changes to sedimentary rock under conditions insufficient for metamorphic (melting) changes (usually <300 °C and <1 kBar), excludes weathering
dialysis – method of ultrafiltration (usually ions) through a semi-permeable membrane
diamond – high-pressure and high-temperature polymorph of carbon (C)
diaspore – α-AlOOH
dickite – $Al_2Si_2O_5(OH)_4$, 1:1 clay mineral, member of the kaolinite–serpentine group

dilatancy – property of dilating or expanding
distal – furthest point
diuranate – a uranate is any UO_x and diurante any U_2O_y
DLVO – theory of colloidal stability after Derjaguin and Landau, Verwey and Overbeek
dolerite (*geology*) – volcanic basalt
dolomite – $CaMg(CO_3)_2$
donbassite – $Al_{4.33}(Si_3AlO_{10})(OH)_8$ can be considered a chlorite group member
Doppler effect or shift (*physics*) – effective frequency change for observer/detector moving relative to the wave source
DSC – differential scanning calorimetry
DSP (*chemistry*) – desilication product
DSTP – deep-sea tailings placement
DTA – differential thermal analysis
dunite (*geology*) – igneous rock >90 per cent olivine
dunitic – dunite-like
duricrust – hard soil crust near to or at the surface
dyke (*geology*) – a tabular or sheet intrusion of one rock in another
EDS – energy-dispersive X-ray spectroscopy, also known as EDX and EDAX
EELS – electron energy loss spectroscopy
Eh – reduction potential, also known as ORP, pE or redox potential
electrolyte (*chemistry*) – chemical compound that gives rise to ions in solution, e.g. NaCl, also refers to the solution that contains the ions
electrophoretic mobility (*chemistry*) – electrophoresis is the motion of particles within a fluid under the influence of an electric field due to net charge on the particles; the charge affects the mobility
electroviscous (*colloid chemistry*) – effects by particle charge on fluid viscosity
electrowinning (*hydrometallurgy*) – metal recovery from solution by means of electrochemical reduction
emulsion (*chemistry*) – mixture of two or more immiscible liquid phases that form a dispersion such that one (or more) of the liquids breaks into small particles (colloids) within the continuous phase of the remaining immiscible liquid. The emulsion may or may not be stable
enantiomorphic (*chemistry*) – a molecular structure whose mirror image cannot be superimposed on itself
endmember (*mineralogy*) – mineral systems can often be described as continuous solid solutions of varying composition; endmembers are the 'pure' forms at the ends of a mineral series
endothermic – reaction that absorbs heat (energy) from its surroundings
entrainment (*mineral processing*) – unwanted particles getting caught in a separation process by other particles, e.g. in the froth for flotation
epidote – $Ca_2Al_2(Fe^{3+},Al)(SiO_4)(Si_2O_7)O(OH)$

epigenetic (*geology*) – ore deposit or formation of a later age than the associated rock formation
epithermal (*geology*) – mineralization formed hydrothermally at shallow depths
epizone (*geology*) – literally zone 'resting upon'
EPR – electron paramagnetic resonance
ESA – edge surface area, in particular reference to clay particles
ESR – electron spin resonance
estuarine (*geology*) – having formed in an estuary
EXAFS – extended X-ray absorption fine structure
exothermic – reaction that releases heat (energy) to its surroundings
facies (*geology*) – characteristic morphology of exposed rock recognizably different from surrounding rock, reflecting differences in process or environment
Faraday – magnitude of charge for one mole of electrons
FBRM® – focused beam reflectance measurement, commercial instrumentation for relative sizing of particles *in situ*
feedwell – entry point for feed going into a thickener (usually at the centre and top) and often taking the form of an open cylinder, although many more complex variations exist on this design. The feedwell ideally serves to dissipate the momentum of the incoming feed, provide conditions that (with addition of a flocculant) favour aggregation and also produce an even discharge of the settling solids into the main body of the thickener.
feldspar – group of rock-forming tectosilicates $KAlSi_3O_8$ – $NaAlSi_3O_8$ – $CaAl_2Si_2O_8$
felsic (*geology*) – igneous rocks rich in feldspar and quartz as opposed to mafic rocks; >65 per cent Si content
ferricrete (*geology*) – duricrust where the sediments are cemented by iron oxide
ferrihydrate [sic] – occasionally used as an alternative to the preferred term ferrihydrite to describe a poorly crystalline or amorphous, hydrous Fe^{3+} oxide. The term ferrihydrate literally means $Fe^{3+}(H_2O)_n$ so is misleading in the context of ferrihydrite.
ferrihydrite – nominally $Fe_2O_3{\cdot}0.5H_2O$ but varies; ferrihydrite is better characterized by its limited XRD as (for example) two, three or six line ferrihydrite depending on the number of main diffraction peaks
ferruginous (*geology*) – containing iron or iron oxides
FFT – fluid fine tailings
FieldSpec – make of portable spectrometer
FIR – far infrared, usually covers the 16 000 nm to 1 mm wavelength range
flagioclase (archaic) – feldspar $NaAlSi_3O_8$ – $CaAl_2Si_2O_8$ more commonly known as plagioclase
floatability (*mineral processing*) – the ease or extent to which a mineral component can be separated by froth flotation
floc – agglomeration of fine/colloidal particles. Also used in industry to refer to the flocculant.
flocculant (*chemistry*) – chemical that promotes flocculation which is the process of contact and adhesion of dispersed particles to form larger clusters (IUPAC). In this sense

flocculation is indistinct from aggregation or coagulation. In mineral processing flocculation is taken as involving bridging molecules (usually polymers).

florencite – rare earth aluminium phosphate, e.g. $(La,Ce)Al_3(PO_4)_2(OH)_6$

flotability (*hydrometallurgy*) – how easily a mineral component 'floats' (see below) depends upon its liberation and surface contact angle (which may or may not involve a collector)

flotation (*hydrometallurgy*) – short for froth flotation in which bubble attachment is used to separate hydrophobic from hydrophyllic minerals

fluvial (*geology*, adj.) – erosion or deposition of river bed sediments

flux (*hydrometallurgy*) – number of particles passing through a unit area per unit of time

foredune (*geology*) – dune ridge that is parallel to a water body

formamide – $HCONH_2$

froth (*hydrometallurgy*) – mineral air bubble mass at the top of a flotation cell

FTIR –Fourier transform infrared

gabbroic (*geology*) – containing gabbro, a coarse-grained, mafic igneous rock chemically the same as basalt

gabbronorite (*geology*) – coarse-grained, mixed rock containing quartz, feldspar, plagioclase, pyroxenes and olivine

gangue (*hydrometallurgy*) – unwanted mineralization in an ore-body that eventually contributes to the bulk of mining/mineral processing waste

garnet – neosilicate of general form $X_3Y_2(SiO_4)_3$ where X is a divalent metal cation and Y a trivalent metal cation

garnierite – poorly defined nickel ore in terms of composition and naming, variously (for example) $(Mg,Ni)_3Si_4O_{10}(OH)_2{\cdot}H_2O$ through to $Ni_3MgSi_6O_{15}(OH)_2{\cdot}6(H_2O)$ (synonyms falcondoite, genthite)

gaspeite – $Ni_{0.6}Mg_{0.3}Fe^{2+}{}_{0.1}(CO_3)$

gel (*chemistry*) – a non-fluid (usually with a low yield stress) formed by a colloid or polymer network

gel point – the specific solids volume fraction at which gel behaviour is first observed (i.e. when a yield stress can be measured)

gelation (*chemistry*) – process of forming a gel or a molecular network

genesis – origin

geochemistry – study of the chemical composition of the earth and the chemical processes that affect the earth's substance

geomorphology – study of physical landforms and the processes that affect them

geotechnical (*engineering*) – branch of civil engineering associated with earth materials (soil mechanics, rock mechanics)

geothermometer – thermometer specifically designed for measuring temperatures at a depth below ground, e.g. in bore holes; not a 'conventional' thermometer by any means and they rely on the thermodynamic behaviour variety of chemical and isotopic systems

gibbsite – γ-$Al(OH)_3$

glaciation (*geology*) – a period of extended glacial activity

glycol – double organic alcohol, e.g. ethylene glycol $HOCH_2CH_2OH$

gmap – generic name for a number of map applications

gneiss (*geology*) – common metamorphic rock characterized by a layered structure reflecting earlier igneous or sedimentary origin

goethite – α-FeOOH

granite (*geology*) – intrusive igneous rock, predominantly quartz, mica and feldspar

granitic or granitoid (*geology*) – coarse-grained, intrusive, igneous rock similar to granite, predominantly feldspar and quartz

granodiorite (*geology*) – intrusive, igneous rock similar to granite, but the feldspar content dominated by plagioclase

granulites (*geology*) – medium- to coarse-grained, high-temperature metamorphized rocks

granulometry – measurement of the size distribution of a group of granular particles

gravimetric (*chemistry*) – mass-based metric used in chemical assessments, similar context to volumetric (volume)

greenstone – generic name for a rock package that consists of a number of different rock types, dominated by mafic and/or ultramafic rocks

gypsum – $CaSO_4{\cdot}2H_2O$

halloysite – $Al_2Si_2O_5(OH)_4$ 1:1 clay mineral, member of the kaolinite–serpentine group

harzburgite (geology) – ultramafic igneous rock, predominantly olivine and pyroxene

hectorite – $Na_{0.3}(Mg,Li)_3Si_4O_{10}(OH)_2$, 2:1 clay mineral, member of the smectite group

hematite – α-Fe_2O_3

hematitic (*geology*) – containing hematite

heterocoagulation (*colloid chemistry*) – coagulation of dissimilar particles

heterogeneous – diverse in nature

heulandite – $(Ca,Na)_{2\text{-}3}Al_3(Al,Si)_2Si_{13}O_{36}{\cdot}12H_2O$ zeolite

hexagonal – one of the seven lattice point or space groups or crystal systems

hexametaphosphate – e.g. sodium hexametaphosphate $(NaPO_3)_6$

hkl (*crystallography*) – Miller indices for lattice plane designation; (hkl) denotes a plane orthoganol to direction (h,k,l); {hkl} denotes a set of planes equivalent to (hkl); [hkl] denotes a lattice vector direction; <*hkl*> denotes a set of directions symmetrically equivalent to [hkl]

hollandite – $Ba(Mn^{4+},Mn^{2+})_8O_{16}$

homionic (*mineralogy*) – applied to clay in which all of the dissociable (exchangeable) cations are of one type

HPAL (*hydrometallurgy*) – high-pressure acid leach

HRTEM – high-resolution transmission electron microscopy

hydration – taking up of water

hydrazine – N_2H_4

hydrobiotite – $[K(Mg,Fe)_3(Al,Fe)Si_3O_{10}(OH,F)_2]{\cdot}[(Mg,Fe,Al)_3(Si,Al)_4O_{10}(OH)_2{\cdot}4(H_2O)]$, 2:1 clay mineral, member of the mica group

hydrocyclone (*mineral processing*) – mineral classifier that sorts suspended particles based either on particle density or size using a cyclonic action

hydrodynamics (*physics*) –study of liquids in motion
hydrogeochemistry – study of chemical characteristics of ground and surface waters in the context of the surrounding geology
hydrology – study of the movement, distribution and quality of water
hydrolysis (*chemistry*) – reaction with water
hydrometallurgy – a field of extractive metallurgy involving the use of aqueous chemistry (as opposed to pyrometallurgy)
hydromorphic – developed in the presence of excess water
hydrophilic – water loving
hydrophobic – water repelling
hydrotalcite – $Mg_6Al_2(CO_3)(OH)_{16}{\cdot}4H_2O$ layered double hydroxide carbonate
hydrothermal (*geology*) – of or pertaining to heated water, to its actions, or to products related to its actions
hydroxamate – salt of hydroxamic acid, RCONHOH
HyLoggerTM – rapid, near-infrared-based system for analysing drill core material
HyMapTM – airborne, hyperspectral imaging sensor (HyVista Corporation)
hypergene (*geology*) – genesis terms: hypogene (processes that occur deep below the Earth's surface), hypergene (processes that occur 'above') and supergene (processes that occur near the Earth's surface). Hypergene and supergene are used equivalently
Hyperion – Hyperion Imaging Spectrometer, hyperspectral satellite in orbit aboard NASA's EO-1 spacecraft
hyperspectral imaging – simultaneous collection and processing of information from a broad part of the electromagnetic spectrum, e.g. visible and infrared
hypochlorite – ClO^-
hysteresis – a lag in one property of a system when a second property is changed
IEP (*chemistry*) – isoelectric point, the pH value at which a particular molecule or surface carries no net charge
igneous (*geology*) – one of the three main rock types. Formed through the cooling/solidification of magma or lava. The other types are sedimentary and metamorphic.
illite – $(K,H_3O)(Al,Mg,Fe)_2(Si,Al)_4O_{10}\,(OH)_2{\cdot}H_2O$, 2:1 clay mineral, member of the mica group, non-IMA name
illitic – composed of or characteristic of illite
ilmenite – $FeTiO_3$
IMA – International Mineralogical Association
imogolite – $Al_2SiO_3(OH)_4$ 1:1 clay mineral
impeller/impellor – any mechanical component that imparts motion to a fluid
indurated (*geology*) – hardened
interbedded (geology) – one lithology lying between beds of different lithology
intercalation (chemistry) – reversible inclusion (without covalent bonding) of one type of molecule or ion within the matrix of (usually) a laminar compound (e.g. clay, graphite). Can also be applied to polymers and non-laminar systems with pre-existing spaces
intergrowth (*mineralogy*) – two or more mineral types intergrown

interstice (*crystallography*) – space between atoms or ions within a crystal lattice. In some structures small atoms can occupy interstitial locations, as opposed to formal substitution
interstitial (*crystallography*) – located at an interstice
interstratified (*crystallography*) – due to their structure, clays can effectively mix layer by layer whereby two or more kinds of clay sheets can stack in a variety of ways to make a new structure
interstratified (*geology*) – see *interbedded*
ironstone – ferruginous sediment that contains an appreciable amount of iron
isoelectric (*chemistry*) – having the same number of electrons, more loosely used in the context of net charge
isomorphic or isomorphous (chemistry) – crystallizing in the same or closely related forms
isotropic – having the same properties in all directions
IUPAC – International Union of Pure and Applied Chemistry
jarosite – $KFe^{3+}_3(OH)_6(SO_4)_2$, the metal cations can be substituted to form a range of jarosite types
Jurassic period (*geology*) – 201.3–145 mya
kaolin – clay rich in kaolinite, also known as China clay
kaolinite – $Al_2Si_2O_5(OH)_4$, 1:1 clay mineral in the kaolinite–serpentine group
karst (*geology*) – landform arising from the dissolution of soluble rocks, e.g. limestone
kerogen (*geology*) – high molecular weight organic matter found in sedimentary rocks, effectively bituminous. When heated, kerogens can degrade and release hydrocarbons such as crude oil and natural gas
kerolite – $(Mg,Ni)_3Si_4O_{10}(OH)_2{\cdot}H_2O$ mixed talc variety
kimberlite (*geology*) – potassic, volcanic rock, can be associated with diamond deposits
komatiite (*geology*) – ultramafic, volcanic rock
Kp (*hydrometallurgy*) – packing factor
kukersite – marine oil shale
kyanite – one of the three common polymorphs of Al_2SiO_5
laminar flow (hydrodynamics) – fluid flow in parallel layers with no disruption between the layers
Landsat – series of Earth resources satellites
lanthanides (*chemistry*) – elements from ^{57}La through to ^{71}Lu
lateral – directed or situated to a side
laterite – high aluminium and iron soil types that evolve due to the long-term leaching removal of silica; can also be considered an ore, e.g. nickel laterites
latex (*colloid chemistry*) – colloidal dispersion resulting from emulsion polymerization, the latex particles are typically 100 nm in size
laumontite – $Ca(AlSi_2O_6)_2{\cdot}4H_2O$
launder (*hydrometallurgy*) – overflow collection receiver for a flotation cell
lepidocrocite – γ-FeOOH
lepidolite – $KLi_2AlSi_4O_{10}F(OH)$, 2:1 clay mineral member of the mica group
leucite – tectosilicate $K[AlSi_2O_6]$

leucoxene – mixture of Fe–Ti oxides, including ilmenite, anatase and rutile
Lewis acid/base (*chemistry*) – a Lewis acid is a molecular species that is an electron pair acceptor and reacts with a Lewis base to form a Lewis adduct; Na^+ and OH^- both fill this definition, but this acid–base concept can be very generally applied to a wide variety of reactions
liberation (*mineral processing*) – the extent to which one mineral is physically associated with another, e.g. the target mineral with the gangue. Highly liberated minerals are generally much easier to process
ligand (*coordination chemistry*) – any ion or molecule that bonds to a central metal ion to form what is known as a coordination complex. Ligand can also be more loosely applied in the context of a surface complex
lignite – brown coal, considered the lowest-ranked coal
limonite – a field descriptive term for iron-rich material consisting of a mix of iron oxyhydroxides, Al-/Si-/Mn-oxides and oxyhydroxides. Previously synonymous with goethite, but now a disregarded term
liquefaction (*geology*) – often applied to certain water-saturated soil types that turn into a liquid under shear (e.g. during an earthquake)
lithocap (*geology*) – literally stone cap
lithology (*geology*) – description of the physical characteristics of a rock
lithiophorite – $Al_{0.7}Li_{0.3}Mn^{4+}{}_{0.7}Mn^{3+}{}_{0.3}O_2(OH)_2$
lixiviant (*hydrometallurgy*) – liquid extractant used to leach (usually selectively) the desired component from an ore
lizardite – $Mg_3Si_2O_5(OH)_4$, serpentine polytype
lucastrine (*geology*) – literally relating to a lake in the context (for example) of a plain or a delta
lutetium – ^{71}Lu, last element in the lanthanide series
mackinawite – $Fe_{0.75}Ni_{0.25}S_{0.9}$
macromolecule (*chemistry*) – any very (single) large molecule, e.g. polymer or protein, usually consists of many thousands of atoms
mafic (*geology*) – magnesium- (**ma**) and iron- (**ferric**) rich rock; <55% Si content
maghemite – γ-Fe_2O_3
magma (*geology*) – molten or semi-molten rock
magnetite – Fe_3O_4
magnetization (or magnetic polarization) – the density of permanent or induced magnetic dipole moments
malachite – $Cu_2CO_3(OH)_2$
marcasite – FeS_2 (iron sulphide with orthorhombic structure, same composition as pyrite)
martite – a variety of hematite, Fe_2O_3, pseudomorph of hematite after magnetite
meq (*chemistry*) – milliequivalent, the given unit divided by 1000; see milliequivalent
Mesozoic era (*geology*) – 252–66 mya
metabauxite – transformed or altered boehmitic bauxite
metakaolin – transformed or altered kaolin, more particularly dehydroxylated kaolinite

metalliferous – containing or yielding metal (not specific to iron)
metamorphic (*geology*) – one of the three main rock types, formed through the action of pressure and temperature on prior rock. The other types are sedimentary and igneous
metamorphism (*geology*) – transformation of one rock type into another; the influence of high pressures and high temperatures, usually >300 °C, >1 kBar
metasomatism (*geology*) – chemical alteration of a rock by simultaneous solution and deposition, usually hydrothermally
MFT – mature fine tailings
mica – 2:1 clay mineral of general formula $X_2Y_{4-6}Z_8O_{20}(OH,F)_4$ where X, Y and Z are various metal ions
micaceous – containing or resembling mica
microcline – $KAlSi_3O_8$ feldspar
microcrystalline – having a crystal dimension only visible under a microscope
micronize – fine grind to produce particles of a few microns in diameter
microplaty (*mineralogy*) – morphological expression of small plates, e.g. hexagonal or pseudo-hexagonal, plate-like hematite where the plates are tens of microns in size
Microtrac – company producing particle analysis equipment
millerite – NiS
milliequivalent (*chemistry/biochemistry*) – equivalent divided by 1000. An equivalent entity (IUPAC) is that corresponding to the transfer of a full charge; the equivalent (eq or Eq) has subsequently been interpreted as a molar equivalent, which is not the intent of the IUPAC equivalent entity. In practice a milliequivalent can be thought of as a millimole. Not used in chemistry.
milling (*mineral processing*) – grinding
minnesotaite – $Fe^{2+}{}_{2.5}Mg_{0.5}Si_4O_{10}(OH)_2$, 2:1 clay mineral in the talc–pyrophyllite group
Miocene epoch (*geology*) – 23.03–5.33 mya
MIR – mid-infrared, usually covering the 2500–5000 nm wavelength range
Mohs (*geology*) – scale of mineral hardness from 1 (talc) through to 10 (diamond)
moiety (*chemistry*) – part of a molecule that may include a whole functional group or part of a functional group, e.g. hydroxyl moiety
monazite – various rare-earth-containing phosphates, e.g. $(Ce,La)PO_4$
monoclinic – one of the seven lattice point or space groups or crystal systems
montmorillonite – $(Na,Ca)_{0.3}(Al,Mg)_2Si_4O_{10}(OH)_2{\cdot}n(H_2O)$ 2:1 clay mineral, member of the smectite group
morenosite – $NiSO_4{\cdot}7H_2O$
morphology – external structure or arrangement
MPVC – mineral processing value chain, a term used in this review to highlight the interconnection between (hydrometallurgical) unit processes and that overall value for mineral extraction is dependent upon each step
Mt – million tonnes
mullite – neosilicate $3Al_2O_3{\cdot}2SiO_2$ or $2Al_2O_3{\cdot}SiO_2$

multispectral imaging (*remote sensing*) – image taken at a number of specific frequencies such that each band images particular aspects, e.g. man-made objects, vegetation, soil moisture, etc. Lower spectral resolution when compared to hyperspectral

muscovite – $KAl_2(AlSi_3O_{10})(F,OH)_2$, 2:1 clay mineral, member of the mica group

mya – millions of years ago

nacrite – $Al_2Si_2O_5(OH)_4$, 1:1 clay mineral, member of the kaolinite–serpentine group, polymorph of kaolinite

nanoparticulate – particle of size range 1–100 nm

natroalunite – $NaAl_3(SO_4)_2(OH)_6$

natrojarosite – $NaFe^{3+}{}_3(OH)_6(SO_4)_2$

network (*chemistry*) – structure in which each unit is connected to each other unit and to the macroscopic phase boundary by multiple paths; examples are colloidal network, polyelectrolyte network, polymer network

Newtonian fluid (*rheology*) – a fluid whose viscosity does not change with the rate of flow or applied shear, e.g. water

NEXAFS – near edge extended X-ray absorption fine structure

NMR – nuclear magnetic resonance

nonionic (*chemistry*) – generally applied to polymers that have no ionic or fully ionizable functional groups

nontronite – e.g. $Na_{0.3}Fe^{3+}{}_2Si_3AlO_{10}(OH)_2.4H_2O$, 2:1 clay mineral member of the smectite group

nordstrandite – $Al(OH)_3$

obduction (*geology*) – overthrusting of continental crust at a convergent point, e.g. building of mountains

ochreous – ochre-like, ochre being a mixture of iron oxide and clay

OES – optical emission spectroscopy

oleic acid – fatty acid $CH_3(CH_2)_7CH{=}CH(CH_2)_7COOH$

Oligocene epoch (*geology*) – 34–23 mya

oligomer (*chemistry*) – a short polymer chain consisting of a few (typically 2–4) monomer units

oligomerization (*chemistry*) – the process of forming an oliogmer

olivine – $(Mg, Fe)_2SiO_4$

ooidal (*geology*, adj.) – term applied to describe iron-rich, fluvial rocks, such as channel iron deposits (CID), comprised of small, rounded, ellipsoidal to spherical, concentrically layered bodies of 0.25–2 mm in diameter (literally egg-shaped)

ophiolite (*geology*) – oceanic crust that has been lifted above sea level and become part of the continental crust

Ordovician period (*geology*) – 485.4–443.4 mya

ore – a mineral deposit of commercial value

orogenesis (*geology*) – formation of mountains by folding/deforming the Earth's crust

orthobauxite (*geology*) – gibbsitic bauxite without concretionary structure and often associated with iron accumulation

orthoclase – $KAlSi_3O_8$ feldspar
orthocumulates (*geology*) – rocks with 75–85 per cent of an accumulated mineral type in the overall rock matrix
orthophosphate – PO_4^{3-}
orthopyroxene (*geology*) – pyroxene crystallized in the orthorhombic system
orthorhombic – one of the seven lattice point or space groups or crystal systems
osmotic pressure (*chemistry*) – pressure required to maintain osmotic equilibrium between a solution and pure solvent by a membrane permeable only by the solvent (IUPAC)
overburden (*mining*) – surface material lying above the ore-body
overshear (*hydrometallurgy*) – a state of excessive shear, usually used in the context of the shear having induced an irreversible change
palaeo- – in the geological past (also paleo)
palaeogeomagnetic – previously recorded direction and intensity of the Earth's magnetic field (occasionally termed palaeomagnetic)
Palaeocene epoch (*geology*) – 66–56 mya
palaeoclimatology – climatic change study through the Earth's geological history
Palaeogene period (*geology*) – 66–23.03 mya
palaeosol (*geology*) – former soil buried under subsequent sedimentary or volcanic deposits
Palaeozoic era (*geology*) – 541–252.2 mya
palygorskite – $(Mg,Al)_2Si_4O_{10}(OH)\cdot4(H_2O)$
parachrysotile – $Mg_3Si_2O_5(OH)_4$ non-IMA name, polytype of chrysotile (asbestos)
paragenesis (*geology*) – co-occurrence/co-formation of different minerals
paragonite – $NaAl_2(Si_3Al)O_{10}(OH)_2$, 2:1 clay mineral, member of the mica group
paramagnetic – materials slightly attracted to an external magnetic field
paste (*hydrometallurgy*) – thickened (dewatered) suspension
PDVF – polyvinylidene fluoride $[–C_2H_2F_2–]_n$
pecoraite – $Ni_3(Si_2O_5)(OH)_4$
pedogenesis (*soil science/geology*) – science and study of soil forming processes
pedolith – synonym for soil
pegmatite (*geology*) – intrusive igneous rock with large (>25 mm) interlocking crystals
pelletoid – pellet-like, not in common use
pentlandite – $(Fe,Ni)_9S_8$
PEO – polyethylene oxide $H–[–OCH_2CH_2–]_n–OH$ also known as polyethylene glycol or PEG
peralkaline (*geology*) – Al-deficient igneous rocks with Na and K levels in excess of what is required for a feldspar
peridotite (*geology*) – coarse-grained igneous rock, mainly olivine and pyroxene
Permian period (*geology*) – 298.9–252.2 mya
permittivity (*physics*) – measure of the resistance to formation, within a medium, of an electric field
petrography (*petrology*) – detailed description of rocks, in particular the mineralogy and textural structure

petrology (*geology*) – origin, composition, distribution and general structure of rocks
PGE – platinum group elements (Ru, Rh, Pd, Os, Ir and Pt)
pH – typically taken as $-\log_{10}[H_3O^+]$ but correctly defined in terms of activity $-\log_{10}[a(H_3O^+)]$
Phanerozoic eon (*geology*) – 542 mya to present (current eon)
phengite – $K(AlMg)_2(SiAl)_4O_{10}(OH)_2$, di-octahedral micas similar to muscovite, non-IMA name
phillipsite – $(Na,K,Ca)_{1-2}(Si,Al)_8O_{16}\cdot 6(H_2O)$, zeolite
phlogopite – $KMg_3(AlSi_3O_{10})(F,OH)_2$, 2:1 clay mineral of the mica group
phosphate – PO_4^{3-}
photometric – usually luminous intensity
phreatic zone (*hydrology*) – zone of water-saturated rock/soil, i.e. an aquifer
phyllic alteration (*geology*) – high-temperature version of argillic alteration in which hydrothermal processes introduce clay minerals into rock
phyllosilicate – large class of silicate minerals characterized by a sheet structure of silica tetrahedral and an underlying Si_2O_5 stoichiometric basis
physiography – synonym for physical geography
physisorption – the adsorption of an atom, ion or molecule on a surface that occurs through the process of physical forces, e.g. dispersion forces; unlike chemisorption, very easily reversed
phytoremediation – remediation of environmental problems through plants
PIMA – Portable Infrared Mineral Analyzer
pimelite – $Ni_3Si_4O_{10}(OH)_2\cdot 4(H_2O)$, 2:1 clay mineral, member of the smectite group
pisoid (*geology*) – concretious grain, primarily limestone
pisolite (*geology*) – sedimentary rock made of pisoids
placer deposit (*geology*) – ore deposit formed by gravity separation during a sedimentary process
plagioclase – solid–solution feldspar mineral with a wide composition range $NaAlSi_3O_8$ to $CaAl_2Si_2O_8$
platelet (*mineralogy*) – a clay plate
platelet (*geology*) – a small tectonic plate
Pliocene epoch (*geology*) – 5.322–2.588 mya
plumbogummite – $PbAl_3(PO_4)_2(OH)_5\cdot(H_2O)$
polyacrylamide – $[-CH_2CHCONH_2-]_n$
polyacrylate – e.g. sodium polyacrylate $[-CH_2CH(COO^-Na^+)-]_n$
polydispersity (*polymer and colloid chemistry*) – a non-uniform distribution, e.g. of particle sizes or molecular weights
polyelectrolyte – polymers that have electrolyte functional groups (take on a charge when dissolved, e.g. in water) on the repeating unit
polyethylene – $[-CH_2CH_2-]_n$
polymorph – different crystal structure form of the same compound
polytype (*crystallography*) – variant in crystal structure of the same compound arising from stacking differences

polyvinylidene – e.g. polyvinylidene chloride $[-CH_2CCl_2-]_n$
polyvinylpyrrolidone (PVP) – $[CH_2CHNC_3H_6CO-]_n$
population balance – a series of mathematical equations that together describe how a population of entities (typically particles) develop particular properties over time. Frequently used in crystallization, cell biology, communition and aerosols, with flocculation (specifically aggregate growth and breakage) being the application discussed in this review
porosity – pore space (volume) in a material
porphyrins (*chemistry*) – heterocyclic organic macrocycles
porphyry (*geology*) – an igneous rock consisting of large crystals (e.g. feldspar or quartz) in a fine-grained matrix
potassic (*chemistry/geology*) – containing potassium or an appreciable quantity of potassium
ppm – parts per million
Precambrian supereon (*geology*) – 4540–541 mya
preg-robbing (*hydrometallurgy*) – or pregnant liquor robbing, a term describing the loss from solution of a valuable component, usually by adsorption or absorption by a co-existing solid phase
propylitic alteration (*geology*) – iron- and magnesium-bearing hydrothermal fluid-altering rock
prospect (*geology*) – a geology with the prospect of containing an ore-body
Proterozoic eon (*geology*) – 2500–542 mya
protolith (*geology*) – precursor rock from which a metamorphic rock will form
protonation – the act of adding a proton H^+ (via transfer from H_3O^+) to another chemical species, within solution or at a surface
protore (*geology*) – literally proto-ore, a mineralization too low in a given component to be considered an ore, but one that might transform to an ore through some secondary enrichment
proximal – nearest point
PSD – particle size distribution
pseudoplastic (*rheology*) – shear thinning fluid
pseudomorph – a particular mineral expression in which one mineral has been replaced by another, and while retaining the original physical form has all other properties determined by the new mineralization, e.g. goethite after pyrite, looks like pyrite form, but is actually goethite
pulp (*hydrometallurgy*) – general term for crushed/ground ore in a liquid processing environment (generally a leach) characterized by a high solids level
PVP – polyvinylpyrrolidone
pyrite – FeS_2 (iron sulphide with cubic structure, same composition as marcasite)
pyroaurite – $Mg_6Fe_2(CO_3)(OH)_{16}\cdot 4(H_2O)$
pyrolusite – MnO_2
pyrometallurgy – a field of extractive metallurgy involving the use of heat (as opposed to hydrometallurgy)

pyrolysis – thermal decomposition of organics

pyrophyllite – $Al_2Si_4O_{10}(OH)_2$, 2:1 clay mineral, member of the talc–pyrophyllite group

pyroxene – general class of rock-forming mineral $XY(Si,Al)_2O_6$ where X and Y are a variety of metal ions and the silica tetrahedra form chains

pyrrhotite – $Fe_{1-x}S$ (x = 0–0.2)

PZC – point of zero charge

QEM*SEM – quantitative evaluation of mineralogy by scanning electron microscope

quartzose (*geology*) – primarily quartz or SiO_2 sandstone

Quaternary period (*geology*) – 2.588 mya to the present

quiescent – inactive

QXRD – quantitative X-ray diffraction (e.g. via Rietveld method)

radiometer – device for measuring radiant flux of EM radiation

rake (*hydrometallurgy*) – applies to the rake at the bottom of a thickener, used to assist the transport of the settled solids within the bed to the underflow take-off point

Raman spectroscopy – method relying on the inelastic scattering of monochromatic (i.e. laser) light (UV–vis–NIR) to access vibrational and rotational states

rectorite – $(Na,Ca)Al_4(Si,Al)_8O_{20}(OH)_4{\cdot}2H_2O$, member of the interstratified chlorite–smectite group

reflectance (*physics*) – fraction of incident electromagnetic power reflected at an interface

refractory (*hydrometallurgical*) – difficult to process

regolith (*geology*) – loose material (e.g. dust, soil, rubble) covering solid rock

remanence (*physics*) – remnant magnetization left behind in a material after an external magnetic field is removed

residuum – residue

Reynolds number (*fluid dynamics*) – Re, dimensionless number that is a ratio of the inertial to viscous forces in a moving fluid. Used to characterize regimes such as laminar and turbulent flow

RH – Rydberg constant

rheogram – plot of shear stress versus shear rate for a fluid

rheology – study of the flow of matter, including high-solids suspensions

rheomalaxis – irreversible decrease in viscosity for a fluid subjected to shear (shear thinning)

Rheomax® – proprietary flocculant

rheometer – device for measuring fluid behaviour; used for fluids with varying viscosity in order to measure the fluid's rheology

rheopexy – time-dependent increase in viscosity for a non-Newtonian fluid subjected to shear

rhodochrosite – $MnCO_3$

Rietveld method – an approach to quantitative analysis of X-ray diffraction patterns

RIR (*crystallography*) – reference intensity ratio of the strongest X-ray diffraction peak of a phase to the strongest peak of a standard

rougher (*hydrometallurgy*) – in froth flotation, if the gangue has some flotation response the first concentrate is diluted and then re-floated in cleaner cells
rozenite – $Fe^{2+}SO_4 \cdot 4(H_2O)$
rpm – revolutions per minute
rutile – polymorph of TiO_2
SAG mill – semi-autogenous grinding mill
sandstone (*geology*) – sedimentary rock composed of sand-sized rock grains (typically quartz and/or feldspar)
saponite – $Ca_{0.25}(Mg,Fe)_3((Si,Al)_4O_{10})(OH)_2 \cdot n(H_2O)$, 2:1 clay mineral, member of smectite group and tri-octahedral subgroup
sapphire – α-Al_2O_3 (corundum)
saprolite (*geology*) – clay-rich decomposed or chemically weathered rock
sauconite – $Na_{0.3}Zn_3(Si,Al)_4O_{10}(OH)_2 \cdot 4H_2O$, 2:1 clay mineral, member of smectite group and tri-octahedral subgroup
scale (*chemical engineering*) – generic term for solid deposits inside processing vessels, heat exchanger pipes, etc.
scattergram – or scatter plot or scatter graph
scavenger cells (*hydrometallurgy*) – the last cells in a froth flotation bank before tailings discharge, used to recover weakly floating material
schist (*geology*) – metamorphic rock with medium to large grains of mica in preferred orientation
schwertmannite – $Fe^{3+}{}_{16}O_{16}(OH)_{12}(SO_4)_2$
scintillometer – device that measures small changes in the refractive index of air
scrubbing (*hydrometallurgy*) – high-speed agitation to remove gangue from ore
sedimentary (*geology*, adj.) – one of the three main rock types, formed through sedimentation. The other types are igneous and metamorphic.
SEM – scanning electron microscope
sepiolite – $Mg_4Si_6O_{15}(OH)_2 \cdot 6H_2O$
sequester – to isolate, separate or set apart
sericite – a fine-grained mica, mainly muscovite
serpentine group – common rock-forming hydrous magnesium iron phyllosilicate $(Mg,Fe)_3Si_2O_5(OH)_4$ minerals found in serpentinite rocks
serpentinite (*geology*) – rock consisting of one or more serpentine minerals
sg – specific gravity; mass per unit volume of a substance, or the ratio of the density of a substance to the density of water
shaly (*geology*) – containing shale
shear stress (*hydrometallurgy*) – force vector component for a fluid moving along a solid boundary
siderite – $FeCO_3$
siderotil – $FeSO_4 \cdot 5(H_2O)$
silanol (*chemistry*) – Si–OH group
silicic (*geology*) – igneous rock rich in silica

siliclastic (*geology*, adj.) – clastic, non-carbonate sedimentary rocks that are primarily silica as quartz or other silicates
sill (*geology*) – tabular sheet intrusion, e.g. volcanic lava, between old layers of sedimentary rock
skarn (*geology*) – metamorphic contact zone around igneous rock intrusion into country rock
slime (*mineral processing*) – fine gangue material that adheres to ore particles
slumping (*geology*) – loosely consolidated material moving down a slope
slurry – solid suspension
smectite – 2:1 clay mineral group with multiple members
Snobrite® – proprietary white kaolin pigment extender
sodalite – $Na_8Al_6Si_6O_{24}Cl_2$, or variants thereof
sol (*chemistry*) – fluid colloidal system of two or more components (IUPAC)
solids, e.g. percentage solids (*hydrometallurgy*) – refers to the quantity of solids in a suspension
sorption – encompasses adsorption and absorption
spectroradiometer (*physics*) – instrument to measure spectral power distribution
sphalerite – ZnS, (Zn,Fe)S
spigot – effectively a tap/faucet, literally a plug for a barrel hole (archaic)
spinel – $MgAl_2O_4$
SSA – specific surface area
staurolite – $(Fe^{2+},Mg)_2Al_9(Si,Al)_4O_{20}(O,OH)_4$
stellarite – archaic term variously applied in the context of oil shales
steric – pertaining to the spatial arrangement of atoms in a molecular structure and how that affects interaction
Stokes flow (*rheology*) – fluid flow with a low Reynolds number, $Re \ll 1$, typical of low fluid velocity and high viscosity
strandline (*geology*) – landscape marking resulting from high water mark tidal action
stratiform (*geology*) – class of ore deposit, where the minerals are deposited concordant with strata
stratigraphy (*geology*) – study of strata, i.e. rock layers and layering
streaming potential (*chemistry*) – potential difference (at zero current) caused by the flow of liquid under a pressure gradient through a membrane, plug or capillary (IUPAC)
supergene (*geology*) – ore deposit where the enrichment has been by precipitation from percolating meteoric water
supernatant (*chemistry*) – fluid above a separated or settled solid or (non-mixing) higher-density fluid
surfactant – a surface active agent or a compound that reduces the surface tension of a fluid, usually water
surficial – relating to the surface of the Earth
suspension (*hydrometallurgy*) – a suspension of non-settling or poorly settling solids in a liquid

SWIR – short wavelength infrared, covering the 1000–2500 nm wavelength range
syncrude – synthetic crude oil made from coal
syngenetic (*geology*) – mineral deposits that formed at the same time as the surrounding rock
synthetic rutile (SR) – >95–96 per cent TiO_2 material produced from the reduction and removal of iron from Fe–Ti oxides
szomolnokite – $FeSO_4.(H_2O)$
tailings (*hydrometallurgy*) – also tails, refers to the waste stream of hydrometallurgical processes. Often these are suspensions that are stored in dams or ponds for evaporation, e.g. tailings storage facility (TSF)
takovite – $Ni_6Al_2(OH)_{16}(CO_3,OH).4(H_2O)$
talc – $Mg_3Si_4O_{10}(OH)_2$, 2:1 clay mineral, member of the talc–pyrophyllite group
tall oil or tallol – by-product of the Kraft process for paper making, complex organic mixture including fatty acids
taxonomy – systematic classification
tectonic (*geology*) – belonging to the Earth's crust
TEM – transmission electron microscope
ternary – composed of three items
terrane (*geology*) – a fragment of tectonic plate crust broken off and subsumed
tetragonal – one of the seven lattice point or space groups or crystal systems
TGA – thermogravimetric analysis
thickener (*hydrometallurgy*) – short for gravity thickener, a unit operation device that treats high volumes of solid suspensions to produce clarified liquid overflows and higher-density underflows. Process is also known as thickening
thiocyanate – SCN^-
thixotropic – a fluid that becomes less viscous when subjected to shear, i.e. shear thinning
TIR – thermal infrared, usually covering the 5000–16 000 nm wavelength range
titania – TiO_2
titaniferous (*geology*) – bearing titanium (often but not exclusively in the context of iron)
todorokite – $(Na,Ca,K,Ba,Sr)_{1-x}$ $(Mn,Mg,Al)_6O_{12}\cdot 3\text{–}4H_2O$
tomography – 3D-imaging by sections
torbanite – variety of fine-grained oil shale
tortuosity (*hydrometallurgy*) – property of a curve being twisted; applies to fluid paths in heap or *in situ* leaching
tosudite – $Na_{0.5}(Al,Mg)_6(Si,Al)_8O_{18}(OH)_{12}\cdot 5H_2O$, member of the interstratified chlorite–smectite group
transgressive dune (geology) – sand dunes migrating inland in the direction of the prevailing wind and wave action
triclinic – one of the seven lattice point or space groups or crystal systems
tridymite – high-temperature polymorph of quartz SiO_2, occurs in seven forms
trigonal – one of the seven lattice point or space groups or crystal systems
trona – $Na_3(CO_3)(HCO_3)\cdot 2H_2O$
TSF – tailings storage facility, also known as a tailings pond

turbidity – cloudiness of a fluid due to suspended particles
TurbiscanTM – instrument range for analysing turbidity of dispersions, emulsions, suspensions, foams, etc.
turbostratic (*crystallography*) – crystal structure where the basal planes are misaligned (e.g. clays, graphites)
ultrafines – particles less than 100 nm in diameter
Ultrafloc – proprietary flocculant and coagulant
ultramafic (*geology*) – low-silica igneous and meta-igneous rock
ultramicrotome – tool to produce thin sections for (for example) TEM; can produce cuts of thickness 30–100 nm
unadsorbed (*chemistry*) – solution species, e.g. ions or polymer molecules that remain in solution and not adsorbed on a solid surface
underflow (*hydrometallurgy*) – thickened solids from a thickener (see thickener)
unthickened (*hydrometallurgy*) – description of feed material for a thickener
uraniferous (*geology*) – bearing uranium (no specific contextual reference to iron)
uranyl – UO_2^{2+}
USGS – United States Geological Survey
UV – ultraviolet
Vadose zone (*hydrology*) – unsaturated zone between the surface and the top of a phreatic zone
van der Waals forces (*chemistry*) – forces between molecular entities not due to bond formation or electrostatic interaction, summation of dipole, induced dipole and London forces (IUPAC)
vanadiferous – (*geology*) – bearing vanadium (no specific contextual reference to iron)
venturi effect (*hydrodynamics*) – reduction of fluid pressure when fluid passes through a constriction; venturi can also refer to the constriction
vermiculite – $(Mg,Fe^{2+},Al)_3(Al,Si)_4O_{10}(OH)_2 \cdot 4(H_2O)$, 2:1 clay mineral, member of the mica group
vestigial – remaining or surviving
yield stress (*rheology*) – point in a stress–strain curve at which the material ceases to behave elastically and becomes plastic
violarite – $Fe^{2+}Ni_2S_4$
viscoelastic – materials that exhibit both viscous and elastic characteristics under deformation
viscometer – device for measuring the viscosity of a fluid
viscosity (*rheology*) – measure of a fluid's resistance to deformation by the application of force
vitreous – composed of glass or having the nature of glass
VMSH (*geology*) – volcanogenic massive sulphide hydrothermal
VNIR – very near-infrared, covering the 400–1000 nm wavelength range
voidage – relative amount or fraction of unoccupied space between packed particles
volatiles – any components (of a mixture) that have a low vapour pressure

volcanogenic (*geology*) – of volcanic origin
volumetric (*chemistry*) – volume-based metric used in chemical assessments, similar context to gravimetric (mass)
vortex – spinning fluid flow about an imaginary axis that may be straight or curved
XANES – X-ray absorption near edge structure
XAS – X-ray absorption spectroscopy
xenotime – YPO_4
xeric – dry, usually applied to habitat, e.g. deserts
XPS – X-ray photoelectron spectroscopy
XRD – X-ray diffraction
XRF – X-ray fluorescence
zeolite – microporous aluminosilicate mineral
zeta potential (*colloid chemistry*) – electric potential in the interfacial double layer around a colloid at the point of transition to bulk fluid; can be considered the effective potential on the colloid particle for all practical purposes
zincite – (Zn, Mn)O
zircon – zirconium silicate $ZrSiO_4$
zirconolite – $CaZrTi_2O_7$

Sources

The glossary has been assembled on the basis of a number of generally available sources, primarily:

Oxford English Dictionary
http://webmineral.com
www.mindat.org
www.bgs.ac.uk
www.britannica.com
http://goldbook.iupac.org
Wills, B.A. 1981. *Mineral Processing Technology*, 2nd edition. New York: Pergamon Press.

In addition, there were numerous minor sources that were impractical to individually cite. A number of words are not in common usage and could be considered archaic; others are very specific to particular research areas and some others have potentially wide meanings. This is especially the case with some mineral names that are not recognized by the IMA but remain in use in narrow areas. The field is given if the word has a possibility of ambiguity. For a comprehensive description of clay-related terms not in this glossary, please refer to www.clays.org/GLOSSARY/GlossIntro.html.

Index

Index entries referring to figures are indicated with an ‘f.’ and to tables with a ‘t.’